Handbook of Radiation Effects

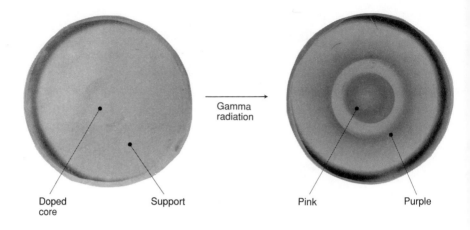

Visible coloration in germanium-doped silica. For full explanation, see p. 252.

Handbook of Radiation Effects

Andrew Holmes-Siedle
*Radiation Experiments and Monitors, Oxford, UK
and Brunel: The University of West London, UK*

and

Len Adams
*Radiation Effects and Analysis Techniques Unit,
European Space Agency–ESTEC, The Netherlands*

Oxford   New York   Tokyo
OXFORD UNIVERSITY PRESS

Oxford University Press, Walton Street, Oxford OX2 6DP

Oxford   New York
Athens   Auckland   Bangkok   Bombay
Calcutta   Cape Town   Dar es Salaam   Delhi
Florence   Hong Kong   Istanbul   Karachi
Kuala Lumpur   Madras   Madrid   Melbourne
Mexico City   Nairobi   Paris   Singapore
Taipei   Tokyo   Toronto
and associated companies in
Berlin   Ibadan

Oxford is a trade mark of Oxford University Press

Published in the United States
by Oxford University Press Inc., New York

© Andrew Holmes-Siedle and Len Adams, 1993
First published 1993
Reprinted with corrections 1994

All rights reserved. No part of this publication may be reproduced, stored in a retrieval system, or transmitted, in any form or by any means, without the prior permission in writing of Oxford University Press. Within the UK, exceptions are allowed in respect of any fair dealing for the purpose of research or private study, or criticism or review, as permitted under the Copyright, Designs and Patents Act, 1988, or in the case of reprographic reproduction in accordance with the terms of licences issued by the Copyright Licensing Agency. Enquiries concerning reproduction outside those terms and in other countries should be sent to the Rights Department, Oxford University Press, at the address above.

A catalogue record for this book is available from the British Library

Library of Congress Cataloging in Publication Data
Holmes-Siedle, Andrew.
Handbook of radiation effects / Andrew Holmes-Siedle and Len Adams.
1. Radiation—Handbooks, manuals, etc.   I. Adams, Len.   II. Title.
QC475.H59   1993   620.1'1228—dc20   92-22567
ISBN   0 19 856347 7

Printed in Great Britain by Biddles Ltd,
Guildford and King's Lynn

Preface

The urge to start compiling a handbook was first stimulated by the European Space Agency when, in 1976, they commissioned the *Radiation effects engineering handbook* prepared by A. Holmes–Siedle and R. Freeman at Fulmer Research Laboratories in the UK with the collaboration of staff at ESTEC in the Netherlands. The welcome given to the Handbook by aerospace contractors and scientific institutes indicated the need to keep updating a reference text and give further detail where required.

The need to keep updating is driven by the rapid pace of development in microelectronics. Furthermore new forms of radiation-induced 'plague' in electronics and optics are always being discovered. One such was the 'Single Event Upset'. Discussions with colleagues in NASA and other centres of radiation effects expertise confirmed the need for official guidelines which the agencies could hand to contractors. This led to the preparation of a guidelines document by the European Space Agency. The present authors supplied most of the material and they also incorporated contributions from many colleagues and consultations with NASA. The 1978 Fulmer handbook, with a 1980 document prepared for a fusion experiment, provided the backbone. A *Radiation design handbook* is issued as an ESA Procedures and Standards document (ESA-PSS-01-609). However, there is also a need for a more 'tutorial' text. In the years since 1976, the authors have had the opportunity to broaden their knowledge, particularly by training a number of young engineers in the discipline of radiation effects and by teaching a number of international seminars and workshops. Both authors have collaborated in research with scientists in diverse non-space disciplines such as cancer treatment, radiological protection, radiation processing, and nuclear fusion technology. The authors were encouraged by many radiation effects practitioners in diverse fields, to broaden the scope of the official space handbooks. Thus, while the space environment and the space industry are widely referred to as examples in this handbook, surveys of the other radiation environments listed above are given and non-space 'hardening' topics which we have worked on are also described. The present work should now be useful to equipment engineers venturing into many fields. For completeness, and since some of the space environment even penetrates to the Earth's surface, we give a brief survey of the terrestrial radiation environment and its risks.

It is no coincidence that space radiation reaches our bodies as well as our equipment on Earth. If we fly high enough in an aircraft, our eyes and our electronics suffer the same single-event upsets as space vehicles. The whole universe is powered by radioactivity and our planet bears various residues of that activity. The overlap between the study of radiation effects in electronic systems and radiation protection for human beings has stimulated lively argument about the appropriate units to use in radiation effects work. Our decision to favour traditional units (rads, curies etc., but adding the Systeme Internationale units in brackets) was not taken lightly. Unlike health physicists, the engineering community continues to use the traditional units, especially in the USA (see e.g. *IEEE Transactions on Nuclear Science*, December 1992). We prefer to give priority to units familiar to engineers and risk the opprobrium of the physicists.

It is strongly hoped that the book will open a door to the world of radiation effects. As well as being a reference tool, we believe that it can act as a bridge between the plentiful, somewhat baffling scientific treatises on radiation effects and the man grappling with the 'real world' — the engineer.

Acknowledgements

The IEEE Nuclear and Space Radiation Effects Conference is the forum for the exchange of ideas. The Transactions of that conference are the convenient 'gospel' of most that is worth knowing in the field. The reader will find that our bibliography bears this out — we have used this resource unstintingly.

The radiation effects community is small, close-knit and highly professional. Over the years we have learnt by discussion with members of this community. If we fail to acknowledge them adequately, we console ourselves by knowing that the true contributors to advances in the field are well recognized and honoured by their peers. We thank them all but would like to give special thanks to a few.

One of the authors (AHS) first took up the profession of radiation effects in the USA. He benefited by leading a group of talented people in jointly reducing their ignorance of the new CMOS device, just invented by RCA. The group at RCA Astro included George Brucker, William Dennehy and Waldemar Poch, aided by Karl Zaininger from the Labs. Back in the UK, Roger Freeman was a close collaborator in organizing material for the 1978 handbook, some of which has been carried over into this book. His clear explanatory style has provided us with a good example. Any obfuscations found have probably crept in since. We have already mentioned the role of ESA document PSS-01-609 in many of the chapters which follow. Below, we further acknowledge some of our fellow contributors to that document. Dr Eamonn Daly has made scientific and engineering advances in the understanding and modelling of the space environment. More directly, he rewrote some of the paragraphs on the space environment and the interaction of matter and space radiation; we have drawn heavily on his description of computer programmes for radiation transport. The dose–depth curves for reference missions (Appendix E) are also due to Dr Daly. Reno Harboe-Sorensen's significant contributions to the field of Single Event Effects are reflected in sections on that topic and much of the tabulated data in Appendix D. Dr Clive Dyer contributed the section on induced radioactivity, and his work on the space environment in general has been of great help to us. To shorten our own text, constant reference has been made to two outstanding books by colleagues in the IEEE community—*Ionizing radiation effects in MOS devices and circuits*, edited by T. P. Ma and P. V. Dressendorfer and *The effects of radiation on electronic systems* by G. C. Messenger and M. S. Ash. Dr John Dennis commented on aspects of radiation protection discussed in Chapters 1 and 2. We also acknowledge radiological texts in our first chapter.

We have had many fruitful discussions with W. E. Price, Dr C. Barnes and Dr P. Robinson of Jet Propulsion Laboratory and Dr E. G. Stassinopoulos of NASA, Dr P. Winokur of Sandia Labs, Prof T. P. Ma of Yale University, Dr G. Messenger and Prof M. Ash. We wish also to acknowledge our European colleagues Prof W. Heinrich, Dr D. Braünig, Dr J. Dennis, Dr J. M. Stephen, and Dr J. Farren. All of these people have contributed in one way or another to this book and we thank them for their ideas and their friendship. It would not be right to end without thanking the managements who gave permission to use technical material and supplied the environment in which thinking and a broad range of work could be done and published: the European Space Agency–ESTEC of Noordwijk, The Netherlands, and Fulmer Research of Stoke Poges, UK.

Oxford, England	A. H.-S.
Noordwijk, The Netherlands	L. A.
December 1992	

Disclaimer

Since this book may be widely used in system design, we deem it prudent to disclaim direct responsibility for the failure of systems outside our control. We must therefore inform readers that, while the information presented is believed to be accurate and reliable, no liability in respect of any use of the material can be accepted by the authors or Oxford University Press.

Contents

Introduction ... 1

1 Radiation, physics, and measurement ... 3
1.1 General ... 3
1.2 Radioactivity ... 4
1.3 Types of radiation ... 5
1.4 Units ... 5
1.5 The measurement of radiation ... 8
    1.5.1 General ... 8
    1.5.2 Techniques ... 9
        1.5.2.1 Ionization chamber ... 9
        1.5.2.2 Calorimeter ... 9
        1.5.2.3 Geiger-Müller tube and proportional counters ... 10
        1.5.2.4 Faraday cup ... 10
        1.5.2.5 Thermoluminescent dosimeters (TLDs) ... 10
        1.5.2.6 Radiophotoluminescent (RPL) glass ... 11
        1.5.2.7 Fricke dosimeter ... 11
        1.5.2.8 Perspex dosimeters ... 12
        1.5.2.9 Colour changes and optical transmission ... 12
        1.5.2.10 Polyethylene and hydrogen pressure ... 12
    1.5.3 Silicon dosimeters ... 12
        1.5.3.1 General ... 12
        1.5.3.2 Dose rate ... 13
        1.5.3.3 Dosimeter diodes ... 13
        1.5.3.4 MOS dosimeters ... 14
References ... 14

2 Radiation environments ... 16
2.1 The space environment ... 16
    2.1.1 General ... 16
    2.1.2 The radiation belts ... 17
    2.1.3 Cosmic rays ... 19
    2.1.4 Geomagnetic shielding ... 22
    2.1.5 Other sources of cosmic rays ... 24
    2.1.6 Solar flares ... 24
    2.1.7 Other planets ... 25
2.2 The nuclear reactor environment ... 26
    2.2.1 General ... 26

	2.2.2	Reactor containment	27
	2.2.3	Reactor vessel and cavity	27
	2.2.4	Mobile nuclear reactors	28

2.3 The radiation processing environment 28
    2.3.1 General 28
    2.3.2 Sources and doses 29
2.4 The weapons environment 29
    2.4.1 General 29
    2.4.2 Radiation output 31
    2.4.3 Electromagnetic pulse 32
2.5 The controlled-fusion environment 32
2.6 The environment of robots 33
2.7 High-energy physics accelerators 36
2.8 Terrestrial and man-made environments 37
    2.8.1 General 37
    2.8.2 Background radiation 37
        2.8.2.1 Extra-terrestrial radiation 37
        2.8.2.2 Environmental radiation 37
        2.8.2.3 Ingested radiation 38
        2.8.2.4 Fallout 39
        2.8.2.5 Internal radiation 40
        2.8.2.6 The monazite survey project, India 40
    2.8.3 The radon hazard 41
    2.8.4 Radiation protection 42
        2.8.4.1 Regulations 42
        2.8.4.2 Risks 43
References 45

3 The response of materials and devices to radiation 48

3.1 Introduction 48
3.2 Degradation processes 48
    3.2.1 General 48
    3.2.2 Atomic displacement 50
        3.2.2.1 General 50
        3.2.2.2 Degradation of transport power 50
        3.2.2.3 Bulk damage at high and low injection levels 54
        3.2.2.4 Annealing of atomic displacement damage 56
        3.2.2.5 Other methods of defect characterization 58
        3.2.2.6 Dependence of damage on particle energy and type 58
        3.2.2.7 Modelling of energy loss in silicon 60
        3.2.2.8 Statistics of particle damage 63
        3.2.2.9 Calculating damage as equivalent particle fluence 63
3.3 Ionization 64
    3.3.1 General 64

	3.3.2	Charge trapping in oxides after ionization	65
	3.3.3	The charge trap system at the surface of oxidized silicon	67
3.4	Induced radioactivity	70
3.5	High dose-rate upsets (transient effects)	71
	3.5.1	General considerations	71
	3.5.2	Bulk semiconductor	73
3.6	Transient photocurrents in p–n junctions	74
3.7	Single-event phenomena	76
	3.7.1	General	76
	3.7.2	Single-event upset (SEU)	76
	3.7.3	Latch-up	79
3.8	Consequences of long-lived degradation	81
	3.8.1	General	81
	3.8.2	Atomic displacement in semiconductors	83
	3.8.3	Ionization	88
	3.8.4	Other effects	88
3.9	Conclusions — an overall view of device response	89
References	90

4 Metal-oxide-semiconductor (MOS) devices — 95

4.1	Introduction	95
4.2	Historical	98
4.3	Charge trapping in MOS devices	100
	4.3.1 Oxide charge trapping and performance degradation — an overview	100
	4.3.2 MOS transistor action and threshold voltage shift	102
	4.3.2.1 General	102
	4.3.2.2 n-Channel transistor (nMOS)	103
	4.3.2.3 Equivalent effects in p-channel devices (pMOS)	105
4.4	Physical model for oxide trapped charge build-up	105
	4.4.1 General	105
	4.4.2 Interface charge trapping	106
4.5	Prediction models for radiation effects in MOS devices	108
	4.5.1 A simple analysis of radiation effects in MOS devices	108
	4.5.2 Terms for gate voltage applied during irradiation	110
	4.5.3 Bias dependence	110
	4.5.3.1 General	110
	4.5.3.2 Zero bias condition: oxide charge vs. interface effects	111
	4.5.4 Intermittent biasing	111
	4.5.5 Gate oxide thickness	113
	4.5.6 Simple graphical model of MOS degradation	115

	4.5.7	A standard graph for assessment of MOS test results	115
	4.5.8	Other prediction programs	115
	4.5.9	Ultra-hard technology versus tolerant commercial technology	117
4.6	Other fundamental effects in MOS devices		117
	4.6.1	Recovery and time-dependent effects (annealing)	117
		4.6.1.1 General	117
		4.6.1.2 High temperature: high recovery rate	119
		4.6.1.3 Room temperature: some recovery	120
		4.6.1.4 Cryogenic temperatures: little recovery	122
	4.6.2	Model of annealing	124
	4.6.3	'Rebound', 'overshoot' or 'super-recovery'	124
	4.6.4	'Reverse annealing'	124
	4.6.5	'Turnaround'	125
	4.6.6	Electron and hole injection; radiation annealing	125
	4.6.7	Low-dose-rate effects	128
	4.6.8	Noise	128
	4.6.9	Resistivity change: heavy damage	128
	4.6.10	Transient effects	128
4.7	Real MOS devices in real environments: a commentary		131
	4.7.1	Effect of total dose on logic circuits	131
		4.7.1.1 Introduction	131
		4.7.1.2 The VNTZ effect and its impact	133
		4.7.1.3 Other effects on logic operation	133
		4.7.1.4 Field oxide leakages	134
		4.7.1.5 Effect of interface states on logic operation – rebound	137
	4.7.2	Classification of 'hard', 'medium' and 'soft' with respect to total dose	138
	4.7.3	Processing variables	138
4.8	Commercial MOS technology: a survey of radiation responses		140
	4.8.1	General	140
	4.8.2	Memory technology	141
	4.8.3	Microprocessor technology	142
		4.8.3.1 General	142
		4.8.3.2 Microprocessor tolerance levels for total dose and upsets	143
		4.8.3.3 Advanced microprocessors	144
		4.8.3.4 Digital signal processors and fast analogue circuits	145
	4.8.4	Logic families	145
	4.8.5	Programmable logic devices (PLDs)	146
4.9	Hardened technologies		147
	4.9.1	General	147

Contents xiii

  4.9.2 Radiation-hardened computer system technology 147
  4.9.3 Examples of hardened technology 148
    4.9.3.1 Bulk silicon MOS ICs 148
    4.9.3.2 Silicon on insulator 149
 4.10 Some advanced MOS structures: potential problems 149
  4.10.1 Introduction 149
  4.10.2 Lightly and moderately doped drains (LDD and MDD) 150
  4.10.3 Non-volatile memories 150
  4.10.4 Heavy metals—disturbance of equilibrium 150
  4.10.5 Ferrite memories 152
  4.10.6 GaAs-on-Si 152
  4.10.7 High-voltage and power CMOS ICs 153
 4.11 Conclusions 153
 References 154
 Recommended background reading 163

5 Bipolar transistors 164
 5.1 Introduction 164
 5.2 Effects of radiation on device function 164
  5.2.1 Gain 164
  5.2.2 Degradation of gain 164
  5.2.3 Other permanent effects 166
  5.2.4 Transient effects 166
 5.3 Bulk damage 168
  5.3.1 General 168
  5.3.2 Influence of base width 169
  5.3.3 Influence of type and energy of radiation 171
  5.3.4 Irradiation results 178
  5.3.5 Prediction of degradation 178
  5.3.6 Selection principles for bipolar transistors 178
 5.4 Surface-linked degradation in gain 180
  5.4.1 Introduction 180
  5.4.2 Statistical prediction of surface damage 182
  5.4.3 Collector–base leakage current 182
  5.4.4 The 'maverick' device 183
  5.4.5 Annealing of surface effects 184
  5.4.6 Slow thermal annealing of bulk damage 184
  5.4.7 Saturation voltage 185
 5.5 Long-lived radiation effects in bipolar integrated circuits 186
  5.5.1 Digital ICs 186
  5.5.2 Emitter-base surface effects in integrated transistors 189
  5.5.3 Analog ICs 189
  5.5.4 Isolation technology 190

xiv Contents

5.6	Transient upsets in bipolar integrated circuits	195
5.7	Summary	197
	References	197

6 Diodes, solar cells, and optoelectronics — 200

6.1	Introduction	200
6.2	Diodes: general	201
	6.2.1 Introduction	201
	6.2.2 Mechanisms	201
6.3	Solar cells	202
	6.3.1 General	202
	6.3.2 Diffusion-length degradation and equivalent fluences	203
	6.3.3 Background	203
	6.3.4 Predicting the degradation of solar cell arrays	205
	6.3.5 Equivalent fluences	206
6.4	Low-power rectifier diodes	207
6.5	High-power rectifier diodes	208
6.6	Zener diodes and diodes in avalanche breakdown	208
6.7	Microwave diodes	208
6.8	Light-detecting devices	209
	6.8.1 Photodiodes	209
	6.8.2 Detector diodes for high-energy physics	210
6.9	Phototransistors	211
6.10	Light-emitting diodes (LEDs) and lasers	212
	6.10.1 General	212
	6.10.2 LEDs	212
	6.10.3 III–V lasers	213
6.11	Optocouplers	213
6.12	Charge-coupled devices (CCDs)	214
	6.12.1 General	214
	6.12.2 Ionization effects in CCD structures	216
	6.12.2.1 General	216
	6.12.2.2 Oxide trapped charge (ot)	217
	6.12.2.3 Interface trapped charge (it)	218
	6.12.2.4 MOS transistor elements in CCDs	218
	6.12.2.5 Charge-transfer channel	219
	6.12.2.6 Voltage shifts	220
	6.12.3 Displacement damage in CCD structures	220
	6.12.3.1 General	220
	6.12.3.2 CTI growth with bulk damage: a model	222
	6.12.3.3 Other CTI growth models	224
	6.12.3.4 Goals for bulk damage limitation	224
	6.12.4 Conclusions on CCD degradation	224
6.13	Electro-optic crystals	225

Contents xv

    6.14 The new optics   225
    6.15 Vacuum devices and extreme environments   226
    6.16 Conclusions   227
    References   227

7 Power devices   233

    7.1 General   233
    7.2 Bipolar power transistors   233
    7.3 Thyristors (silicon-controlled rectifiers)   235
    7.4 Power MOSFETs   236
        7.4.1 Parameter changes under radiation   236
        7.4.2 Radiation-tolerant power MOS circuits   237
        7.4.3 Transient and heavy-ion-induced burn-out   237
    7.5 Static induction transistor (SIT)   238
    7.6 'Smart power' devices   238
    7.7 Conclusions   238
    References   239

8 Optical media   240

    8.1 General   240
    8.2 Window materials   242
        8.2.1 General   242
        8.2.2 Colour centres in halides and oxides   242
        8.2.3 Silicate glasses   243
        8.2.4 Particle-induced defects   247
    8.3 Coatings   247
    8.4 Optical light guides   248
        8.4.1 Introduction   248
        8.4.2 Sources of radiation-sensitivity in silica and glasses   248
        8.4.3 Prediction models for optical fibre loss versus dose   248
            8.4.3.1 Fundamentals   248
            8.4.3.2 Simple mathematical model   249
        8.4.4 Vapour-deposited fibre technology   251
        8.4.5 Fibres drawn from Suprasil rods   253
        8.4.6 Fibre luminescence   253
        8.4.7 Polymer optical fibres   256
    8.5 Scintillators   256
    8.6 Conclusions   257
    References   257

9 Other components   260

    9.1 Junction field-effect transistors   260
        9.1.1 Introduction   260
        9.1.2 Mechanisms of degradation of FETs   260

Contents

9.2 Transducers	262
9.2.1 General	262
9.2.2 Previous transducer studies	263
9.3 Temperature sensors	266
9.4 Magnetics	266
9.5 Superconductors	267
9.6 Mechanical sensors	268
9.6.1 General	268
9.6.2 Silicon micromechanisms	268
9.7 Miscellaneous electronic components	268
9.7.1 Capacitors	269
9.7.1.1 Total-dose effects	269
9.7.1.2 Dose-rate effects	270
9.7.2 Resistors and conductors	270
9.7.3 Quartz crystals	270
9.7.4 Vacuum tubes	271
9.7.5 Semiconductor microwave devices	271
9.7.6 Miscellaneous hardware	272
References	272

10 Polymers and other organics   275

10.1 Introduction	275
10.2 Radiolytic reactions	276
10.3 Radiation tolerance of polymers and organics according to application	277
10.3.1 General	277
10.3.2 Polymers in electronics	278
10.3.3 Remote handling	278
10.3.4 Accelerator parts	279
10.3.5 Optical fibres, windows and scintillators	279
10.3.6 Lubricants	280
10.3.7 Bombardment of coatings in space	280
10.4 Radiation processing	280
10.4.1 Sterilization of products	280
10.4.2 Irradiation of foods	280
10.4.3 Radiation curing of plastics	281
10.4.4 Resists	281
10.5 Long-lived degradation in polymers	281
10.5.1 Relative sensitivity	281
10.5.2 Effects of additives and fillers	282
10.5.3 Combined effects of stress (fields, vacuum, and temperature) and ageing with irradiation	283
10.6 Radiation-tolerance of plastics according to technology	283

	10.6.1	General	283
	10.6.2	Thermoplastics	284
		10.6.2.1 Structural plastics	284
		10.6.2.2 Plastics films as dielectrics and coatings	284
	10.6.3	Thermosetting plastics	284
	10.6.4	Elastomers	285
10.7	Radiation-induced conductivity in insulators		286
10.8	The space environment		287
10.9	Conclusions		288
References			288

11 The interaction of space radiation with shielding materials — 291

11.1 Introduction — 291
11.2 Particle radiation transport and range — 292
    11.2.1 General — 292
    11.2.2 Range — 292
11.3 Transport of electrons — 293
    11.3.1 Transmission coefficients for electrons — 295
    11.3.2 Stopping power — 296
    11.3.3 Internal spectrum — 297
11.4 Transport of protons and other heavy particles — 299
    11.4.1 Interactions — 299
    11.4.2 Energy loss and attenuation — 299
11.5 Electromagnetic radiation: bremsstrahlung, X- and gamma-rays — 299
    11.5.1 General — 299
    11.5.2 Bremsstrahlung — 300
    11.5.3 Other electromagnetic radiation — 300
    11.5.4 Production and attenuation of electromagnetic radiation — 302
        11.5.4.1 Production — 302
        11.5.4.2 Attenuation — 303
        11.5.4.3 'Build-up' — 306
    11.5.5 Soft X-rays and vacuum ultraviolet: generation and special effects of long-wavelength X-rays — 306
        11.5.5.1 Introduction — 306
        11.5.5.2 Machines and optics — 307
        11.5.5.3 Absorption constants — 307
        11.5.5.4 Effects — 308
11.6 Radiation attenuation by shielding; deposition of dose in targets — 310
    11.6.1 Dose versus depth — 310

xviii Contents

      11.6.2 Shielding: relation between space radiation flux and deposited dose   311
  11.7 Atomic displacement damage versus depth   316
  11.8 Influence of material type on radiation stopping   318
      11.8.1 Deposition of dose   318
      11.8.2 Shielding materials   318
      11.8.3 Routine calculation of particle transmission   324
  11.9 Conclusions   324
  References   325

12 Computer methods for particle transport   327

  12.1 Introduction   327
  12.2 Environment calculations   328
  12.3 Dose computation   329
      12.3.1 Space particle types   329
      12.3.2 Monte Carlo techniques   330
      12.3.3 Methods using a dose 'look-up table': SHIELDOSE   331
      12.3.4 Methods using straight-ahead approximation   331
      12.3.5 CHARGE program   333
      12.3.6 Sector analyses   334
      12.3.7 Comparisons   335
  12.4 Single-event upset prediction   336
  12.5 Neutrons, gamma-rays, and X-rays   337
  12.6 Conclusions   338
  References   339

13 Radiation testing   341

  13.1 Introduction   341
  13.2 Radiation sources   341
      13.2.1 Simulation of radiation environments   341
      13.2.2 Gamma-rays   343
      13.2.3 X-rays: steady-state and pulsed   344
      13.2.4 Electrons: steady-state and pulsed   346
      13.2.5 Protons   349
      13.2.6 Neutrons: steady-state and pulsed   350
      13.2.7 UV photon beams and other advanced oxide-injection methods   351
      13.2.8 Summary of requirements for steady-state radiation sources   351
  13.3 Cosmic ray upset simulation: heavy ions   352
  13.4 Dosimetry for testing   355
  13.5 Test procedures for semiconductor devices   356
      13.5.1 Introduction   356

Contents xix

	13.5.2 Objectives	357
13.5.3 Comparison of space with military requirements		358
13.6	Radiation response specifications	359
	13.6.1 General	359
	13.6.2 Product assurance techniques and special radiation assessment	359
	13.6.3 ESA/SCC specification (Europe)	360
	13.6.4 BS 9000 specification and CECC (Europe)	360
	13.6.5 MIL specifications (USA)	360
	13.6.6 ASTM specifications (USA)	360
13.7	Device parameter measurements	361
	13.7.1 MOS threshold voltage	362
	13.7.2 MOS flatband voltage (V_{FB}) and C-V plot	364
	13.7.3 Quiescent current (I_{SS}) in CMOS logic	364
	13.7.4 Leakage currents	364
	13.7.5 Current gain	364
	13.7.6 Input offset in analog ICs	365
	13.7.7 Noise-immunity and DC switching of logic gates	365
	13.7.8 AC and functional testing	365
	13.7.9 Single-event upset testing	365
	13.7.10 Measurement of transient photocurrent	366
13.8	Engineering materials	366
13.9	Time-dependent effects and post-irradiation effects	366
13.10 Conclusions		368
References		368
Data compilations		371

14 Radiation-hardening of semiconductor parts 372

14.1 General 372
14.2 Methodology of total-dose hardening 372
14.3 Hardening of a process 375
    14.3.1 Introduction 375
    14.3.2 Material preparation and cleaning 376
    14.3.3 Oxide growth 376
    14.3.4 Oxide anneal 377
    14.3.5 Gate electrode 377
    14.3.6 Modified gate insulators 378
    14.3.7 Field oxide hardening 378
    14.3.8 Other processing steps 378
14.4 Hardening for total dose by 'layout' 379
14.5 Hardening against transient radiation 380
    14.5.1 Pulsed gamma rays 380
    14.5.2 Single-event upsets 380

	14.6 Hardening of parts other than silicon	382
	14.7 Conclusions	382
	References	383

15 Equipment hardening and hardness assurance — 385

15.1 Introduction — 385
15.2 Elementary rules of hardening — 386
    15.2.1 General — 386
    15.2.2 Measures at systems level — 387
        15.2.2.1 Prediction and statistics — 387
        15.2.2.2 System theory — 387
15.3 Robots, diagnostics and military vehicles in penetrating radiation — 387
    15.3.1 Manipulators for nuclear plant — 387
    15.3.2 Hardening of a robotic vehicle — 390
    15.3.3 Preventive replacement and fault detection — 390
    15.3.4 Remotely controlled maintenance of fusion reactors — 392
    15.3.5 Instruments and detectors in power reactor and accelerator facilities — 393
    15.3.6 Military systems — 394
    15.3.7 General guidelines for hardening against pulsed gamma rays and neutrons — 396
    15.3.8 When only vacuum electronics will do — 397
15.4 Equipment in non-penetrating radiation: space, X-rays, and beta-rays — 397
    15.4.1 Introduction — 397
    15.4.2 Typical spacecraft configurations and materials — 399
        15.4.2.1 General — 399
        15.4.2.2 Properties of typical spacecraft materials — 402
        15.4.2.3 Spacecraft structure as a radiation stopper — 402
    15.4.3 Add-on shielding — 402
        15.4.3.1 Introduction — 402
        15.4.3.2 An example of shield weight trade-off — 403
15.5 On-board radiation monitoring — 405
15.6 Hardness assurance — 406
    15.6.1 General — 406
    15.6.2 Hardness assurance defined — 406
    15.6.3 Management of hardness assurance — 407
    15.6.4 Databases — 408
    15.6.5 Parts procurement and radiation design margins — 408
    15.6.6 Economics of hardness assurance — 410
        15.6.6.1 Programme costs — 410
        15.6.6.2 Testing costs — 413
        15.6.6.3 Sample sizes — 414

15.7 Conclusions	415
References	415

16 Conclusions 419

Appendix 421

A Useful general and geophysical data 421
    Table A1 Conversion factors, physical properties, and constants 421
    Table A2 Frequency, wavelength, and energy 422
    Table A3 Geophysical and orbital parameters and conversion factors 422

B Useful radiation data 423
    Table B1 Radiation units and data 423
    Table B2 Energy absorption versus photon energy for air 424
    Table B3 Typical performance figures for high-energy radiation sources 424
    Table B4 Typical photon energies and wavelengths 425
    Table B5 Radioisotopes useful in radiation experiments: main emission energies 426
    Table B6 Practical ranges of electrons in aluminium 427
    Table B7 Selected values of range of protons in aluminium 428
    Table B8 Range of alpha particles in silicon 428
    Table B9 Total mass attenuation coefficients of selected materials 429

C Useful data on materials used in electronic equipment 430
    Table C1 Densities and chemical formulae of commercial plastics 430
    Table C2 Radiation absorption effectiveness of various materials 431

D Test data: radiation response of electronic components 433
    Table D1 Degradation under total ionizing dose exposure: transistors 433
    Table D2 Dose for onset of malfunctions under total ionizing dose exposure: memories 434
    Table D3 Observations of single-event upset and latch-up under beams of heavy ions 437
    Table D4 Observation of single-event upset and latch-up under beams of protons 440
    Table D5 Total ionizing doze and single event upset data on some microprocessors 442

E Dose–depth curves for representative satellite orbits 444

Fig. E1	Geostationary transfer orbit	444
Fig. E2	Geostationary orbit	445
Fig. E3	Low Earth orbit: polar	446
Fig. E4	Low Earth orbit: space station	447
Fig. E5	'Molniya' orbit	448
Fig. E6	Typical interplanetary mission	449

F	Degradation in polymers	450
Fig. F1	Radiation-tolerance of elastomers	450
Fig. F2	Radiation-tolerance of thermoplastic resins	451
Fig. F3	Radiation-tolerance of thermosetting resins	452
Table F1	Analysis of data on plastics from Figs F1–F3	453

The appendices are issued as technical information, are not a licence to use information, and not part of a contract. In particular, while all information furnished is believed to be accurate and reliable, no liability in respect of any use of the material is accepted by the authors or Oxford University Press.

Author index 455
Subject index 465

Introduction

This handbook is intended to serve as a tool for designers of equipment and scientific instruments in cases where they are required to ensure the survival of the equipment in radiation environments. High-technology materials, especially semiconductors and optics, tend to degrade on exposure to radiation in many different ways. Hence the need for a guidebook to that set of 'radiation effects'. As for the radiation environments, first, space has a naturally high radiation level. Secondly, the surface of the earth is radioactive at a low level. When we go down a uranium mine or explore space, we run into much higher levels. Man concentrates radiation energy for a multitude of purposes. Intense high-energy radiation environments are found in nuclear reactors and accelerators, machines for radiation therapy, and industrial sterilization. Some engineers have to build equipment which will survive a nuclear explosion from a hostile source. There is the problem of machining or handling intensely radioactive isotopes such as spent nuclear fuel rods or radiotherapy capsules.... Finally, we have to continue to plan for the unthinkable. A disastrous explosion, dispersing radioactivity too intense for human survival, occurred at Chernobyl in the Ukraine, in 1986. Proper handling of a disaster with radioactive materials requires equipment which depends utterly on semiconductor microelectronics and imaging devices. Thus the technology of radiation-tolerant electronics is an instrument for good in social spheres as diverse as disaster planning and the exploration of Mars.

If we are to design equipment for intense environments like those described above, then degradation from high-energy irradiation must be seen as a BASIC DESIGN PARAMETER. It must be given as much thought as temperature, humidity or shock waves; engineering specifications for radiation exposure and the monitoring of degradation have to be written. The aim of this handbook is to assist the engineer or student in that thought; to make it possible to write intelligent specifications; to offer some understanding of the complex variety of effects which occur when high-technology components encounter high-energy radiation; and to go thoroughly into the balance of choices of how to alleviate the effects and hence achieve the design aims of a project. We go into detail on the more sensitive technologies such as solid-state devices, optics and organic materials; we discuss in depth the best solutions for spacecraft, terrestrial robots and imagers. However, to understand these, the reader will soon see that it is also necessary to have a 'running knowledge' of the main types of radiation and how they interact with solids. We also attempt to supply this knowledge.

2 Introduction

The discussion covers radiation physics, solid-state physics, and electronics, but our explanations are pitched so that advanced scientific training is not needed to follow them. An appreciation of engineering terms and methods is assumed but it is hoped that the book will act as a reference tool and a bridge between the plentiful, somewhat baffling scientific treatises on radiation effects and the potential user, the engineer. The explanatory mode should also make the book useful to students, although it is not designed to fit any formal course of training that exists today.

The authors have both worked primarily with the space engineering and nuclear fusion community but have also collaborated with workers in the fields of microelectronics, defence, radiotherapy, accelerators, sterilization, reactors, and robotics; while the space environment and the space industry are widely referred to as examples in this handbook, other radiation environments, noted in our opening paragraph, are also described. Discussion of the intricate difficulties and the ingenious solutions which have been used in space should have something to offer engineers in other fields considering similar environments. The authors point out where problems and solutions are common to more than one field; they aim to offer complex information as clearly as possible.

If a project requires equipment to operate in a radiation environment, it is important that guidance on radiation effects reaches the designer at the right phase of the work. The authors maintain that there is only one appropriate time—right at the start. Radiation effects should reside in the BASIC DESIGN PARAMETERS of the project from the start. The authors hope to increase the understanding of this managerial point among students as well as professionals. It is hoped that these workers use the knowledge we impart to evolve engineering techniques, develop novel approaches to further research, and devise better management policies.

1
Radiation, physics, and measurement

1.1 General

Radiation effects is an interdisciplinary subject which contains radiological physics and solid-state physics at its core. This first is the science of high-energy particles and photons and their interaction with matter. This science began with the discoveries, in the 1890s, of X-rays and radioactivity by Röntgen, of radioactivity by Becquerel, and of radium by the Curies (see Roesch and Attix 1968). The effect of natural radiation on silver chloride – a mixture of a physical and a chemical radiation effect – was the phenomenon which ushered in Becquerel's discovery of radioactivity. The damage that radiation can do to human tissues – a biological radiation effect – was discovered, at some personal cost, by the users of Röntgen's tubes. The radiation effects which concern us here occur predominantly in solids and require considerable appreciation of solid-state physics. The need for this understanding arose as a matter of urgency only in the late 1950s, when semiconductor technology began to be used in space and military equipment. These two 'aerospace' environments – nuclear weapons and space – provide exposure to radiation which leaves many active semiconductor devices 'stone dead'.

The workers in the field – and the authors of this book – have been engaged since, in collecting and using the parts of radiological and solid-state physics which are relevant to understanding and predicting the many phenomena which occur when radiation interacts with an active semiconductor device. A few other classes of high-technology materials, such as polymers, naturally fall within the same sphere of interest.

This first chapter gives a brief review of radiological physics, units, and measurement methods, before launching into radiation environments and the effects they cause. Several outstandingly clear radiological course books will help the reader at this stage, including Attix (1986), Johns and Cunningham (1969), Knoll (1989), and Profio (1979).

We live in a universe which is constantly bombarding us from inside and outside (Adams and Holmes-Siedle 1991). In all matter, the protons and neutrons within a nucleus are in continual motion. As a result of this motion, energy is transferred between particles. In a stable nucleus no particle acquires enough energy to escape. In a radioactive nucleus it is possible for a particle to gain high energy and to escape from the nucleus. As a result, we are bombarded by high-energy particles and photons from radioactive nuclei. We can measure environmental dose from any sample of earth, rock or air. As will be described in Chapters 2 and 3, we are also bombarded from space by particles accelerated by dynamic magnetic fields in and near the sun, other stars and planets. Space and nuclear hazards are only intensified versions of what we are already exposed to in Nature.

1.2 Radioactivity

We learnt above that some atomic nuclei are less stable than others. The lighter atoms are all stable with a neutron/proton ratio of 1. This ratio gradually increases to 1.5 with the heavier elements. All elements with Z between 1 and 92 (uranium) exist naturally and those from 93 to 106 have been produced artificially. All elements with $Z > 82$ (lead) are radioactive and undergo nuclear rearrangement with emission of sub-atomic particles and gamma radiation until a stable configuration is reached. Nuclides of high atomic number (as are the majority of fission products) emit negative beta particles. Proton-rich nuclides emit positive beta particles.

The SI unit of *radioactivity* is the becquerel, which is a rate of 1 disintegration per second. The traditional, but still commonly used, unit of activity, the Curie, was derived from the number of disintegrations in 1 gram of radium and is a rate of 3.7×10^{10} becquerels.

The *decay rate* of a nuclide is defined in terms of its 'half-life'. The *half-life* is defined as the time taken for the radioactivity to fall to half its original value. The number of disintegrations occurring at any one time is proportional to the number of unstable nuclides present, hence

$$t_{1/2} = \frac{\log 2}{\tau} = \frac{0.693}{\tau},$$

where τ is the decay constant for the nuclide. The half-life of any nuclide is peculiar to the nuclide and independent of the activity present.

The *specific activity* of a nuclide is the amount of activity associated with one gram. The SI unit is becquerels per gram, although Curies per gram is still used. As noted earlier, the specific activity of radium (used as a reference) is 1 Curie per gram (3.7×10^{10} becquerels per gram).

The change in neutron/proton ratio during nuclear rearrangement does not necessarily result in a stable nuclide. The daughter nucleus will also be unstable and undergo rearrangement and decay which continues until the final nuclide is stable. Such a sequence is known as a decay series or chain.

1.3 Types of radiation

Gamma rays and X-rays. These are short-wavelength forms of photon or electromagnetic radiation. The different names derive from discovery at different times. A gamma ray has its origin in a nuclear interaction, whereas an X-ray originates from electronic or charged-particle collisions. The ways in which these photons interact with matter are identical. They are lightly ionizing and highly penetrating and leave no activity in the material irradiated.

Alpha particles. Alpha particles are the nuclei of helium atoms. They have a mass of 4 and a positive charge of 2 units. Normally of high energy (in the MeV range), they interact strongly with matter and are heavily ionizing. They have low penetrating power and travel in straight lines. A typical alpha particle energy is 5 MeV with a typical range of 50 mm in air and 23 μm in silicon.

Beta particles, electrons, positrons. Beta particles have the same mass as an electron but may be either negatively or positively charged. With their small size and charge they penetrate matter more easily than alpha particles (see Section 11.2 and Appendix B6) but are more easily deflected. Their high velocity, normally approaching that of light, means they are lightly ionizing.

Neutrons. A neutron has the same mass as a proton but has no charge and consequently is difficult to stop. The neutron can be slowed down by hydrogenous material. The capture of a neutron results in the emission of a gamma ray. Neutrons are classified according to their energy: thermal (<1 eV); intermediate; and fast (>100 keV). Water is an especially effective shield for neutrons (Profio 1979).

Protons. The proton is the nucleus of a hydrogen atom and carries a charge of 1 unit. The proton has a mass some 1800 times that of an electron, and consequently is more difficult to deflect, with a typical range of several centimetres in air and tens of micrometres in aluminium at energies in the MeV range (see Section 11.2 and Appendix B7).

1.4 Units (see also Appendix A)

Energy units. The SI unit of energy is the joule (J). However the electron volt (eV) is more frequently used in radiation technology. One eV is the energy gained by one electron in accelerating through a potential difference of 1 volt. The conversion factor is: $1 \text{ eV} = 1.6 \times 10^{-19}$ J. Energies in nuclear reactions are usually quoted in MeV (10^6 eV) or keV (10^3 eV). A traditional unit of energy, the erg, is 10^{-7} joules.

Flux. Also known as the fluence rate or flux density, this is the number of particles passing through some defined zone per unit time. To allow for isotropic as well as anisotropic radiation, the zone is usually taken as a sphere with a cross section of 1 cm^2. The unit of flux is cm^{-2}s^{-1}; the symbol for the particle is normally added.

Fluence. This is the time-integrated flux of photons or particles. The unit is cm^{-2}. As for flux, the type of particle is also usually stated.

Roentgens, rads, and grays. Until recently, the gas ionization chamber was the only means to measure electrically the dose derived from a radiation beam. This chamber is merely a pair of electrodes arranged to collect the air ions created in a certain volume, but the values thus measured still serve as a standard to which to relate other units, and formed the basis on which the roentgen unit has been defined. The latter is that unit of exposure to radiation which creates air ions to the level of 2.58×10^{-4} coulombs per kilogram (previously defined as 1 esu cm^{-3} in air of density 0.001293 g cm^{-3} at stp). This corresponds to the deposition of energy in air at the rate of 87 erg per gram.

The rad and gray (Gy) are units of energy deposition; a rad has been absorbed by the sample of interest when 100 ergs per gram, and a gray when 1 joule per kilogram, has been deposited. One rad thus equals 10^{-2} gray or 1 centigray (cGy).

The roentgen was abandoned in the 1960s as a standard of radiation quantity but is used as a measure of radiation exposure, convertible to 'air kerma' in rads (Gy) if we know the relative mass energy absorption coefficients for the radiation in air and in the material in question. For 1 MeV photons, the factor for converting roentgens to rads in water is 0.965 rad roentgen^{-1},

TABLE 1.1 Relative photon energy absorption coefficients (water–material) for various materials. The physical state of the elements is not important

	1 MeV	100 keV
H	0.557	0.631
C	1.11	1.19
O	1.11	1.08
Al	1.15	0.663
Fe	1.18	0.117
Cu	1.20	0.0848
Pb	0.82	0.0112
Perspex (PMMA)	1.03	1.08
Polyethylene	0.97	1.05
LiF	1.20	1.14
Glass	1.12	0.788

i.e. the flux of 1 MeV photons which yields 0.87 rads in air yields 0.965 rads in water. Some useful figures on relative photon energy absorption coefficients (water/material) are given in Table 1.1. The values for glass and Al are very similar at 1 MeV (from which one may reasonably assume that the figures for SiO_2 and Si would be only marginally different, say 1.12 and 1.15 respectively). These figures are useful for calculating relative doses for radiation testing.

Even though the gray is the SI unit, the rad is used in this book because this is still the working unit for most published papers in radiation effects (see e.g. IEEE Transactions) and also for current medical practice. Some workers continue to work in rads but write cGy instead. Since the SI unit may well become more widely used, we attempt to note SI units in brackets beside traditional units for reference.

Biological units. In radiobiology the dose-absorbing media of interest are tissue, bone, muscle, and blood-forming organs. All ionizing radiation produces similar biological effects but the actual damage effect for a given absorbed dose differs from one radiation type to another. The difference is expressed as the Relative Biological Effectiveness (RBE):

$$\text{RBE} = \frac{\text{Absorbed dose of reference radiation}}{\text{Absorbed dose of given radiation with same effect}}$$

Alternative units for estimating biological effects are Quality Factor, Q, based on the linear energy transfer of the radiation concerned; and 'radiation weighting factors' based on the risks of doses absorbed in given patterns in a living body (ICRP 1966). Some typical quality factors are given in Table 1.2.

The biological unit corresponding to the gray is the sievert (Sv):

$$\text{biological dose} = \text{dose}\,(\text{rad or gray}) \times \text{RBE}.$$

In a mixed radiation field the integrated biological dose can be obtained by summing the doses from each radiation type, each with its own RBE.

Kerma. The deposition of dose by a photon beam in an absorbing material is a two-step process. The first step is when the kinetic energy of the photons

TABLE 1.2 Quality factors for different types of radiation

Radiation	Quality factor
X-ray, gamma, beta	1
Slow neutrons	5
Fast neutrons and protons	10
Alpha particles	20
HZE particles (cosmic rays)	20

HZE = high atomic weight and energy

is converted to the production of energetic electrons. The second step is when the electrons deposit energy by ionization in the material.

In 1962 the ICRU introduced a unit to describe the first step: the Kinetic Energy Released in Matter, or kerma K, which for photons is:

$$K = \Psi \cdot \mu/\rho,$$

where Ψ is the energy fluence and μ/p the mass energy transfer coefficient for the material (Attix 1986).

If all of the energy absorbed in the material is converted into dose, then kerma and dose are equal and electronic equilibrium is said to exist. Equilibrium is also defined as when the same number of electrons are set in motion in a given volume as come to rest in the same volume. This is when the material is at least as thick as the range of the Compton electrons. For the majority of conditions considered in this book, electronic equilibrium exists. An extreme case may exist in the testing of thin samples such as Mylar or gold films. 'Build-up' material is placed in front of the test sample. In the case of Co-60 gamma radiation a typical build-up material would be 2 mm of aluminium. For another case, see Section 4.10.4.

1.5 The measurement of radiation

1.5.1 General

Dosimetry is the science of measuring the amount of energy absorbed in a sample when exposed in a given radiation beam. It is also taken here to cover the measurement of particle or photon fluxes and of the absorption or deposition of energy in the radiation-sensitive sample of interest. Dosimetry grew up in the fields of radiobiology and medicine, where energy deposition in the form of ionization is of interest and takes place primarily in the aqueous or organic media of living tissue.

Classical dosimetry methods and concepts have been developed mainly for the calculation of ionization in organic materials, water, and air (the medium used in ionization gauges). In the field of radiation effects in electronic components, we are interested mainly in

(a) deposition of ionization energy in silicon dioxide, silicon and a few other materials;

(b) deposition of energy in the form of atomic displacements in crystalline lattices.

Thus, dosimetry for the aerospace field covers certain concepts and apparatus not found in the classical dosimetry textbooks.

A wide range of dosimeters is available and the choice of dosimetric technique is determined by the type of radiation, the accuracy required, the dose range of interest, and the physical constraints of the experiment. Dosimetric techniques may be broadly divided into physical and chemical and are briefly reviewed in the following sections. For a more detailed treatment the reader is referred to books by Johns and Cunningham (1969), Attix (1986), McLaughlin *et al.* (1989), Paić (1988), and Greening (1981), especially the bibliography in the last. For discussion of pulse-counting detectors, see Knoll (1989) and Delaney *et al.* (1992).

1.5.2 Techniques

1.5.2.1 *Ionization chamber*

Ionization chambers are used as primary and secondary reference standards. Miniaturized and ruggedized versions (thimble chambers) are also employed in routine applications in therapy, radiation protection, and radiation testing. The theory of the thimble chamber is clearly discussed in Greening (1981). The essential features of an ionization chamber are a gas-filled cavity surrounded by electrically conducting walls. A collecting electrode is connected to a sensitive, high-impedance, current-measuring instrument such as an electrometer. A potential difference (several hundred volts) is maintained between the collecting electrode and the chamber walls. When the chamber is exposed in a radiation field, secondary electrons are produced mainly in the walls of the chamber. The secondary electrons produce ions in the gas which are collected by the central electrode and cause a current to flow in the external circuit.

The operation of an ionization chamber is governed by Bragg–Gray theory, which states that 'The energy absorbed per unit volume of the wall material is equal to the product of the ionization current produced per unit mass of gas in the cavity'.

The condenser chamber consists of a thimble ionization chamber connected to a capacitor which is built into the stem of the chamber. This is then charged to several hundred volts. When the thimble portion is irradiated, the reduction in, the voltage across the capacitor is proportional to the charge collected in the ionization chamber. This chamber can be carried about and irradiated without connection to a power supply.

The 'electrometer' personnel dosimeter is derived from the gold leaf electroscope, where a central electrode (generally quartz) is deflected by the charge accumulated on parallel electrodes connected to the ionization chamber. The mechanical deflection, measured optically, is proportional to the absorbed dose.

1.5.2.2 *Calorimeter*

The calorimeter, based on the measurement of energy deposited as heat in a thermally isolated mass of a chosen target material such as silicon or carbon, is a primary standard dosimeter. A calorimeter may be used in all radiation fields and is energy-, dose-rate-, and total-dose-independent. It is therefore used as a primary standard, despite the difficulties of the method. The detailed design of a calorimeter varies according to application. The temperature rise is measured with a thermistor or thermocouple and calibration is carried out using an embedded resistor. Special care is required in calorimeter design to minimize heat loss, for example by evacuating the enclosure or using plastic insulating foam. Although it is a simple device, the calibration and use of a calorimeter are quite complex, involving dynamic energy absorption, heat transfer, and heat loss.

1.5.2.3 *Geiger–Müller and proportional counters*
Any ionization chamber, operated at a high potential, will exhibit *gas multiplication*, in which a single ionization event leads to an 'avalanche' of ionization events. The current pulse from the avalanche is collected and registered as a count. Counting tubes typically consist of coaxial electrodes enclosed in an argon gas-filled volume. The central electrode is usually a thin tungsten wire (anode), and the cathode a conductive coating on a thin-walled coaxial cylinder of glass. The tube operation is governed by the gas pressure and the potential applied between the electrodes. The principle of operation is based on the ionization of the gas and multiplication by impact ionization. Various regimes of operation exist, the main regimes being a plateau in the I–V response where the tube acts as an ionization chamber; a 'proportional' regime where pulse height is proportional to anode voltage and the size of the initiating event; and finally the Geiger–Müller regime where pulse heights are all approximately the same. Each regime has its own specialized design and geometry. Geiger–Müller tubes require a mixture of an inert gas and a quenching agent. They are energy-dependent but well characterized for this dependence and are frequently used in miniaturized and ruggedized forms as personnel and area dosimeters for radiological protection.

1.5.2.4 *Faraday cup*
For beams of charged particles, a block of metal of the appropriate thickness will stop all the particles and the flow of the resultant charge can, of course, be measured. This arrangement is known as a Faraday cup and refined forms are used for controlling many accelerators in conjunction with single-particle counters collecting scattered radiation. The main refinements are the evacuation of air around the cup electrode, the shaping of the cup and the addition of biased grids so that secondary electrons do not escape.

1.5.2.5 *Thermoluminescent dosimeters (TLDs)*
Many crystalline materials exhibit thermoluminescence (Attix 1986). Light is emitted when the material is heated after exposure to radiation. Electron and hole traps are filled during radiation exposure (this is a solid-state ionization effect); heating frees electrons and holes from these traps. Light is emitted when the electrons and holes recombine. The light emitted is

characteristic of the material and a function of temperature. The plot of light output versus temperature is known as the 'glow curve' for that particular material. The integrated light output in the glow curve is a measure of the radiation dose. Since the traps are emptied by heating, TLDs can be reused after a further high-temperature annealing step, so long as residual bulk damage is not too great (i.e. so long as the doses are low).

A TLD 'reader' consists of a heat source (a lamp or a platinum 'planchette') and a photomultiplier with a light-sensitive phosphor. The output of the photomultiplier is taken to a current integrator and readout system.

TLD materials commonly used are lithium fluoride (LiF), calcium fluoride (CaF_2), calcium sulphate ($CaSO_4$), lithium borate ($Li_2B_4O_7$), beryllia (BeO), and alumina (Al_2O_3). These materials are activated with the addition of rare earths or transition metals. Many other materials exhibit useful thermoluminescence, including silica and some natural materials used in archaeological dating and sterilization dosimetry.

TLDs are available in many forms: powder; hot-pressed pellets; powder combined with carbon or Teflon; rods; and tapes. A typical TLD pellet is 4 mm × 4 mm × 1 mm, hence is very convenient for radiological protection, therapy, and radiation testing. The smallest dose that can be measured with TLDs is in the range of a few millirads. The maximum dose is generally quoted as being in the range of 10^4–10^5 rad (10^2–10^3 Gy), but as these levels are approached there is some loss of linearity. Repeated exposures to high dose levels may result in a permanent loss of sensitivity. TLDs are subject to 'fading'; this is quite small in the case of LiF (a few per cent over a period of months) but can be quite high in the case of $CaSO_4$ and CaF_2 (10 to 50 per cent in a matter of hours). TLDs should not be exposed to light or elevated temperatures between exposure and readout.

TLDs can be made sensitive to neutrons as well as gamma rays; they also exhibit energy-dependence according to the material, so with careful selection and combination of TLDs, and the use of different shielding materials, it is possible to discriminate between types of radiation and provide energy information.

1.5.2.6 *Radiophotoluminescent (RPL) glass*

Certain defects in glass increase their ability to luminesce linearly with ionizing dose. This effect can be calibrated to register doses with an accuracy of 10 per cent, and RPL chips are used in accident dosimetry lockets, and in high-dose measurements on nuclear accelerators (Coche *et al.* 1988). The reader is a UV flash-lamp and photomultiplier.

1.5.2.7 *Fricke dosimeter*

The ferrous sulphate, or Fricke, dosimeter is a secondary standard chemical dosimeter based on the oxidation of ferrous ions to ferric ions. The Fricke dosimeter is very accurate (1–2 per cent) and the radiation chemistry is well understood. The dosimeter is usually in the form of a sealed glass ampoule of a few cm^3 volume containing an aqueous solution. Scrupulous cleanliness and purity are necessary in the preparation of the dosimeters. Readout is performed using a spectrophotometer, and the light absorbance at 303 nm (absorption maximum of aqueous ferric ions) is a measure of the radiation dose.

1.5.2.8 *Perspex dosimeters*
Perspex (polymethyl methacrylate, or PMMA) is a very simple dosimeter, widely used for high dose applications, 10^5–10^8 rad (10^3–10^6 Gy). Perspex darkens under radiation, and readout is performed by measuring the change in optical density at a wavelength near 310 nm. The response is reasonably linear up to about 10^6 rad, after which it becomes sub-linear. An accuracy of about 5 per cent can be achieved, and Perspex has the advantage of being water- and muscle-equivalent.

Dyed Perspex dosimeters (red, amber, and Gammachrome-YR) have been developed to improve sensitivity at lower doses. Gammachrome-YR, for example, covers the range 10–300 krad (0.1–3 kGy). Absorption measurements for dyed Perspex are made at around 640 nm.

Perspex dosimeters are subject to fading (particularly at elevated temperatures) and can be affected by relative humidity.

1.5.2.9 *Colour changes and optical transmission*
Radiochromic dosimeters are plastic films containing dyes which become deeply coloured under radiation, within a dose range of 10^4–10^8 rad (10^2–10^6 Gy). Nylon-based films are frequently used as dosimeters. The change in absorbance of these films is usually measured at about 600–630 nm. Radiochromic films can be affected by temperature and relative humidity. Colour changes in other dyes and pigments under irradiation have been investigated. Some media can be painted on to surfaces (see Coche *et al.* 1988). For a discussion of optical transmission, see Chapter 8.

1.5.2.10 *Polyethylene and hydrogen pressure*
For high doses (megarad to megagray), the pressure of hydrogen developed by a weighed quantity of polyethylene in a glass ampoule can be measured and calibrated against ionizing dose. This has been used on accelerators of the Rutherford Laboratory, England, and CERN, Geneva (Coche *et al.* 1988).

1.5.3 Silicon dosimeters
1.5.3.1 *General*
Silicon has already been mentioned with respect to calorimetry. Here we discuss various junction and MOS (metal-oxide semiconductor) devices in

1.5 The measurement of radiation 13

sawn chip form. Silicon p-n junction detectors can be used for both dose rate and total dose measurements. These detectors have high sensitivity and provide 'real-time' information. For radiobiological applications (Attix 1986) there are difficulties related to 'tissue equivalence' (energy dependence compared with water or muscle), but for radiation testing of the type discussed in this book, the silicon device has the advantage of being of the same material as most of the electronic devices under test.

1.5.3.2 *Dose rate*
For dose-rate measurements a p-i-n (p-intrinsic-n) reverse-biased silicon diode is used. The large intrinsic region plus the depletion region acts as a solid-state ionization chamber. Electron–hole pairs generated by the radiation will be separated by the applied field and collected by p and n regions as an 'ionization current'. A particle, or photon, passing through the detector will generate an electrical pulse and the amplitude of the pulse will be proportional to the charge deposited. This property means that this type of p-i-n detector (called a dE/dX detector) can be used as an energy spectrometer. If the thickness of the intrinsic region and depletion layer is greater than the range of the particle, then the charge collected is proportional to the total energy of the particle. In order to improve the sensitivity to low-LET particles (e.g. alpha particles), lithium may be diffused into the silicon to increase resistivity (and hence the size of the depletion region for a given bias voltage). Unfortunately lithium can diffuse in silicon at room temperature; consequently 'lithium-drifted' detectors must be cooled to avoid a steady degradation in resolution at room temperature.

P-i-n detectors are frequently used to monitor accelerator beams and for instruments measuring the space radiation environment (Knoll 1989).

1.5.3.3 *Dosimeter diodes*
Silicon detectors will suffer permanent degradation as a function of total dose (particularly at the higher dose levels in the range of 10^6 rad). This effect can be used to provide an integrating dosimeter for high-dose applications. For the measurement of the integrated energy deposition from massive particles, the electrical changes produced by atomic displacement in semiconductors can be used. The forward voltage drop of a semiconductor diode is affected by the defect states produced near the junction (see Chapter 6), and this effect is used in dosimeters designed to measure high-energy neutrons without registering dose from high-energy gamma rays or electrons (McCall *et al.* 1978). The effect per rad of neutron tissue dose in roughly constant for neutron energies between 0.3 and 14 MeV. The voltage drop for a constant test current increases linearly versus dose and is large enough to measure neutron doses in the region of 1 tissue rad (10 mGy) and so is useful in accident dosimetry.

1.5.3.4 *MOS dosimeters*

For integrated ionizing dose measurements a different response mechanism is used. Silicon dosimeters for this application (RADFETs) are MOS structures which exploit the fact that charge generation and trapping in the gate insulator of an MOS device will cause a shift in electrical characteristics, notably the threshold voltage, and this shift is proportional to the radiation dose. The physical mechanisms are explained in Chapter 4, and reviews of RADFET principles and applications have been published by the present authors (Holmes-Siedle and Adams 1986; Holmes-Siedle *et al.* 1992). The present status of RADFET technology is such that the minimum measurable dose is about 1 rad and the maximum about 10^6 rad. The main applications so far found for RADFETs are cases where remotely monitored, real-time dose information is required, with minimal demands on voltage and detector head size, such as space vehicles, high-energy accelerators, nuclear facilities, hostile environment robots, and radiotherapy. These devices have achieved considerable success as in-orbit monitors of doses of 1–10 rad per day (0.01–0.1 Gy per day) as discussed in Sections 15.5 and 15.6. More sensitive forms are being investigated as probes for radiotherapy (Gladstone *et al.* 1991*a*, *b*).

References

Adams, L. and Holmes-Siedle, A. G. (1991). La survie des composants dans l'espace. *La Recherche*, **22**(236), 1182–9.

Attix, F. H. (1986). *Introduction to radiological physics and radiation dosimetry*. Wiley, New York.

Coche, M., Coninckx, F., Schönbacher, H, Bartolotta, A., Onori, S. and Rosati, A. (1988). *Comparison of high-dose (megarad to megagray) dosimetry systems in accelerator radiation environments*. Report No. TIS-RP/205 CERN, Geneva.

Delaney, C. F. G. and Finch, E. C. (1992). *Radiation detectors: physical principles and applications*. Oxford University Press.

Gladstone, D. J., Chin, L. M. and Svensson, G. K. (1991*a*). Automated data collection and analysis system for MOSFET radiation detectors. *Medical Physics*, **18**, 542–8.

Gladstone, D. J., Chin, L. M. and Holmes-Siedle, A. G. (1991*b*). MOSFET radiation detectors used as patient radiation dose monitors during radiotherapy. *Medical Physics*, Abstr. S3, **18**, 593. (May/June 1991), presented at 33rd Annual Meeting of the American Association of Physicists in Medicine, San Francisco, July 21–25, 1991.

Greening, J. R. (1981). *Fundamentals of radiation dosimetry*. Adam Hilger, Bristol.

Holmes-Siedle, A. G. (1980). *Radiation effects in the Joint European Torus experiment: guidelines for preliminary design*. Report No. R857/2 Fulmer Research laboratories, Stoke Poges UK.

Holmes-Siedle, A. G. and Adams, L. (1986). RADFET — a review of the use of metal-oxide-silicon devices as integrating dosimeters. *International Journal of Radiation Physics and Chemistry*, **28**(2), 235–44.

Holmes-Siedle, A. G., Adams, L., Leffler, S. and Lindgren, S. R. (1992). The RADFET system for real-time dosimetry in nuclear facilities. In *Proceedings of the 7th Annual ASTM–EURATOM Symposium on Reactor Dosimetry, Strasbourg*. Kluwer, Dordrecht, pp. 851–9.

ICRP (1966). International Commission on Radiation Protection, Report No. 9. Pergamon Press, Oxford.

Johns, H. E. and Cunningham, J. R. (1969). *The physics of radiology*. Charles C Thomas, Springfield, Illinois.

Knoll, G. F. (1989). *Radiation detection and measurement* (2nd edn.). Wiley, New York.

McCall, R. C., Jenkins, T. M. and Oliver, G. D., Jr. (1978). Photon and electron response of silicon-diode neutron detectors *Med. Phys.* **5**, 37.

McLaughlin, W. L., Boyd, A. W., Chadwick, K. H., McDonald, J. C. and Miller, A. (1989). *Dosimetry for radiation processing*. Taylor and Francis, New York.

Paić, G. (1988). *Ionizing radiation: protection and dosimetry*. CRC Press, Boca Raton, Florida.

Profio, A. E. (1979). *Radiation shielding and dosimetry*. Wiley, New York.

Roesch, W. C. and Attix, F. H. (1968). Basic concepts of dosimetry in *Radiation dosimetry*, (eds. F. H. Attix and W. C. Roesch), Vol. 1, Ch. 1. Academic Press, New York.

2
Radiation environments

In this chapter we present a brief overview of the various environments likely to have a degrading effect on electronic devices and systems: space, nuclear reactors, processing, weapons, and controlled fusion. We also provide a description of the terrestrial environment, partly because of public concern regarding such events as the Chernobyl nuclear accident, and partly to place in context the various levels of radiation exposure which may be experienced by humans and equipment in these widely differing environments.

2.1 The space environment

2.1.1 General

The space radiation environment is composed of a variety of energetic particles with energies ranging from kev to GeV and beyond. These particles are either trapped by the Earth's magnetic field or are passing through the solar system.

The main elements of the radiation environment are:

1. Trapped radiation. This consists of a very broad spectrum of energetic charged particles trapped in the earth's magnetic field, forming the *radiation belts*.
2. Cosmic rays. These are low fluxes of energetic heavy ions extending to energies beyond TeV and including all ions in the periodic table.
3. Solar flares. Solar eruptions produce energetic protons with a minor contribution of alpha particles, heavy ions, and electrons. Energies range to hundreds of MeV.

In addition, space is pervaded by a plasma of electrons and protons with energies up to about 100 keV. The fluxes associated with this plasma are as high as $10^{12}\,\text{cm}^{-2}\,\text{s}^{-1}$. Within the trapped radiation belts these particles merely represent the low-energy extremes of the trapped electron and proton populations. In the outer zones of the magnetosphere and in interplanetary space, these particles are associated with the solar wind, and considerable fluxes will be encountered at very high altitudes. The low-energy particles are easily stopped by very thin layers of material and hence only the outermost surfaces such as thermal control material and solar cell cover slips are affected. The low-energy plasma can cause spacecraft charging, and

2.1 The space environment

the internal electronics may be affected by this charging and subsequent discharging. The interaction of the above environments with structural and shielding materials can result in the generation of secondary radiation, including highly penetrating bremsstrahlung and neutrons.

2.1.2 The radiation belts

The Earth's radiation belts consist mainly of electrons of energy up to a few MeV and protons up to several hundred MeV which are trapped in the Earth's magnetic field. The field is basically that of a magnetic dipole and in those regions where the field lines are 'closed', charged particles become trapped in the magnetosphere. The field is not geographically symmetrical; local distortions are caused by an offset and tilt of the magnetic axis and by geological influences; one important distortion is known as the South Atlantic anomaly. The sun also heavily distorts the magnetosphere to a great degree. The resulting form of the field has characteristics comparable to the 'bow wave' and 'wake' of a solid object moving through a fluid. This effect is shown in Fig. 2.1 (Daly 1989). Field lines are not shown specifically, but

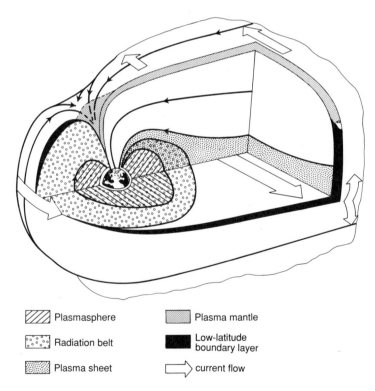

Fig. 2.1 The magnetosphere and radiation belts surrounding the Earth. The sun is to the left of the figure (Daly 1989). (ESA with permission.)

18 Radiation environments

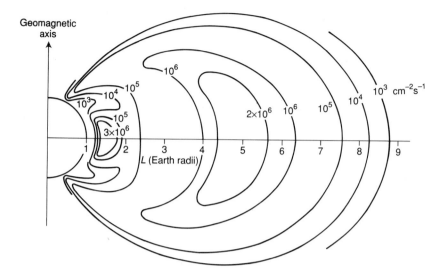

FIG. 2.2 Trapped-electron radiation belts (Daly 1989). Cross-section in the plane of the Earth's magnetic axis plotting contours of equal electron flux of energy above 1 MeV; the radius of the Earth is 6371 km. (ESA with permission.)

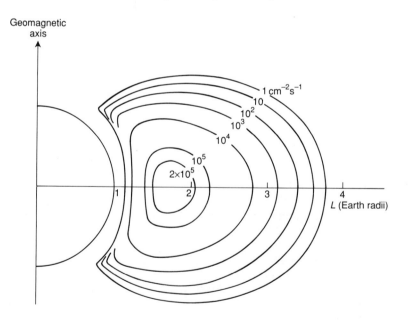

FIG. 2.3 Trapped-proton radiation belts (Daly 1989). Cross-section in the plane of the Earth's magnetic axis plotting contours of equal proton flux of energy above 10 MeV; the radius of the Earth is 6371 km. (ESA with permission.)

the magnetosphere here is divided into regions of different trapping capability. The sun is to the left of the figure. The structure of the trapped electron and proton belts is shown in Figs 2.2 and 2.3 (Daly 1989). The contours join points of equal average particle flux.

The electron environment consists of particles up to 7 MeV with the most energetic particles occurring in the 'outer zone' described below. In contrast, proton energies extend to several hundred MeV, with the most energetic found at lower altitudes. The electron environment shows two flux maxima, referred to as the 'inner' and 'outer' zones. The inner zone extends to about 2.4 earth radii (R_e) and the outer zone from 2.8 to 12 R_e. The gap between 2.5 and 2.8 R_e is referred to as the 'slot'. The outer zone envelops the inner zone, its contours extending towards the earth in cusps of relatively high flux (the 'polar horns').

The proton environment does not exhibit 'inner' and 'outer' zones nor 'polar horns'. The flux varies with distance from earth inversely and monotonically with energy, the outer boundary being at about 3.8 R_e (Stassinopoulos 1988). A 'cross section' of the radiation belts is shown in Fig. 2.4, which shows the radial flux profiles in an equatorial orbit. The South Atlantic anomaly and polar horns are shown in Fig. 2.5 as contour plots on a world map (Stassinopoulos 1970).

The models shown in Figs 2.2 and 2.3 are compilations of large numbers of scientific detector readings in orbit. These were combined into rational models (e.g. AP8 for protons and AE8 for electrons). Pioneers in this field were Vette and Stassinopoulos. References to the work can be found in Holmes-Siedle and Freeman (1978) and Stassinopoulos (1988).

2.1.3 Cosmic rays

There are three sources of cosmic rays: galactic, solar, and terrestrial.

Galactic cosmic rays. These are 'primary' cosmic rays which originate outside the solar system but are associated with the galaxy and provide a continuous, low-flux component of the radiation environment. They comprise about 85 per cent protons, 14 per cent alpha particles and 1 per cent heavier nuclei with energies extending to 1 GeV. Figure 2.6 shows the elemental abundances of galactic cosmic rays together with solar system abundances.

Solar cosmic rays. In addition to producing an intense burst of both UV and X-rays, solar flares accelerate solar material to high velocities. These solar particles are similar to galactic cosmic rays but, owing to their different origin, are not identical in composition. During a flare, cosmic rays are dominated at low and medium energies by solar material of low atomic weight. In addition, the extragalactic overall cosmic radiation in an earth orbit may increase dramatically. This is because solar flares cause geomagnetic disturbances resulting in considerable lowering of the geomagnetic barrier. 'Geomagnetic shielding' is discussed below.

20 Radiation environments

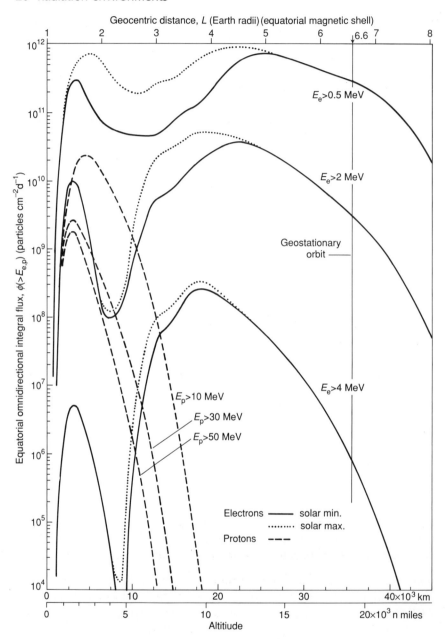

FIG. 2.4 Radial flux profiles in space around the Earth: trapped electron and proton integral flux (orbit-integrated) as a function of altitude in circular equatorial orbits. Data should be regarded as qualitative only, being derived from models current in the 1970s (Holmes-Siedle and Freeman 1978).

2.1 The space environment 21

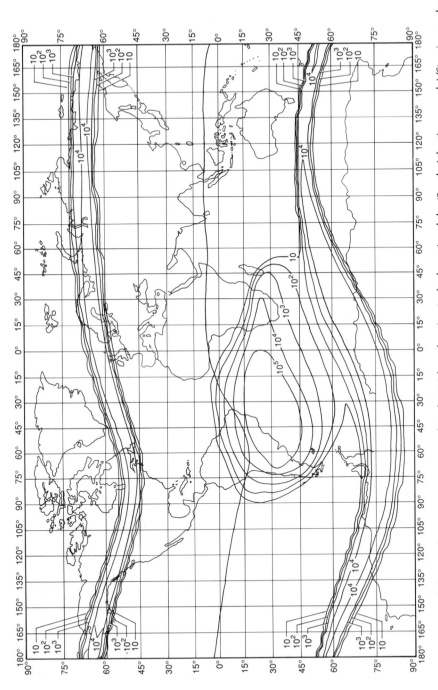

Fig. 2.5 Contours of electron fluxes at >1 MeV, at altitude 500 km, showing 'polar horns' and the 'South Atlantic anomaly' (Stassinopoulos, 1970, © 1970 IEEE. Reprinted with permission).

22 Radiation environments

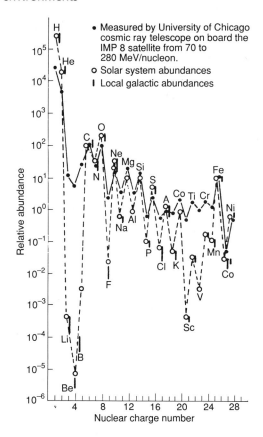

FIG. 2.6 Elemental abundance in galactic cosmic rays compared with solar system abundance (J. Adams 1981).

Terrestrial cosmic rays. The primary cosmic radiation which penetrates the earth's atmosphere is rapidly transformed by interactions which produce a cascade of secondary radiation. These cascades take place in the main body of the atmosphere and the secondary rays produced are the principal components of cosmic radiation at the Earth's surface.

2.1.4 Geomagnetic shielding

For cosmic rays to reach a spacecraft in earth orbit or the earth's surface, they must penetrate the earth's magnetic field. Since they are moving charged particles, they will tend to be deflected by the magnetic field. However, this tendency is opposed by the energy of the particles as they move at high velocity towards the earth. A particle's penetrating ability is determined by its momentum divided by its charge, and this quotient is referred to as its 'magnetic rigidity'.

2.1 The space environment   23

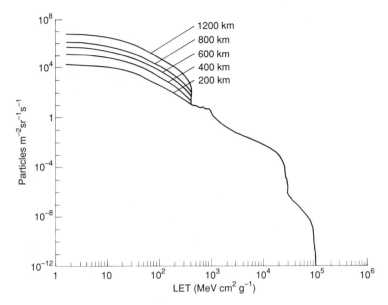

FIG. 2.7 Integral spectra of flux versus linear energy transfer (LET) inside a spacecraft in a circular orbit of 60° inclination for the altitudes indicated (J. Adams 1982).

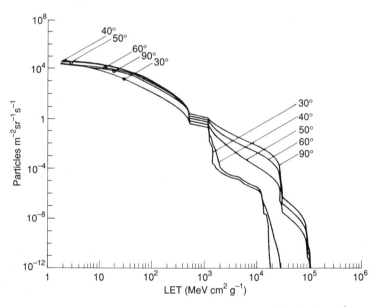

FIG. 2.8 Integral spectra of flux versus linear energy transfer (LET) inside a spacecraft in a circular orbit at 400 km altitude for the orbital inclinations indicated (J. Adams 1982).

24 Radiation environments

For each point within the earth's magnetic field there is a minimum magnetic rigidity which a cosmic ray requires to reach that point. Particles below the minimum will be deflected and this minimum is called the geomagnetic cutoff value. As mentioned earlier, geomagnetic disturbances may affect the cutoff, usually lowering it. The cutoff value falls to zero at the edges of the magnetosphere and at the earth's magnetic poles. Since the cosmic ray flux is highest at low energies, a satellite in earth orbit will be protected to some extent from cosmic rays by the magnetic field. The degree of protection will depend on the altitude and inclination of the orbit. Figures 2.7 and 2.8 (Adams 1982) show the effect of both altitude and inclination on geomagnetic shielding. The geostationary orbit, at an altitude of 35 860 km and crucially important for communications satellites, is afforded virtually no geomagnetic shielding against cosmic rays; polar orbits, important for earth observation satellites, are also significantly exposed.

2.1.5 Other sources of cosmic rays

Adams (1981) referred to two other sources of energetic charged particles in the interplanetary medium. 'Co-rotating events' give rise to infrequent modest increases in particle flux in the energy range up to 20 MeV. The particles are thought to be from the high-energy part of the solar wind, and the streams are correlated with phenomena co-rotating with the sun.

The 'anomalous cosmic ray component' appears infrequently near solar maxima. The anomalous spectra are observed for helium, nitrogen, and oxygen and it is thought that the particles involved are singly charged. Their origin could be clouds of neutral interstellar gas entering the solar system. The energies of the nuclei are mainly in the range 1 to 30 MeV. An important point is that, if the anomalous component is only singly charged, the particles will penetrate deeper into the magnetosphere than cosmic rays of similar mass and energy.

2.1.6 Solar flares

Solar flare protons, together with electrons and alphas in smaller quantities, are emitted by the sun in bursts during solar storms. Their fluxes, besides being intermittent, vary overall with the solar cycle. The energy spectra of solar protons are likely to be softer than those associated with trapped protons, but a spacecraft may nevertheless be exposed to considerable total fluence levels. A single flare in August 1972 completely dominated the solar cycle 20 in terms of fluence and total dose. Figure 2.9 (Goswami *et al.* 1988) shows the solar flare proton events for cycles 19, 20 and 21. More recently a major flare in October 1989 caused significant damage to a number of spacecraft solar arrays (Goldhammer 1990), single-event upsets in electronics (Harboe-Sorensen *et al.* 1990), and measurable increases in absorbed dose (Holmes-Siedle *et al.* 1990a).

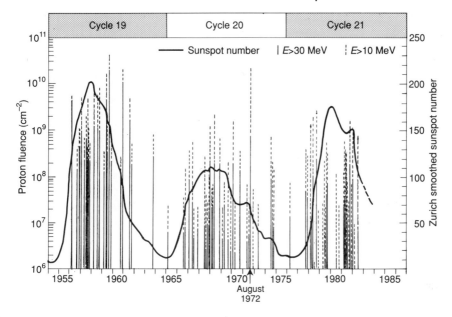

FIG. 2.9 Occurrence of large solar flare fluences (vertical lines) during solar cycles 19, 20, and 21, superimposed on the curve of sunspot numbers. The anomalous size of the August 1972 event is apparent (Goswami et al. 1988).

The degree of exposure to such effects is highly dependent on orbital parameters. The Earth's magnetic field exhibits a shielding effect in the equatorial regions, but allows the proton flux to be funnelled in towards the magnetic poles (these protons produce the well-known polar absorption phenomena in the ionosphere which has a strong disturbing effect on radio communications). Thus, equatorial orbits will be shielded from solar protons except at high altitudes, whereas polar or highly inclined orbits may be greatly exposed even at low altitudes.

Geomagnetic shielding exhibits significant east–west asymmetry, with protons able to penetrate to lower altitudes from the west than from the east. Disturbances in the magnetosphere also affect geomagnetic shielding, and the shielding has been seen to weaken significantly following geomagnetic storms (Daly 1988).

2.1.7 Other planets

The subject is of special importance when long-term missions to remote parts of the solar system are under way. The giant planets are believed to possess trapped-radiation belts similar to, but more intense than, Earth's. The Jupiter environment is considerably more intense and extensive than that of

Earth. A great deal of work has been carried out to prepare spacecraft electronics for survival in that environment. There are authoritative reports suggesting that Jovian electrons are transported to the earth's magnetosphere (Baker et al.1986). Saturn and Uranus are also known to have energetic radiation environments, and measurements made by the short-lived Phobos probe showed Mars to have a radiation environment (Shutte et al. 1989). However, in these cases, the intensities are much lower than for Earth and do not provide a threat to normal electronics.

2.2 The nuclear reactor environment

2.2.1 General

Three levels of environment severity need to be considered for nuclear fission power plants:

(1) within the reactor core and cavity;
(2) in containment,
(3) in containment under accident conditions;

These environments are compared in Table 2.1 for three types of reactor (Kakuta and Yagi 1986). For a description of types of containment see Lamarsh (1983, ch. 11).

TABLE 2.1 Reactor environments (Kakuta and Yagi 1986)

Type of reactor	In core		In containment	
	Neutrons (n cm^{-2} s^{-1})	Gamma (rad h^{-1})	Normal (Mrad)	Accident (Mrad)
PWR	10^{12}–10^{14} (thermal) 10^{10}–10^{14} (fast)	10^{5}–10^{10}	50 (40 years)	150
BWR	10^{12}–10^{14} (thermal) 10^{10}–10^{14} (fast)	10^{5}–10^{10}	50 (40 years)	26
FBR	10^{12}–10^{14} (thermal) 10^{10}–10^{14} (fast)	10^{5}–10^{10}	10 (30 years)	10^{4}

PWR = pressurized water reactor; BWR = boiling water reactor; FBR = fast breeder reactor

2.2.2 Reactor containment

The most important environment for equipment and components is 'in containment' and, while the gamma dose rate (10^{-3}–10^2 rad h^{-1}) and neutron flux (1–10^5 n cm^2 s^{-1}) are moderate, the requirement for 40 years' operating life results in significant accumulated levels. Safety equipment is designated 'Class 1E' and is required to operate at the end of a specified lifetime. Such equipment must also operate during, and after, a radioactive accident. The question of active dosimetry for Class 1E equipment has been addressed by Holmes-Siedle *et al.* (1992), and this could be an important technique for the monitoring of safety-related equipment.

Johnson *et al.* (1983) pointed out that other stress factors (ageing, temperature, etc.) must be taken into account in conjunction with radiation effects in order to arrive at a true estimate of life expectancy and to define adequate qualification tests. Since radiation damage is accumulated over a very long period of time, dose-rate effects (discussed in Chapter 4) become important, particularly since a number of modern semiconductor technologies have been shown to degrade more severely in a low-dose-rate environment ('reverse annealing' or 'rebound').

2.2.3 Reactor vessel and cavity

In pressurized water reactors, the steel pressure vessel is surrounded by a concrete shield–support. There is an air cavity between them which men can enter and install instruments in during shutdown. Neutronic instruments are also placed within the core (Knoll 1989).

The fuel rods and core matrix are exposed to very high levels of radiation damage from neutrons. The pressure vessels are also heavily irradiated, and damage effects in the steel, including embrittlement, can be significant towards the end of the reactor life. Research is proceeding at most reactors to estimate the precise damage rates and totals in the pressure vessels and other core structures and to decrease neutron leakage by the use of dummy rods and other shielding. The discipline of 'reactor dosimetry' has developed to serve the above aims. This includes the design and exposure of 'cavity dosimeter' capsules, which are carried into reactor vessels when the reactor is shut down, and extensive modelling of the reactor structure and neutron/gamma fluxes within them and calculation of the damage caused in various steels and other parts (see for example Maerker 1992).

In a large reactor at full power, the fluxes in the fuel elements may be over 10^{12} n cm^{-2} s^{-1} (E > 10 keV) – see Table 2.1 – with roughly the same flux of gamma photons. At the cavity walls, the rates may be over ten times lower than this. Since the life of a core part may be 40 years, cumulative neutron fluxes (using an 'up' time of say 10^9 seconds) may be as high as 10^{21} n cm^{-2} (E > 10 keV). These very high damage levels (in which each atom of the material has been displaced several times) will probably be exceeded in the first wall of power fusion reactors (Holmes-Siedle *et al.* 1984). Neutron energy spectra for reactor cores are compiled by the International Atomic Energy Agency (Zsolnay *et al.* 1992).

Similar issues arise in the design and maintenance of other intense radiation sources, such as weapon simulators and high-energy physics machines. Here the range of materials of interest may be broader, extending for example to insulators, sensors, and optical systems.

2.2.4 Mobile nuclear reactors

As well as large nuclear electrical power generating plants, mobile nuclear power plants such as marine propulsion systems are now quite common. The consequences of an accident in a marine dockyard in a densely populated area could be serious, and adequate accident warning and monitoring systems are required. Interplanetary satellite probes frequently carry isotope power sources and public concern was seen in the legal efforts made to prevent the launch of the Galileo and Ulysses interplanetary satellites in 1990. A spaceborne nuclear reactor, SP-100, has been in the design stage for some years, while a Russian reactor, Topaz, has had several orbital flights.

2.3 The radiation processing environment

2.3.1 General

Radiation processing is a branch of radiation technology which involves the deliberate introduction of radiation damage into materials for beneficial purposes. The aspect of radiation processing which has caught the attention of the public in recent years concerns the irradiation of food products in order to eradicate harmful organisms and extend shelf-life. This is a small part of what is now becoming a major industry world-wide.

Examples include the following:

1. Modern consumer-oriented society generates a large amount of waste, and radiation processing is a potential solution for many of the problems involved in waste treatment and disposal and the cleansing of water supplies.

2. Radiation processing is used in the medical field for the sterilization of a number of products such as dressings, hypodermic needles, and catheters.

3. Industrial applications are increasing, particularly in the field of materials modification. The polymer industry uses radiation for the cross-linking of polymers to produce durable insulation for wires and cables particularly for submarine use. Current investigations centre around polymer grafting, with particular reference to the controlled delivery of drugs. A major benefit to be obtained here may be obtained from the development of an oral drug for diabetes.

4. The semiconductor industry is beginning to use radiation processing for the modification of starting materials, particularly silicon.

5. Many developing nations benefit from radiation processing. The economy of such nations frequently depends on single, perishable crops. Radiation processing is used to delay ripening and extend the period during which they may be brought to the market-place.

2.3.2 Sources and doses

A wide range of radiation sources are used in the processing industry. Gamma sources range from 0.3 to 3 megacuries (11–111 PBq), comprising an array of Co-60 pencils in a water tank. Electron accelerators, pulsed and continuous, operate at 5–20 MeV and a power of 20–250 kW. Current research is directed to 150–300 keV machines for surface treatment. Continuous processing facilities (e.g. for cable insulation) use a vertical 'curtain' of radiation or linear scanning, with a continuous feed of the material under treatment. A typical processing rate for these applications is 15 000 Mrad m min^{-1} for a filamentary product. Typical processing doses are:

(1) food products – up to 1 Mrad with most doses in the range 30–500 krad;
(2) medical supplies – 2.5 Mrad;
(3) sewage processing – up to 1 Mrad;
(4) polymer modification – 10–15 Mrad.

For further reading see the bi-annual proceedings of the International Radiation Processing Symposium (RPC 1990).

2.4 The weapons environment

2.4.1 General

A nuclear weapon based on fission is constructed with a configuration of fissionable material (generally plutonium or U-235) slightly below the critical point. A nuclear detonation is triggered when the configuration is made supercritical. This may be achieved either by driving two pieces of subcritical material together as in the Hiroshima weapon, or by imploding a spherical shell of material as used in the Nagasaki weapon.

A thermonuclear weapon operates on the fission–fusion principle whereby a small fission weapon in the centre of the device is surrounded by a shell of lithium deuteride, which is in turn surrounded by a case of U-238. The prompt neutrons from the fission device interact with the lithium to form tritium. The energy released heats the tritium and deuterium to the point where D–T fusion can take place. The function of the U-238 case is to reflect prompt neutrons to enhance the fission process and to add to the fission yield by fissioning itself when absorbing fast neutrons.

The energy release associated with a nuclear weapon requires a special classification scheme. Nuclear yields are expressed in equivalent kilotons (kt) or megatons (Mt) of TNT explosive. 1 kt will generate 10^{12} cal, which should be compared with the requirement of only 56 g of U-235 to release the same amount of nuclear energy. In less than 1 microsecond the detonation energy of a weapon has escaped into an air mass many hundreds of times larger than the mass of the device itself.

The energy from the weapon is transferred by four mechanisms:

thermal (fireball)
blast
nuclear radiation
electromagnetic pulse (EMP).

The thermal and blast components are independent of the weapon construction; the radiation and the EMP component are determined by the weapon construction and materials and are independent of the external environment.

Immediately following detonation, the soft X-rays of a few keV are absorbed by the air, which is excited into the creation of a luminous fireball reaching millions of degrees Celsius in a fraction of a microsecond.

The energy partition of a typical nuclear weapon detonated in the atmosphere is:

blast/shock	50 per cent
thermal radiation	35 per cent
delayed nuclear radiation	10 per cent
prompt nuclear radiation	5 per cent

The time division between prompt and delayed radiation is taken as 1 minute.

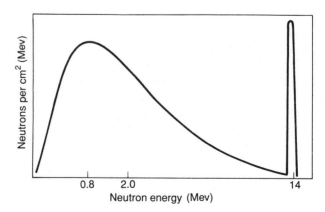

FIG. 2.10 Neutron energy spectrum for a thermonuclear detonation reproduced by permission of authors and Wiley (Messenger and Ash 1986 © Van Nostrand Reinhold, New York. Reprinted with permission).

2.4.2 Radiation output

The radiation consists of neutrons, X-rays, gamma rays, alpha particles, beta particles and secondary particles. Some 90 per cent of the neutrons generated by the fission–fusion reactions are absorbed within the bomb case. The neutron spectrum shown in Fig. 2.10 (Messenger and Ash 1986) peaks at 0.8 MeV with an average energy of 2 MeV and a fusion peak at 14 MeV. Neutron fluence is roughly proportional to yield. The neutrons undergo a $1/r^2$ geometrical attenuation as well as an exponential attenuation in any medium in their path.

The neutron yield is maximized in 'enhanced radiation' weapon, which are low-yield tactical weapons. For comparison the 500 rad radius is 1000 ft for a normal low-yield (0.1 kt) weapon and 2860 ft for an 'enhanced radiation' weapon.

The gamma rays originate from two principal sources:

(1) prompt gammas generated in the first few milliseconds which come from a nucleus excited by capturing a neutron and then falling back to ground state;

(2) delayed gammas generated by the decay of radioactive fission fragments.

The gamma spectrum extends to 12 MeV with most of the fluence in the range up to 0.75 MeV. As with neutrons, the gammas undergo geometrical attenuation as well as exponential attenuation in materials. The gamma dose can be enhanced by the lowered density of the air during the blast.

The altitude at which an explosion takes place will modify the energy partition. In a high-altitude or exo-atmospheric burst the majority of the energy will be given off as X-rays. For an underground detonation, the thermal component is suppressed and 95 per cent of the energy is dissipated in the shock wave.

Typical radiation levels to be expected from a nuclear weapon are, for prompt radiation at 1 mile (air density $1 \, \text{g l}^{-1}$):

1 kt	9 rad gamma
	2.1 rad neutrons
100 kt	960 rad gamma
	210 rad neutrons
1 Mt	62 000 rad gamma
	2 100 rad neutrons

(1 rad tissue = $4.4 \times 10^8 \, \text{n cm}^{-2}$).

Comparing atmospheric with exoatmospheric bursts, a dose of 1 krad in a silicon target would be obtained at 2.5 km in the atmosphere and 30 km outside the atmosphere.

2.4.3 Electromagnetic pulse

The electromagnetic pulse (EMP) effect is somewhat outside the scope of this book but merits brief discussion, owing to the important consequences of a weapon burst causing disruption of communications and upset to computers.

EMP is a result of the generation of Compton electrons by the primary radiation, which interact with the earth's magnetic field, travel the field lines, and emit synchrotron radiation. The delayed gammas feed the generation of electrons. EMP has a frequency spectrum extending from DC to GHz and can travel a significant distance by means of earth–ionosphere waveguide effects. EMP effects are most severe for high-altitude bursts and can cover an area of 2000 miles or so. The effects are confined to about 1 mile for a sea-level burst and non-existent for subsea or underground bursts.

2.5 The controlled-fusion environment

It is predicted that the generation of electrical power by nuclear fusion reactors will become practical in the second decade of the 21st century. Although this may seem a distant prospect, detailed investigations of the radiation generated by fusion reactors have already been carried out. This is because some of the control devices for fusion reactors will be immersed in the neutron fields and hence will be very severely damaged by the environment.

A magnetic-confinement fusion reactor operates on the basis of an extremely high-temperature plasma (equivalent to that which may be found in the centre of a star), confined within a tubular ring structure, or torus, by means of a magnetic field. The aim of such a reactor is to approach the plasma conditions under which deuterium and tritium may fuse.

A comprehensive study was carried out for the Joint European Torus (JET) by Holmes-Siedle (1980) with particular reference to the radiation levels for various proposed locations for electronic equipment in the torus hall and in the basement. The fusion experiments to be carried out by JET are still in their exploratory phase; however, the worst case radiation environment can be based on the success of the experiment running for 2 years in the D-T ignition phase.

The primary radiation source is 14 MeV neutrons from the D-T reaction. Taking into account the shielding afforded by the structural elements of the machine, walls and floor, the levels were estimated for a number of different locations and are shown in Table 2.2. In general the radiation field is mixed neutron and gamma, except in certain specific locations where large masses of iron result in gamma ray absorption.

Diagnostic instruments are required to operate in penetrations through the shields. The penetration ducts may incorporate several bends to reduce neutron 'streaming'. A study carried out by Holmes-Siedle *et al.* (1984)

TABLE 2.2 Fusion reactor environments (Holmes-Siedle 1980)

Location	1 Mev (equiv) neutron fluence (n cm^{-2})	Total dose (rad Si)
On the torus, within the structure	4.7×10^{16}	3.4×10^{7}
On the torus, outside the steel magnetic yokes	9.8×10^{15}	1×10^{6}
Average on the floor of the torus hall	7.8×10^{15}	1×10^{7}
Typical locations of support electronics	3.2×10^{12}	3.4×10^{3}
Basement of torus hall	1.5×10^{13} (thermal)	3×10^{3}

examined the exposure levels anticipated at different locations in different shapes of penetration. Typical levels were in the range 10^{15}–10^{22} equivalent 1 MeV n cm^{-2}. Ionizing doses were estimated to be in the range 10^{3}–10^{7} rad (Si) s^{-1}.

Sensor survival in such a severe environment will require a number of techniques; strong limitations on exposure time; frequent replacement of various elements; ingenious design of local shields and shape of penetrations; and the development of 'super-hard' technologies. As fusion research progresses, serious thought will have to given to radiation effects on electronic piece parts and instruments, and appropriate hardness assurance programmes will need to be developed.

2.6 The environment of robots

In space and the nuclear industry, the presence of man in certain locations can be tolerated only for a very short time. In many cases (see e.g. Chapter

TABLE 2.3 Gamma ray does rates at various stages of spent fuel reprocessing. Any neutron component of the dose rate is less than 2 per cent of the gamma rate

	Maximum gamma ray dose rate (rad(Si) h^{-1})	Maximum neutron dose rate (rad(Si) h^{-1})
Spent fuel dismantling facility	10^{5}	0.1
Spent fuel storage pond	10^{5}	0.1
Fuel element shearing	10^{5}	0.1
Fuel dissolution	10^{5}	0.1
Solvent extraction	10^{4}	0.1
Pu finishing	2.0^{a}	0.04
U finishing	0.5^{a}	0.04
Vitrification	10^{5}	0

[a] Photons of low energy about 0.15 MeV; in other cases the spectrum is of higher average energy (0.2 to 1 MeV)

34 Radiation environments

TABLE 2.4 Selected information on gamma radiation environment and operational lives of equipment in various nuclear facilities

	Typical operational life (h)	Typical dose rates (rad h^{-1})		Radiation tolerance requirement (rad)	Comments
		Peak	Common		
Fuel fabrication (and plutonium materials handling)	10^4 to 3×10^5	10	10^{-1}	10^3 to 3×10^4	Low-energy gamma photons plus some neutrons
In-reactor, on load					
In core	10^4 to 3×10^5		10^9	10^{13} to 3×10^{14}	High neutron fluxes
Outside radial shield			10^4	10^8 to 3×10^9	
Fusion-first wall		—	10^{10}	10^{14} to 3×10^{15}	
Fusion-shield inner		—	10^7	10^{11} to 3×10^{12}	
Mission-oriented					
Fission reactor inspection (off-load)	10^4	10^6	10^4	10^8	Close-to-core inspection, neutrons negligible
		—	10^2	10^6	
Tokamak maintenance	10^3	10^3	10^2	10^6	

2.6 The environment of robots

Decontamination of cells and equipment	10^3 to 10^4	10^6	10^3	10^6 to 10^7	Post-reactor cells
Decommissioning of reactors	10^4	3×10^3	30	3×10^5	Gas reactor
Reactor incident inspection	—		—	3×10^8	TMI[a] model recommendations
Post-reactor (fuel processing and materials handling)					
Fuel handling	10^4 to 3×10^5	10^6	10^4	10^8 to 3×10^9	
PIE[b]		10^6	10^4	10^8 to 3×10^9	
Process plant		10^5	10^3	10^7 to 3×10^8	
Underground storage					
Fixed systems	10^4 to 10^6	10^5	10^3	10^7 to 10^8	Waste cooling reduces long-term dose
Mobile systems	10^3 to 10^4	10^4	10^3	10^6 to 10^7	

[a] Three Mile Island, USA
[b] Post-irradiation full examination of fuel

36 Radiation environments

TABLE 2.5 EC TELEMAN-2 Programme—radiation requirement

	Dose rate $\dot{D}$		Accumulated dose[a] D		Comments
	(Gy h^{-1})	(rad h^{-1})	Gy	rad	
Light mobile	10	10^3	10^3	10^5	1/2 ton only
Robust mobile	10	10^3	10^6	10^8	Several tons
Long arm	1/2	50	10^3	10^5	
Pipe crawler	1	10^2	10^3_+	10^5_+	Small camera

[a] All gamma ray

15), remote operation of machines would be preferable; the technology for the purpose, started in the 1960–90 period, is still being developed in the 1990s. The cost and ease of operation will be controlled largely by the length of time for which the solid-state and optical elements can survive the radiation environment without replacement.

Table 2.3 gives some examples of the dose rates and total mission doses encountered in a spent-fuel-element processing plant. Table 2.4 is a broader list of situations envisaged in the nuclear field, including fusion and the mission models worked out after the Three Mile Island accident in the USA. Table 2.5 gives the estimates used in the preliminary stages of a research programme of the European Community, the aim of which was to develop teleoperators and robots specifically for hostile environments in the nuclear industry. The highest dose is 10^8 rad (10^6 Gy). The requirements vary, owing to the variety of missions. Distances from sources will vary; the intervening shields will vary; and sources will vary from the slightly radioactive steel in a powered-down reactor duct to 'hot' debris expelled during a reactor fire. Doses possibly exceeding those shown could further be found in the case of a robotic machine servicing a reactor in space. In large accelerators, some autonomous equipment will also be needed.

2.7 High-energy physics accelerators

High-energy physics research is carried out on beams of electrons or protons having energies as high as 500 GeV, often operated so that two beams collide, producing short-lived particles of interest in determining atomic structure. The equipment for accelerating the particle beams is large, often several miles in length, such as the Super Proton Synchrotron at CERN in Geneva. Many magnets and RF power sources are used to guide and accelerate, and the target areas are surrounded by very large arrays of radiation detectors of the most advanced type, including high-speed, ultrasensitive electronic detection circuitry (Heijne, 1983; Anghinolfi 1991; Stevenson 1990; Groom 1990).

The secondary radiation which escapes from the accelerator tube consists mainly of photons and neutrons of high energy. The dose rates are of the order annually of 10^7 rad (10^5 Gy) (Schönbacher et al. 1990; Tavlet 1991). The consequent need for some parts and materials to withstand 10^8 rads (10^6 Gy) opens up a new range of requirements generally higher than those needed in space radiation or the military environment.

2.8 Terrestrial and man-made environments

2.8.1 General

This is a major subject in its own right and can be only briefly treated in a book of this nature. Furthermore the levels of background radiation are such that the effects are only radiobiological in nature, which is not the major thrust of this book. Books such as those by Turner (1986), Lamarsh (1983), and Dragnić et al. (1990) give a more comprehensive account of the terrestrial environment. The United Nations recently issued a concise, impartial guide (UNEP 1991).

2.8.2 Background radiation

The 'background' radiation dose is generally taken to be in the region of 2000 μSv yr^{-1} (200 mrem yr^{-1}); Fig. 2.11 shows the various contributions to this background level. However, Dennis (personal communication) points out that if the effect of radon in the lung is weighted as recommended (ICRP 60), then the above dose value is almost doubled.

2.8.2.1 *Extra-terrestrial radiation*

The sources of extra-terrestrial radiation have been described in Section 2.1. At sea level the radiation consists of 'muons' resulting from interactions of the primary radiation with the earth's atmosphere. These give a soft-tissue dose in mid-latitudes of 280 μSv (28 mrem) per year. The dose rate varies with geographical location, in both altitude and magnetic latitude. The cosmic ray dose increases by roughly a factor of 10 in ascending to 6000 metres. Significant dose rates may be encountered by air travellers, particularly in supersonic transport aircraft, but of course the exposure periods are comparatively short.

2.8.2.2 *Environmental radiation*

This concerns gamma rays from rocks and soil and varies significantly depending on geographical location. Granites are generally highest in activity, limestones and sandstones are low, and chalk is very low. Building bricks can be very high (owing to the thorium content), depending on the source of the raw material. Environmental dose rates may vary seasonally owing to precipitation, humus and vegetation. Typical dose rates at 1 metre height in air vary from 200 μSv (20 mrem) per year over limestone to 1.5 mSv (150 mrem) per year over granite (Attix 1969). Living in a granite built house

38 Radiation environments

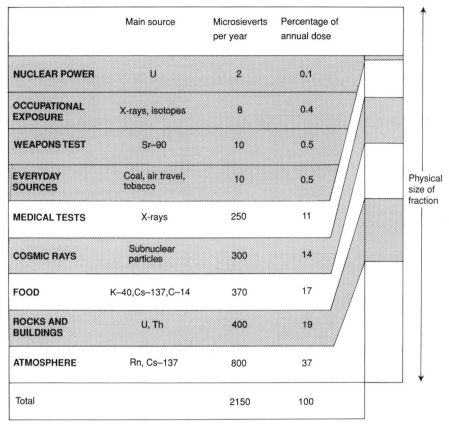

	Main source	Microsieverts per year	Percentage of annual dose
NUCLEAR POWER	U	2	0.1
OCCUPATIONAL EXPOSURE	X-rays, isotopes	8	0.4
WEAPONS TEST	Sr–90	10	0.5
EVERYDAY SOURCES	Coal, air travel, tobacco	10	0.5
MEDICAL TESTS	X-rays	250	11
COSMIC RAYS	Subnuclear particles	300	14
FOOD	K–40, Cs–137, C–14	370	17
ROCKS AND BUILDINGS	U, Th	400	19
ATMOSPHERE	Rn, Cs–137	800	37
Total		2150	100

Traditional units: 1 microsievert = 0.1 mrem, thus total = 215 mrem

FIG. 2.11 Sources of radiation in the terrestrial environment, averaged for the British population (UKAEA 1986, quoting NRPB data).

may provide up to 5 mSv (500 mrem) per year. Certain specific areas in China, India, and Brazil have reported levels in the range of 100–200 mSv (10–20 rem) per year (compared with the legal limit of 50 mSv (5 rem) per year for radiation workers).

2.8.2.3 *Ingested radiation*
Naturally occurring radionuclides are mainly ingested via food and water (apart from inhaled radon gas discussed later). The human body generally establishes equilibrium with radioactivity by means of normal excretion processes.

Uranium is to be found in drinking water, together with its daughter products. Concentrations vary widely up to a few $pCi\,l^{-1}$. Spa waters can have extremely high concentrations in the $100\,000\,pCi\,l^{-1}$ range.

Radium is to be found in foodstuffs, as well as polonium and potassium-40 (the latter being present particularly in green vegetables). A typical western diet would generally contain a few pCi per day from foodstuffs, particularly cereals.

2.8.2.4 *Fallout*

This includes radioactivity discharged into the atmosphere by the nuclear industry as well as the residual effects of nuclear weapons tests. Fallout contains a wide range of artifical radionuclides resulting from fission and fusion reactions as well as the results of interactions of these nuclides with the atmosphere or with soil. Radioactive material in the atmosphere is generally carried back to the ground in precipitation and is taken up by the soil. From the soil it enters the food chain via crops and meat and milk from animals grazing in contaminated areas. The most common radionuclides in fallout are caesium-137, strontium-90 and -89, uranium-238, plutonium-239 and -240, and iodine-131. The iodine has only a short half-life (8 days) but is particularly dangerous, as it accumulates in the thyroid gland.

The major burden of radioactive accidents such as Chernobyl is radio-activity in milk and meat. In the Chernobyl disaster a fission reactor pressure vessel in the Ukraine exploded, uranium fuel rods melted, and the burning core distributed volatile radioactive products such as Cs-137 and I-131 over Europe and Russia. Radiation levels near the reactor were over 100 Roentgen hr^{-1} (Medvedev 1991). Workers attempting to contain the radioactivity received doses ranging from lethal to sublethal. The population near to the reactor received from the plume deposits dose levels which were much larger than those permitted to the general public. Wolfson (1991) notes that the 23 000 people most exposed near Chernobyl received on average 43 rem (430 mSv). This can be compared with a whole-body dose limit of 5 rem (50 mSv) quoted in Table 2.6. In countries outside Russia, the radiation levels were enhanced but only in the region of 10 per cent (Lange *et al.* 1988). Ingestion of Cs-137 from crops and meat was the main hazard requiring

TABLE 2.6 Annual radiation dose limits (mSv) specified by ICRP (ICRP 1990)

Members of the public:	5
Non-classified workers:	15
Classified workers:	
Whole body:	50
Feet:	500
Eyes:	150
Hands:	500

measurement and control. The maximum surface air concentrations in northern Europe were in the region of 10 Bq m^{-3} of I-131 and 1 Bq m^{-3} of Cs-137. No enhancements were observable in England on Geiger counters sensitive to the normal background level of 0.1 μSv h^{-1}. Troops and other personnel sent into the reactor area to tidy up the debris were allowed a measured exposure of 20 roentgens before replacement. An area of over 1000 km^2 had levels of contamination in the range 10^5–10^6 Bq m^{-2}. This would give a dose rate some ten times the natural background. Some vegetation near the facility received a dose of 30 krad (300 Gy). The conclusion five years after the disaster is that, for Europe and the USA, the predicted effect on public health is a small percentage increase in cancer incidence (UNEP 1991; Wolfson 1991).

2.8.2.5 *Internal radiation*
Some 200 μSv yr^{-1} (20 mrem yr^{-1}) of natural 'background' is contributed by sources internal to the body, which contains a number of naturally occurring radionuclides. These include potassium-40, radium, lead-210, and polonium. The actual content varies according to a number of factors including age and sex.

2.8.2.6 *The monazite survey project, India*
Sandy strips on the coast of Kerala, in the south of India, contain monazite sand having the highest concentration of thorium in the world. The area is densely populated and areas of low and high background radiation levels alternate (giving a control population). The state government and the Bhabha Atomic Research Centre in Bombay have performed intensive epidemiological and medical investigations over the past 20 years in an attempt to discern radiation-related disease in the population (Gopal-Ayengar *et al.* 1975; George *et al.* 1989).

Investigations included chromosome and fertility studies. No significant differences were found in the incidence of any diseases or chromosomal abnormalities. One report of an increased incidence of Down's syndrome was later rebutted (Sundaram 1977). The investigation showed that the average annual exposure for gamma rays alone was over 1.3 Roentgen (compared with under 0.1 Roentgen in south–east UK). In one area (Chavara), thousands of of dosimetry measurements proved that many households received 1 to 2 rad yr^{-1} (20 000 μSv yr^{-1}), which is ten times the worldwide figure. The above studies are probably the best continuing, controlled studies on which to base judgements of the risks inherent in human occupational exposures and in background radiation environment levels raised by man-made radiation.

2.8.3 The radon hazard

Radon accounts for at least 50 per cent of the natural 'background' dose and is now becoming a source of major concern. The build-up of radon in houses has been exacerbated by current efforts to build energy-efficient homes with minimal ventilation.

Radon originates from radium and goes through a transition from gas to solid progeny. The decay chain for radium is shown in Fig. 2.12. The gas is inert, whereas the progeny are are solid, charged particles, with sizes in the region of 1 nm. The progeny give off radioactive alpha particles in the continuing decay process. The progeny can remain unattached, attach themselves to aerosols, or plate out on to available surfaces; see Fig. 2.13 (Thomson and Singmin 1988). Radon diffuses into the atmosphere and its progeny attach themselves to particles and return to the ground in precipitation. The human risk is that, during normal breathing, the progeny attached to aerosols may be drawn into the lungs and expose the cell linings to alpha radiation. The action level, at which corrective action is legally required, is set by the US Environmental Protection Agency at 4 pCi l^{-1} (148 Bq m^{-3}). In the UK the NRPB recommend remedial action at the slightly higher level of 5.4 pCi l^{-1}

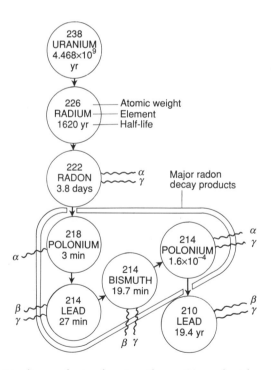

FIG. 2.12 Radium decay chart, showing the origins of radon (Thomson and Singmin 1988. Reprinted with permission).

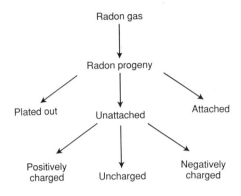

FIG. 2.13 Radon decay chart (Thomson and Singmin 1988). (Reprinted with permission.)

(200 Bq m^{-3}). The average natural level outdoors is about 0.2 pCi l^{-1} (7.4 Bq m^{-3}) or less. Levels vary widely according to geographical location, with highest levels being found in granite-bearing areas. A great deal can be done to reduce the accumulation of radon in buildings by means of surveys followed by additional ventilation, and the use of 'air ionizers' (Thomson and Singmin 1988).

2.8.4 Radiation protection

2.8.4.1 *Regulations*

The International Commission on Radiation Protection (ICRP 1966) and various national bodies such as NIST in the USA and NRPB in the UK have established detailed regulations for the maximum tolerable doses for the public at large and for occupationally exposed persons. A list of tolerable doses is given in Table 2.6. For astronauts a similar list has been proposed by the NCRP and is given in Table 2.7 (see also Cucinotta *et al.* 1991).

TABLE 2.7 Radiation dose limits for astronauts: NCRP tentative proposal (Sinclair 1986)

	Annual (Sv)	Career (Sv)
Blood-forming organs (BFO)	0.5	According to age[a]
Eyes	2	4
Skin	3	6
Testes	0.5	1.5

[a] BFO career dose is calculated according to: 200 + 7.5 (age−30) rem for males; 200 + 7.5 (age−38) rem for females. (Divide by 100 for Sv.)

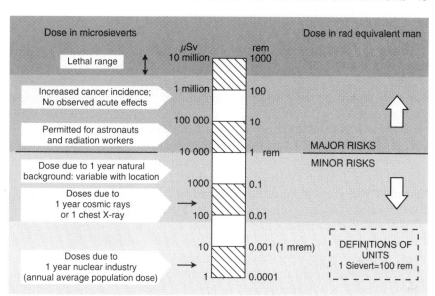

FIG. 2.14 Magnitudes of radiation dose and associated risks: the graduations of the vertical bar are logarithmic and cover seven orders of magnitude from 'apparently harmless' to 'lethal'. For clarity, the risks are arbitrarily divided into 'major' and 'minor' at a dose of 1 rem (0.01 Sv).

2.8.4.2 Risks

Figure 2.14 gives a conversion between the old unit, rem, and the SI unit, sievert (UKAEA 1986). The microsievert is sometimes called the 'unit' of population exposure. Background dose values are then in easily remembered and handled units, say '2000 units' for the annual background dose and '7.5 units per hour' for the allowed exposure of the public. The figure also compares the magnitudes of various radiation exposure levels; the range is so wide that a logarithmic scale is necessary. A suggested dividing line is shown between major and minor hazards due to radiation. Occupations which involve hazardous encounters with radiation tend to receive annual doses somewhere near this line. The definition of 'minor' is arbitrary but the meaning is similar to Brewin's description (1992) of certain radiation dose regimes as 'dilute' (see later in this section).

The question of risk in exposure to radiation and chemicals is complicated by the dual risks of acute exposure in accidents and long-term effects from chronic doses. Regulations for radiation workers, who include astronauts, must aim to maintain the risks from chronic doses at about the level of those for workers in other hazardous occupations such as miners, aircrew and toxic chemical workers. Figure 2.14 is intended to bring out the large quantitative differences between the very low chronic man-made environments (of a few microsievert units), the 'medium' natural background (thousands of units), and the acute and lethal levels (over a million units).

In assessing the radiobiological risks of radiation, the primary database comprises some 95 000 survivors of the Hiroshima and Nagasaki weapons of 6 and 9 August 1945. In addition, the history of some 50 000 patients given X-ray therapy many years ago has been studied. Both groups were exposed to large doses of radiation delivered in a short time. This is in contrast to the natural environment situation of a low dose rate spread over a very long period of time. An additional problem has been in estimating the yield of the Hiroshima and Nagasaki bombs and, in particular, the ratio of neutron to gamma doses. The very high Quality Factor (20) attached to neutrons means that small variations in the estimated neutron fluence have a major effect on the estimated absorbed biological dose. In the early 1980s both events underwent a major reassessment and the results of this reassessment had a profound impact on our approach to the effects of low-level ionizing radiation. First, it was found that the neutron yield was considerably less than originally thought, and second, the gamma dose in the case of one of the cities had been overestimated by a factor of 2. Both of these findings led to a major reduction in the estimated absorbed dose for the survivors and a complete revison of the models for biological effects (BEIR 1990).

The new models provide a lifetime risk factor for all types of cancer of 0.8 per cent for a single exposure of 100 mSv (10 rem). This means that in a population of 100 000 people exposed to 100 mSv, roughly 21 000 would die of cancer, and 800 of these deaths could be blamed on radiation. Furthermore it was decided that there was no 'threshold' dose below which effects were negligible. Previously this 'threshold' was thought to be in the region of 100 mSv.

The radiation to which the general population is exposed is seen to be largely from natural sources which are not under our control, with fallout and emission from the nuclear industry together providing only about 0.5 per cent. Medical doses can be said to be under our control but given voluntarily in cases where the benefit to our health outweighs the perceived radiation risk.

Public concern over the effects of radiation on health is intense. Except for engineers and a few with mathematical minds, it is difficult to get a proper feeling for the relative magnitudes of the risk to health of sensible uses of radiation. Brewin (1992) attempts to draw some lessons from Chernobyl, to help us in the future management of disasters. Reactions due to a wrongly perceived risk can sometimes cause more sickness, hardship, and economic breakdown than the true hazard. The object of education is to help people balance the possible disruption of excessive measures (closing down power stations, panic evacuations, destroying food, etc.) against the true hazard of radiation-induced disease. Brewin suggests the intuitive concept of 'dilute' radiation for cases where there is 'probably very small risk of carrying on as normal'. Wolfson (1991) provides a balanced account of our energy needs and the risks associated with them. Cohen (1983) gives a more partisan case for nuclear power. All agree that a better job of public education is needed.

References

Adams, J. H. (1981). Cosmic ray effects on microelectronics. Part 1. The near earth particle environment. *Naval Research Laboratories*, Memo Report 4506.

Adams, J. H. (1982). The ionizing particle environment near earth. AIAA Aerospace Sciences Meeting.

Anghinolfi, F., Aspell, P., Campbell, M., Heijne, E., Jarron, P., Meddeler, G. and Santiard, J. C. (1991). *ICON, a current mode preamplifier in CMOS technology for use with high-rate particle detectors*. IEEE Nuclear Science Symposium, Santa Fe, NM, 5-9 November.

Attix, F. H. (ed). (1969). *Topics in radiation dosimetry*. Academic Press, New York.

Baker, D. N., Blake, J. B., Klebesadel, R. W. and Higbie, P. R. (1986). Relatavistic electrons in the earth's outer magnetosphere. Lifetimes and temporal history 1979-1984. *Journal of Geophysical Research*, **91**, 4265.

BEIR Report. (1990). *Biological effects of ionizing radiation*. National Academy Press, Washington, DC.

Brewin, T. B. (1992). Excessive fear of dilute radiation. *Journal of the Royal Society of Medicine*, **85**, 311-13.

Cohen, B. L. (1983). *Before it's too late—a scientist's case for nuclear energy*. Plenum Press, New York.

Cucinotta, F. A. *et al*. (1991). Radiation risk predictions for Space Station Freedom orbits. NASA Technical Paper 3098. NASA Office of Management, Washington, DC.

Daly, E. J. (1988). Evaluation of the space radiation environment for ESA projects. *ESA Journal*, **88**(12), 229.

Daly, E. J. (1989). The radiation environment; in ESA (1989).

Dragnić, I., Dragnić, Z. and Adloff, J-P. (1990). *Radiation and radioactivity; on Earth and beyond*. CRC Press, Boca Ralon, FL.

ESA (1989). European Space Agency, *Radiation design handbook*. ESA Document PSS-01-609. ESTEC, Noordwijk.

George, K. P., Sathy, N. and Sarvandam, K. V. (1989). Investigations on the biological effects of high background radiation: frequency estimates for congenital abnormalities among newborns. In DAE Symposium on Human Genetics, Ahmedabad, p. 49-51.

Goldhammer, L. J. (1990). Recent solar flare activity and its effect on in orbit solar array. In *Proceedings of IEEE Photovoltaic Specialists Conference*, Kissimmee, Florida.

Gopal-Ayengar, A. R., Sundaram, K., Mistry, K. B. and George, K. P. (1975). Current status of investigation on biological effects of high background radioactivity in the monazite bearing areas of Kerala coasts in south-west India. In International Symposium on Areas of High Natural Radioactivity, Pocos de Caldas, Brazil.

Goswami, J. N., McGuire, R. E., Reddy, R. C., Lai, D. and Jha, R. (1988). Solar flare protons and alpha particles during the last 3 solar cycles. *Journal of Geophysical Research*, **93**, 7195.

Groom, D. E. (1990). *Radiation levels in detectors at SSC*. Report No. CERN 89-10, vol. I pp. 96 and 103, (CERN, Geneva), and SSCL-SR-1054 (Superconducting Supercollider Laboratory, Dallas, Texas).

Harboe-Sorensen, R., Daly, E. J., Underwood, C. I., Ward, J. and Adams, L. (1990). The behaviour of measured SEU at low altitude during periods of high solar activity. *IEEE Transactions on Nuclear Science*, **37**.

Heijne, E. H. M. (1983). *Muon flux measurement with silicon detectors in the CERN neutrino beams*. Report No. CERN 83-06, (CERN, Geneva).

Holmes-Siedle, A. G. (1980). Radiation effects in the Joint European Torus experiment – guidelines for preliminary design, Final report R857/2. Fulmer Research, Stoke Poges, UK.

Holmes-Siedle, A. G., Engholm, B. A., Battaglia, J. F. and Baur, J. M. (1984). Damage calculations for devices in the diagnostic penetration of a fusion reactor. *IEEE Transactions on Nuclear Science*, **31**, 1106-14.

Holmes-Siedle, A. G., Ward, A. K., Bull, R., Blower, N. and Adams, L. (1990). The Meteosat-3 dosimeter experiment: observation of radiation surges during solar flares in geostationary orbit. In *Proceedings of ESA Space Environment Analysis Workshop*, ESTEC Report No. WPP-23.

Holmes-Siedle, A. G., Leffler, J. S., Lindgren, S. R. and Adams, L. (1992). The RADFET system for real time dosimetry in nuclear facilities. In *Proceedings of 7th Annual ASTM-Euratom Symposium on Reactor Dosimetry,* Strasbourg. Kluwer, Dordrecht. pp. 851-60.

ICRP (1966) (International Commission on Radiation Protection) *ICRP Publication No 9*, Pergamon Press, Oxford.

Johnson, R. T. *et al.* (1983). A survey of aging of electronics with application to nuclear plant instrumentation. *IEEE Transactions on Nuclear Science*, **30**, 4358-62.

Kakuta, T. and Yaqi, H. (1986). Irradiation tests of electronic components and materials. In *3rd International Workshop on Future Electron Devices*, RDA/FED, Tokyo.

Knoll, G. F. (1989). *Radiation detection and measurement*. Wiley, New York.

Lamarsh, J. R. (1983). *Introduction to nuclear engineering*. Addison-Wesley, Reading, MA.

Lange, R., Dickerson, M. H. and Gudiksen, P. H. (1988). Dose estimates from the Chernobyl accident. *Nuclear Technology*, **82**, 311-23.

Maerker, R. E. *et al.* (1992). Analysis of Venus-3 experiments [and other papers]. In *Proceedings of 7th ASTM-EURATOM Symposium on Reactor Dosimetry*, Strasbourg. Kluwer, Dordrecht, pp. 627-34. (See also 5th Symposium, Geesthacht, FRG. Reidel, Dordrecht, 1987.)

Medvedev, G. (1991). *The truth about Chernobyl.* (I. B Tauris Co. Ltd., London).

Messenger, G. C. and Ash, M. S. (1986). *The effects of radiation on electronic systems*. Van Nostrand Reinhold, New York.

RPC (1990). *Radiation processing: state of the art.* (1990). Proceedings of 7th International Meeting on Radiation Processing. *Radiation Physics and*

Chemistry, *International Journal of Radiation Applications and Instrumentation, Part C,* **35**(1-6).

Schönbacher, H. and Coninckx, F. (1990). Dose to the CERN 450 GeV super proton synchrotron and an estimate of radiation damage. *Nuclear Instruments and Methods in Physics Research,* **A288**, 612-8.

Shutte, N. M. *et al.* (1989). Observation of electron and ion fluxes in the vicinity of Mars with the HARP spectrometer. *Nature,* **341**, 614-6.

Sinclair, W. K. (1986). Radiation protection standards in space. *Advances in Space Research,* **6**, 335-43.

Stassinopoulos, E. G. (1970). *World maps of constant B, L and flux contours.* Report No. SP-3054 (National Acronautics and Space Administration, Washington, DC).

Stassinopoulos, E. G. (1988). The space radiation environment for electronics. *IEEE Proceedings,* **76**, 1423-42.

Sundaram, K. (1977). Down's syndrome in Kerala. *Nature,* **267**, 728.

Stevenson, G. R. (1990). *New dose calculations for the LHC detectors.* Report No. CERN 90-10, (CERN, Geneva).

Tavlet, M. and Florian, M. E. L. (1990). *PSAIF: Facilite d'irradiation au PS-ACOL du CERN. Proc.* RADECS '91, Montpellier, Sept 9-12. IEEE Catalog No. 91 THO 400 2 Vol. 15, pp. 582-5.

Thomson, I. and Singmin, A. (1988). Applications of a portable continuous working level radon monitor. In *Proceedings of Health Physics Society Annual Meeting,* Boston, MA.

Turner, J. E. (1986). *Atoms, radiation and radiation protection.* Pergamon Press, Oxford.

UKAEA (1986). *Radiation and you.* Information Services Branch, UK Atomic Energy Authority, London.

UNEP (1991). *Radiation doses, effects, risks. United Nations Environment Programme.* Blackwell, Oxford.

Wolfson, R. (1991). Nuclear choices. A citizen's guide to nuclear technology. MIT Press, Cambridge, MA and London.

Zsolnay, E. M., Nolthenius, H. J., Greenwood, L. R. and Szondi, E. J. (1992). Reference data file for neutron spectrum adjustment and related damage calculations. In *Proceedings of 7th ASTM-EURATOM Symposium on Reactor Dosimetry,* Strasbourg. Kluwer, Dordrecht. pp. 299-306. (See also 5th Symposium, Geesthacht, FRG. Reidel, Dordrecht 1987.)

3
The response of materials and devices to radiation

3.1 Introduction

This chapter gives a general description of the effects of radiation – pulsed and steady-state – in materials and devices. It is intended to introduce the reader to the range and variety of effects to be expected, before it is discussed in later chapters how these effects strike at specific device structures. We describe first the two main types of interaction of radiation with materials (atomic displacement and ionization) and the degrading effects thereof. We describe other responses to radiation which are important but do not fit the description of 'degradation', including two classes of transient effect and activation from space radiation. Finally, the consequences in general of these degradation effects for the individual parts of a device are described. The description is extended to a bar chart and a table of about 60 varieties of device structure and material. The classification of these varieties is given in greater detail by Holmes-Siedle (1974).

The effect of irradiating an electronic material and the consequent degradation in performance of devices made from such material can follow a number of courses. The final result will depend upon the type of radiation, its mode and rate of interaction with the material, the type of material, its particular contribution to the device function, and the physical principles upon which the function of the device is based. The flow chart, Fig. 3.1, gives an impression of the variety of effects which radiation can have on devices and materials.

3.2 Degradation processes

3.2.1 General

Energetic particles or photons passing through matter lose energy through a variety of interactions and scattering mechanisms. It is not within the scope of this book to discuss these complex mechanisms in detail. We discuss some key aspects below and refer heavily to authoritative texts on the theory. We are concerned primarily with the two major consequences of energy transfer from radiation to electronic materials, namely ionization and atomic displacement, and the serious degradation effects in device parameters which can be caused thereby.

3.2 Degradation processes

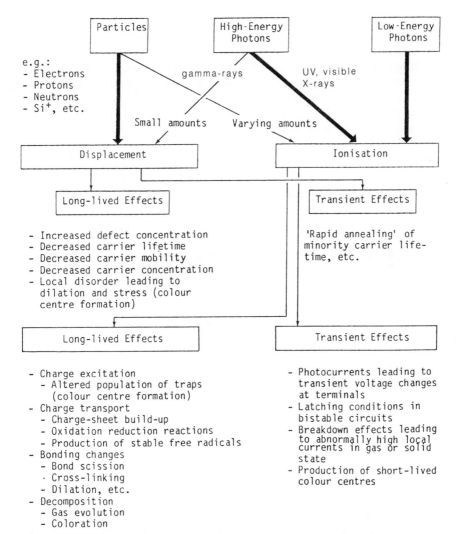

FIG. 3.1 Flow chart showing disturbances induced by high-energy radiation in solid-state devices and materials, and their relationships.

The alteration of metals due to atomic displacement, for example radiation hardening and embrittlement, is only noticeable at very high particle fluence levels. Gamma and X-rays have negligible effects. No detailed information on radiation damage in metals is presented here. Instead, see Agullo-Lopez *et al.* (1988) Thompson (1969) and Hrabal (1992).

3.2.2 Atomic displacement

3.2.2.1 *General*

The displacement of an atom from its equilibrium position in an otherwise cohering lattice of atoms (i.e. a lattice well below its melting point) is of such fundamental interest that scientists have been studying ways of creating such displacements with energy sources of all kinds, from a stylus to a cyclotron beam. The main interest is in determining the process by which the atom is moved and in studying the sequel, which may be a straight return (recombination) or a complex migration to some alternative point of stability in the lattice such as a surface or a second defect.

For some materials, the motivation of the study is both fundamental and practical. The control of the engineering properties of materials often relies heavily on very well-ordered crystalline lattices. In both metals and semiconductors, irradiation can produce lattice defects which degrade the performance of the material. The selection of structural materials for nuclear reactors initiated much early work on neutron-induced defects in metals (Dienes and Vineyard 1957; Alberts and Schneider 1989; Hrabal 1990); the radiation-induced changes in the resistance of magnetic coils under severe radiation environments was also of concern; the need for high crystallinity in early designs of germanium and silicon rectifiers led workers to the study of damage in semiconductors, using electron and proton accelerators. The work benefited from an earlier interest, mainly theoretical, in atom displacement in alkali halides, which produce 'colour centres' (Schulman and Compton 1963; Crawford and Slifkin 1975). The device theory needed to obtain good predictions of displacement damage and its degrading effects in silicon devices is described authoritatively by Larin (1968), van Lint *et al.* (1980), and Messenger and Ash (1986). For the surface effects which later became important and for transient effects, Ma and Dressendorfer (1989) present an advanced treatise.

3.2.2.2 *Degradation of transport power*

1. *General* When a momentum exchange occurs between a high-energy particle and an atom in a solid, the target atom is likely to leave its position at high velocity and is known as the 'knock-on'. The atom will leave its site if it receives energy greater than the displacement energy E_d. Its departure leaves a vacancy (by definition, a defect in the lattice). The removed atom may collide many times with other atoms and produce a 'cascade' of displacements. Each displaced atom will either meet a vacancy and 'recombine', or lodge in an interstitial position in the lattice. Vacancies are usually mobile and either combine with impurity atoms or cluster with other vacancies. In semiconductors, the resulting complexes are usually electronically active, but the interstitial atoms are less active. Both the displacement process and its consequences are complex but amenable to modelling, especially in crystalline materials. The atomic displacement process is conventionally

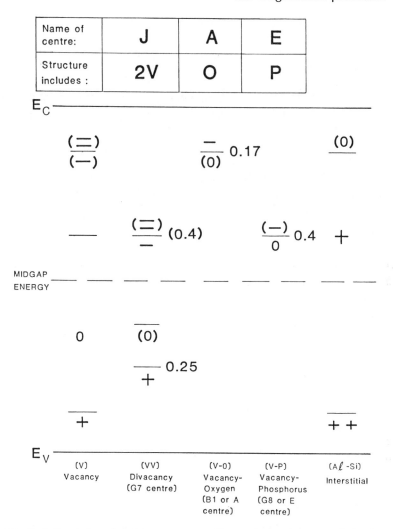

FIG. 3.2 Radiation-induced defect states: energies and charge states of the better-known radiation-induced defect centres in the forbidden gap of silicon. The levels indicated, in eV, are measured from E_C or E_V, depending on position. The charge states are shown as $=$, $-$, 0, $+$, and $++$. The position of the symbol relative to the thin line indicates where the Fermi level must be to attain that state (adapted from van Lint *et al.* 1980. Reprinted with permission of John Wiley & Sons, Inc.).

termed "bulk damage". The details of defect creation and structure in ionic crystals, semiconductors, molecular crystals and metals is dealt with in three volumes edited by Crawford and Slifkin (1975). In Volume II, Corbett and Bourgoin describe the many types of defects in semiconductors, including diamond, silicon and III-V compounds. Figure 3.2 shows the major defects found in silicon. Similar lists have been made for most crystalline materials.

52 The response of materials and devices to radiation

The defects were described above as electronically active. That is, they act in the same way as dopants in that they are a source of carriers in the semiconductor; they also act as traps for carriers and increase the number of collisions experienced by a moving charge.

2. *Minority carriers* In many silicon devices, the most important action of traps is to provide a site for recombination of minority carriers, which carry the electronic signal. The amount of recombination can be expressed in the transport parameters of the carrier, namely lifetime and diffusion length (Messenger and Ash 1986).

Shockley–Read–Hall recombination theory predicts that, when few excess carriers are present, diffusion lengths and minority-carrier lifetimes in silicon should change according to

$$1/\tau - 1/\tau_0 = K_\tau \phi, \tag{3.1}$$

$$1/L^2 - 1/L_0^2 = K_L \phi, \tag{3.2}$$

where τ is the minority-carrier lifetime after irradiation, τ_0 the initial minority-carrier lifetime, K_τ the minority-carrier lifetime damage constant (dependent on material, type of particle, injection level, and temperature), L the diffusion length after irradiation, L_0 the initial diffusion length, K_L the minority-carrier diffusion-length damage constant (dependent on

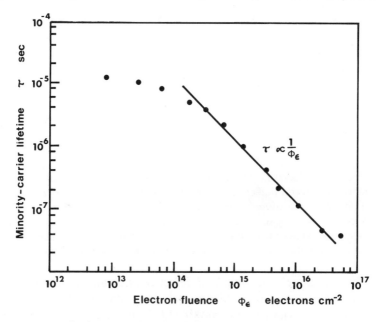

FIG. 3.3 Atomic displacement damage: the effect of irradiation by electrons of energy 1 MeV on n-type silicon, showing minority-carrier lifetime vs. fluence. The solid line indicates the linear dependence expressed by the equation shown (Wertheim 1957).

3.2 Degradation processes

material, type of particle, injection level, and temperature), and ϕ the particle fluence.

The damage constants are practical factors which assume that, if the right parameter is chosen, then the degradation is linear over the range of interest and can be used to normalize test results for different particle fluence values. The fact that linearity applies only to certain ranges is shown in Fig. 3.3, which shows the minority-carrier lifetime of silicon as a function of 1 MeV electron irradiation (Wertheim 1957). Figure 3.4 shows diffusion length plotted against fluence for neutrons in a form which permits comparison with eqn (3.2) and shows linearity.

Damage constants for 1 MeV electrons and neutrons are often taken as the standard with which the effect of all other particles is compared. An electron of higher energy will displace more atoms per unit volume. A neutron

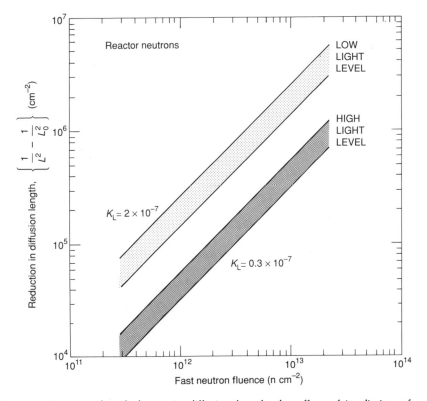

FIG. 3.4 Neutron-induced change in diffusion length: the effect of irradiation of a solar cell with fast neutrons. The solid lines indicate the linear dependences expressed by the equations. The effect of injection level (excess carrier density) on the damage constant can be seen. This is a trend plot of results by several authors (see Chapter 6). Values of K_L are calculated using equation 3.2. The effects of light level are discussed in Section 3.2.2.3.

or proton of the same kinetic energy will displace over a thousand times more because it has far greater momentum. A detailed review of neutron damage in devices can be found in Larin (1968) and in Messenger and Ash (1986). Further discussion of standard irradiation energies is given later in this chapter.

For junction devices such as transistors and solar cells, the damage constants will vary with the design of the device and the conditions of application. This is because the recombination of minority carriers and their re-emission from traps is affected by the level of excess minority carriers ("injection level") in the active regions, as well as the effect of resistivity changes in the radiation-sensitive regions. These two effects may even work in opposite directions in affecting device operation. The uses of the above minority-carrier formulae are further discussed in the chapters on individual devices.

3. *Majority carriers* The best collections of data on changes in majority carrier transport in silicon, induced by particle irradiation are by Van Lint and co-workers (1975, 1980), and Stein and co-workers (1968). The latter also shows the recovery of transport properties as the defects anneal on heating. Curtis (1966, 1975) correlates majority–carrier and minority-carrier effects. Messenger and Ash (1986) give a mathematical treatment which assists in the correlation of carrier densities, resistivity and other transport properties and the tracking of transport properties with time after irradiation (fast and slow annealing). These authors employ the useful resistivity predictions of Buehler (1968). Carrier density changes have been useful as a tool for measuring basic defect properties of silicon in simple (Hall bar) form (see e.g. Corbett and Burgoin 1975).

3.2.2.3 Bulk damage at high and low injection levels

Figure 3.4 is a case where the excess carrier level ("injection level") can be seen clearly to affect the damage constant. It is lower at high light levels. Sunlight levels excite "medium high" excess carrier levels in silicon cells (concentrations of the order of 10^{14} cm^{-3} giving rise to tenths of amperes per cm^2). Small-signal transistors are intended to work at "low" levels (mA cm^{-2}); power devices work at very high levels (A cm^{-2}). TREE effects (see below) generate very high levels. A dose rate of 10^{10} rad s^{-1} excites over 10^{18} carriers per cm^3. Low-level injection (LLI) is taken to be the case when excess concentration is lower than the doping; the opposite holds for high-level injection (HLI).

The explanation of the effect of injection level lies in the availability to the excess carriers of several routes for recombination. The first is Shockley–Read–Hall recombination, which occurs via the defect centres described in many places in this handbook. Unfortunately, the models are simple only when a single defect energy level is present. Many recombination

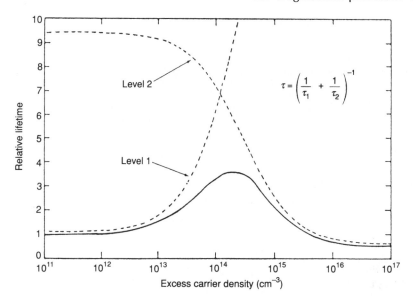

FIG. 3.5 Two-level model for recombination: a model for the dependence of carrier recombination lifetime on excess carrier density. The broken lines show a simple dependence for individual defect levels; the solid line shows the observed result in gamma-irradiated silicon (after Srour 1984. Reprinted by permission of Robert Noyes, Park Ridge, N.J.).

effects appear to follow a behaviour which indicates recombination through at least two trapping states (Srour *et al.* 1984). As shown in Fig. 3.5, a deep level dominates at LLI and a shallow one at HLI (Curtis 1975), even for the simplest form of defect (e.g. Co-60 damage), leading to a maximum, as shown, in the dependence of lifetime on excess carrier level. Curtis likens the slowing of recombination in the middle range (increase of lifetime) to a 'traffic jam' as the recombination defects get overcrowded with carriers. When the shallower level comes into play, the jam is relieved. For damage by heavy particles, the clustered nature of the defects gives even larger changes with injection level (Srour *et al.* 1975). Examples of excess carrier effects in neutron-damaged devices are to be found in Figs 3.4 and 6.1a. In most cases it is advisable to rely on measurement to confirm the expected dependence of lifetime or diffusion-length effects on excess carrier concentration.

A second mechanism for recombination is the Auger mechanism, in which two electrons and a hole are necessary but no defect is required. This is explained in detail by Kerns (1989). Naturally, the Auger route dominates only when the concentration of carriers is very large, as in accelerators or nuclear pulses. Messenger (1979) has developed models for the variation in effective minority carrier lifetime as a function of excess carrier density.

3.2.2.4 *Annealing of atomic displacement damage*

A method of distinguishing the various defects which can appear in one material is annealing, which is the determination of the temperature at which each type of defect spontaneously dissociates. Figure 3.6 shows a simplified scheme of the annealing of silicon defects, in which the material is subjected to temperature increments at regular intervals (an isochronal anneal). The resulting spontaneous decrease in defect concentration takes place over a small temperature range, so that the effect concerned can often be identified unambiguously by monitoring one of the optical or electrical properties of the device. Moreover, the life of the device under radiation can sometimes be extended by heat treatment.

Experimental observations are sometimes less clearcut than the schematic representation shown in Fig. 3.6. Figure 3.7 shows an actual case of residual carrier removal rate, plotted as a fraction of the original ('unannealed') value of silicon samples irradiated with electrons and neutrons at about 80 K. Vacancies are virtually frozen where formed at this temperature (Stein 1971; Vook and Stein 1978). On warming, the vacancies migrate to form a series

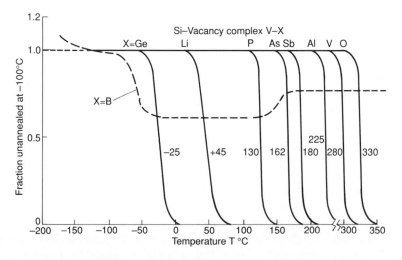

FIG. 3.6 Anneal by stepping temperature: schematic representation of the anneal of certain radiation-induced defect centres in silicon. The curves represent the reduction in concentration of various defect centres as temperature is increased in steps of equal time (e.g. 10 min). The compounds or complexes contained in the defect are denoted by V–X, where V is a vacancy in the lattice and X is an impurity atom bound to it. Data for Ge,P,As,Sb,Al,V, and O from Corbett and Bourgoin (1975); data for Li from Wysocki (1974); data for B from Srour (1970). The temperatures shown by each curve are those for 50 per cent anneal. (Reproduced with permission.)

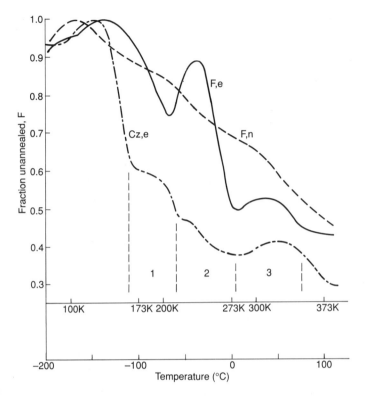

FIG. 3.7 Anneal of carrier removal: actual isochronal annealing of carrier removal rate in irradiated silicon of various types (Stein 1969, 1971; Vook and Stein 1969) **F, e**: p-type float-zone Si (oxygen-lean), 1.8 MeV electrons; **Cz, e**: p-type Czochralski Si (oxygen-rich), 1.8 MeV electrons; **F, n**: p-type float-zone (oxygen-lean) Si under reactor neutrons. Common peaks or shoulders in regions 1, 2, and 3 can be distinguished but concentration is much lower in the oxygen-rich sample.

of complexes. The result is a series of peaks and valleys. If the new complex has a larger electrical effect there is an increase in carrier removal rate or "reverse annealing" as is evident in the peak in region 2 of Fig. 3.7. The differences between curves C and Z is thought to be due to the higher concentration of oxygen in the silicon. Figure 3.8 shows a simpler case, the anneal of a complex of a vacancy and a phosphorus atom, the E centre, on heating gamma−irradiated n−type silicon at a constant temperature (isothermal anneal) (Hirata et al. 1965). Isothermal annealing may be useful in effecting the recovery of certain devices in place, in irradiated equipment, including emergency robots and unmanned spacecraft (Holmes-Siedle et al. 1991).

58 The response of materials and devices to radiation

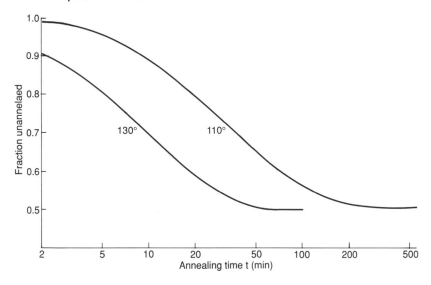

FIG. 3.8 Anneal of lifetime damage: actual isothermal annealing of minority-carrier lifetime in gamma-irradiated, phosphorus-doped silicon at temperatures 110 and 130°C. Float-zone silicon with 10^{14} phosphorus atoms cm^{-3}. Gamma dose about 1 megarad (10 kGy) (after Hirata *et al.* 1967, reprinted by permission).

3.2.2.5 *Other methods of defect characterization*

Radiation-induced defects in silicon can be identified by transient capacitance measurements (Walker and Sah 1973; Mooney *et al.* 1977). Electronic control systems have made possible a convenient form of defect spectroscopy using thermal ramps, known as Deep Level Transient Spectroscopy which is suitable for use on many semiconductor device structures (Lang 1974; Kimerling *et al.* 1979; Miller *et al.* 1977). The power of defect spectroscopy by electron spin resonance is brought out in the classic work of Corbett (1966), and Watkins whose joint research (e.g. Watkins and Corbett 1964), elucidated the main electrically-active defects in silicon.

3.2.2.6 *Dependence of damage on particle energy and type*

Introduction. A broad spectrum of particles forms a major element in the internal environment of many electronic systems exposed to radiation. In the calculation of silicon degradation, especially for solar cells, transistors and charge-coupled devices in space or in nuclear environments, the exact values for the energy-dependence curve are very important. We review the energy-dependence question here in its general aspects but mention other aspects in later sections on devices.

3.2 Degradation processes

The practical need in radiation effects engineering is to find a simple but accurate method of calculating the net electrical, mechanical or optical effects of a spectrum of particles, such as that found within a reactor or unmanned spacecraft, on the material of interest. Given a nominal curve relating energy and damage, test results for only a few energies can be extended to fulfil this task.

Various workers have derived curves for the energy-dependence of particle damage, based on experiments on devices at various energies and on the modelling of collisions in solids. We will discuss the more important of these; more comprehensive descriptions and reviews are given in Burke (1985, 1986) and Norgett *et al.* (1975). Classic examples of the determination and use of energy dependence curves and particle equivalence are Brucker (1967) for planar bipolar transistors; Summers *et al.* (1987) for various devices, and Brown, Gabbe, and Rozenzweig (BGR) (1963) for solar cells. The BGR curves have been widely used in solar array damage prediction programmes. The curves, which specify incident electron and proton energy and coverglass ('front-shield') thickness, can be used to arrive at a damage value in terms of 'damage-equivalent, normally-incident 1-MeV electrons cm^{-2}, sometimes abbreviated to DENI.

A similar system for predicting bulk damage with all-round shielding of greater depth is in its formative stages (see e.g. Section 6.12; Holmes-Siedle 1991). Research to standardize damage effects in III-V materials are in progress. For example, Griffin *et al.* (1991) for neutrons, and Pons *et al.* (1980, 1985) for electrons. Recently, interest has risen again owing to the discovery of proton effects in charge-coupled devices (see Chapter 6) and the steel embrittlement problem as a bar to extending reactor lifetimes (Lowe and Snidow 1990; Kelly *et al.* 1991).

In nuclear environments (see e.g. Srour *et al.* 1984; Raymond 1985), the possibility of bulk damage caused by gamma rays or X-rays of the same energy must be remembered. Photons possess negligibly small momentum so that they themselves cannot interact with, and displace, an atom in a silicon crystal. They may however generate an appreciable quantity of high-energy Compton electrons which, in turn, produce damage (1 rad of Co-60 gamma rays produces about 10^7 Compton electrons per cm^2 of energies up to about 1.3 MeV in the steel cap of a transistor). Thus, high doses of electromagnetic radiation may produce displacement damage, the amount depending upon details of the component's shielding configuration (see Section 5.3).

Standard energies for bulk damage equivalence. In making calculations of the effect of a broad particle spectrum, it is necessary to select a standard particle energy and assign to this a damage effectiveness of unity. Particles at other energies have a higher or lower relative damage effectiveness factor. If K_S is the bulk damage constant for the chosen standard value of energy, then K/K_S will usually have a value between 0.1 and 20 (except perhaps for some of the resonances in neutrons). If the standard particle and energy chosen is 1 MeV electrons, then the functional damage due to a particle spectrum can be expressed in simple terms, using the unit 'equivalent 1-MeV electrons per cm^2.

Some workers have used the following standard energies: 20 MeV for protons, 1 MeV for neutrons (Conrad 1971; Lowe and Snidow 1990; Heimbach 1992; Kelly *et al.* 1991) and either 1 or 3 MeV for electrons (see Chapters 5 and 6 on solar cells, bipolar transistors, CCDs and optoelectronic devices).

3.2.2.7 *Modelling of energy loss in silicon*

Primary knock-on atoms (PKAS). In this section, we explain the parameters which control the formation of electrical defects during proton irradiation. When a high-energy electron, proton, deuteron, neutron or helium atom bombards a silicon lattice, collisions lead to two types of energy loss:

1. Ionization in the silicon. This is electronic energy loss in the semiconductors (not to be confused with ionization effects in oxide layers (see Chapter 4). Electrons are stripped from nuclei and the energy released is subsequently converted into lower energy forms such as lattice vibration, light or electrical current.

2. Non-ionizing energy loss (NIEL). The major element is atomic displacement (there are other minor losses); calculations of NIEL from first principles are a major tool for understanding and predicting bulk damage in semiconductor devices (see e.g. Burke 1986; Dale *et al.* 1988; Marshall *et al.* 1990).

The dose values deposited in these two forms are referred to as ionization kerma and non-ionizing kerma. The first displaced atoms are called primary knock-on atoms (PKAs), which have a very high energy. These carry on to collide with many more atoms, many of which are themselves displaced. In other words there is a 'displacement cascade'. The key calculations required are:

(a) the energy spectrum of the PKAs;
(b) the result of the secondary collision of the PKAs with lattice atoms in terms of degree of displacement: are they displaced far enough to leave the lattice site and not to return at once?

Computer codes are now available which simulate this sequence with accuracy. These show that different particles give quite a variety of PKA spectra. It might be expected that the nature of the damage in the active region of a silicon device would differ for each case and that the correlation of damage for each particle type and energy might not be possible. On the contrary, it has been found by experiment that this expectation is incorrect.

3.2 Degradation processes 61

The fortunate fact which makes such calculations useful to device physicists is that, despite the differences in PKA from one type of particle to another, there is a *very good* linear relationship between NIEL (expressed in energy deposited per unit mass) and device degradation, for example degradation expressed in reduction of minority-carrier currents in transistors or emission from recombination centres in several types of charge-coupled imaging devices, as discussed in Section 6.12.

In seeking the explanation for this fortunate result, the investigators (see e.g. Dale *et al*. 1991) find that, during the many collisions between atoms in the lattice, the individuality of the original PKA spectrum is lost and the final product in terms of stable defects is fairly similar for neutrons, protons, and electrons over a wide range of energies. This finding is extremely useful when attempting to predict the behaviour of a modern device exposed to a broad, mixed particle spectrum, using the results of tests with limited particle energies and types. It also opens the door to the use of a large amount of earlier research on silicon damage effects which was performed on bulk silicon or on now outdated devices using reactor neutrons and accelerators (Kelly and Griffin 1991; Raymond and Petersen 1987).

Secondary processes. We will now follow the physical events which occur after the displacement cascade. Many atoms have been displaced, leaving lattice atom vacancies. Of these pairs, the silicon atom is rarely heard from again, but the vacancy is electronically and chemically active *and* physically mobile. For many both devices it is appropriate to consider the 'microvolume' in which the electronic action takes place. Both the shape of the microvolume and the chemical doping present can influence the result. If the vacancy migrates to a surface of that volume, it is lost to the complex of electronic events going on within it. If it meets another vacancy or an impurity atom, it frequently forms a stable recombination centre, which is observed in long-term physical measurements. It is important to understand the latter processes because the chemical composition of modern silicon devices varies grossly. Implanted layers, nitrided surfaces or epitaxial layers lead to sharp discontinuities.

Choice of model. In order to justify the choice of a particular energy-dependence curve when making our predictions, it is necessary to choose detailed theories of the origin of PKA spectra in the silicon (nuclear vs. classical scattering) to fit the spectrum in question. Three processes dominate the form of the silicon atom PKA spectrum given by a particle of energy E_p. The energy available for displacement damage will be called damage energy, E_{dam}. This is not the same as the displacement energy E_d – see above.

1. Coulombic elastic scattering. Interactions between the charge on the

proton and the charges in the silicon nucleus are the classical effects which lead to Rutherford scattering. These are elastic interactions, that is, no energy is lost in the collision. It is straightforward to calculate the PKA spectra for this type of collision. Dale *et al.* (1989, 1990) have shown that the energies typically vary as $1/E_p$ for protons and are of the order of hundreds of eV at a proton energy of 10 MeV. For example, the calculated values of E_{dam} for coulombic interactions are as follows: 176 eV for 12 MeV protons, 213 eV at 22 MeV, and 287 eV at 63 MeV (Dale 1990). For a displacement energy of 10 eV, this means that a 12 MeV proton should produce about 17 vacancy–interstitial pairs in the lattice. In the present discussion, this can be regarded as a small cluster of defects.

TABLE 3.1 Values of non-ionizing energy loss (NIEL) in silicon (Burke, reprinted with permission)

Proton energy (MeV)	NIEL in Si (keV cm^2 g^{-1})	$\dfrac{\text{NIEL}(E)}{\text{NIEL}(10)}$
0.01	2000	290
0.02	1300	188
0.03	1000	145
0.05	690	100
0.07	540	78.2
0.1	410	59.4
0.2	240	34.8
0.3	170	24.6
0.5	110	15.9
0.7	85	12.3
1.0	63	9.13
2.0	32	4.64
3.0	22	3.10
5.0	14	2.03
7.0	9.8	1.42
10	6.9	1.0
20	4.7	0.681
30	4.3	0.623
50	3.7	0.536
70	3.3	0.478
100	3.0	0.434
200	2.4	0.348
300	2.2	0.319
500	2.0	0.290
1000	1.7	0.246

3.2 Degradation processes   63

2. Nuclear elastic scattering. Some particles collide, so that a nuclear force is exerted upon the silicon atom. This produces a different range of damage energies. For energies below 10 MeV, E_{dam} is lower than the Coulombic values, but it becomes significant above this energy. For example, E_{dam} is 76.5 eV at 12 MeV, 111 eV at 22 MeV, and 152 at 63 MeV. That is, these interactions, when they occur, add about 50 per cent to the damage energy at 22 MeV, and this affects in some way the number of vacancies finally formed after secondary energy partitioning in the lattice.

3. Nuclear inelastic scattering. This implies breakup of the nucleus after the particle interacts. The key process for the formation of inelastically formed PKAs is the 'evaporation' of nucleons from a highly excited fragment of the original silicon atom. The resulting PKAs have damage energies of over 100 000 eV. However, these processes do not come into strong operation below an E_p value of 60 MeV for protons (Dale 1991).

3.2.2.8 Statistics of particle damage
Burke (1986) calculated the energy-dependence of the NIEL for protons, alpha particles, deuterons, and neutrons, and more recently calculations were extended to lower energies (Dale 1991). Table 3.1 records the values of NIEL for a number of energies which interest us (E. A. Burke, personal communication). The authors have used this table for preliminary calculations of equivalent 10 MeV proton damage inside spacecraft in various orbits (Holmes-Siedle *et al.* 1991; see also Section 6.12.3).

Dale *et al.* (1990) have described the consequences in certain small semiconductor structures of the statistical variance of the damage energy and its effect on the physical form of the resultant stable defects. While the defects are largely simple combinations of a vacancy and oxygen or dopant atoms, the environments of the defects appear to differ. The result is that, if a high electrical field is present in the microvolume around a given signal storage electrode and the majority carrier concentration in that volume is depleted, then the emission varies significantly from one electrode to another. The study of the statistics of the emissive characteristics of defects in such microvolumes represents one of the important lines of research in radiation effects in silicon in recent years (Dale *et al.* 1988, 1989; Srour 1986; Hopkinson 1989). One effect of practical importance is the generation of 'dark current' in CCDs (see Section 6.12). The nature of the differences in defect environment from one electrode to another is not clear.

3.2.2.9 Calculating damage as equivalent particle fluence
The theoretical conversion of NIEL per particle to a number of displacements per particle is straightforward. However, the migration and chemical combination processes described above remove about 90 per cent of the displacements which originate in the silicon microvolume. For the reasons

stated, the percentage loss must be device-dependent. Usually, the precise percentage loss is best found out by means of experiment. Measurements which determine defect concentration, such as diode rectifying properties, photovoltaic efficiency, IR absorption, and transient spectroscopy, give absolute concentrations and levels in the energy gap. Other measurements can then determine the precise effects of that defect concentration on device function, using experiments such as CCD dark current and CTI measurement. If all of the above elements of the analysis are performed with consistency, we can have some confidence in applying the resulting damage prediction to an engineering plan.

In summary, the required elements of the damage analysis necessary to calculate equivalent 10 MeV proton fluences (or other standard particle fluences) are:

(1) appropriate curve vs. E_p of

$$\frac{\text{silicon damage at } E_p}{\text{silicon damage at } E_s}$$

(theoretical calculation from PKA spectra in silicon);
(2) differential spectrum of protons at the surface of the spacecraft (orbital integration from model);
(3) appropriate differential spectrum of protons within the CCD chip enclosure (proton transport programme);
(4) experimental results of degradation of the required CCD type and parameter (e.g. CTI, I_D) versus appropriate fluences of protons at a standard enerqy E_s and for appropriate measurement and anneal temperatures.

3.3 Ionization

3.3.1 General

The primary interactions between energetic radiation and the electronic structure of atoms are more complex and varied than the simple transfer of momentum to nuclei of atoms described earlier. Despite the initial variety of interaction (Johns and Cunningham 1971), much of this loss of energy to the electrons in semiconductors and inorganic insulators is ultimately converted to the form of electron–hole pairs. In this process, known as ionization, the valence band electrons in the solid are excited to the conduction band and are highly mobile if an electric field is applied. Any solid — even an insulator — thus conducts for a time at a level higher than is normal. The positively charged holes are also mobile, but to a different degree. The production and subsequent trapping of the holes in oxide films cause serious degradation in MOS and bipolar devices. This effect is the subject of several later sections.

In polymers, the main result of ionization may be the breaking of chemical bonds and the creation of new ones. In this case, conduction may also result, but, in the space environment, long-lived forms of physical breakdown may be more apparent. Other sequelae to ionization include luminescence (scintillation) and thermoluminescence (see Sections 1.5 and 8.4).

It is worth noting that the net energy required to create an electron–hole pair is relatively small (e.g. 18 eV in SiO_2). Therefore, since no momentum transfer to atoms is involved, the energy of the radiation causing ionization is not so critically important as in atomic displacement (of course, the number of pairs created depends upon the particle or photon energy). Thus, ionization effects as produced by megavolt particles in space may often be simulated by much lower energy X-rays, electron beams, or even ultraviolet light. For displacement damage, freedom of choice of simulation methods is much more limited. The deposition of energy in a material by means of ionization is conventionally termed 'dose' and measured in rads or grays (see Chapter 1).

It is often necessary to specify the dose rate at which radiation is delivered. For example, a spacecraft orbiting in the van Allen belts is said to be exposed to a low dose rates, and a sample close to a reactor core to a high dose rate. A nuclear weapon pulse delivers an even higher rate. We can specify this rate in terms of the average energy absorbed per unit mass and time (e.g. rads per second). This method of specifying rates is adequate for high-energy electrons and photons. However, local ionization effects on very dense electronic components exposed to certain particles may lead to strong transient electrical response. Special questions of dose-rate effects such as these will be dealt with as required.

3.3.2 Charge trapping in oxides after ionization

Insulator films are used in solid-state devices, often acting as stand-off layers which act as barriers, to arrest charge motion between two layers of semiconductor or conductor. A prime hazard of radiation is that it may (a) temporarily lower the barrier; (b) catch some of the charge travelling across the oxide and freeze it in place, producing a semi-permanent charge sheet (this will have its own built-in field, which will bend bands and so affect conductivity in charge-sensitive layers around it); and (c) disturb the labile bonds which often occur at interfaces, especially insulator–semiconductor interfaces. Effects (b) and (c) constitute two of the most serious ionization effects in microelectronic semiconductor devices; see especially Chapters 4 and 5. In this section, we describe the structures of the traps which produce charge sheets in oxide films under irradiation.

The effects described are so diverse for a number of reasons. First, the variety of dielectrics employed in device technology is much greater than the variety of semiconductors. Second, the gap between valence and conduction bands of a dielectric is large. Dielectrics are usually covalent compounds of several chemical elements. Thus, a large variety of trapping levels and excitations is possible, while polarization effects may be long-lived. Third, impurity concentrations tend to be higher because, unlike semiconductors, processes analogous to 'doping' are not commonly used.

Charge trapping sites are localized defects at which electrons or holes may be captured. The trapped charges can be excited back into the conduction band of the solid concerned. In insulators, however, the energies required to do so are often large. For example, in amorphous SiO_2, the insulator of main interest in Si devices, the band gap is about 9 eV and trapped holes lie at a level of the order of 2.5 eV above the conduction band. At this depth, thermal processes are unlikely to excite the charges from the traps, although electron injection processes can annihilate them. Often, shallower traps also exist, which permit holes to perform a 'random walk' through the oxide. The state of the charge sheets (and hence the degrading effect) is thus best referred to as 'semipermanent', but the degradation can be very long-lived. For example, the present authors have remeasured metal-oxide semiconductor devices 20 years after irradiation and found a large fraction of the trapped charge to persist in the oxide.

It is not widely realized that amorphous insulators have a well-established order and hence an electronic structure which can be closely defined in

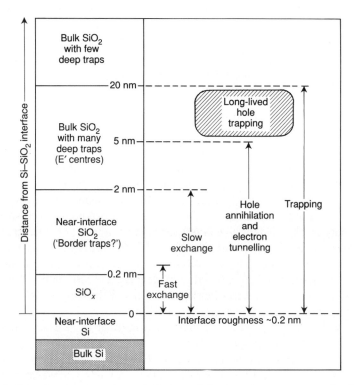

FIG. 3.9 Trapping zones: a general picture of the trapping zones near the interface of the oxidized silicon surface and of the metal-oxide–silicon device structure (see text).

quantum terms and investigated by all types of spectroscopy, and possess well-defined conduction states for electrons and holes. Helms and Deal (1988) introduce this subject and Helms (1988) further reviews it in a book describing the advanced state of knowledge of the Si–SiO$_2$ system.

3.3.3 The charge trap system at the surface of oxidized silicon

The processes of charge trapping are reviewed by Svensson (1988) and the methods of measurement by Lefevre and Schulz (1988). When silicon is placed in an oxidizing gas (e.g. a mixture of oxygen and water vapour) it is possible to grow a well-controlled form of amorphous SiO$_2$ which, except for the interface regions, is stoichiometric and pure except for small amounts of hydrogen. This bulk layer bears a strong resemblance in X-ray structure and electronic properties to fused silica. Charge trapping in the oxide film is concentrated near the silicon. The description by Helms (1988) of regions near the Si–SiO$_2$ interface includes five regions, which we sketch in Fig. 3.9. On the left of the figure are the subdivisions given by Helms, based on microscopy. On the right are shown the positions and types of process believed by the present authors to be involved in the generation and decay of oxide trapped charge (Q_{ot}) and interface trapped charge (Q_{it}). Trapping processes also occur at the oxide–metal surface but are not shown. When

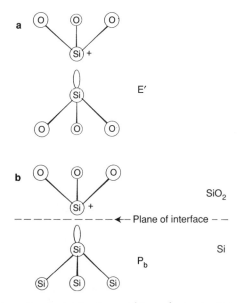

Fig. 3.10 Charge trap structure: structures of two charge traps in SiO$_2$ films. (a) The broken-bond defect believed to be associated with the E' centre in SiO$_2$, shown 'filled' with a hole charge; (b) the analogous defect at the Si–SiO$_2$ interface, which structure is found in ESR spectra as the P_b centre. In (b) only, the charge state is dependent on the Fermi level in the silicon (after Helms 1988. Reprinted by permission of Elsevier Science Publishers).

radiation-generated holes drift into this region from the bulk, they are trapped; but those within tunnelling distance of the silicon are quickly annihilated by electrons. The net result is a charge sheet located as suggested by the shaded area in the figure. Electronic exchange of interface trapped charge with the silicon is fast within an atomic bond length and slow at greater distances (see Holmes-Siedle and Adams 1983).

In the 'interfacial SiO_2' region, a high level of strain may exist and it is likely that charge may be exchanged as shown with the silicon by a tunnelling process. Traps in this region appear to have a mixed character, bearing some features of 'bulk' states and some of 'interface' states. Based on his studies of noise from these states (and the historical analogy of 'Border States' in the Civil War in America 1861–5), Fleetwood (1992) proposes the name 'border traps' for states of mixed allegiance. One criterion proposed for these traps is that they can be filled with charge from the silicon within about 1 minute. The 'SiO_x' region probably contains the 'fast interface states', in which the charge state changes rapidly with the Fermi level, including changes caused by bias on the MOS gate electrode.

The bulk SiO_2 probably contains both shallow and deep hole traps which dominate the trapping of holes at different temperatures. This charge constitutes 'oxide trapped charge'. A picture of the possible difference in the charge traps in the bulk and interfacial region (Q_{ot} and Q_{it} repectively) is given in Fig. 3.10 (Helms 1988). The E' centres shown are present in the oxide as grown on silicon, that is before irradiation. The P_b centres may be latent electrically before irradiation and then be triggered by irradiation or charge injection. The consequences of the appearance of charge-bearing forms of these two traps, Q_{ot} and Q_{it}, are discussed in the chapters dealing with devices of various types. This duality of ionization effects in MOS structures manifests itself in almost all oxide-passivated devices, including bipolar transistors and charge-coupled devices.

Defects like the E' centre also occur in transient and permanent effects in optical media. Cognate forms of defect are found in other compounds (Henderson and Wertz 1977). The trap in the E' centre is thought to be a non-binding sp^3 orbital on a tetrahedral silicon atom linked to three oxygen atoms, sometimes represented as $O_3 \equiv Si\bullet$ (Feigl *et al.* 1974; Griscom and Friebele 1982). The non-binding orbital projects into a vacancy caused by an oxygen deficiency in the network. Other forms for the hole trap have been proposed (Griscom 1989), including oxygen excess centres such as the peroxide radical (an Si–O–O–Si structure is broken between Si and O) or the non-bonding oxygen hole centre. The NBOHC is inactive when in the form (–Si–OH . . . HO–Si–). It becomes a free radical when it traps a hole and loses a hydrogen atom, becoming (Si–OH . . . + •O–Si). All these species are probably present in oxidized silicon surfaces, which, more than bulk silica, also possess a high content of strained chemical bonds. Griscom *et al.* (1988) gave an overall picture of the oxidized silicon surface under irradiation. Fig. 3.11 shows the proposed flow of reacting species which they

3.3 Ionization

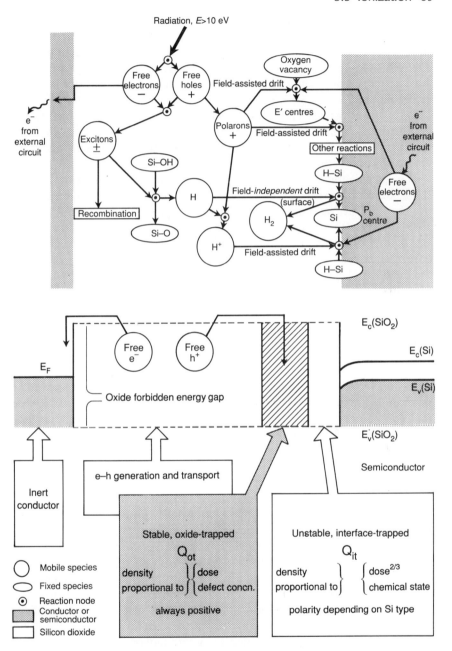

FIG. 3.11 Dynamic processes in oxides: the upper part of the picture shows the main excitation and trapping processes proposed by Griscom et al. (1988) for the oxide film of a metal–oxide–silicon structure when it is irradiated under a positive gate voltage (reprinted with permission). The lower part shows a simplified form of the processes, noting the energy levels, dose dependence and generic names of the charges built up.

proposed. Those in circles can move; those in ellipses are fixed. The main interactions and motions are shown by reaction arrows passing through small circles, or 'reaction nodes'. A polaron is an electronic charge—here a mobile hole—which is stabilized by the relaxation of the lattice around it. An exciton is a hole-and-electron, able to move as a pair without instantly recombining.

In the lower part of Fig. 3.11 we show a simpler charge flow picture but also distinguish the trapping zones, previously explained apropos Fig. 3.6, the energy levels and other properties of the traps determined by recent device physics research.

The microscopic structure of hole and electron traps in oxide films is still a subject of research. For example, it is known that the effects of hole trapping can be 'annihilated' by electron injection (Di Maria and Stasiak 1989; Feigl 1983; Ning 1976) or by hydrogen treatment (Witham and Lenahan 1987). This does not fit the popular 'bond strain gradient' model (Grunthaner *et al.* 1982), which demands irreversibility. A proper answer to the question of reversibility or irreversibility of ionization effects in oxide films has not yet been obtained.

3.4 Induced radioactivity

Any material exposed to an energetic-particle environment becomes radioactive. Protons, neutrons, nuclei, and pions are all capable of transforming stable nuclei of any material into radioactive nuclei by the removal of nucleons or, in the case of low-energy neutrons, by neutron capture. A wide range of radioactive nuclei is produced and these decay at a later time according to their characteristic half-lives.

The majority decay by emitting a positron (beta$^+$) or by capture of an orbital electron, often accompanied by gamma-ray emission, while in a few cases—such as neutron capture—emission of an electron (beta$^-$) occurs. Such products have typical energies in the range 0.01 to 10 MeV. In space, primary protons are the most abundant particles producing radioactivity, while in thick materials activation by secondary protons, pions, and especially neutrons becomes increasingly important. The total interaction cross section for protons increases monotonically with energy, reaching a peak at 30 MeV, after which it falls slightly to a constant level for energies greater than 200 MeV. While total cross sections are known with an error of 5 per cent, the spallation cross sections for production of particular radionuclides are mostly unknown but can be estimated to within a factor of two using semi-empirical formulae. For high-Z target materials, some 200 radionuclides contribute significantly to the induced radioactivity.

In space, the most intense proton fluxes are to be found in the heart of the inner radiation belt where radioactivity is induced by the energetic proton component in the energy range 30 to 400 MeV. At altitudes below or above this trapped-proton regime, the more energetic (GeV) cosmic rays dominate the induced radioactivity, and the precise flux and level of induced activity vary with the geomagnetic latitude. At high altitudes (e.g. geosynchronous or interplanetary space), solar flare protons provide the major component of activation for some 20 days per year at solar maximum.

Doses due to induced radioactivity are small compared with the dose produced directly by the inducing radiation. However, such radioactivity is a major source of background in the performance of highly sensitive X-ray and gamma radiation astronomy and remote-sensing spectroscopy. Induced radioactivity due to high-energy protons and cosmic rays has been directly observed by a number of instruments on various spacecraft (Peterson 1965; Dyer *et al.* 1980).

Calculations of induced radioactivity in thin homogeneous materials can be made using methods and formulas given by Dyer *et al.* (1980) and Barbier (1969), and based either on semi-empirical cross sections or on ground-based irradiations using representative particle beams. To a first approximation, one radioactive decay results from each proton interaction, so that a saturated activity of some 4–6 decays per second per kilogram of material is attained in a cosmic ray flux of 1 particle per cm^2 per second.

Trapped protons can be stopped by ionization without interacting and, in general, are at least a factor of three less efficient in inducing a decay even under minimal shielding. The detailed decay behaviour and spectrum of emissions is a complex superposition of many radionuclides and requires methods such as those described by Dyer *et al.* (1980). For heavy spacecraft, activation by secondary neutrons becomes important and particle transport calculations must be performed. Activation by neutron capture is very dependent on the spectrum of neutrons and the detailed cross section of the material of interest.

3.5 High dose-rate upsets (transient effects)

3.5.1 General considerations

Transient radiation effects in electronics (TREE) are also called dose-rate or gamma-dot phenomena ($\dot{\gamma}$, the expression sometimes used for dose rate; we prefer the term d). The main source is the detonation of a nuclear weapon (Raymond 1985; Sartori 1991; Glasstone 1977). Gamma rays, X-rays and neutrons are emitted in times varying from nanoseconds to microseconds (see Chapter 2), along with degrees of air blast and thermal fluxes which vary greatly with the surroundings of the burst. In all equipment exposed to the burst, one thing in common is that a uniform flux of penetrating radiation is produced and that the transient effects of this irradiation on the semiconductor parts are usually significant for exposure levels well below those leading to structural damage (melting or fracture). The electron-hole pair concentrations generated by a pulse of, say, 10^9 rad(Si) s^{-1} are as high as 10^{18} cm^{-3}, well above the doping densities of

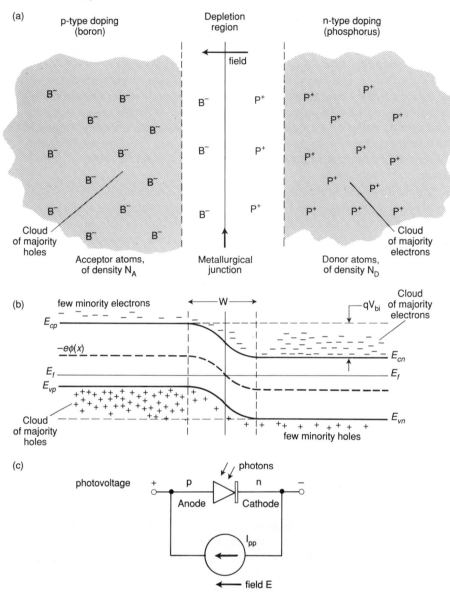

FIG. 3.12 Diode transient photocurrent: (a) Space charge depletion layers form an electric dipole layer field; (b) the electric dipole forms a potential barrier qV_{bi} and a depletion region of width W; (c) the primary photocurrent, I_{pp}, moves in the direction of the field e.g. minority holes flow to the left. Convention and symbols used for the photocurrents and photovoltages appearing across an unbiased p–n junction.

3.5 High dose-rate upsets (transient effects)

most active semiconductor elements. The effect is thus to 'swamp' many junctions. As shown in Fig. 3.12, currents flow in the direction which is normally 'blocked' and voltages larger than those being used as signals are generated. Pulses from small junctions, such as the drain junction of the MOS transistor shown in Fig. 3.13 (a radiation pulse from a LINAC is used here to simulate a nuclear burst), are generated simultaneously at many nodes in the circuit. The proper method for the analysis of a complex circuit exposed to TREE is by a computer code describing the photocurrents and their joint effects at nodes (see Chapters 4 and 14). A survey of transient effects in simple structures follows below; see also Thatcher *et al.* 1965.

3.5.2 Bulk semiconductor

A disturbance of minority-carrier concentration by high-energy radiation or light follows the relation

$$dp'/dt = G_L + G_{th} - R,$$

where p' is the excess carrier concentration, G_L the rate of electron–hole pair generation by light or a radiation beam throughout the semiconductor, G_{th} the rate of thermal generation, and R the rate of recombination of

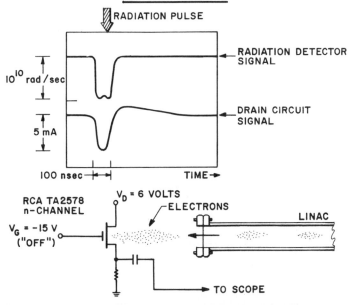

FIG. 3.13 Typical transient response of an MOS device: pulsed exposure of part of a CMOS circuit to a 100 ns pulse from the beam of a 10 MeV LINAC accelerator, simulating a nuclear burst. The curves show the oscilloscope traces for the dose rate detector (not shown) and the output of the drain junction.

excess carriers, which is the reciprocal of the minority-carrier lifetime, τ. From this generation equation, the excess carrier concentration can be calculated. For steady-state irradiation, the rates of recombination and generation are in equilibrium, so that

$$p' = G_L/R = G_L \cdot \tau = g_o \cdot d \cdot \tau,$$

where g_o is the number of carriers generated per unit dose, being 4.2×10^{13} electron–hole pairs per rad(Si) (1×10^{15} pairs per Gy(Si)), and d is the dose rate. The carrier concentration generated by a pulse of radiation decays with time as follows:

$$p'(t) = p'(0)\exp(-t/\tau),$$

where t is the time after the pulse and $p'(0)$ is the concentration during the pulse.

3.6 Transient photocurrents in p–n junctions

The effects of photocurrents on the electrical behaviour of devices reflects the complexity of the device structure itself. Transistors and logic circuits

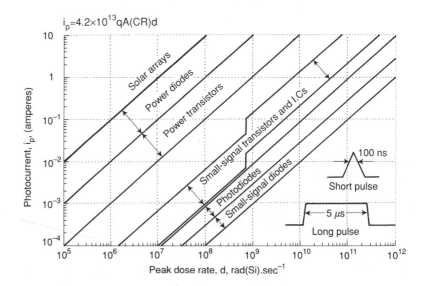

FIG. 3.14 Typical ranges for the transient response of silicon junction devices to pulsed ionizing radiation. The wide spread of the ranges shown derives from the wide range of junction areas in each class of device and a reduction in response to short (e.g. 100 ns) compared to long (e.g. 5 μs) pulses (see text for explanation). The discontinuity exhibited by some classes of device is caused by the onset of parasitic effects such as 'secondary photocurrent' and 'latch-up' in complex junction geometries.

3.6 Transient photocurrents in p–n junctions

give complex responses and will be discussed device by device. A moderately simple case, the p–n junction diode, is discussed here. Much of the early work on radiation hardening of silicon devices was an attempt to understand and reduce the transient effects which intense gamma-ray pulses (typically 10^6 to 10^{12} rad(Si). sec^{-1} for a duration of less than 1 m sec) produce in circuits containing junction devices. Figure 3.14 shows the response of a wide range of silicon devices as a function of dose rate. The magnitudes are explained below.

The instantaneous value of the photocurrent i_p in a p–n junction diode is:

$$i_p = 4.2 \times 10^{13} \, q \, A \, (CR) \, d(t) \tag{3.3}$$

where q is the charge on an electron, A is the junction area, in cm^2 (CR) is the depth of the collection region in cm and d(t) is the dose rate in rad(Si).sec^{-1} or cGy(Si).sec^{-1}. In the case of a pulse of radiation, the dose-rate is time-varying and the current waveform initially follows the radiation pulse. The arrow in Fig. 3.12c indicates the direction of positive charge movement.

The carriers in the depletion region of the junction are collected very rapidly often in less than 1 ns, giving a prompt photocurrent for which the (CR) term has a value, W, the width of the depletion region of the junction. A delayed photocurrent is supplied by diffusion of carriers from a zone roughly within a diffusion length of the junction. This may persist after the prompt current has died down, in the form of a current decaying over a time controlled by the minority-carrier lifetime in the material. The amplitude of the delayed term in the current equation is dependent upon the length of the pulse. For very long pulses, the delayed current contribution is at a maximum and, in equation (3.3), the collection region term is (W + Lp + Ln), where the L terms are the values of the minority-carrier diffusion lengths – potentially hundreds of micrometres – in the n and p regions respectively. In other words, for long pulses, not only is the photocurrent prolonged but maximum amplitude is produced because the effective collection region has been widened considerably. For a collection region of 10 micrometres and a p–n junction area of 1 cm^2, the value of the current is a few amperes at a dose rate of 10^{10} rad.sec^{-1}. This can be expressed as a responsivity of 6.2×10^{-9} amp.sec.rad (Si)$^{-1}$. Figure 3.13 shows the response of the drain junction of an MOS logic transistor. The responsivity is about 5×10^{-13} amp.sec.rad(Si)$^{-1}$. Figure 3.14 shows that power diodes and solar cells give the largest responses. Small-signal diodes are physically very small (i.e. have a small value of A) and thus appear at the other end of the response scale.

Bipolar transistors and integrated circuits are intermediate in size. The curves for these devices show a discontinuity. They not only generate photocurrent but their junctions are configured so as give secondary currents and latch-up (see Chapters 4 and 5). Unsophisticated, junction-isolated

76 The response of materials and devices to radiation

integrated circuits (early TTL types) can exhibit latch-up at a gamma dose rate value as low as 10^6 rad(Si).sec^{-1} (10^4 Gy(Si).sec^{-1}).

The engineer can use Fig. 3.14 to estimate the order of magnitude of photocurrent generated by pulsed or steady-state radiation or light in a given environment. More precise calculations require values for the junction areas and the length of the radiation pulse. Larin (1960) provides a useful framework for calculating photocurrents from typical solid-state devices such as a model bipolar transistor. Responsivity tables for a wide range of device types are supplied by Rudie (1976) and Messenger (1986). Transient conductivity in cable insulators is discussed by van Lint *et al.* (1980). The output pulses may lead to logic upset (Kerns 1989), latch-up (see below) or burnout of associated semiconductor devices.

3.7 Single-event phenomena

3.7.1 General

This is a special case of ionization effect which is of increasing concern to the space community. For this topic we have adopted the terminology of NASA–JPL whereby the term single-event phenomena (SEP) covers both single-event upset (SEU), or 'soft error', and latch-up. There is as yet no generally accepted terminology in this field of radiation effects, and the reader may well find other publications in which the term SEU is considered to cover latch-up as well as 'soft error'.

SEU is not a phenomenon unique to space. Cosmic rays have been shown to cause upsets in the memories of computers on Earth (Ziegler and Lanford 1981). Effects in aircraft computers were christened 'homotrons' by the Airline Pilots' Association (Hazard Prevention 1981). The computer industry has been concerned since 1978 with soft errors due to alpha particles from radioactive materials in the device packaging (see for example Carter and Wilkins 1984). It has been suggested that an irreducible minimum rate of soft errors is imposed by radioactivity in the silicon itself.

3.7.2 Single-event upset (SEU)

An SEU, or 'soft error', is the change of state of a bistable element caused by the impact of an energetic heavy ion (cosmic ray) or proton. The effect is non-destructive and may be corrected by 'rewriting' the affected element (triggering it back to its intended state). An SEU is the result of ionization by a single energetic particle, or the nuclear reaction products of an energetic proton.

The ionization induces a current pulse in a p–n junction. The charge injected by the current pulse at a sensitive ('off') node of a bistable data storage element may exceed the critical charge, Q_c, required to change the logic state of the element. The mechanism of the current pulse generation and time constants involved are shown in Fig. 3.15 (Pickel 1983). The resulting change of state is often known as a 'bit-flip'. The SEU mechanism

3.7 Single-event phenomena 77

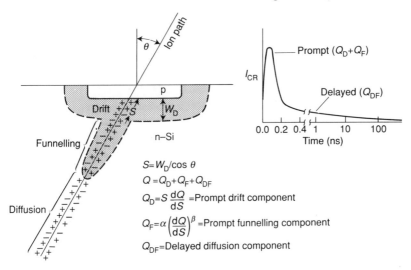

FIG. 3.15 Current pulse in a p–n junction from a single-ion event: cross section and time curve (after Pickel 1983. Reprinted by permission).

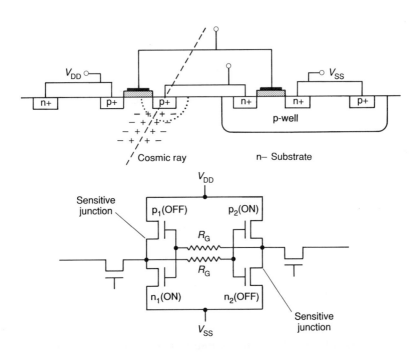

FIG. 3.16 Mechanism of a single-event upset (SEU) in a typical CMOS bistable memory cell: cross section and circuit (after Dawes 1985). The cross-section is similar to Fig 4.1(a). (Reprinted by permission).

78 The response of materials and devices to radiation

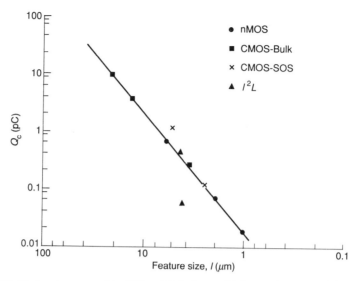

FIG. 3.17 Technology trend in SEU: critical charge (Q_c) as a function of feature size for various technologies (after Burke 1985. Reprinted by permission).

affects both bipolar and MOS technologies. Figure 3.16 (Dawes 1985) shows the mechanism in a typical CMOS bistable memory cell. The reason for the increasing importance of the SEU mechanism is clearly illustrated in Fig. 3.17 (Burke 1985), which shows that the technology evolution towards smaller device geometries results in a decrease in critical charge and consequently an increase in vulnerability to SEU. Vulnerability does not have a linear relationship to critical charge, as there is some compensation from the fact that the ion path, and consequently the charge deposited, is smaller. It has been suggested that although SEU may become a limiting factor in space application, we may approach some form of saturation in VLSI error rates, as a result of the path-length versus critical-charge compensation mechanism. As geometries become smaller, other mechanisms may cause problems, in particular multiple errors as the result of a single ion strike and new, potentially destructive mechanisms such as rupture of oxide layers.

Charge deposition is a function of a particle's LET (linear energy transfer), or energy loss which has the dimensions of MeV g^{-1} cm^{-2}. This in turn can be related to charge deposited over a given distance in picocoulombs per micrometre (3.6 MeV energy loss produces 0.16 picocoulombs of charge in silicon).

The vulnerability of a device to SEU is defined by two parameters:

1. Threshold LET. This is the minimum LET required to produce upset and corresponds to a charge deposition comparable with the critical charge Q_c. Critical charge is related to the volume of a sensitive node and the degree of charge funnelling.

3.7 Single-event phenomena

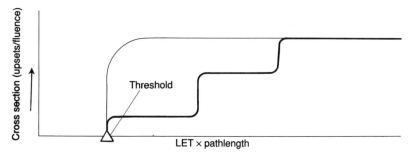

FIG. 3.18 Typical SEU responses (Robinson 1987. Reprinted by permission).

2. Saturated cross section. This is when all incident ions are capable of producing an upset and no increase in upset rate is seen for an increase in LET. The cross section has the dimensions of cm² and is a function of the surface area of the sensitive nodes.

Figure 3.18 (Robinson 1987) shows these two parameters on an idealized SEU response curve. Also shown is a more complex response which is the result of the different sensitivities of different designs of bistable within a particular device (Zoutendyk et al. 1984).

3.7.3 Latch-up

Bulk CMOS structures contain parasitic vertical and lateral bipolar p–n–p and n–p–n transistors. In a typical p-well technology these parasitic transistors can form a p–n–p–n thyristor or SCR structure. Figure 3.19 (Robinson 1987) shows a cross section of a p-well device and the parasitic transistors. Under normal operating conditions the SCR structure is 'off' and in a state with high impedance between anode and cathode. The SCR may be triggered into an 'on' condition by a pulse of ionizing radiation. Under these circumstances excess minority carriers are generated in the base region of the parasitic transistors, 'forcing' them into conduction. Positive feedback around the p–n–p–n loop causes the structure to remain in a low impedance state, when it can be held 'on' by a comparatively low holding voltage. This is classical SCR behaviour, shown in Fig. 3.20 (Srour 1983). Certain conditions must be satisfied for latch-up to occur:

(1) the gain product $B_{pnp} \cdot B_{npn}$ must be greater than 1 to allow positive feedback to occur;

(2) Both transistors must have their emitter–base junctions forward-biased so that injection may occur;

(3) Power supplies must be able to sink or source a current greater than the holding current required to maintain the latch.

Single-event-induced latch-up is a serious occurrence, since it is long-lived and potentially destructive.

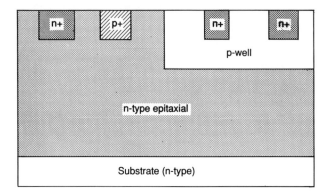

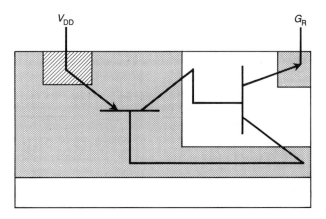

Fig. 3.19 P–n–p–n effects in CMOS: vertical cross-section of circuit in which latch-up is possible (after Robinson 1987. Reprinted by permission).

A condition known as 'snapback' can affect n-channel devices (Ochoa *et al*. 1983). This is not an SCR condition but a three-layer avalanche effect sustained by injection from the source of the device. There is no feedback loop, so the 'snapback' condition is recoverable under normal operating conditions and is unlikely to be destructive.

A condition known as a single-event-induced 'soft latch' has been detected in complex devices. Under these conditions a device 'locks up' and fails to respond to control signals. Normal operation can be restored only by removing and then reapplying the power supply voltage. This is not latch-up as described earlier; device currents have been seen to drop in a 'soft latch' condition. The exact causes of the phenomenon are not known, but it is possible that an SEU at an internal node may cause the internal logic to enter an illegal 'blocked' condition. It may also be that an internal n-channel 'snapback' blocks operation of the device. A destructive sequel to latch up,

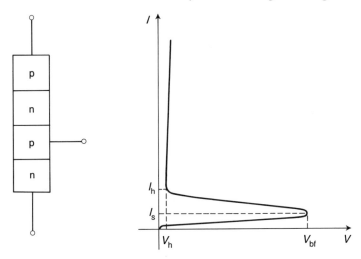

FIG. 3.20 P–n–p–n effects in CMOS: structure and I–V characteristic (after Srour 1983. Reprinted by permission).

constituting 'burn out', has been observed in power MOS and bipolar transistors (See Chapters 5 and 7).

3.8 Consequences of long-lived degradation

3.8.1 General

The preceding sections have discussed important radiation response mechanisms without specific reference to devices and their structures. In this section, we attempt to focus more on device structures, that is the way a *collection* of subelements responds. This requires classification according to the following points:

(1) degradation may be caused by either ionization or atomic displacement;
(2) the effect may be long-lived or transient;
(3) degradation may be associated with one or more 'subelements' of a device (Gregory 1973);
(4) each subelement can be classified as being made of dielectric, semiconductor, or conductor.

It can be seen that there is scope for an extremely complex set of processes leading to long- or short-lived loss of performance. The more sophisticated the microelectronic device structure, the greater the possibilities. Nevertheless, long-lived effects can still be summarized fairly simply in a bar chart such as Fig. 3.21. The dominant effect—bulk damage (b) or ionization (i)—can usually be predicted with some confidence. The authors were final

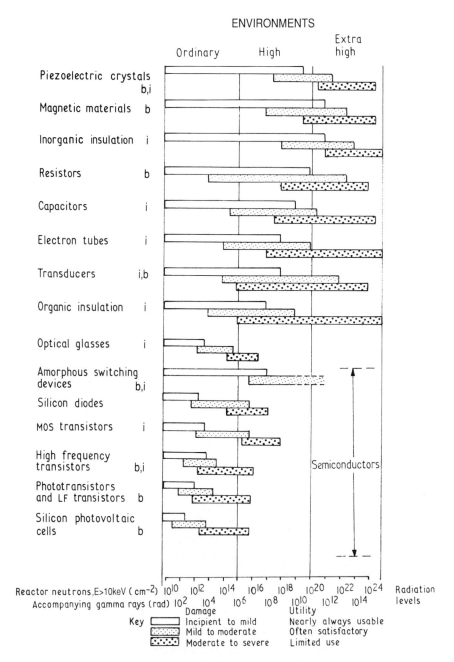

Fig. 3.21 Bar chart of materials and devices: comparison of typical long-lived degradation under mixed neutron–gamma irradiation typical of a nuclear reactor. Symbols 'b' and 'i' indicate the most likely mode of failure: 'bulk damage' or 'ionization' (Holmes–Siedle 1980).

arbiters of the positioning of the bars, but of course the choices made derive from a very large number of physical experiments and radiation tests, published and unpublished, performed on devices in the USA and Europe during the design of electronics for space vehicles, military equipment, and reactor components, plus some extrapolation by one of the authors (Holmes-Siedle 1974, 1980). The fact that the bar indicating, say, 'mild to moderate' degradation extends over perhaps six decades of radiation fluence is because one member of a class may degrade much more rapidly than another. For instance, a capacitor employing an organic dielectric (such as Mylar) may degrade more quickly than one employing an inorganic dielectric (such as mica). Even the mechanisms can be different and multiple; the Mylar device may develop leakage via the degraded dielectric, change capacitance due to dimensional change, or rupture due to gas release, while the mica would probably not manifest any of these effects.

The choice of scale in Fig. 3.21 is arbitrary; other particles could have been chosen. Many workers have compared the effect of reactor radiation with that given by other radiation. They found that, in general, so long as dose or damage was expressed in the correct units, the same positions of the bars resulted, no matter which source had been used for irradiation. This suggests that, given the correct interpretation, tests performed with one radiation environment in mind may also provide useful data for other environments. This point is expanded in discussions elsewhere of 'equivalent fluences', fluence–rad conversion and the equivalence of X-rays and cobalt-60 gamma rays in testing.

This chapter has so far described modes of degradation which are likely to be induced by ionization and displacement in various device subelements. Tables 3.2 and 3.3 separate these two effects and name specific device structures most sensitive to one effect or the other. The implications of these tables are discussed briefly below. The purpose of the tables is to show the large degree of commonality of effects which extends across device and materials technology and, conversely, the range and variety of the effects. The number of 'X' symbols gives a preliminary idea of the degree of sensitivity of a device to ionization induced in its dielectric subelement and indicates the problems which may arise when the device is used in a severe radiation environment. However, the magnitude of the effect is not indicated. Methods for calculating magnitudes in particular cases are given in later chapters.

3.8.2 Atomic displacement in semiconductors

Table 3.2 reviews displacement effects in semiconductor devices. Bipolar transistors, solar cells, opto-electronic devices, and silicon-controlled rectifiers all require a high minority-carrier lifetime for efficient operation. Field-effect transistors (FETs) withstand significantly higher particle fluences because majority-carrier transport parameters such as density and

TABLE 3.2 Atomic displacement effects in semiconductor devices (updated from Holmes-Siedle 1974)[a]

Device	Carrier lifetime reduction	Carrier removal by trapping	Carrier mobility decrease	Slow emptying of traps	Other
p-n JUNCTION DEVICES					
(a) Low reverse fields					
Bipolar transistors & SCRs	XX	X	—	—	—
MIS (MOS) field-effect transistor	—	XX	X	—	—
Variable threshold MIS transistor	—	XX	X	—	—
Junction FET, MESFET, static induction transistor	—	XX	X	—	—
Rectifying/blocking diode	XX	X	—	—	—
Tunnel diode	—	—	—	—	X
Schottky-barrier diode	XX	X	—	—	—
Junction photosensor	XX	X	—	X	—
Optoisolator	XX	X	—	X	—
Junction electroluminescent diode (LED)	—	X	—	XX	—
MIS electroluminescent diode	—	X	—	XX	—
Single-crystal solar cell	XX	X	—	—	—

3.8 Consequences of long-lived degradation

Device				
(b) Avalanche devices				
Zener and IMPATT diode	X	XX	–	–
Surface-controlled avalanche diode	–	XX	–	X
(c) Other				
Charge-coupled device	XX	X	–	XX
Hall-effect device	–	X	XX	–
OTHER DEVICES				
Transferred-electron device	–	X	–	XX
Photoconductive photosensor	–	–	X	X
Storage photosensor	–	–	X	X
Mechanical transducer	–	XX	X	–
Ovonic threshold switch	–	–	–	XX
Ovonic memory cell	–	–	–	XX
Amorphous Si solar cell	–	–	X	X
Amorphous tunnel triode	–	–	–	XX
Photostructural switch	–	–	X	XX
Cold-cathode electron emitter	X	X	XX	–

[a] XX, primary failure mode; X, secondary failure mode; –, negligible effect.

86 The response of materials and devices to radiation

TABLE 3.3 Ionization in dielectrics: long-lived effects characterized (updated from Holmes-Siedle, 1974)[a]

Device	Charge excitation		Structural change	
	Retrapped locally[a]	Charge transport	Bonding changes	Decomposition
(a) ACTIVE ELECTRICAL FUNCTION				
Charge storage				
Variable threshold transistor	x	x	–	–
Storage photosensor	–	x	–	x
Ferroelectric memory	x	–	x	–
Charge emission				
Photoemitter, photomultiplier, phototube, channel multiplier, etc.	–	x	–	–
Charge transport				
Tunnel emission cathode	–	–	–	–
Filamentary switch	–	–	x	–
Other transduction				
Pyroelectric detector	–	–	–	x
Piezoelectric detector	–	–	–	x
Surface acoustic wave device	–	–	x	–
(b) PASSIVE ELECTRICAL FUNCTION				
Passivation or gate layers in bipolar, MOS and CCD devices	–	x	–	–
Josephson device	–	–	–	–
Spacers in ICs.				
–Field oxide	x	x	–	–
–Surface encapsulation layer	–	x	x	–
–Dielectric substrate (SOS, SOI)	x	x	–	–
–Trench isolation	–	x	x	–
Capacitor insulator	–	x	x	x

3.8 Consequences of long-lived degradation 87

Standoff insulator (cable sheath, high-voltage stack)	x	–	x	x
(c) ACTIVE OPTICAL FUNCTION				
Light-beam modulator	x x	– –	– –	x x
Light-beam deflector	x x	– –	– –	x x
Storage medium				
Photostructural switch	– –	– –	x x x	– –
Photochromic memory	x –	x –	x x x	– –
Thermoplastic memory	– –	x –	x x x	– x
Phosphors and display media				
Laser medium	x x	– –	x x x	– – – –
Cathodoluminescent phosphors	x x	– x	x x x	– – – –
Electroluminescent phosphors	x x	x x	x x –	– – – –
Thermoluminescent phosphors	x x	x x	– –	– – – –
Scintillator phosphors	x x	– –	x x	x x –
Dark-field displays	x x	– –	x x	– –
(d) PASSIVE OPTICAL FUNCTION				
Lenses, filters & gratings	x x	– –	x x x x	– –
Interference and antireflection coating	x x	– –	x x x x	– –
Light guide	x x	– –	x x x x	x x
Thermal control coating	x x	– –	x x x x	x x
Glass dosimeter	x	–	x x	–
(e) PASSIVE MECHANICAL OR THERMAL FUNCTION				
Corrosion protection coating	– –	x –	x x x	x x x
Refractory layer	– –	– –	x x x	x x x
Nuclear fuel encapsulant	– –	– –	x x x	x x x
Thermal insulator	x –	x –	x x x	x x x
Solid lubricant system	– –	– –	x x x	x x x

[a] xx, primary failure mode; x, secondary failure mode; –, negligible effect.

mobility control performance. A few types of devices, including charge-coupled devices and high-electron-mobility transistors, are sensitive to the slow emptying of displacement-induced charge traps. This effect may produce inconvenient 'tails' on the otherwise sharply falling edge of electrical pulses or could increase the dark current of an imaging device. Only at higher fluences would other properties (e.g. the electromechanical constant of a transducer) be affected. Such effects are unlikely in the space environment but may be encountered, for example, in reactor environments.

3.8.3 Ionization

Table 3.3 summarizes the long-lived effects of ionization in the dielectric subelements of a number of devices. It is divided according to whether the dielectric has an active or passive electrical or optical function.

The significance of ionization in a dielectric is that charge is produced and also often transported in an element whose whole purpose is to block such transport. If the block is removed — even temporarily — unfortunate results can ensue. Fortunately, charge motion usually persists only while excitation continues. However, the *long-lived effects* resulting from the displacement of a charge sheet can be drastic, particularly space-charge build-up, in MOS gate oxides and passivation layers. They may also be accompanied by the alteration of chemical bonds in the dielectric — examples are interface-state production and the cross-linking of polymers. Chemical decomposition is an advanced form of bonding change, in which the material 'falls apart'. Devices incorporating dielectrics may be disturbed by charge transport and trapping. The MOS field-effect transistor and especially the variable-threshold transistor memory device depend on a well-controlled field at the surface.

In semiconductors, ionization produces electron–hole pairs, leading to spurious photocurrents. Upset and latch-up due to these are dealt with in detail in later sections. In glasses, optical and mechanical effects occur (see Chapter 8).

3.8.4 Other effects

Other effects in materials, theoretically interesting, do not command as much interest here as those discussed above, since they do not have much impact on electronic equipment. For example, atomic displacement in dielectrics is well known in optical materials and metal halide salts. The vacancies so produced are known as 'colour centres' (see later). However, displacement defects in dielectrics rarely interfere with the function of electronic devices. This is because dielectric materials do not require crystalline perfection in order to be good dielectrics, and the colour centres which cause darkening of glass are not created by the radiation, only repopulated thereby. Many good dielectrics have no crystalline structure at all, such as the widely used SiO_2 layers used in silicon processing as a separator. Consequently, the displacement of nuclei from their original sites has little effect on dielectric properties. Neither the capacitance nor the leakage of a capacitor is sensitive to radiation-induced displacements, and the function of a stand-off insulator is not impaired by such displacement events. Radiation-induced atomic displacements in dielectrics are of significance only in special cases where the material is used for purposes other than insulating, e.g. in thermoluminescent dosimeters (TLDS), where the number of defect sites must be controlled. In such a case, displacement effects might limit the useful dose range to a value less than that desired.

Little is said in this handbook about radiation effects in conductors. Silicon gates of MOS devices, being highly doped, are little affected. Changes in electrical properties of metals under radiation may occasionally be important in reactor applications, but rarely in other applications. In some circumstances, atomic displacement could conceivably cause small changes in conductivity which might affect a carefully balanced resistive network in a microelectronic circuit. Some handbooks mention neutron tests on the electrical properties of metals useful in electronics (Harper 1970) but in-depth studies have not yet been justified. In large metal structures in reactors, atom displacement may lead to serious changes in resistance and mechanical strength, important examples being magnetic coil windings and pressure vessels. Ionization is not likely to create problems in metals, except in so far as photo-emission and charge-scattering may cause unwanted electrical currents in vacuum devices. Steel embrittlement has already been discussed briefly.

3.9 Conclusions—an overall view of device response

The intention of this chapter has been to review general aspects of the response of solid-state devices and materials to radiation and to show what aspects there are in common. Tables 3.2 and 3.3 summarize the worst effects in a whole range of device technology. Figure 3.21 offers another bird's-eye view of technology. These overviews tend to confirm that, given the correct interpretation, tests performed with one radiation environment in mind may also provide useful data for other environments. Finally, Fig. 3.21 brings out the fact that the radiation sensitivity of the semiconductor group is markedly greater than all others. Semiconductors are materials engineered so precisely that any disturbance is bad. Complexity in the structures provides the computational plenty we all enjoy. It is tempting to say that the more complex the geometry and operational principle of a solid-state device, the more vulnerable it is to radiation-induced changes; since this is not always so, we will say instead that complexity creates more intricate possibilities for degradation problems. Further, as we move from the purely electronic type of device to the type in which light is used to produce electronic effects in the solid (or which itself generates light by such effects), the possibilities of radiation-induced degradation are even more varied; few semiconductor devices survive far into the 'high' range of radiation fluence. It is the severe practical problems produced by this group in certain applications which leads to the need for a handbook such as the present one.

References

Agullo-Lopez, F., Catlow, C. R. A. and Townsend, P. D. (1988). *Point defects in materials*. Academic Press, London.

Alberts, W. G. and Schneider, W. (1989). Reactor dosimetry: standardization of procedures and considerations on further developments for LWR PV surveillance. In *Reactor dosimetry*, Proceedings of the 6th ASTM-Euratom Symposium, (ed. H. Farrar IV and E. P. Lippincott), STP 1001. ASTM, Philadelphia.

Barbier, M. (1969). *Induced radioactivity*. North Holland, Amsterdam.

Brown, W. L., Gabbe, J. D. and Rosenzweig, W. (1963). *Bell System Technical Journal*, 42, 1505-59.

Buehler, M. G. (1968). Design curves for predicting fast neutron induced resistivity changes in silicon. *Proceedings of the IEEE*, October, p. 1741.

Burke, E. A. (1985). The impact of component technology trends on the performance and radiation vulnerability of microelectronic systems. *IEEE NSREC Short Course*, Monterey, CA. IEEE, New York.

Burke, E. A. (1986). Energy dependence of proton-induced displacement damage in silicon. *IEEE Transactions on Nuclear Science*, NS33, 1276-81.

Brucker, G. J. (1967). Correlation of radiation damage in silicon transistors bombarded by electrons, protons and neutrons. In Symposium on Radiation Effects in Semiconductor Components, Toulouse.

Carter, P. M. and Wilkins, B. R. (1984). Alpha-particle-induced failure modes in dynamic rams. *Electronics Letters*, 21(1), 38-9.

Conrad, E. E. (1971). Considerations in establishing a standard for neutron displacement energy effects in semiconductors. *IEEE Transactions on Nuclear Science*, NS18, 200-6.

Corbett, J. W. (1966). *Electron radiation damage in semiconductors and metals*. Academic Press, New York.

Corbett, J. W. and Bourgoin, J. C. (1975). Defect creation in semiconductors. In Crawford and Slifkin, Vol. 2. 1-147.

Crawford, J. H., Jr and Slifkin, L. M. (eds) (1975). *Point defects in solids*, Vols 1-3. Plenum Press, New York.

Curtis, O. (1975). Effect of point defects on electrical and optical properties of semiconductors. In Crawford and Slifkin, Vol. 2, 257-332.

Curtis, O. (1966). Effects of oxygen and dopant on lifetime and carrier concentration in neutron-irradiated silicon. *IEEE Transactions on Nuclear Science*, NS13, 33-40.

Dale, C. J. and Marshall, P. W. (1991). Displacement damage in Si imagers for space applications. *SPIE Proceedings*, Vol. 1447, 70-87.

Dale, C. J., Marshall, P. W. and Burke, E. A. (1990). Particle-induced spatial dark current fluctuations in focal plane aways. *IEEE Transactions on Nuclear Science*, NS37, 1784-91.

Dale, C. J., Marshall, P. W. and Burke, E. A., Summers, G. P. and Bender, G. E.

(1989). The generation lifetime damage factor and its variance in Si. *IEEE Transactions on Nuclear Science*, **NS36**, 1872-81.
Dale, C. J., Marshall, P. W., Burke, E. A., Summers, G. P. and Wolicki, E. A. (1988). High energy electron induced displacement damage in silicon. *IEEE Transactions on Nuclear Science*, **NS35**, 1208-14.
Dawes, W. R. (1985). Radiation effects hardening techniques. IEEE NSREC Short Course, Monterey, CA.
Deutsches Institut für Normung (1990). *Neutron fluence measurement*, DIN 25456, Parts 1 to 6. Beuth Verlag, Berlin.
Dienes, G. J. and Vineyard, G. H. (1957). *Radiation effects in solids*. Interscience, New York.
DiMaria, D. J. and Stasiak, J. W. (1989). Trap creation in silicon dioxide produced by hot electrons. *Journal of Applied Physics*, **65**, 2342.
Dyer, C. S. and Hammond, N. D. A. (1985). Neutron capture induced radioactive background in spaceborne, large volume, NaI gamma-ray spectrometers. *IEEE Transactions on Nuclear Science*, **NS32**, 4421-4.
Dyer, C. S., Trombka, J. I., Seltzer, S. M. and Evans, L. G. (1980). Calculation of radioactivity induced in gamma-ray spectrometers during spaceflight. *Nuclear Instruments and Methods*, **173**, 585.
Feigl, F. J. (1983). Characterization of dielectric films. In *VLSI electronics*: *microstructure science*, (ed. Einspruch, N. G. and Larabee, G. B.), 147. Academic Press, New York, USA
Feigl, F. J., Fowler, W. B. and Yip, K. L. (1974). Oxygen vacancy model for the E' center in SiO_2. *Solid State Communications*, **14**, 225.
Fleetwood, D. M. (1992). Border states in MOS devices. *IEEE Transactions on Nuclear Science*, **NS-39**, 269-70.
Glasstone, S. (ed.) (1977). *The effects of atomic weapons*. US Department of Defense, Defense Nuclear Agency; US government Printing Office, Washington.
Gregory, B. L. (1973). Radiation defects in devices. In *Radiation damage and defects in semiconductors*, (ed. J. E. Whitehouse), Institute of Physics Conference Series No. 16, pp. 289-94. Institute of Physics, London.
Griffin, P. J., Kelly, J. G., Luera, T. L. and Lazo, M. S. (1991). Neutron damage equivalence in GaAs. *IEEE Transactions on Nuclear Science*, **NS38**, 1216-25.
Griscom, D. L. (1989). *Physical Review*, **B40**, 4224.
Griscom, D. L. and Friebele, E. J. (1982). Effects of ionizing radiation on amorphous insulators. *Radiation Effects*, **65**, 63.
Griscom, D. L., Brown, D. B. and Saks, N. S. (1988). Nature of radiation induced point defects in amorphous SiO_2 and their role in $Si-SiO_2$ structures. In Helms and Deal, pp. 287-97.
Grunthaner, F. J., Grunthaner, P. J. and Maserjian, J. (1982). Radiation-induced defects in SiO_2 as determined with XPS. *IEEE Transactions on Nuclear Science*, **NS29**, 1462-66.
Harper, C. A. (ed.) (1970). *Handbook of materials and processes for electronics*. McGraw-Hill, New York.
Hazard Prevention (1981). A new enemy—homotrons. (Reprinted from International Federation of Airline Pilots' Association Newsletter, September 1980.)

Heimbach, C. (1992). The use of neutron–sensitive diodes as damage monitors for 1 MeV equivalent neutrons. In 7th ASTM–Euratom Symposium on Reactor Dosimetry, Strasbourg, pp. 789-96. Kluwer, Dordrecht.

Helms, C. R. (1988). Physical structure and chemical nature of the Si–SiO$_2$ interface. In *The Si-SiO$_2$ system*, (ed. P. Balk), pp. 77-128. Elsevier, Amsterdam.

Helms, C. R. and Deal, B. E. (eds) (1988). *Physics and chemistry of SiO$_2$ and the Si–SiO$_2$ interface.* (Plenum Press, New York.)

Henderson, B. and Wertz, J. E. (1977). *Defects in the alkaline earth oxides.* Taylor and Francis, London.

Hirata, M., Hirata, M., Saito, H. and Crawford, J. H., Jr (1967). Effect of impurities on the annealing behavior of irradiated silicon. *Journal of Applied Physics*, **38**, 2433-8.

Holmes-Siedle, A. G. (1974). Radiation sensitivity and amorphous materials, present and future. *Reports on Progress in Physics*, **37**, 699-769.

Holmes-Siedle, A. G. (1980). Radiation effects in the Joint European Torus experiment: guidelines for preliminary design, Fulmer Report R857/2. Fulmer Research Ltd, Stoke Poges, UK.

Holmes-Siedle, A. G. and Adams, L. (1983). The mechanisms of small instabilities in irradiated MOS transistors. *IEEE Transactions on Nuclear Science*, **NS30**, 4135-40.

Holmes-Siedle, A. G., Holland, A., Johlander, B. and Adams, L. (1991). Limiting the effects of radiation damage in charge-coupled devices by control of operating conditions. Proc. RADECS'91 (1991). Montpellier, France, Sept 9-12 1991. *IEEE Conf. Record*, Cat. No. 91 THO 400 2, Vol. 15, 338-42.

Hopkinson, G. R. (1989). Proton damage effects in a CCD imager. *IEEE Transactions on Nuclear Science*, **NS36** 1865-71.

Hrabal, C. A. (1992). Modified damage parameters applied to a typical light water reactor's pressure vessel supports. In 7th ASTM–Euratom Symposium on Reactor Dosimetry, Strasbourg, pp. 739-46. Kluwer, Dordrecht.

Johns, H. E. and Cunningham, J. R. (1971). *The physics of radiology.* Thomas, C. C. Springfield, IL.

Kelly, J. G., Griffin, P. J. and Luera, T. F. (1991). Use of silicon bipolar transistors as sensors for neutron energy spectra determination. *Transactions on Nuclear Science*, **NS38**, 1180-6.

Kerns, S. E. (1989). Transient-ionization and single-event phenomena in *Ionizing radiation effects in MOS devices and circuits*, (eds. T. P. Ma and P. V. Dressendorfer), Wiley Interscience, New York, pp. 485-576.

Kimerling, L. C., Blood, P. and Gibson, W. (1979). Defect states in proton bombarded silicon at T < 300K. Institute of Physics Conference Series No. 46, pp. 273-280.

Lang, D. V. (1974). Deep-level transient spectroscopy: a new method to characterize traps in semiconductors. *Journal of Applied Physics*, **45**, 3023-32.

Larin, F. (1968). *Radiation effects in semiconductor devices.* Wiley, New York.

Lefevre, H. and Schulz, M. (1988). Electrical evaluation of the Si–SiO$_2$ system in Balk, pp. 273-348.

Lowe, A. L., Jr and and Snidow, N. L. (1990). DPA versus fluence (E > 1 MeV) for reactor vessel wall attenuation. In 7th ASTM–Euratom Symposium on Reactor Dosimetry, Strasbourg, pp. 699-710. Kluwer, Dordrecht.

Ma, T. P. and Dresserdofer, P. V. (1989). *Ionizing radiation effects in MOS devices and circuits*. Wiley Interscience, New York.

Marshall, P. W., Dale, C. J., Summers, G. P. and Bender, G. E. (1990). Proton induced displacement damage distributions and extremes in silicon microvolumes. *IEEE Transactions on Nuclear Science*, NS37, 1776–83.

Messenger, G. C. (1979). Conductivity modulation effects in diffused resistors at very high dose rate levels. *IEEE Transactions on Nuclear Science*, NS26, 4725–9.

Messenger, G. C. and Ash, M. S. (1986). *The effects of radiation on electronic systems*. Van Nostrand Reinhold, New York.

Miller, D. L., Lang, D. V. and Kimerling, L. C. (1977). Capacitance transient spectroscopy. *Annual Review of Materials Science*, 1977, 377–448.

Mooney, P. M., Cheng, L. J., Suli, M., Gerson, J. D. and Corbett, J. W. (1977). Defect energy levels in boron-doped silicon irradiated with 1-MeV electrons. *Physical Review* B, 15, 3836–43.

Ning, T-H. (1976). High-field capture of electrons by coulomb-attractive centers in silicon dioxide. *Journal of Applied Physics*, 47, 3202.

Norgett, M. J., Robinson, M. T. and Torrens, I. M. (1985). A proposed method of calculating displacement dose rates. *Nuclear Engineering and Design*, 33, 50–4.

Ochoa, A., Sexton, F. W., Wrobel, T. F., Hash, G. L. and Sokel, R. J. (1983). Snapback: a stable regenerative breakdown mode of MOS devices. *IEEE Transactions on Nuclear Science*, NS30, 4127–30.

Peterson, E. L. (1965). Radioactivity induced in sodium iodide by trapped protons. *Journal of Geophysical Research*, 70, 1762.

Pickel, J. C. (1983). Single event upset mechanisms and predictions. IEEE NSREC Short Course, Gatlinburg, IEEE, New York.

Pons, D., Mooney, P. M. and Bourgoin, J. C. (1980). Energy dependence of deep level introduction in electron irradiated gallium arsenide. *Journal of Applied Physics*, 51, 2038–42.

Pons, D. and Bourgoin, J. C. (1985). Irradiation induced defect in GaAs. *Journal of Physics C: Solid State Physics*, 18, 3839.

Raymond, J. P. (1985). Environments, failure mechanisms and simulation of radiation effects on microelectronic components. Short Course Notes, NSREC'85 Conference, Monterey, CA. IEEE, New York.

Raymond, J. P. and Petersen, E. L. (1987). Comparison of neutron, proton and gamma rays effects in semiconductor devices. *IEEE Transactions on Nuclear Science*, NS34, 1622–8.

Robinson, P. A. (1987). Packaging testing and hardness assurance. IEEE NSREC Short Course, Snowmass, CO. IEEE, New York.

Rudie, N. J. (1976). *Principles and techniques of radiation hardening*. (Vols 1 and 2). Western Periodicals Co., North Hollywood, CA.

Sartori, L. (1991). Effects of nuclear weapons. In *Physics and nuclear arms today* (ed. D. Haefemeister). American Institute of Physics, New York.

Schulman, J. H. and Compton, W. D. (1963). *Color centers in solids*. (Pergamon, Oxford).

Srour, J. (1983). Basic mechanisms of radiation effects on electronic materials, devices and integrated circuits. *Tutorial Short Course, IEEE Nuclear and Space Radiation Effects Conference*. Gatlinburg, Tennessee.

Srour, J. R., Hartman, J. A. and Kitazaki, K. S. (1986). Permanent damage produced by single proton interactions in silicon devices. *IEEE Transactions on Nuclear Science*, **NS33**, 1597–1604.

Srour, J. R., Long, D. M., Millward, D. G., Fitzwilson, R. L. and Chadsey, W. L. (1984). *Radiation effects on and dose enhancement of electronic materials*. Noyes, Park Ridge, N.J.

Srour, J. R., Othmer, S. and Chiu, K. Y. (1975). Electron and proton damage coefficients in low-resistivity silicon, *IEEE Transactions on Nuclear Science*, **NS22**, 2656–62.

Stein, H. J. (1969). In Corbett, J. W. and Watkins, G. D. (eds) *Radiation effects in Semiconductors*. Gordon and Breach, Philadelphia, P.A.

Stein, H. J. (1969). Electrical properties of neutron-irradiated silicon at 76°K; Hall effect and electrical conductivity. *IEEE Transactions on Nuclear Science*, **NS15**, 69.

Stein, H. J. and Gereth, R. (1968). Introduction rates of electrically active defects in n- and p-type silicon by electron and neutron irradiation. *Journal of Applied Physics*, **39**, 2890.

Summers, G. P., Burke, E. A., Dale, C. J., Wolicki, E. A., Marshall, P. W. and Gehlhausen, M. (1987). Correlation of particle induced displacement damage in silicon. *IEEE Transactions on Nuclear Science*, **NS34**, 1134–9.

Svensson, C. (1988). Charge trapping in silicon dioxide and at the silicon dioxide-silicon interface, in Balk, pp. 221–72.

Thatcher, R. K. and Kalinowski, J. J. (1969). *TREE handbook*, DASA Report. Battelle Memorial Institute, Columbus OH.

Thompson, M. W. (1969). *Defects and radiation damage in metals*. Cambridge University Press, Cambridge.

Van Lint, V. A. J., Gigas, G. and Barengoltz, J. (1975). *IEEE Transactions on Nuclear Science*, **NS22** 2663–8.

Van Lint, V. A. J., Flanagan, T. M., Leadon, R. E., Naber, J. A. and Rogers, V. C. (1980). *Mechanisms of radiation effects in electronic materials*, Vol. I. Wiley, New York. (No Vol. II published to date)

Vook, F. L. and Stein, H. J. (1968). in Vook, F. L. (ed) *Radiation effects in semiconductors*. Plenum Press, New York. pp. 99–114.

Walker, J. W. and Sah, C.-T. (1973). Properties of 1 MeV electron irradiated defect centers in silicon. *Physical Review* B, **7**, 4587–605.

Watkins, G. D. and Corbett, J. W. (1964). Defects in irradiated silicon: electron paramagnetic resonance and electron–nuclear double resonance of the Si-E center. *Physical Review* A, **134**, 1359–77.

Wertheim, G. K. (1957). *Physical Review*, **105**, 1730.

Witham, H. S. and Lenahan, P. M. (1987). The nature of the deep hole trap in MOS oxides. *IEEE Transactions on Nuclear Science*, **NS34**, 1147–51.

Ziegler, J.F. and Lanford, W.A. (1981). The effect of sea-level cosmic rays on electronic devices. *Journal of Applied Physics*, **52**, 4305–12.

Zoutendyk, J. A., Malone, C. J. and Smith, L. S. (1984). Experimental determination of single event upset (SEU) as a function of collected charge in bipolar integrated circuits. *IEEE Transactions on Nuclear Science*, **NS31**, 1167–74.

4
Metal-oxide-semiconductor (MOS) devices

4.1 Introduction

MOS integrated circuits, particularly those of the complementary form (CMOS), are very suitable for use in high-performance electronics such as timers, battery-powered computers, robots, missiles, and space vehicles. The power consumed by the CMOS logic element is extraordinarily low compared with nMOS and bipolar circuits. In addition, the CMOS inverter employs voltage signals which can be made highly immune to noise. These features are uniquely suitable for advanced data-handling and control systems in severe, remote environments. It is unfortunate therefore that many MOS devices show strong, variable and long-lived response to total-dose radiation. In any radiation environment exceeding about one thousand rads (10 Gy), it is necessary to consider the effect of oxide charge trapping and interface-state generation; heavy ions and intense pulses of radiation can also cause the upset of logic states.

If long-term degradation of MOS devices is to be avoided in severe environments, then special attention has to be paid to the design of the MOS structure, the modelling of the device physics, its processing and procurement. This chapter will discuss in detail the physical problem of total-dose effects and how to predict the electrical changes caused, and will also indicate some of the best solutions to the radiation problem. The topic is so large that it will be necessary to refer heavily to several authoritative treatises on the subject (Ma and Dressendorfer 1989; Grove 1967; Sze 1969). The insulating oxide layers have the main influence upon how an MOS device changes in the presence of radiation. The designer must know the factors which affect the oxides. Analysis methods are described here. Probably the the most important factors concern the *material* parameters of the oxide. These are strongly affected by the processing conditions during and after oxide growth. Equally important is the *applied electrical field* in the oxide—hence the differences in MOS behaviour for various powered and unpowered states.

The mechanisms which are the most important in the degradation of MOS devices are caused by ionization (hole–electron pair creation) within the silicon dioxide layers and photocurrent generation in the source and

96 Metal-oxide-semiconductor (MOS) devices

drain junctions. This chapter is mainly concerned with the effect of ionization in the oxide and the consequent effects upon function. The effects of ion tracks and photocurrents are discussed in relation to design factors in MOS devices, such as in the development of components tolerant to single-event upsets or weapon bursts.

The term 'MOS' is normally extended to devices in which the oxide, SiO_2 in the simplest case, is modified to contain other elements (e.g. nitrogen) and in which the metal gate may be replaced by a polycrystalline silicon electrode (Sze 1969). The latter type is distinguished by the name 'silicon-gate MOS devices'. Reference is made in the following pages to both discrete

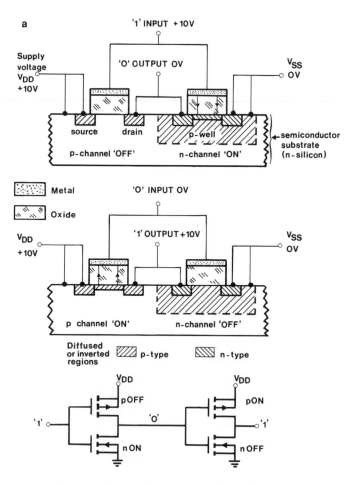

FIG. 4.1 Typical MOS structures. (a) Structure of an integrated complementary-symmetry metal-oxide-semiconductor (CMOS) inverter showing typical bias and logic levels. The lower part shows two inverters in cascade.

MOS devices and integrated circuits, particularly the complementary MOS network (CMOS) which incorporates a number of p- and n-channel transistors on a single chip. Figure 4.1(a) shows an example of a CMOS circuit configuration (two CMOS inverters in cascade) and a schematic representation of the device. We will later refer to this sketch in our discussion of transient response to radiation pulses. The source and drain regions are heavily doped 'islands' in the substrate or body, and the channels are the surface regions lying between source and drain. Material of n-type is typically produced by doping with phosphorus, while p-type may be obtained by doping with boron. The gate oxide layer, mainly thermally grown SiO_2, is typically less than 0.1 μm thick.

Figure 4.1(b) shows more realistic detail of a CMOS chip. The device shown is a large-scale integrated (LSI) circuit. The dimensions of subelements may be as small as 0.5 μm, the gates are made of polycrystalline silicon, and the gate oxide thickness may be as small as 0.03 μm. The 'field oxides' are an essential feature in radiation response. Designed as inert spacers to carry signal tracks, they present severe problems of radiation-induced charging. As noted later, heavy-metal layers may well be embedded in the upper surface of the chip.

A source of confusion in nomenclature should be noted here. The term 'radiation-hardness' is often used to describe several different forms of tolerance to radiation in MOS and other devices. This term has a military connotation. Originally, 'hard' meant 'invulnerable to attack'. Where possible, we will use the term 'tolerance (or sensitivity) to radiation' instead of 'hardness'. However, the term 'hardness assurance' is widely used in

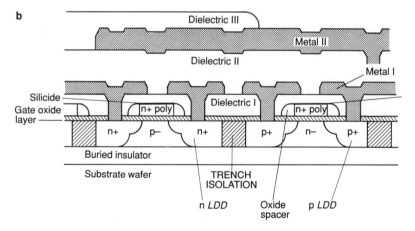

FIG. 4.1 (b) Silicon-gate CMOS Structure on one-micrometre scale: includes new features such as lightly-doped drain (LDD), buried oxide layer, and trench isolation (Van Vonno, personal communication). Reproduced by permission of Harris Semiconductor.

98 Metal-oxide-semiconductor (MOS) devices

TABLE 4.1 Comparison between typical radiation environments in a military 'weapon burst' and space radiation belts

	Military	Space
Damage: neutron fluence (cm^{-2})	10^{14}	none
electron/proton fluence	none	$10^{12}/10^{9}$
equiv. 1 Mev electron fluence (cm^{-2})	10^{17}	10^{12}
Total dose: rad(Si)	10^{6}	10^{5}
Dose rate: rad(Si) s^{-1}	10^{11}	10^{-3}

the USA. 'Megarad-hard' is used to signify 'capable of withstanding one megarad of ionizing radiation'. The term 'SEU-hard' is also used for single-event upset. Confusion also arises on account of the military requirement to survive (a) the transient effects of an intense burst of gamma rays and neutrons at a dose rate at least 10 billion times higher than that experienced in space, and (b) the long-lived effects or 'total doses' received from this burst. The extent of (b) may be of the same order as that accumulated in a space mission, but neutron damage will also be received in the weapon burst. From typical figures for the space and military environments, given in Table 4.1, it can be seen that transient effects, usually the dominant problem in military devices, are negligible in the case of space radiation belts or most exposures in and around reactors or isotope sources. The space doses are for a typical internal location within a spacecraft structure. Because of the 'logic upset' produced in the military case, special device structures, which minimize the photocurrents produced, have been developed. These may *not* have good tolerance to total-dose effects. They are often not exploitable commercially. Thus, in the procurement of devices, the term 'radiation-hardened' should be investigated and quantified precisely before its relevance to the environment in question can be established. Similar distinctions should be drawn when considering testing methods.

Much radiation test data for commercial MOS devices is available, but is too voluminous for inclusion here. In Sections 4.8 and 4.9, we draw on specialized knowledge of technology assessed for the space environment. Data banks include the ESA database developed 1988–1991 (see also Braünig *et al.* 1977). Other data compilations are described in Chapter 13. Appendix D contains a listing of test results from ESA programmes.

4.2 Historical

The effect of radiation on the performance of MOS devices was first noted by H. L. Hughes and Giroux (1964). This observation contradicted the earlier belief that, because surface channel devices would not be affected by the production of minority-carrier recombination centres in silicon, they would not undergo strong changes as a result of irradiation. Even though radiation-induced charge-trapping had long been recognized as, for example, the source of colour centre formation in bulk silica and other insulators, it was not appreciated that similar effects in the insulator layer of any MOS device would strongly influence performance. In fact, some early workers were misled by an early test in which developmental MOS transistors were placed in a nuclear reactor core without a bias circuit (Christian 1966). Since little change in performance was observed, it must be assumed that the heating induced by the neutron–gamma environment was sufficient to anneal out the trapped charge and interface states produced, so that the workers concerned appear to have missed a discovery.

During 1965 and 1966, much experimental evidence was produced to support the theory that radiation-induced degradation in MOS devices was caused by the trapping of charge (holes) in the oxide layer (Zaininger 1966a,b; Holmes-Siedle and Zaininger 1968; Dennehy et al. 1967). This theory gained support from work proceeding at the same time on surface degradation effects in planar bipolar transistors (Brucker et al. 1965). It had already been proposed that these effects were not due solely to the well-known atomic displacement damage but were due to the trapping of charge in the passivating oxide layer. H. L. Hughes, who was the discoverer of the oxide trapping effects in MOS devices, has given an authoritative review of this period (H. L. Hughes 1989)

In MOS devices, it was found that the amount of charge trapped depended strongly upon the voltage across the oxide during irradiation. It was also found that a secondary effect, involving the rearrangement of atomic bonds at the oxide–silicon interface, was contributing to the degradation; this is the production of new 'interface states', including 'slow-state instabilities' (Dennehy et al. 1966; Holmes-Siedle and Adams, 1983)

Figure 4.2 illustrates the two effects mentioned above. Note that the term 'irradiation bias' (V_I) is used to specify the gate voltage applied to a device during irradiation. This is commonly a steady voltage which induces the electric field in the oxide (it may however be variable during real device operation). The influence of this field is very important in determining the rate of device degradation under irradiation. Atomic displacement damage in either the silicon or the insulator does not appear to play a significant role in these phenomena. The trapping of holes appears to occur to some extent in all types of insulator, in which the holes are usually much less mobile than electrons. Preliminary attempts were made to discover the origin of the hole-trapping and to suppress it, including the use of different insulators (Zaininger and Waxman, 1969), but these did not meet with immediate practical success. Users concerned with radiation environments (including equipment for some military environments, space, nuclear industry, and nuclear physics) then settled down to try and understand, control, and accommodate the changes produced by radiation in MOS devices (see e.g. Dennehy et al. 1969b). Work was funded by the US Government, through

100 Metal-oxide-semiconductor (MOS) devices

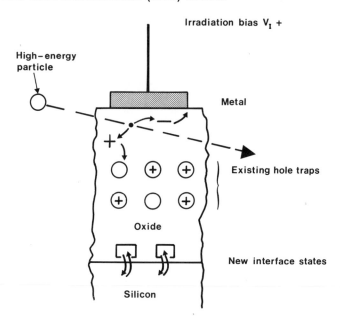

FIG. 4.2 MOS degradation mechanisms: ionization, charge transport, hole-trapping, and interface states in the gate insulator of a metal-oxide semiconductor structure.

the US Naval Research Laboratories, NASA and other agencies (Aubuchon 1971; Holmes-Siedle and Harari 1971). Semiconductor laboratories leading the effort included Sandia National Laboratory, RCA, National Semiconductor, and Harris. European governments funded research at a lower level, for example in the 1970s at GEC Hirst Research Centre, Wembley (Emms *et al*. 1974), SGS, Milan, and CEN, Grenoble. Progress reports of the above work can be found in the annual proceedings of the IEEE Nuclear and Space Radiation Effects Conference, published regularly as the December issue of the *IEEE Transactions on Nuclear Science*, while Ma and Dressendorfer (1989) have edited an authoritative account of the field.

4.3 Charge trapping in MOS devices

4.3.1 Oxide charge trapping and performance degradation—an overview

The electrical consequences of radiation-induced physical changes in MOS devices are progressive loss of function and eventual 'failure' of an MOS circuit. Figure 4.3 shows a series of parallel I_D-V_G curves for an n-channel device, resulting from successive increases in dose. Each of the curves corresponds to the onset of particular malfunction in the CMOS logic gate shown in Fig. 4.1. Table 4.2 lists the four major performance degradation effects

4.3 Charge trapping in MOS devices

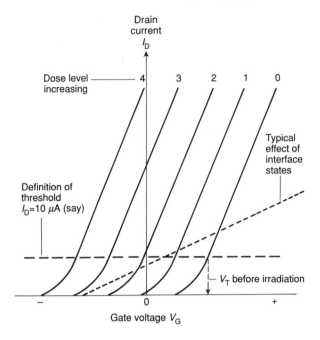

FIG. 4.3 nMOS degradation and failure points: typical drain current–gate-voltage curves for an n-channel MOS device, showing the progressive effect of increasing radiation dose. Curves 1 to 4 correspond roughly to the points at which the four different 'failure mechanisms' described in Table 4.2 occur. The solid curves show the ideal case, where positive charge only is introduced into existing traps; the further effect of the production of interface states on the shape of the curves is shown by the broken line.

which are likely to appear at the four dose levels indicated in Fig. 4.3. The table also gives an indication of the real dose and threshold shift associated with each failure mode in a typical commercial CMOS circuit.

It is clear that if one were designing CMOS circuits *ab initio*, one would use circuit techniques to build-in some tolerance to these effects. However, most LSI circuits are designed with a commercial purpose in view. The internal circuit design is predetermined and may have poor tolerance to one of the four effects noted above. Thus, 'failure' cannot be defined by the performance of a single device element, but is rather the point at which a particular network of devices no longer tolerates the effects. Thus, in this chapter, we separate the fairly predictable 'device effects' from the less easily predictable joint or 'system' effects. We can however define a radiation dose limit or 'maximum acceptable dose', $D_{A(max)}$, for any CMOS circuit based on the dose at which some limitation such as the power available is reached. Power drain usually becomes too large by virtue of an

102 Metal-oxide-semiconductor (MOS) devices

TABLE 4.2 Four types of 'failure' in CMOS circuits from successive irradiations; no. 2 (VNTZ) is the most common cause of system malfunction

Dose level in Fig. 4.3	Failure mechanism number	Main degradation effect	Symbol	Typical values for CMOS LSI	
				Dose (rad(Si))	$-\Delta V_T$ (V)
1	1	Minor 'noise immunity reduction'; possibly minor loss in switching speed; device 'out of specification'	NIR	8×10^2	0.2
2	2	Sharp quiescent current increase due to 'V_T of n-channel crossing zero'	**VTNZ**	5×10^3	1
3	3	Switching speed reduction	SSR	1×10^4	2
4	4	Change of logic state impossible: 'Logic failure'	LF	3×10^4	4

effect noted in Table 4.2, namely 'V_T of the n-channel element crossing zero (VTNZ)'. This leads to a large increase of quiescent current. The levels of current increase acceptable in battery-operated devices may be very small.

4.3.2 MOS transistor action and threshold voltage shift

4.3.2.1 *General*

Details of the solid-state physics of MOS transistors can be found in the books by Grove, Nicollian and Brews, and Sze, listed under 'Recommended background reading'. A detailed treatise on radiation effects in MOS devices by a number of authorities has been edited by Ma and Dressendorfer (1989). The threshold voltage of an MOS transistor is ascertained by measuring the channel current (usually termed 'drain current'), I_D, as a function of the gate voltage, V_G, at a constant supply voltage, V_{DD}. The characteristic curve for an n-type channel in a p-type substrate is shown in Fig. 4.4(b). The minimum in the C–V curve, Fig. 4.4(a) reflects the transition from 'depletion' to 'inversion' conditions. Comparing Figs 4.4(a) and (b), it is seen that the value marked as V_T on the I–V characteristic is always aligned with V_i, a point near this minimum at which 'inversion' is established. The flatband condition, at voltage V_{FB}, occurs when the ratio C/C_0 is approximately 0.8. Note that the measurement of C does not involve the measurement of channel current and, in fact, can be accomplished using an MOS capacitor with appropriately doped silicon substrate; source and drain diffusions are not required. MOS capacitors therefore simplify the

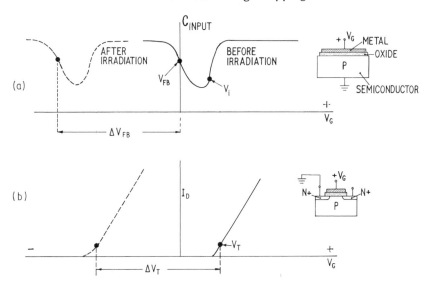

FIG. 4.4 MOS characteristics: typical variation of (a) capacitance and (b) drain current with gate voltage, showing the shifts in flatband and threshold voltages due to trapped charge (no interface states).

laboratory investigation of radiation effects (see e.g. Winokur *et al.* 1984 Dressendorfer 1989).

The shifts observed in irradiated MOS devices can conveniently be divided into two largely independent elements: the voltage shifts caused by *oxide trapped charge* (ΔV_{ot}) and that caused by *'interface trapped charge'* (ΔV_{it}). These are explained later, but the adoption of these terms derives from a terminology derived by Deal and co-workers (Deal 1980). Winokur and his colleagues at Sandia (Winokur *et al.* 1983) have demonstrated practical methods of separating the two species.

Trapped charge, Q_{ot}, causes simple translation of the I–V characteristic. Interface charge, Q_{it}, cause shift *and distortion* of the I–V characteristic. In this chapter, we give a quantitative treatment of the effects of ΔV_{ot} and a semiquantitative treatment of the effects of ΔV_{it}. This is probably valid for all except the most thoroughly radiation-hardened MOS devices, in which ΔV_{ot} has been so strongly suppressed by 'hardening', that ΔV_{it} dominates.

4.3.2.2 *n-Channel transistor (nMOS)*

We will now describe the effect of Q_{ot} and Q_{it} on an MOS transistor having an n-type channel in a p-type substrate. Trapped holes induced by ionization in the oxide layer of a MOS device (Q_{ot}) will have the same qualitative effect on the potentials in the silicon as the application of a positive gate voltage it will bend the energy bands further downwards and tend to induce inversion in the p-type substrate. An alternative interpretation of this effect, without involving energy bands, is to say that the trapped positive charge induces negative image charge in both the gate and the substrate; the charge in the substrate therefore increases the n-type conductivity in an n-channel device. As a result, a smaller threshold voltage will be registered. Given sufficient trapped charge, inversion may be established by the trapped charge alone and a 'leakage current' will flow in the channel even in the absence of a gate voltage. This is the VTNZ effect in Table 4.2. The existence of trapped charge also means that a greater negative gate voltage will be required to achieve the flatband condition. Thus, simple trapped charge accumulation results in an essentially parallel shift of the C-V and I_D-V_G characteristics. Discussion of these shifts in threshold, ΔV_T, and flatband voltages, ΔV_{FB}, forms the major part of this section. If we ignore various distorting effects, especially ΔV_{it}, then ΔV_T is equal to ΔV_{FB}, is equal to ΔV_{ot}. As described later, the degree of parallel shift observed in MOS devices varies very widely with the material parameters of the insulator layer. In the 'real world', n-channel transistors are usually integrated into a network. For this, the device may be surrounded by a thicker 'field oxide' which may give leakage failure before the VTNZ effect occurs. See Section 4.7 and later sections for discussion of 'real world' considerations.

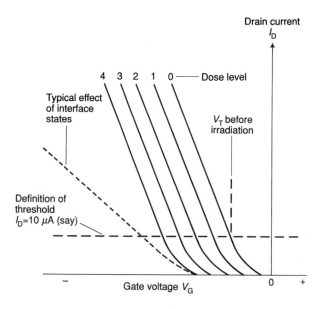

FIG. 4.5 pMOS degradation: curves for a p-channel device, corresponding to Fig. 4.3. In this case, V_T does not cross zero, but progressively higher gate voltages will be required to invert the channel region.

4.3.2.3 Equivalent effects in p-channel devices (pMOS)

Whereas, in the absence of Q_{ot}, an n-channel MOS requires a positive gate voltage to make the channel conduct, a p-channel devices requires a negative gate bias. As shown in Fig. 4.5, the parallel shift in the I_D-V_G characteristic caused by Q_{ot} means that an increasing negative bias is required to operate the device. At high radiation doses, it may therefore become impossible to switch the device 'ON'. Additionally, the creation of interface states, resulting in a typical characteristic as shown by the dotted curve, must always – in contrast to n-channel devices – make matters worse.

4.4 Physical model for oxide trapped charge build-up

4.1.1 General

A detailed physical model for the transport and eventual trapping of holes in a silicon dioxide film was developed by McLean, McGarrity, Boesch, Ausman, and Oldham from 1975 to 1986 (see e.g. McLean et al. 1976) and is summarized by those authors in Ma and Dressendorfer (1989). This dose-dependence model is confined to Q_{ot} build-up mainly because, as Winokur (1989) notes, the dose dependence of Q_{it} build-up is quite variable. For low doses,

$$\Delta V_{ot} = -(q/\varepsilon_{ox}) \cdot K_g \cdot f_y \cdot f_t \cdot d_{ox}^2 \cdot D, \qquad (4.1)$$

where q is the charge on the electron, ε_{ox} is the permittivity of the oxide, K_g is the energy-dependent charge generation coefficient, f_y is the field-dependent fractional free charge yield, f_t is the field-dependent fraction of radiation-generated holes which are trapped, d_{ox} is the oxide thickness, and D is the radiation dose.

It should be noted that, in the model, oxide trapped charge, Q_{ot}, is converted into a voltage shift, ΔV_{ot}, by means of the factor $-(q/\varepsilon_{ox})$. This term incorporates the capacitative effects of the MOS structure in a convenient way (the smaller the capacitance, the larger the voltage effect of a given charge). The thickness element of capacitance is included in the d_{ox} term. Equation (4.1) gives a 'thickness squared' dependence of ΔV_{ot}. This applies for oxides thicker than 30 nm.

Figure 4.6 shows a general picture of charge build-up processes which are going on simultaneously in an oxide under a steady dose rate. The local oxide field varies with the amount of charge trapped in the oxide and hence with dose and position. Thus, f_y and f_t both vary in a complex manner. The change in V_T is thus not perfectly linear with dose, but is nearly so over some dose ranges. The practical consequences of this are noted later in discussion of a simple prediction model. Here we will note only that, for unhardened discrete MOS transistors or MSI CMOS logic, the equal increments of ΔV_T in Figs 4.4 and 4.5 can properly be assumed to be caused by equal increments of dose.

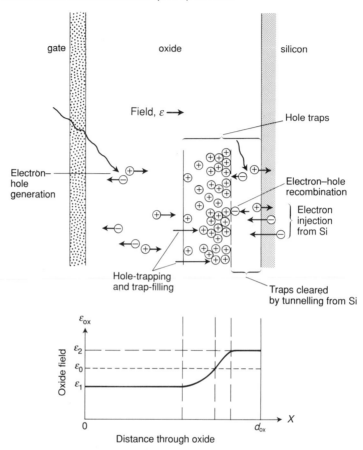

FIG. 4.6 Charge build-up mechanisms: charge build-up and removal processes in an MOS structure. (a) Sketch of processes taking place under irradiation. (b) The strong variation of the electric field in the region of the sheet of trapped charge; as expressed by McLean et al., in Ma and Dressendorfer (1989); compare with Figs 3.6 and 3.8. (Reprinted with permission.)

4.4.2 Interface charge trapping

In many CMOS devices, the effects of Q_{it} near oxide charge saturation has profound effects on device action, including 'turnaround' in the n-channel device (see, for example, Freeman et al. (1978), Dressendorfer (1989), and Section 4.6). Also, if some of the Q_{ot} is annihilated, then 'rebound' of V_T may be significant. As noted by Freeman et al. (1978), these effects must be recognized in any practical prediction for modern CMOS devices. They are discussed further in section 4.6.

The effects of radiation-induced interface states on MOS devices are

4.4 Physical model for oxide trapped charge build-up

strong, especially in high-technology ICs having thin gate oxides. The main effects are:

(1) lowering of transconductance;
(2) distortion of I_D-V_G characteristics;
(3) additional threshold voltage shifts, ΔV_{it}; these shifts are in a negative direction for p-channel and positive for n-channel transistors ('turn-around', see below);
(4) Generation of 'slow states' and a resulting slow drift of V_T with time (Holmes-Siedle and Adams 1983).

Some of the effects of (1), (2) and (3) have been incorporated in the simple models discussed below. A particularly important effect of (3) is 'rebound' in n-channel transistors. Effect (4) is not normally important in logic circuits but is important in instrumentation applications.

At the interface between silicon and a thermally grown oxide layer it is inevitable that there should be a mismatch in the chemical bonds passing across the plane of that interface. Uncompleted or dangling silicon bonds are electronically active; in particular, these defect centres or interface states are able to alter their charge state according to the availability of electrons or holes; for example they depend on the Fermi level in the silicon. The resulting interface trapped charge sheets, Q_{it}, appear either as a fixed charge or an instability in the I-V and C-V characteristics (see Section 4.3). Modern processing technology has been successful in reducing the observable defects, N_{it}, to a density of the order of 10^{10} cm^{-2}, at which value they have little effect on device performance. However, several kinds of stress, including excitations within the oxide caused by radiation, tend to disrupt many more bonds, creating new interface states which often have a different character from the as-processed states. It is fair to say that MOS device processing may have minimized or 'passivated' the defect states but leaves a distinctly metastable condition near the interface which is disturbed by physical excitation. This is assisted by the presence of chemical species other than SiO$_2$ in the oxide layer, notably hydrogen. Winokur (1989) reviews the nature and formation of interface traps by irradiation. The role of hole and hydrogen transport from the bulk to the interface after pulsed radiation exposure is currently under investigation and a 'hole trapping/hydrogen transport' theory has been proposed (Saks *et al.* 1991; Shaneyfelt *et al.* 1990).

Since 1980, the use of electron spin resonance and X-ray photoelectron spectroscopy has contributed to the understanding of the structure of interface states (Lenahan *et al.* 1984; Grunthaner *et al.* 1982). The interface state appears to be a dangling silicon bond, known as the P_b center, which has energy states broadly distributed throughout the silicon forbidden gap and becomes paramagnetic when it accepts an electron into an otherwise unsatisfied s-type orbital. The use of a technique which is a hybrid of magnetic resonance and device electrical measurement, spin-dependent recombination (Vranch *et al.* 1988; Jupina and Lenahan 1990), is assisting the correlation of atomic-scale measurements and those gross electrical effects of interface states which operate in devices.

4.5 Prediction models for radiation effects in MOS devices

4.5.1 A simple analysis of radiation effects in MOS devices

Prediction models have been developed to deal with threshold voltage shift in MOS devices as a function of dose and electrical stress applied during irradiation. An example of an early simple analysis is that of Freeman and Holmes-Siedle (1978). The analysis adopts straightforward assumptions with respect to trap location in the oxide and yield of the available charge. It employs a 'worst-case' (linear dependence) of charge build-up on dose at low dose values, making a transition to a saturated condition at high doses. The result is an analytical method suitable for codifying the behaviour of commercial MOS devices such as simple CMOS logic elements without field oxides (CD4000 series and successors) under laboratory radiation test and the use of the data to predict degradation with time in a radiation environment. The analysis was developed for commercial devices with thickness values generally above 60 nm (600 Å), available widely, and used in European space programmes.

The starting point is laboratory test data from which the oxide charge trapping characteristics of an individual transistor are found and assigned an 'A-value'. This is a charge-trapping probability, and the most sensitive devices have an A-value of 1.0. The prediction of threshold voltage shift, ΔV_T, is made using the formula:

$$\Delta V_T = R.A.D. \qquad (4.2)$$

R is calculated from oxide thickness and other parameters as explained below. Further detail is given in Freeman and Holmes-Siedle (1978).

The simple model assumes that the charge is a thin positive sheet of density Q_{ot} separated from the silicon by a distance x_1, as shown in Fig. 4.7. With a positive voltage on the gate, radiation-induced charge is collected from Region 2 (of length x_2) and deposited in the trap sheet centred at a distance x_1 from the silicon. With a negative voltage, the charge is collected from Region 1 and deposited in the same place. If we call the depth of the collection region x_c, then the number density of holes generated in the oxide is $g \cdot x_c$, where g is the generation rate for electron-hole pairs. The number trapped in the charge sheet to form Q_{ot} is this number, modified by the trapping probability, A, and the fractional charge yield, f_E, which escapes recombination. x_1 is usually much smaller than x_2.

4.5 Prediction models for radiation effects in MOS devices

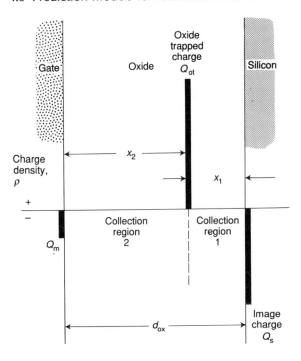

FIG. 4.7 Simple charge sheet model: a very thin sheet of charge, labelled Q_{ot}, produced when holes generated in collection regions 1 or 2 are swept towards it by the applied field. Q_s is the image charge in the semiconductor, which determines the threshold voltage. Co-ordinates are used in the text (Freeman and Holmes-Siedle 1978).

By electrostatics, it can be shown that

$$Q_s = \frac{x_2}{d_{ox}} \cdot Q_{ot} \tag{4.3}$$

and, converting charge into voltage,

$$\Delta V_T = \frac{x_2}{d_{ox}} \cdot \frac{Q_{ot}}{C_{ox}} \tag{4.4}$$

where C_{ox} is the capacitance of the oxide.

Calculating Q_{ot} from the values of trapping probability and charge collection mentioned, we obtain our final equation:

$$\Delta V_T = (q \cdot g \cdot x_2 \cdot x_c \cdot f_E / \varepsilon_{ox} \cdot \varepsilon_o) A \cdot D = R.A.D, \tag{4.5}$$

where R collects all the terms characteristic of the experiment conditions and MOS structure. A is the key variable, the trapping capability of the oxide, through which we can define the radiation tolerance of variants of a given series of fixed oxide thickness. We will quote 'A-values' in future when discussing this subject. This equation does not embrace the generation of interface states.

D is the mission dose predicted for a device in a given environment. If the measurement of original V_T values are not possible in the laboratory, the manufacturer can supply the original value of V_T for the devices as used in integrated circuits and an estimate of the V_T value at which the circuit will fail. Given these data, we can calculate the earliest time at which the device may malfunction in a given environment.

Prediction programmes for densely packed, LSI MOS devices are subject to several pitfalls:

(1) failure modes may involve either the gate oxide or the much thicker field oxide;

(2) with small geometry comes lower gate oxide thickness, which enhances the distorting effects of Q_{it} versus Q_{ot}.

Refinements of the model can be introduced to deal with the above effects and also late or post-irradiation effects (Freeman and Holmes-Siedle, 1978; Brown *et al.* 1989).

4.5.2 Terms for gate voltage applied during irradiation

The terminology recommended for the gate bias values applied during irradiation is as follows:

$V_I +$: positive bias applied to gate during irradiation;

$V_I -$: negative bias applied to gate during irradiation;

$V_I\ 0$: zero bias (gate shorted to body) during irradiation.

Thus, for example, 'irradiation at $V_I + 5$' would describe the application during irradiation of $+5$ V to the gate relative to the silicon body or substrate of the device. The oxide field, E_I, created thereby in an oxide thickness of 0.1 μm would be 50 V μm^{-1}. In a CMOS gate in the '1' condition with $V_{DD} = 5$ V, the n-channel device is in the $V_I + 5$ condition, but the p-channel device is in a different condition, not unlike '$V_I 0$', see Fig. 4.1a.

4.5.3 Bias dependence

4.5.3.1 *General*

Since the beginning of radiation experiments on MOS devices (see e.g. Holmes-Siedle and Zaininger 1968) it was noted that oxide charge buildup per unit dose increased steeply as a function of positive bias during irradiation and much less steeply for negative bias. The oxide trapped-charge model (Fig. 4.7) predicts this because of the larger collection region, x_2, when bias is positive. In Fig. 4.8(a), broken lines show the predicted curve. The

4.5 Prediction models for radiation effects in MOS devices

minimum is zero and linear curves are expected for both n- and p-type devices. For a number of reasons, explained below, the experimental curve is rarely as simple as this. The solid line shows an irradiation of type CD4007 commercial CMOS devices. Negative bias was applied to the p-channel side and positive bias to the n-channel. The shifts at zero bias are finite and the curves are not linear.

4.5.3.2 *Zero bias condition: oxide charge vs. interface effects*

A growth curve for commercial CMOS devices is given in Fig. 4.8(b). It again shows that V_T shift at zero bias increases at a finite rate with dose. The curve is different for n- and p-MOSFETs on the same chip. Up to a dose of 30 krad (300 Gy), the charge build-up is seen to be the same for the two FETS, suggesting that the charge trapped in this range is Q_{ot}. At higher dose values, the curves diverge, suggesting that, after this point, the generation of Q_{it} overtakes that of Q_{ot}. The n-channel transistor exhibits the well-known 'turnaround' effect. Similar curves are to be found for harder oxides in Dressendorfer's account of device phenomena (Dressendorfer 1989).

The zero-bias response has to be taken into account if, for example, a CMOS device has to withstand a large dose in a passive condition (for example a 'cold spare' circuit in space or munitions in store) before operation under radiation commences (Holmes-Siedle and Adams 1986). In the ISO spacecraft instrument, ISOCAM, redundant 8OC86 microprocessors are installed and operation is switched from one to the other according to the readings of a co-located dosimeter.

The amplitude of the zero bias effect is large enough to be used in RADFET dosimetry, where the use of the passive condition is often convenient (see Holmes-Siedle *et al.* 1991). R. C. Hughes (1985) has argued that, since there is no applied field to drive the radiation-generated electrons and holes apart, the zero-bias response must be due to the diffusion of electrons out of the oxide.

4.5.4 Intermittent biasing

It has been known for some time that when the electric field across the oxide layer of a MOS device is released *during* irradiation, the trapped charge begins to decline. This is due to the neutralizing effect of photoelectrons emitted into the oxide by the electrodes. This mechanism is effective even if the rate of cycling of the field is many times per second. Thus, for devices which are subject to cycling, the maximum acceptable dose is likely to be higher than if the irradiation bias were applied continuously.

An illustration of the effect of intermittent biasing is shown in Fig. 4.9(a). Here, irradiation biases of zero and -9 V are applied to a typical p-channel MOS device. If the cycling procedure is interrupted and the bias thereafter held constant, the resulting 'growth curve' should gradually approach the upper and lower limit curves. There is scope for introducing a 'resting'

112 Metal-oxide-semiconductor (MOS) devices

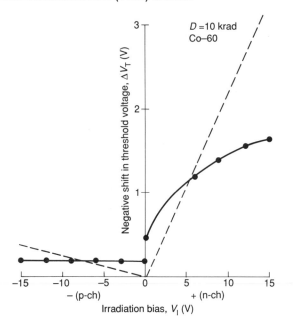

(a)

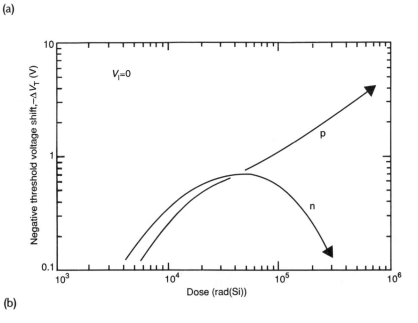

(b)

FIG. 4.8 Irradiation bias effects in real CMOS: (a) Bias-dependence curve: threshold voltage shift vs. gate voltage applied during irradiation. For our simple prediction model (broken line), a minimum near zero would be expected and the same ΔV_T values would be derived for n- and p-type devices. In the CD4007 CMOS devices

4.5 Prediction models for radiation effects in MOS devices

policy, either within the chip or over the whole subsystem. Such redundancy would significantly prolong the useful life of all MOS components in space. However, significant build-up of interface states continues during irradiation at zero bias; therefore, caution should be exercised in applying a resting technique. Stanley *et al.* (1985) observed this as a significant effect in hardened oxides.

Figure 4.9(b) shows the marked effect of '50% cycling' (V_I applied for half the time) on the onset of the VTNZ effect. We take this point as the 'maximum acceptable dose'. Low bias values are not only better—this is the expected bias dependence—but the 50% curve also shows that cycling is even more beneficial at low bias. It should be noted that the broken sections of the curves in this figure tend towards a very high dose value. That is, devices which are unbiased would not fail until megarad doses are received.

Fleetwood *et al.* (1990) studied cycled bias patterns in Sandia hardened oxides (thickness less than 50 nm), separating trapped charge and interface states (V_{ot} and V_{it}) for doses in the megarad range. Prediction schemes for cycled, hardened oxides have been developed (Fleetwood and Dressendorfer 1987). In such oxides, recovery of V_T shift occurred when bias was released during irradiation. However, the n-channel devices exhibited 'turnaround' within this dose range (see Section 4.6), which makes the patterns more complex than in the early growth stages of soft oxides, as shown above. A different effect was discovered by Okabe *et al.* (1990), namely a strong effect of AC gate bias applied after exposure. This may be due to the different ability of conduction electrons and holes in the excited oxide to follow very high-frequency fields.

Given the above variety of the effects of cycling, the best engineering approach for MOS device users is to operate the devices during laboratory irradiation tests (a) in an electrical mode which simulates that expected during use and (b) in the worst case of DC bias conditions.

4.5.5 Gate oxide thickness

Experimental work on oxide thickness dependence, described by Fossum *et al.* (1975), indicated that radiation-induced threshold shifts varied as the cube of the oxide thickness d_{ox}. G. W. Hughes and Powell (1976) disagreed

---

used, negative bias was applied to the p-channel side and positive bias to the n-channel side. (b) Zero-bias effects: growth of V_T versus dose for a CMOS device with zero applied bias during irradiation; p- and n-channel elements both shift in a negative direction until the 'conventional' effect of interface states (positive shift for n-channel) takes over, causing the curve for the n-channel element to reverse. Experimental data for an early (1971) RCA CD4007A 'hard' device (A-value 0.05) Freeman and Holmes-Siedle 1978.

114 Metal-oxide-semiconductor (MOS) devices

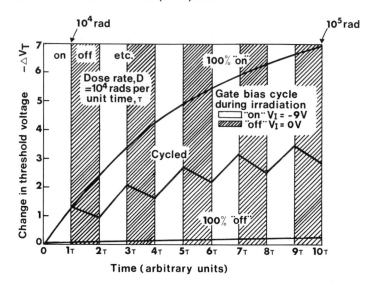

(a)

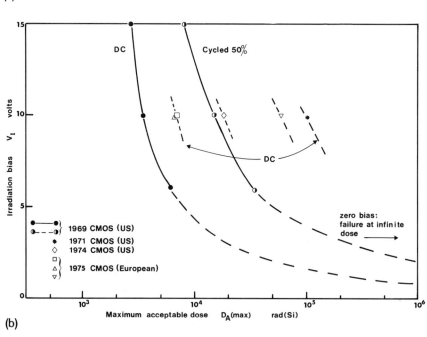

(b)

FIG. 4.9 Effect of intermittent bias: (a) on threshold voltage shifts as a function of time (equivalent to dose) for a typical p-channel MOS device subject to constant −9V bias, zero bias or 50% cycling bias during irradiation; (b) on dose value for failure due to VTNZ ('maximum acceptable dose') as a function of irradiation bias. The beneficial effect of 50% cycling in these cases is clear.

4.5 Prediction models for radiation effects in MOS devices

with this conclusion and demonstrated dependence on the square of d_{ox}. In the very simple model for threshold shift, described earlier in this chapter, threshold voltage shift is effectively dependent upon the square of the oxide thickness. The majority of studies have found the square rule to be adequate, and this can be used in prediction studies as a preliminary assumption.

4.5.6 Simple graphical model of MOS degradation

The equations derived in the simple analytical model described in Section 4.5.1 have been used to construct the set of curves shown in Figs 4.10(a) and (b). These figures show examples of a number of possible sets of curves derived from the model. As shown in the insets, some typical values of oxide thickness and charge distribution have been selected. In modern technologies, the oxide thickness values may be lower than those shown.

By comparing experimentally determined growth curves with theoretical sets such as those illustrated, it is possible to obtain a statement of the radiation sensitivity of an MOS technology in terms of the trap density and 'trapping probability' (A-value) of the oxide. Experimental data at low doses will indicate the A-value; those at very high doses indicate the trap density. Equation (4.4) and others show, however, that, for a complete statement, one also needs the value of oxide thickness, not usually available on data sheets.

4.5.7 A standard graph for assessment of MOS test results

The simple engineering model described above can be used to make a graph for assessing test results. Given the gate oxide thickness value for a device under test (a value sometimes available from the manufacturer with fair accuracy), and making assumptions about the location of the charge sheet (e.g. distance x_1) and the fraction, f_E, of charge collected, one can use eqn (4.5) to draw the growth curve for a device having an A-value of 1. This is shown in Fig. 4.11. We can then draw in a 'band' of ΔV_T values which are thought to cause failure in the circuit in question. Where the curve intersects the band is, of course, the trouble spot. Furthermore, plotting experimental values on the same diagram allows us to estimate the 'A-value' or 'hardness' of the actual device and the dose at which failure problems may start to occur in practice. The graph is recommended as a standard and explicit method of assessing MOS transistors. The simple form shown of course breaks down in the higher dose ranges where saturation and, even more strongly, 'turnaround' begin to be important.

4.5.8 Other prediction programs

Computer models have been developed for the modelling of ΔV_T of MOS devices as a function of radiation dose. R. C. Hughes (1985) developed models of the generation and transport of charge created by pulsed X-rays. Boesch *et al.* (1986) reported a model for the saturation of ΔV_{midgap}

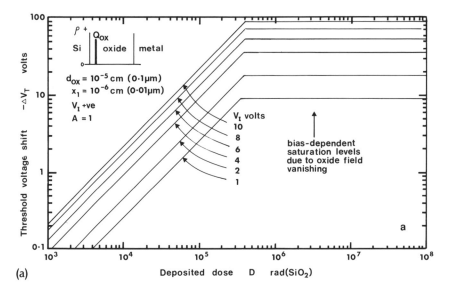

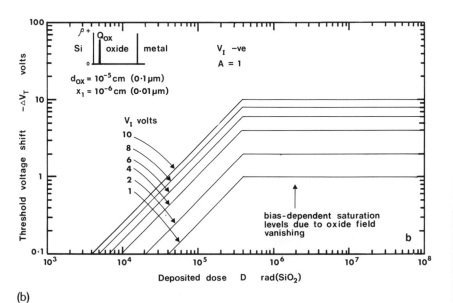

FIG. 4.10 Engineering growth curves: (a) V_I + ve, simple model of threshold shift as a function of dose for several values of positive irradiation bias, with trapping probability equal to unity; (b) V_I − ve, as (a), but for several values of negative irradiation bias. In case (a), the field which is noted as vanishing is ε_1 in Fig. 4.6.

of MOS capacitors which employs Poisson's equation to calculate the internal fields created by the trapped charge sheets and can set the number and cross section of the hole traps. Fossum *et al.* (1975) reported a programme to optimize oxide thickness and p-well surface doping density that will give the maximum post-irradiation voltage (i.e. delay the VTNZ effect to the highest possible dose). A detailed physical model of charge build-up has been collected and explained by McLean *et al.* (1989). Brown *et al.* (1989) have characterized the growth curve of commercial and hardened devices by parameters which denote not only charge build-up and interface state build-up but also the time-dependence of these parameters (i.e. annealing).

4.5.9 Ultra-hard versus tolerant commercial technology

The above models are more complex than the analysis and 'graphical engineering model' given in this section but do not conflict with it in principle. The main difference is that the models quoted in 4.5.8 were developed mainly for *very thin oxides*. The crucial experiments concerned the behaviour of a limited, specialized range of oxide types, mainly laboratory specimens or devices specially developed for megarad-hard equipment. Unlike commercial devices (typical value 80 nm), thickness values were rarely above 60 nm. Ultra-hard oxide films are designed, with great effort, to encourage room-temperature recovery of V_T with time (by loss of Q_{ot}). As Lelis *et al.* (1989, 1991) found, annealing of ultra-hard oxides requires that the sheets of trapped holes should lie very close to the Si–SiO$_2$ interface, i.e. the value of the term x_1 in Fig. 4.7 is very small compared with our 'commercial' model of Section 4.5.1. Therefore, ultra-hard oxides may constitute an unusual type of oxide, not necessarily suitable for modelling commercial types. For example, the sensitivity of the oxides used in the study by Boesch *et al.* was 0.0002 mV rad^{-1}, which is 500 times harder than the normal commercial oxide, which is of the order of 0.1 mV rad^{-1}. It will not be surprising if it is found that the ultra-hard oxides have a different distribution of traps, in both energy and position, and it has yet to be proved that recovery models developed for ultra-hard devices can be accepted for soft types. At the same time, the latter models have proved their use in Sandia's outstanding system for production and quality control of megarad-hard devices and systems (see e.g. Winokur *et al.* 1991; and Dressendorfer 1989).

4.6 Other fundamental effects in MOS devices

4.6.1 Recovery and time-dependent effects (annealing)

4.6.1.1 *General*

Positive charges (holes) which have been generated by radiation in the oxide layers of an MOS device may migrate to the region near the silicon, in which case they are usually trapped very firmly. However, trapped holes are not

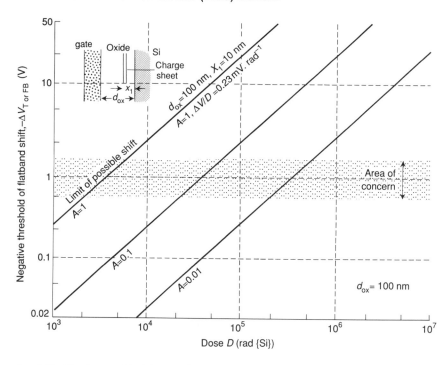

Fig. 4.11 Simple graphical format for experimental data on the degradation of MOS transistors. Predicted curve for ΔV_T vs. dose for 100% trapping of holes in the charge sheet ($A = 1$) as shown; the shaded area is the band of values of ΔV_T which will cause problems in the application in question. When experimental data are plotted, the diagram graphically demonstrates the comparison of the device under test with the worst case.

immovable in all cases, especially if the oxide structure is designed to aid the relaxation of the charge sheet. Relaxation is a natural tendency, because the sheet possesses a 'built-in field' which drives the charges away from it. The rate of this relaxation will depend on temperature, on trap depth, and on the possibility of electron injection into the region of the sheet and the availability of conduction mechanisms in the insulator. As a result, the rate of relaxation is extremely variable, taken across the field of MOSFET technology. Electrons injected into the oxide annihilate any trapped hole they find (see section 4.6.6.). This can be done with a simple UV lamp such as an EPROM eraser.

The terminology for recovery is varied. The scientific term is 'annealing', which can include a worsening of an effect. 'Relaxation' can be used to describe specific changes of bonds or charges. Many of the rebound and reverse effects are embraced in the overall terms 'Post-irradiation effects' (PIE), 'Time-dependent effects' (TDE) or — preferred by the present

4.6 Other fundamental effects in MOS devices

authors—'late effects', which is widely used in biology. 'Fading', a term already used in dosimetry, has been used for radiation-sensitive MOSFET dosimeters (Holmes-Siedle *et al.* 1991). The key problem of late radiation effects is how best to characterize various technologies with respect to late radiation effects. This may involve research into the improved measurement of device parameters as well as into the development of new sample preparation and irradiation procedures (see e.g. Shaw *et al.* 1991)

US and European agencies are now attempting in introduce into their testing techniques some recognition that such recoveries occur in the real environment (Brown, *et al.* 1989; Barnes *et al.* 1991). An example is the inclusion of a 100 °C anneal following exposure (see Chapter 13.6). The above effects are not the same as 'Turnaround', which is discussed elsewhere.

4.6.1.2 *High temperature: high recovery rate*

Early in the investigation of hole-trapping in MOS systems, it was discovered that the trapped charge could be removed ('annealed') without trace by heating the device to a temperature above 200 °C. The 'irradiate–anneal' cycle could be repeated many times with little alteration. However, these results were puzzling because they conflicted with the evidence from physics which pointed to a high activation energy required for the release of holes. The exact mechanism of thermal annealing has not been established, but

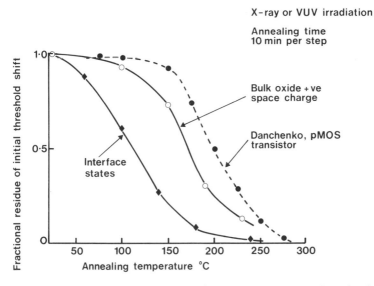

FIG. 4.12 Thermal annealing of threshold voltage shift: experimental results showing the fractional residue of radiation-induced threshold shift as a function of isochronal annealing temperature. Comparison of data of Danchenko *et al.* (1980, 1981) (broken line) and of Holmes-Siedle and Groombridge (1975) (solid lines, capacitors).

120   Metal-oxide semiconductor (MOS) devices

Danchenko *et al.* (1980) have proposed that the effect is brought about by electrons being injected into the Si–SiO$_2$ interface from the silicon. Figure 4.12 shows a comparison between Danchenko's results and some by Holmes-Siedle and Groombridge (1975) obtained from MOS capacitors irradiated with X-rays or vacuum UV light. Also shown are results by the same authors of the annealing of interface states. The effect of a varying thermal environment was studied by McWhorter *et al.* (1990).

4.6.1.3 *Room temperature: some recovery*

Brucker *et al.* (1982, 1983) and the present authors studied the recovery of V_T shift in irradiated MOS devices at room temperature. They and others found that, with oxides in the range 60 to 100 nm, p-channel types are often slower to recover than n-channel devices. Figure 4.13 (Winokur *et al.* 1983) shows an experiment in which CMOS inverters from five manufacturers were irradiated rapidly and then observed for the next 11 days. Some n-channel devices recovered less than 10 per cent; others exhibited a large recovery. In some, V_T even *overshot* the original value. This overshoot is known as 'rebound'. The p-channel transistors degraded further with time after exposure. This is known as 'reverse annealing'. A lower dose rate would clearly have an effect on the result, since significant recovery would occur during exposure. In contrast to the fast recoveries shown in some cases, in a 15-year study of ultra-soft p-channel devices used as space radiation

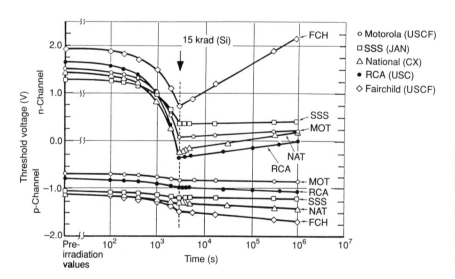

FIG. 4.13 Recovery of CMOS devices: threshold voltage curves for n- and p-channels of commercial 4007 inverters during and after a 5 rad(Si) s^{-1} Co-60 irradiation, showing recovery after irradiation ends at 3000 s (after Winokur *et al.* 1983) (10^6 s ≃ 2 weeks).

dosimeters, a typical recovery over 10 years was between zero and 20 per cent, depending on the dose and irradiation bias condition (Blamires *et al.* 1986; Holmes-Siedle *et al.* 1991). To illustrate the wide variety of possible annealing rates, Fig. 4.14 shows some of the results quoted above compared in terms of fractional recovery with log(time). The annealing curves are straight lines on this scale, and slopes vary from 3 to 25 per cent per decade of time. Also shown on this chart are p-channel devices exhibiting increased degradation with time (curve d). Both this 'reverse annealing' effect and the 'rebound' of n-channel devices, mentioned earlier, are explained by the build-up of new interface states with time.

The mechanism of recovery of n-channel devices under positive bias has been studied by Harry Diamond Laboratories (see for example McLean *et al.* 1976; Lelis *et al.* 1989, 1991; Oldham *et al.* 1989). A positive bias on the gate of thinner oxides (20 to 40 nm) is said to draw electrons into traps in the oxide which are very close to the interface (Walters and Reisman 1991). Negative bias can cause the electrons to tunnel out again (reverse-anneal). In the softer samples, Oldham and co-authors predicted that, at temperatures below 125 °C, full recovery will require about 3 billion years. However, since the recovery is a function of the logarithm of time (a tunnelling effect), then the percentage recovery in the early stages is quite noticeable

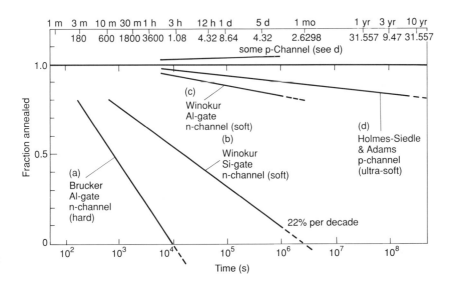

FIG. 4.14 Room-temperature annealing: typical post-irradiation changes in MOS devices as a function of log(time) at room temperature. Trapped charge in the oxide may relax rapidly, especially in 'hard' n-channel transistors (a), or very slowly as in the ultra-soft p-channel (RADFET dosimeter) devices (d). Lines (b) and (c) are examples from an analysis of CMOS devices by Winokur *et al.* (1983). (3×10^8 s $\approx$ 10 years).

and should always be considered when predicting the end of life of a thin-oxide MOS device at low dose rates.

A NASA-sponsored group of investigators (Brucker *et al.* 1982, 1983; Stassinopoulos *et al.* 1983a, 1983b, 1984, 1991; Danchenko *et al.* 1980, 1981; Barnes *et al.* 1991; Shaw *et al.* 1991) have made extensive studies of the time-dependence of radiation effects in MOS devices. This work included a detailed investigation of the relative effects of electrons, protons and gamma radiation on a uniform batch of CMOS inverters of the 4007 type. This batch, containing a fairly 'soft' type of oxide, exhibited recoveries of V_T shift in the region of 35% in a year. Particle type had some influence on the mode of recovery.

The impact of room-temperature annealing on engineering work such as life prediction and radiation testing specifications is strong for many modern MOS ICs. The physics of room-temperature annealing is only partly understood, but sufficient testing has been done to develop mathematical models for the main recovery effects. If further work proves this model to be reliable, we may be able in due course to apply 'annealing coefficients' to devices which are to be operated in space at a very low rate, and evolve methods for handling recovery and rebound effects in the development of aerospace devices and equipment.

4.6.1.4 *Cryogenic temperatures: little recovery*

Both hard and fairly soft MOS devices, irradiated at cryogenic temperatures, exhibit larger radiation-induced shifts than at room temperature. At room temperature, holes migrate quickly to the interface, where they are discharged or partially trapped. At very low temperature, virtually all of the holes are trapped where they are formed, in shallow traps which permeate the oxide. The net effect may be a much *worse* case than for irradiation at room temperature and the damage can be said to be frozen-in. The charge escapes in the normal way if the device is then warmed to room temperature.

McLean *et al.* (1976) and Boesch *et al.* (1978) have characterized the above processes for ultra-hard, very thin oxides. The holes proceed to the interface at a rate determined by the temperature and a 'random walk' process, suggesting a polaron form (see Section 3.3). Holmes-Siedle (1989) has calculated shifts for 'soft' oxides in the middle to low temperature range. A model was required for the recovery of cooled sensors over several years in a radiation environment. Figure 4.15 shows the temperature-sensitivity of the recovery of an MOS capacitor (or transistor) with a A-value of 0.25 from +7 to −130 °C. Figure 4.16 shows a similar calculation for a still softer oxide ($A = 0.5$), using a linear time scale. For holes placed in the oxide all at one time, at a temperature of −100 °C, it will take over a year for the curve to flatten out (e.g at 0.25 × initial post-pulse V_T value in Fig. 4.15). The latter state occurs when most of the holes have 'walked' to the sheet of

4.6 Other fundamental effects in MOS devices

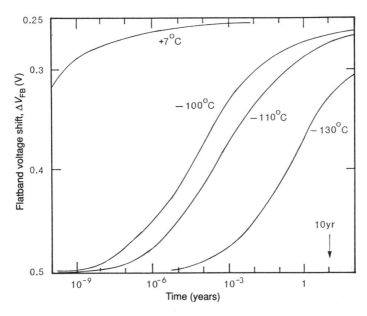

FIG. 4.15 Holes walking at low temperature (logarithmic time scale): response of a 'soft' MOS capacitor (A = 0.25) to a pulse of radiation at various temperatures. On a log time scale. Calculated from model of Boesch *et al.* by Holmes-Siedle (1989).

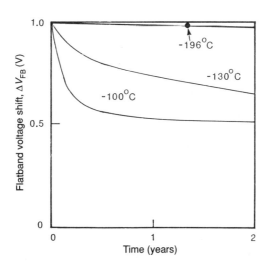

FIG. 4.16 Holes walking at low temperature (linear time scale): response of a 'soft' MOS capacitor (A = 0.5) to a pulse of radiation at various temperatures. Calculated from model of Boesch *et al.* by Holmes-Siedle (1989).

deep traps near the Si–SiO$_2$ interface. It is clear from the figure that, at $-130\,°C$, the same point will not be reached under 10 years, while at $-196\,°C$ (cryogenic nitrogen) there is practically no motion.

Note that these low-temperature 'random walk' effects are not the same as the 'recovery' described above for high- and room-temperature annealing. The 'random walk' effect is from slow *population* of the trap sheet from the bulk of the oxide; the latter is from slow *depopulation* (or electron compensation) of that sheet.

4.6.2 Model of annealing

Winokur *et al.* (1983) have produced an analysis of room-temperature recovery effects which is useful for very low dose rates. It allows us to perform tests in a reasonably short time and to extrapolate the results to long times, say several years in space. By the use of simple linear system theory, a mathematical value — the 'annealing slope' — is extracted from the test data. A high slope value implies a rapid room-temperature recovery characteristic in the laboratory test and predicts a milder degradation in space at low dose rates. If this technique is used in projects, caution has to be exercised in device selection, because batches with different recovery characteristics are produced from the same source.

4.6.3 'Rebound', 'overshoot' or 'super-recovery'

We have described above how, in n-channel MOSFETs, the build-up of oxide trapped and interface charge counteract each other, having opposite effects on the threshold voltage. If the oxide trapped charge now begins to relax with time at room temperature, it is quite possible for the threshold voltage of the device to overshoot its original value (Schwank 1984). This 'rebound' of the threshold voltage has caused some serious failures of ICs, installed in good condition in equipment after irradiation tests and later failing 'spontaneously' due to rebound (Johnston 1984). Dressendorfer (1989) explains the effect and describes the influence of process factors and the bias applied after irradiation. Rebound effects led to the need for rework on two spacecraft (NASA's Galileo and ESA's Ulysses) after pre-irradiated microcircuits passed a radiation acceptance test on delivery but failed electrically during re-test some months later.

4.6.4 'Reverse annealing'

It is possible for a damage species to develop *well after* the irradiation of the sample is stopped. This 'reverse annealing' is well known in atomic displacement damage at low temperatures (see Chapter 3). For MOS devices, 'reverse annealing' is due to the development of new interface states with time after irradiation. Winokur observed this in 1977 using LINAC pulses and Saks and Brown confirmed it later for thin oxides (see e.g. Dressendorfer 1989; Saks *et al.* 1991). The effect may be beneficial, if it counteracts the

4.6 Other fundamental effects in MOS devices

effects of oxide trapped charge, or neutral, as in the case where the threshold voltage of RADFET dosimeters continues to increase for a few days after exposure.

4.6.5 'Turnaround'

'Turnaround' derives from the same cause as 'rebound'. It was noted in the 1970s that n-Channel MOS transistors exhibited a 'turnaround' effect, but that p-channels never did. This was explained by Freeman and Holmes-Siedle (1978) in terms of different rates of buildup of fast states, N_{FS} and oxide trapped holes, N_T. Figure 4.17 shows the resultant threshold shifts calculated in that work. Since that time, the terminology of Section 4.3.2 has been adapted to this case and relative contributions of the two charges are represented by ΔV_{ot} and ΔV_{it} (see e.g. Winokur 1989). Similar curves are to be found for harder oxides in Dressendorfer's account of device phenomena (Dressendorfer 1989). The turnaround is caused by the continuing growth of negative Q_{it} as a function of dose after Q_{ot} has begun to saturate. Eventually Q_{it} dominates the growth of V_T (see also Schwank et al. 1989). If turnaround occurs before VTNZ occurs (see section 4.2) then a CMOS device may remain in specification up to a very high total dose. If the oxide trapped charge decreases with time after irradiation, then recovery of leakage and VTNZ effects will be assisted by the turnaround effect. In other words, serious quiescent current (VTNZ effect) leakage via the channel may be avoided (Holmes-Siedle and Zaininger 1968; Dressendorfer 1989a, 1989b; Brown et al. 1989; Schwank et al. 1989). Dressendorfer's reviews show many examples of V_{Tn} turning round before the onset of serious leakage effects. This does not, however, save a device from field oxide leakage (section 4.7).

4.6.6 Electron and hole injection; radiation annealing

The internal emission of carriers in insulators has achieved some importance in the field of characterizing MOS devices for trap content. The processs can be used to charge or discharge traps in oxides or other insulator layers and so provide a partial simulation of high-energy radiation. In internal photoemission (Williams 1965; Nicollian and Brews 1982), light is the exciting agent. Electrons from the conduction band of the silicon are excited over the Si–SiO$_2$ barrier and travel through the oxide. This effect is not to be confused with short-wavelength UV irradiation, which excites electron-hole pairs in the oxide itself (Powell and Derbenwick 1971; Holmes-Siedle and Groombridge, 1975). It was shown by Nicollian and co-workers (see e.g. Nicollian and Brews 1982) that AC fields in the silicon could induce avalanche processes in it. Of the resulting 'hot electrons' and 'hot holes', some have the energy to surmount the Si–SiO$_2$ barrier and move through the oxide. This technique proved a powerful method of establishing the types and cross sections of traps in the oxide. Later, charge injection

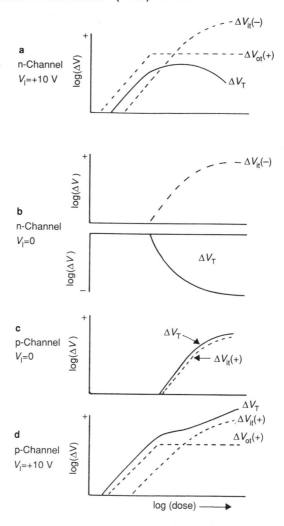

FIG. 4.17 Explanation of differences between n- and p-MOS behaviour: curves marked V_{ot} represent the voltage charge due to build-up of oxide trapped charge, and V_{it} similarly for interface trapped charge. The models explain the distorting effects of interface states on the simple charge-sheet model for V_T shifts in MOS devices under irradiation. These effects vary with channel polarity and irradiation bias value as shown in (a)-(d); in (b) and (c), ΔV_{ot} is not shown, being assumed small. (Freeman and Holmes-Siedle 1978).

4.6 Other fundamental effects in MOS devices

techniques were studied for their suitability as a simulation of high-energy irradiation (Boesch et al. 1979; H. L. Hughes and Razouk 1981). Figure 4.18 compares the effect on an MOS capacitor of 200 krad of gamma rays and an equivalent flux of hot holes injected from the silicon. It can be seen that the voltage shift and distortion due to interface states are very similar (Holmes-Siedle 1982). Practical difficulties experienced in routine simulation of gamma rays by avalanche injection (Vandenbroeck et al. 1986) were due to three factors: process-induced variation in doping at the silicon surface; low hole injection currents are difficult to control; holes stopping in the first few nanometres of oxide anneal rapidly.

Photoemission of electrons (holes are not easily injected by light) has been used extensively for the erasure of radiation-induced charge ('photo-annealing') (see e.g. Walters and Reisman 1991). Gamma-ray-induced electron injection is probably the main driver in so-called 'radiation annealing' and cycling effects (Poch and Holmes-Siedle 1970).

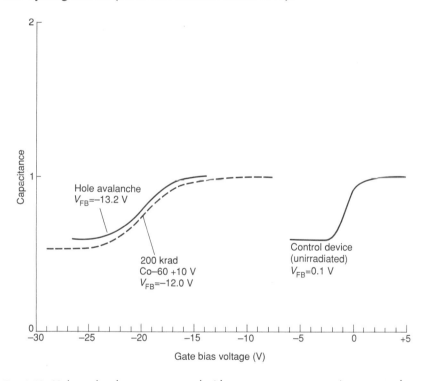

FIG. 4.18 Hole avalanche test compared with gamma-ray test: capacitance vs. voltage curves for a 'soft' metal-oxide semiconductor capacitor at 1 MHz. The C–V curves of the devices exposed to gamma rays and to hole injection are practically identical, indicating good simulation of radiation by hole injection in soft oxides (Holmes-Siedle, 1982 assisted by R. Lawrence and R. Hevey; capacitors prepared for US Naval Research Laboratories by R. Razouk—see Hughes and Razouk 1981).

4.6.7 Low-dose-rate effects

It was explained earlier that if an MOS device is prone to 'room-temperature annealing', then the radiation-induced shift of V_T will be affected by the dose rate used, so long as the shifts are measured immediately after irradiation. It has also been explained that this is not a true dose-rate effect, but the effect of concurrent trapping and relaxation of charge. Winokur and co-workers sought methods for the calculation of end-of-life state of an MOS device. Their work shows that, if a correction is made for the relaxation, then the dose rate used in gamma irradiations can be varied between a few rads per hour and thousands of rads per second without affecting the predicted value. In other words, no fundamentally different physical processes occur during accelerated testing compared with the slow rate expected in space (Fleetwood, *et al.* 1989). These findings are enshrined in new proposals for accelerated testing (see Chapter 13). The existence of the 'rebound' effect makes it important to use an accurate correction for dose rate (See section 4.6.3).

4.6.8 Noise

$1/f$ Noise in MOS transistors is sensitive to the presence of chargeable defects near the interface. The cause of pre-irradiation noise has not been well established, but recently it has been proved that, in thin MOS oxide layers, additional noise induced by exposure to radiation is correlated with the amount of oxide trapped charge generated by radiation. It is suggested that pre-irradiation noise is due to the precursors of the oxide charge traps filled by radiation (Meisenheimer and Fleetwood 1990).

4.6.9 Resistivity change: heavy damage

Transport of carriers in the channel of an MOS device is by majority-carrier conduction. The effect of atomic displacement damage on this transport is slight at levels below the megarad range (such as those which are likely in space) and any such changes are almost certain to be swamped by the ionization effects already described. Applications requiring consideration of displacement damage are the use of MOS devices in long-term, intense gamma-ray fields (damage is due to Compton electrons) or within reactors (say controls in fusion reactors or sensors in gamma dosimetry for fission reactors). For sensors used in power reactors, the exposure to fast neutrons may be of the order 10^{16} cm^{-2}, which would affect carrier density strongly, altering source–drain characteristics. The factors which affect the resistivity of bulk silicon are discussed in Chapter 3.

4.6.10 Transient effects

Since the MOS transistor contains two diodes, the main transient effect of pulsed radiation is diode primary photocurrent. For connected accounts

4.6 Other fundamental effects in MOS devices

of the theory and practical prediction of these currents, see Chapters 3 and 5. Figure 4.19 and 4.20a shows the transient response and equivalent circuit for one of the earliest examples of a transient radiation test on a CMOS device by Dennehy, Holmes–Siedle and Leopold (1969a). The source junctions do not produce external effects because, in this circuit, they are 'shorted out'. The currents from the two drain junctions of the device flow directly to the inverter output node after a radiation pulse. Figure 4.20(b) is a simplification of this circuit for the logic '1' state. The voltage pulse seen by the next circuit element (in this case an oscilloscope) is in theory proportional to the net photocurrent at the output node, the dose rate, and the value of the load resistor. Here, the load resistor is the n-channel, which is 'ON'. The positive-going pulse at the output is relatively small because the current from the three junctions involved (two drains and the p-well) nearly cancel each other. Figure 4.21 shows the dependence on dose rate of the peak voltage in the '1' state. Above about 10^9 rad s^{-1} (10^7 Gy s^{-1}), the linearity breaks down because drain, well and substrate, which form a parasitic bipolar transistor, amplify the photocurrent pulse. This 'parasitic' action gives rise to the self-sustaining current effect shown in Fig. 4.19(c). This exploratory experiment made it clear to investigators at the time that, to make CMOS integrated circuits which would tolerate pulsed

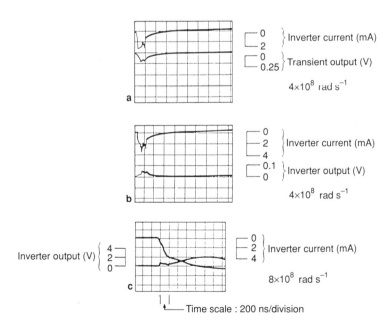

FIG. 4.19 Waveforms from radiation pulses: waveforms in the power supply leg ('inverter current') and at the output node for peak dose rates above 10^8 rad s^{-1} (10^6 Gy s^{-1}) for 100 ns pulses on a CD4007D inverter (after Dennehy et al. 1969a).

130 Metal-oxide semiconductor (MOS) devices

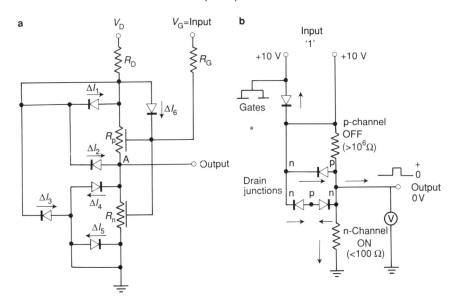

FIG. 4.20 Equivalent circuits for transients: models for a CMOS inverter exposed to pulsed radiation (after Dennehy et al. 1969a). (a) Full circuit: the photocurrents contributing to the flow of charge out of the output node are ΔI_2, ΔI_3 and ΔI_4; the gate protection diode provides ΔI_6. (b) Simplified equivalent circuit model: the diodes and photocurrents shown are the ones important in producing signals at the terminals when the input is in the logic '1' state and in producing parasitic 'latch-up'; the possibility of a pnpn path can be seen.

radiation, the parasitic action had to be suppressed and, although less important, the minority-carrier collection region around the drain junctions should be reduced. This was done in subsequent years by several methods, including epitaxial layering, lifetime killing (Friedman et al. 1989), and dielectric isolation. The first use of the latter was by RCA, who grew single-crystal silicon films on sapphire and other oxide crystals (Dennehy et al. 1969b; Schlesier 1974).

Tests by many workers on bulk CMOS ICs confirmed that, as in the above experiments, the upset of logic operation usually occurs in the range 10^8 to 10^9 rad(Si) s^{-1} (10^6 to 10^7 Gy(Si) s^{-1}), usually by the parasitic mechanism. The use of silicon-on-sapphire regularly increases this threshold to about 10^{10} rad s^{-1} (10^8 Gy s^{-1}) (see Long's and Gover's bar charts, collected in Messenger and Ash 1989). Success in reducing transient responses in MOS logic circuits has also been achieved by reducing the area of the drain junctions, a method named 'topological adjustment' by Dennehy et al. (1970).

Circumvention (see Chapter 15) has been implemented at chip level, for example in a design for a 'shadow RAM' (Murray 1991). In this chip, a

4.7 Real MOS devices in real environments: a commentary

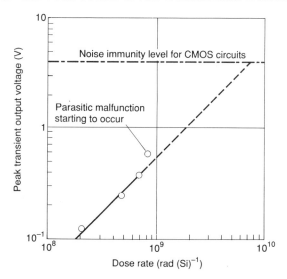

FIG. 4.21 Dose-rate dependence and latch-up threshold: peak transient output voltage vs. dose rate; the output voltage constant for the '1' state is 6×10^{-10} V per rad s^{-1}. The onset of logic upset is due not to crossing the noise immunity line but to the onset of a parasitic mode of upset. (After Dennehy et al. 1969a.)

non-volatile SNOS memory (see Section 4.10) retains at any one time a large fraction of (i.e. it shadows) the information passing through the faster, volatile conventional random-access memory. The circuit was functional after 1.2 Mrad (12 kGy) and is expected to surpass the tested data retention level of 10^{11} rad s^{-1} (10^9 Gy s^{-1}). Logic upset, as expected, was at 5×10^8 rad s^{-1} (5×10^6 Gy s^{-1}) because it was not built in SOI technology.

4.7 Real MOS devices in real environments: a commentary

4.7.1 Effect of total dose on logic circuits

4.7.1.1 *Introduction*

The problem of high and varying sensitivity of oxide films in MOS and other commercial devices to total dose is a theme repeated many times in this and other works. The practical problems are at their most acute in the field of 'real-world' equipment design for nuclear-industrial, space or military equipment or high-energy physics instrumentation. General discussions of the total-dose problems of circuits for space are given in Adams and Holmes-Siedle (1991, 1992). Stevenson (1990) and Heijne et al. (1990) note the special problems of circuits used in experiments near accelerator collision zones. This section enlarges on the total-dose problems of integrated circuits

132 Metal-oxide semiconductor (MOS) devices

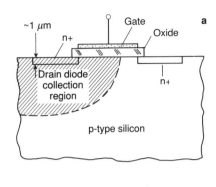

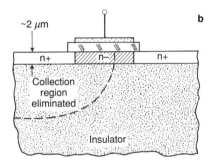

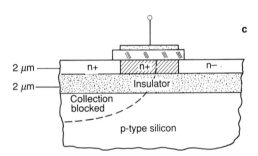

FIG. 4.22 Hardening a MOSFET against transient effects: growing a silicon film on sapphire or other insulator has the effect of cutting off most of the minority-carrier collection region around the junction from which the photocurrents derive and disrupting latch-up paths (a) Bulk silicon; (b) silicon film on a bulk insulator; (c) silicon film on a thin-film insulator.

4.7 Real MOS devices in real environments: a commentary

and reviews the extensive available test results and comments on their significance. A collection of test data in tabular form is given in Appendix D. We will attempt to show, by reviewing the effects of radiation on commercial devices and relating it to our knowledge of radiation response mechanisms, that device types selected from the commercial manufacturer can often be used successfully in radiation environments below 100 krad (1 kGy).

We stated earlier in this chapter that the electrical consequences of radiation-induced physical changes in MOS circuits are the progressive loss of function and eventual 'failure'. Successive increases in dose cause parallel shifts of the I_D-V_G curves of the transistors which make up the logic gates. We listed four types of malfunction in Table 4.2. The table also gave an indication of the real dose and threshold shift associated with each failure mode in a typical CMOS circuit.

Although most of the physical research and the prediction models have been carried out on MOS transistors or capacitors, the system designer has to go further, and to predict the point at which a particular network of devices no longer tolerates the multiple effects occurring in the components. In this section, we discuss the total-dose effects liable to occur in networks of MOS devices, as illustrated by the CMOS logic circuit. Early work was done by Poch and Holmes-Siedle, 1970 and reviewed by Oldham *et al*. 1984. Discussions of transient radiation effects in networks are given later.

4.7.1.2 *The VTNZ effect and its impact*

The growth of leakage, or quiescent current (I_{ss}), in CMOS devices as a function of dose is illustrated in Figure 4.23. This is the most important functional consequence of 'V_T of the n-channel crossing zero (VTNZ)' is often the most easily monitored parameter in MSI or LSI circuits. The 'maximum acceptable I_{ss} value' has been arbitrarily set at 0.5 mA. The figure shows that the maximum acceptable dose value can vary by over a hundredfold from type to type. This means that the time to failure of the system can vary by a hundredfold depending on the material parameters of the CMOS gate oxides. It is this uncertainty which makes extensive radiation testing necessary.

4.7.1.3 *Other effects on logic operation*

A CMOS inverter may be regarded as a type of 'potential divider' with the potential, V_{DD}, being dropped across various sections of the device according to its logic state. Clearly, a pronounced leakage current in a transistor which is meant to be 'OFF' will reduce the potential difference across that channel. Thus, the difference in output voltage between the '0' and '1' states will be reduced, correspondingly degrading the input (gate) voltages to the next inverter pair. This is the noise immunity reduction (NIR) noted in Table 4.2.

The effect of this is illustrated in Fig. 4.24. The full curves show the

134 Metal-oxide semiconductor (MOS) devices

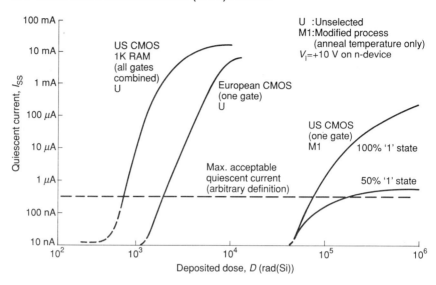

FIG. 4.23 Leakage current in CMOS devices: quiescent current (supply current when the gates are not changing state) as a function of dose in typical commercial CMOS integrated circuits. Growth of this current is one important functional effect of the shift of threshold voltage in the n-channel element ('VTNZ effect') (Holmes-Siedle and Freeman 1978).

voltage waveform associated with CMOS gate-switching before and after various doses of radiation. The broken curves show the relevant channel current which flows during the switching phase. In the experiment shown, V_{DD} was +10 V and the irradiation bias was +10 V for the n-channel and zero for the p-channel. In other logic gate structures, such as the transfer gate (Hatano *et al.* 1984), different fields and switching waveforms will apply.

4.7.1.4 *Field oxide leakages*

Virtually all modern CMOS IC circuits employ 'field oxides', which are thick oxide layers covering all parts of the chip except the functional elements. The layer acts as a spacer to keep the interconnects from interacting with the silicon. Charge in the region lying over the edge of the source and drain will tend to cause inversion and hence an n-type channel which shunts the main channel. On irradiation of CMOS ICs, it is very often found that the first mode of failure is the onset of source–drain leakage in the n-channel element, caused by the field oxide structure (Kelleher 1991; Vandenbroeck *et al.* 1986). An example is shown in Fig. 4.25(a) and (b) (Kelleher 1990), which show I–V curves for an n-channel silicon gate device on linear and

4.7 Real MOS devices in real environments: a commentary

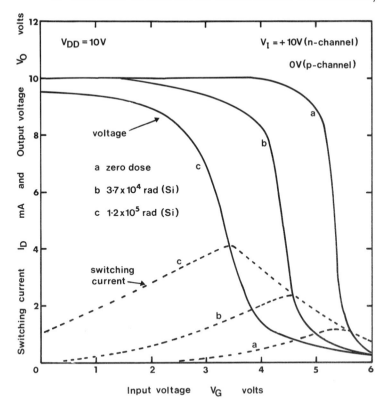

FIG. 4.24 CMOS switching characteristics: radiation-induced changes in logic transfer characteristics of a CMOS inverter. Increasing dose causes progressive loss of a logic definition and decrease in height of the output voltage step, with corresponding increase in power consumption. The left-hand side of the switching-current waveform is controlled by the n-channel component, large shift observable, and the right-hand side by the p-channel which did not shift (Holmes-Siedle and Freeman 1978).

log scales of current. The shunt effect gives prodigious increases in the I_{ss} value and may cause logic failure, even though the n-channel gate oxide itself has not crossed zero V_T. This effect would not be observed in a simple discrete MOSFET, and is due to oxide trapped charge in the field oxide (for examples of field oxides, see Figs 4.27 to 4.29).

In commercial CMOS ICs, the silicon under the field oxides usually receives extra doping (a 'guard-band'), but this may not be heavy enough to counteract the very large radiation-induced shifts in threshold voltage of the 'field oxide transistor' (see e.g. Fig. 4.25a–c). Dressendorfer (1989) describes the effect in hardened circuits and proposes IC layouts which move the onset of the effect to higher dose levels. The French atomic energy

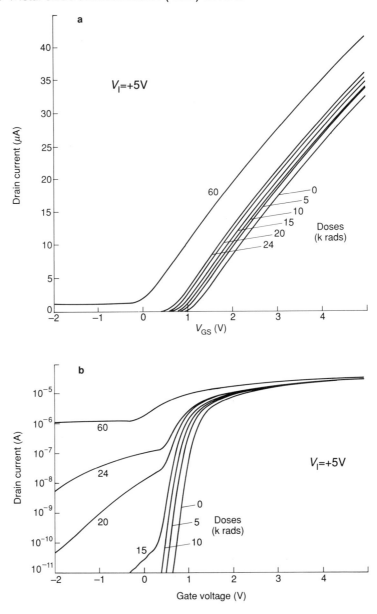

FIG. 4.25a–b I_D–V_G current curves for an n-channel, Si-gate CMOS device at an irradiation bias, V_I value of +5V(a), (b) unhardened technology with gate oxide thickness of about 50 nm, irradiated at 0–60 (0–0.6 kGy) krad; (a) linear current scale (b) logarithmic current scale, showing sub-threshold currents (after Kelleher, unpublished work. Reprinted with permission of National Microelectronic Research Centre, University College, Cork, Eire).

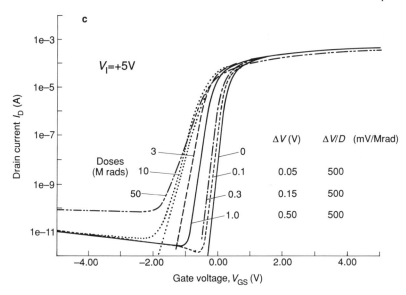

FIG. 4.25c Sub-threshold current curves for an ultra-hard, CMOS-SOI n-channel device with a gate oxide thickness of 28 nm, measured at 0–50 Mrad (0–0.5 MGy) (after Dupont-Nivet et al. 1991. Reprinted with permission).

agency (see Fig. 4.25(c), Dupont-Nivet et al. 1991 and Leray et al. 1990) are developing layouts which will withstand megagray doses without field oxide leakage (see also Section 14.3).

4.7.1.5 Effect of interface states on logic operation—rebound

The qualitative effect of interface states (the complex rearrangements of atomic bonding in the SiO_2– Si interface caused by ionization) on MOS transistors is to distort both the I_D-V_G and C-V characteristics. The common result is that the I_D-V_G curves of both n and pMOS devices are reduced in slope, as shown by the broken curve in Figures 4.3 and 4.5. In an n-channel device, this effect is to a degree beneficial in that the threshold shift is effectively reduced. A high degree of interface state production can decrease the I_D-V_G slope so much that the threshold voltage, V_{TN}, becomes even higher than its pre-irradiation value (rebound). The transconductance of the channel (the rate of change of I_D with V_G at given gate and supply voltages) is reduced and this may lead to further degradation of switching speed in a CMOS device.

4.7.2 Classification of 'hard', 'medium' and 'soft' with respect to total dose

Figure 4.26 shows a classification scheme for total-dose tolerance based on the 'growth curve' format presented earlier. The figure is for device irradiation under a positive gate bias of 10 V. An oxide thickness of 0.1 μm is taken. The format is divided into three areas: 'hard', 'medium' and 'soft' to total-dose effects. The position of a device test result in this framework is a useful indication of the suitability of that device for use in a given environment. Experimental results generally follow the 'corridors' shown when medium dose rates are used. Pulsed radiation or very low dose rates may need different treatment (see Section 4.6).

4.7.3 Processing variables

The key to the radiation-sensitivity of MOS devices lies in the preparation of the gate insulator and field oxide layers, at present grown by various methods of thermal oxidation of the silicon wafer. This sensitivity varies with the concentration of hole traps in this layer and, although research has been intensive, there is still controversy about:

(1) the exact structure of the hole traps;

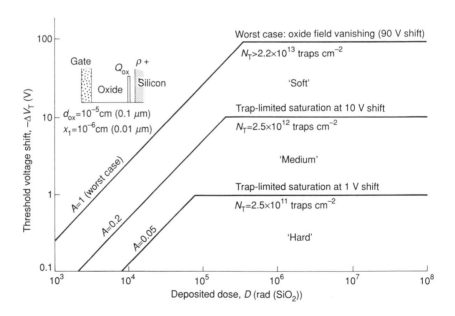

FIG. 4.26 'Hard', 'Medium', and 'Soft' defined graphically: the 'simple' model of V_T growth vs. dose, showing three proposed regions of radiation sensitivity; devices under +10V irradiation bias with trapped charge sheet as shown in the inset.

4.7 Real MOS devices in real environments: a commentary

(2) the exact mechanism of trap generation during processing;

(3) the best methods of controlling the generation of hole traps.

The important point is that, apparently, small variations in growth temperature, annealing temperature, subsequent metallization, silicon quality and cleanliness, and oxidation atmosphere make a large difference in hole-trap concentration (see Chapter 14). Although considerable attention is paid to these 'process parameters' in commercial MOS devices, it still appears that the control exercised is often not close enough to maintain a uniform level of radiation sensitivity. V_T shifts for a given dose vary from unit to unit and from batch to batch, giving a serious problem of 'scatter'.

For a cost-conscious small purchaser of devices, a still more unfortunate fact is that the MOS process parameters chosen by manufacturers on grounds of profitability are well removed from those which produce low radiation sensitivity. For example, the current 'recipe' for the best tolerance to radiation calls for such features as slow oxide growth, limited thickness in specially purged dry oxygen furnaces at low temperature, low-temperature anneal, deposition of metal without electron beam heating, and 'tailoring' of V_T. By contrast, commercial LSI circuits are often made to a recipe which includes fast oxide growth in non-purged furnaces (for high throughput), thicker oxides (for fewer pinhole defects), silicon gate electrodes (for self-alignment and low V_T) and high-temperature anneal (for low V_T and high yield). Furthermore, competition in the area of switching speed at low V_{DD} values causes the manufacturer to 'tailor' V_T to as low a value as possible.

For commercial reasons, manufacturers do not hold large stocks of devices produced to the 'hard' recipe. That is, they are barely definable as being in the 'commercial' product line. The term 'commercial' is sometimes used in a pejorative sense. However, for semiconductor devices, this is certainly not the case. For a semiconductor manufacturer to be profitable with, say, a sub-micrometre CMOS microprocessor, strong process and quality control giving good repeatability, homogeneity and yield is essential to commercial success. All of these are positive points from the radiation hardness point of view. The negative aspect of a commercial process is that it may be subject to significant process adjustment as time goes by, with the object of maintaining or improving process yield. The details of such activities are usually proprietary, are normally not visible to the user, and may seem trivial to the manufacturer. However, they may have a profound impact on radiation response.

On the other hand, because of the positive factors noted above, the user constrained to use commercial devices may be fortunate. A designer may say: 'I have discovered this great Japanese CMOS microprocessor in a plastic package; can I use it?'. In the space community, we are tempted to apply all that we have learnt about MOS radiation response in an

140 Metal-oxide semiconductor (MOS) devices

imaginative fashion. In fact, given a rigorous examination of the device and closely defined conditions for its use, it is often found that the 'great microprocessor' is acceptable; see for example the successful use of microprocessors in spacecraft, described in Section 4.8.2.

If one can be sure that a given batch of devices is homogeneous, then one can start an examination with reasonable confidence. The authors have found that commercial manufacturers will often provide the necessary traceability data, provided it is clearly explained why such data is required, and not too much proprietary detail is sought. An initial radiation assessment covering a wide range of conditions and, preferably, using samples covering as wide a production period as possible will provide the bulk of the data needed for a decision. For proper rigour, 'mask revision' data should be sought from the manufacturer, die photographs should be made and critical dimensions recorded as part of the test procedure. The result is a 'radiation-characterized' (not hardened) device.

The picture in the field of military hardware and the silicon technology developed for it is very different. In military radiation effects, research is in the hands of a a small number of government-funded laboratories. Military laboratories exert considerable control on the design and processing of MOS devices for military hardware (see for example Winokur *et al.* 1991) and cost is not always a prime consideration. Many of the products concerned are, for security reasons, not available outside the USA, although the presence of the technology, once released, may spur later commercial exploitation, as in the case of silicon-on-insulator technology (see Section 4.6). The physical models are biased towards the hardened product (thin, low-temperature oxides; see e.g. the chapters by McLean and by Winokur and two chapters by Dressendorfer in Ma and Dressendorfer 1989).

In summary, the problem of high and varying sensitivity of commercial devices to total dose may remain for some time. We must accommodate it by a mixture of (a) pre-selection, (b) batch monitoring, (c) protection of components from the external environment, and (d), given time and funds, the procurement of special alterations, by the manufacturer, of his process (see Chapters 14 and 15). The technological tricks which make devices 'hard' with respect to total dose (oxides with reduced trapping) and transient effects (photocurrent reduction or latchup suppression) should not be confused.

4.8 Commercial MOS technology: a survey of radiation responses

4.8.1 General

We will attempt to show, by reviewing the effects of total dose on commercial devices and relating it to our knowledge of radiation response mechanisms, that device types selected from the commercial manufacturer can often be used successfully for radiation environments in the kilorad range.

4.8 Commercial MOS technology: a survey of radiation responses

4.8.2 Memory technology

For data storage in spacecraft and military systems (here described as aerospace systems) it is most likely that CMOS static RAMs will continue to be the most widely used form of semiconductor memory. First, the static memory configuration is less prone to upset than are bipolar or dynamic MOS forms, and second, the low power drain of CMOS meets the low power requirements of spacecraft. In aerospace circuitry, strong emphasis is placed on process reliability and tolerance to noise and various radiation effects. CMOS circuitry is preferable in most of these respects so long as 'latch-up' effects are suppressed. Nevertheless aerospace technology is subject to some of the same pressures as the commercial world, and dynamic memories are now under serious consideration for space use (particularly as replacements for mechanical tape recorders).

In recent years it has been recognized that the memory field is developing so rapidly that conventional space technology evaluation and qualification techniques cannot keep pace. The initial reaction was to 'freeze' memory technology for space application at a comparatively low density (16 kbits) in order to allow radiation and reliability approval programs to 'catch up'. In certain highly critical applications this approach is still valid but generally concerns 'non-commercial' technologies such as SOS. For a broad range of applications it has been found that the application 'pull' and semiconductor manufacturing 'push' have combined to make a more rapid response. Particularly in the field of 'non-hardened' CMOS memories there is little point in embarking on extended approval programmes when the product is on the point of obsolescence from the manufacturing and application points of view by the time it is approved.

Certain laboratories, notably ESA, Sandia and Aerospace Corporation, have now developed a 'rapid response' approach which allows a quick and cost effective radiation evaluation which at least allows candidate devices and technologies to reach the prescreened stage for space use with minimum delay. A comprehensive memory survey was carried out by Harboe-Sorensen *et al.* (1991a) which covered 256 kbit SRAMs as well as 1 and 4 Mbit DRAMs. As may be anticipated, a wide range of total dose functional failure levels was found. For the SRAMs the range was 6 krad to 70 krad, for 1 Mbit DRAMs, 5 krad to 60 krad, and for 4 Mbit DRAMs, 8 krad to 80 krad. This tends to belie the popular belief that DRAMs are, by definition, less tolerant than SRAMs.

A number of SEU surveys of memories have been performed. Harboe-Sorensen *et al.* (1990, 1991a), using Cf-252, found that SRAMs showed saturated cross sections in the range from 3×10^{-8} cm^2 per bit to 2×10^{-6} cm^2 per bit. Similar results were obtained by Koga *et al.* (1991) using heavy-ion accelerators with threshold LETs in the range 2 to 10 MeV mg^{-1} cm^{-2}. Latch-up cross sections were in the range 3×10^{-6} to

1×10^{-3} cm² per device, with threshold LETs in the range 15 to 72 MeV mg^{-1} cm^{-2}. Interestingly enough, two commercial devices did not latch up. Harboe-Sorensen's DRAM survey gave Cf-252 saturated cross sections in the range 3×10^{-7} to 1×10^{-6} cm² per bit. Heavy-ion accelerator testing showed threshold LETs lower than 3 MeV mg^{-1} cm^{-2}. The conclusions to be drawn from these two significant surveys are that there are no clear rules for technology selection for commercial high-density memories. Total dose tolerances have been found which are more than adequate for a wide range of space applications. A word of caution regarding total dose tolerance concerns leakage currents. Many reports refer only to the level at which functional failure occurred. Leakage currents, particularly in DRAMs, may increase significantly, well before the functional failure level. The onset of increased leakage has been noted at about 50 per cent of the functional failure dose. In large mass memories the increase of leakage current is likely to be the limiting factor.

Generally, DRAMs do not suffer from single-event latch-up and, although lower SEU thresholds are found, it is quite possible that the best DRAM may have better orbital performance than the worst SRAM. Certain high-density SRAMs have been found that do not suffer from single-event latch-up, and these offer the possibility of implementing a low-power, high-density, mass memory.

The concept of identifying appropriate candidate devices from application and performance aspects and then performing a radiation pre-screen has been shown to be valid. A final word of caution concerns traceability, which is extremely difficult to establish in the high-volume commercial world. We tend to recommend that all candidate devices be decapped and the chip examined and photographed. In the field of high-density memories, it is not unknown for packages from manufacturers X and Y to contain a chip supplied by manufacturer Z.

4.8.3 Microprocessor technology

4.8.3.1 *General*

A microprocessor is essentially a computer system on a chip. Being such highly individual circuits, there are only a few high-performance types available and even fewer are tolerant to radiation. Some of these are available only to the military or the cost is extremely high. While memories may be considered as standard, general-purpose components, processors are much more application-specific and it is difficult for the parts specialist to constrain the choice of processor. Many laboratories tend to 'standardize' on a particular processor and develop a significant infrastructure of software and hardware development tools. The development of the military 'ADA' language and the MIL-1750A processor protocol has had a beneficial effect on standardization, and these standards are imposed on many projects. However, there will always be many aerospace and nuclear industry systems

4.8 Commercial MOS technology: a survey of radiation responses

based on commercial processors, for the reasons stated earlier and because of steadily increasing requirements in speed and processing power.

A number of commercial processors have been evaluated for aerospace use and are indeed being selected for space equipment in Europe and the USA. In the 'general-purpose processor' category, a popular candidate has been the 80C86 (a low-power CMOS version of the Intel 8086). In the readily available 'military' form, the total-dose tolerance was found to lie in the range 7–10 krad. Early versions were sensitive to latch-up, but transfer to epitaxial technology and optimization of epitaxial thickness solved this problem. Even with a rather low total-dose tolerance, the 80C86 can be used in a comparatively severe environment, with the appropriate countermeasures. One current European application (ISO) is in a scientific instrument for use in a highly elliptical orbit. The countermeasures in this case include switching off during transit of the radiation belts and heavy local shielding to bring the dose at component level down to 5 krad. Additional assurance is obtained by the provision of a second processor in 'cold redundancy' and co-located active 'RADFET' dosimetry to allow routine monitoring of absorbed dose to provide adequate warning of the need to bring the redundant processor into operation. Soft errors (SEU) have been taken into account in hardware and software design. The NS-800 has been successfully used in the polar orbit for several years in the University of Surrey satellite UOSAT-2, and the 80C186 is used in the recently launched UOSAT-5, also in the polar orbit. The Z-80, 8086 and 80C86 were evaluated for polar orbit application by Harboe-Sorensen *et al.* (1986). The Z-80 and NS-800 were studied for SEU by Cusick *et al.* (1985); the NS-800 showed latch-up but not the Z-80.

There is a major thrust amongst designers to adopt 32-bit general-purpose and reduced instruction set (RISC) processors. Complex manned space projects such as Space Station Freedom, Columbus, and the European 'spaceplane' Hermes have significant requirements in this area. For these low-earth-orbiting manned programmes, the total dose tolerance is not a major issue, although a certain low dose (generally in the range 2.5–7 krad) is specified. The single-event requirement, both for SEU and latch-up, is more demanding, particularly for high energy proton exposure in the South Atlantic Anomaly.

4.8.3.2 *Microprocessor tolerance levels for total dose and upsets*

Mattsson (1991) evaluated types 80386 and 68020, 32-bit general-purpose microprocessors, for total dose and SEU (using Cf-252). For the CHMOS-IV technology, the 80386 was found to have a total dose tolerance in the region of 15 krad. The SEU cross section using Cf-252 was found to be 7.8×10^{-7} cm^2 per bit. Using a 'cycled' total dose test of 2 krad exposure followed by a 24 hour anneal, Scott (1990) found a 20–24 krad tolerance for the same technology. He also performed Cf-252 and accelerator heavy-ion and proton testing. The threshold LET for SEU was found to be 6 MeV mg^{-1} cm^{-2} with a saturated cross section of 10^{-7} to 10^{-4} cm^2 per bit, depending on the register tested. Latch-up threshold LET was in the region of 28 MeV mg^{-1} cm^{-2} with a saturated cross section per device of 10^{-6} to 10^{-4} cm^2. Proton testing was carried out on the earlier CHMOS-III technology at 148 MeV and no upsets were seen up to a fluence of 2×10^{10} p cm^{-2}. For the 68020, Mattsson (1991) recorded total dose functional failures around 7 krad and, using Cf-252, found a range of SEU cross sections from 1×10^{-3} to 6×10^{-3} cm^2 per device depending on the operating mode. Chapuis (1991) performed accelerator heavy-ion testing on the 68020 and 68882 co-processor in epitaxial and bulk versions. For the 68020 (epitaxial) no latch-ups were detected. The threshold LET for SEU was about 12 MeV mg^{-1} cm^{-2} and the saturated cross section was about 2×10^{-3} cm^2 per device. Similar results were found for the 68882 co-processor, but for this device latch-up was detected with a threshold LET of about 27 MeV mg^{-1} cm^{-2}. For the bulk versions, high sensitivity to latch-up was found from LET 12.7 MeV mg^{-1} cm^{-2} upwards. The SEU threshold LET was significantly lower at 1.7 MeV mg^{-1} cm^{-2} and saturated cross sections were about an order of magnitude higher than for the epitaxial version.

4.8.3.3 *Advanced microprocessors*
There is continued interest in the unique INMOS Transputer designed for parallel processing, and with potential application in fault-tolerant computing. Thomlinson *et al.* (1987) evaluated the T414 16-bit machine and found a total dose tolerance of 15 krad at 2000 rad min^{-1} and 50 krad at 500 rad min^{-1}. Accelerator heavy-ion testing showed the on-chip RAM to be the dominating factor with a threshold LET of 2–3 MeV mg^{-1} cm^{-2} and a saturated cross section of 2×10^{-6} cm^2 per bit. By using external RAM the 'per device' cross section was reduced from 2×10^{-1} to 5×10^{-5}. No latch-up was seen on epitaxial versions. The Transputer is a rapidly evolving product and a later evaluation in 1989–90 showed a significant loss of radiation tolerance, particularly for the on-chip RAM, which now reached only 5 krad. The new 32-bit T800 showed similar behaviour, but disabling of the internal RAM improved tolerance to between 25 and 40 krad. INMOS co-operated in a programme to determine how normal production could be used in a radiation environment and the degree of 'spread' of radiation tolerance to be anticipated. A large 'spread' in tolerance was found from lot to lot and the 'spread' across a wafer was found to be very small. The conclusions of this programme were that for a given tolerance, devices could be selected from a given wafer with high confidence. At the end of the programme it became known that the T800 would shortly be discontinued and replaced by the T805, a 'shrunk' version of the T800 with a floating-point unit. The object lesson here is that advanced processor technology is

4.8 Commercial MOS technology: a survey of radiation responses

evolving extremely rapidly, and to take advantage of such technology requires well-formulated and rapidly implemented radiation assessment programmes carried out with the co-operation of the manufacturer.

The R-3000 RISC processor is currently under evaluation in the USA and Europe and seems to be a reasonable candidate for space use with latch-up protection and SEU mitigation (Kaschmitter *et al.* 1991).

4.8.3.4 *Digital signal processors and fast analogue circuits*

Another category of processor is the digital signal processor (DSP), optimized to perform matrix and vector operations at high speed and used in image processing, data compression and digital filtering. At this time the two main candidates for space use are the TMS320-C25 and the ADSP-2100A. The TMS320-C25 has performed well in geostationary orbit for 3 years (Adams *et al.* 1991). Total dose tolerance is 5 krad, the SEU threshold LET for the RAM is about $12\,\text{MeV}\,\text{mg}^{-1}\,\text{cm}^{-2}$, and the saturated cross section is $10^{-6}\,\text{cm}^2$ per bit.

The ADSP-2100A has been evaluated by Kinnison *et al.* (1991) and Harboe-Sorensen *et al.* (1991b). Both found a total dose tolerance in excess of 25 krad. Kinnison *et al.* found a heavy-ion latch-up threshold of $13-14\,\text{MeV}\,\text{mg}^{-1}\,\text{cm}^{-2}$ and a latch-up cross section of $10^{-4}\,\text{cm}^2$ per device for commercial devices. No latch-up was detected on epitaxial versions. No proton upsets were noted until significant total dose damage had been accumulated at $4.5 \times 10^{11}\,\text{p}\,\text{cm}^{-2}$ at 63 MeV. Harboe-Sorensen *et al.*, also using epitaxial devices, found an SEU threshold LET of about $8\,\text{MeV}\,\text{mg}^{-1}\,\text{cm}^{-2}$ with a saturated cross section of $2 \times 10^{-6}\,\text{cm}^2$ per bit. They also reported latch-up with a threshold LET of $12-19\,\text{MeV}\,\text{mg}^{-1}\,\text{cm}^{-2}$ with a cross section of about $10^{-5}\,\text{cm}^2$ per device. Latch-up was also seen on proton testing from 200 to 800 MeV and the proton SEU cross section was $5 \times 10^{-14}\,\text{cm}^2$ per bit at 200 MeV.

CERN, Geneva, along with international experimental groups collaborating with CERN, are advancing the development of ultra-fast experiment-specific circuitry which will survive about $10^7\,\text{rad}(\text{Si})$, ($10^5\,\text{Gy}(\text{Si})$). This includes fast analogue circuitry for the front end of experiments, which include preamplifiers, analogue pipelines and analogue memories, operating at over 50 MHz (Jarron 1989; Anghinolfi 1990; Faccio *et al.* 1992). SOS versions have been tested by Bingefors *et al.* (1992) to a dose of $3 \times 10^6\,\text{rad}(\text{Si})$. CEA, France are collaborating in this experimental circuit-specific technology (Dupont-Nivet *et al.* 1991; Leray *et al.* 1990).

4.8.4 Logic families

Standard logic families are often available in several forms, such as fully hardened (1 Mrad), tolerant (100 krad), or commercial/military (10 krad). As noted in Chapter 14, there is significant interest in developing processes and designs which are tolerant enough to radiation for a wide range of space applications (30–50 krad) but are still an attractive commercial product. An example of this is the 54HC High-Speed CMOS series, for which tolerant processes were developed under ESA sponsorship. The same process is used for mass production. The SGS 4000 B series was developed in radiation-tolerant (100 krad) form in Europe to met the needs of the ESA Ulysses project which will be exposed to the severe radiation environment of Jupiter. The tolerant process was developed from a commercial process, but in this case the changes are sufficiently profound that the radiation-tolerant and commercial processes are run separately.

In the standard logic field, Advanced-CMOS (AC/ACT) is of great interest and has been the subject of 'pre-screening' programs. Koga *et al.* (1990) performed accelerator heavy-ion testing on the National Semiconductor 'FACT' series which had been shown by the manufacturer to have a total dose tolerance in the 100–300 krad range. No latch-ups were seen on the epitaxial versions, and SEU threshold LETs were in the range 40–60 MeV mg^{-1} cm^{-2} with saturated cross sections in the range 2×10^{-6} to 9×10^{-5} cm^2 per device. Johlander (1991) performed a survey of total-dose effects for devices from three manufacturers as a pre-screen prior to qualification. He found parametric failures ('out of specification') at 10 krad for one manufacturer and between 25 and 50 krad for the other manufacturers. All devices remained functional at the maximum dose used of 100 krad, since the radiation-induced leakage had not affected the basic logic functions. One type exhibited 'rebound' during the annealing tests (see Section 4.6.3). In his report, Johlander makes an important observation regarding the influence of design and complexity: that, when performing an assessment of logic 'families' which may include a wide range of designs and levels of complexity, it is dangerous to extrapolate results on a few device types to a whole family. The pre-screen test shows whether it is worth further, more thorough radiation assessment. Maurel *et al.* (1991) reported an assessment programme performed on the tolerant 54HC series mentioned earlier. This programme is a good example of test matrices covering the influence of design, function, and complexity as well as variability in response from lot to lot, wafer to wafer and across a wafer.

4.8.5 Programmable logic devices (PLDs)

PLDs are a class of technology which offer significant advantages in any application seeking to minimize mass, volume, and power consumption. The PLD is classed as an application-specific integrated circuit (ASIC) and is based on a large array of cells capable of performaing logic but reprogrammable like the well-known memory systems. 'Macro cells' are available covering most standard logic functions. A comprehensive survey of CMOS PLDs was carried out by Lopez-Cotarelo (1991); she identified a number of suitable candidates with good radiation tolerance.

One field programmable gate array (FPGA) was shown to have a total dose tolerance in excess of 400 krad, good SEU performance, and no latch-up. This particular device by ACTEL has already been adopted for a 'Microsat' space project and for other flight instruments.

The two main subfamilies within PLD technology are 'standard' and 'complex'. Within the 'complex' group we find the FPGA, which has a gate-array type of architecture and offers a high degree of flexibility. Typical 'gate counts' range from 1000 to 10 000 equivalent gates.

PLDs are evolving very rapidly. Many 'manufacturers' are, in fact, design houses, with the actual silicon chips being supplied by silicon foundries. This is therefore a good example of the complications of the 'real MOS technology'. Any user who has to know radiation tolerance must 'track' the technology very closely. This requires both contact with the manufacturer and independent test and analysis. It is all too easy for an unsuspecting user to be caught out by a change in technology.

4.9 Hardened technologies

4.9.1 General

The previous section is not intended to give the impression that the best, or only, choice of technology lies in the non-hardened commercial/military field. The main application of such technology is in non-critical subsystems such as industrial equipment and low-cost space projects. For equipment which is crucial to the operation of an aircraft, a nuclear reactor or a major physics experiment, clearly time and funds should be spent to ensure lasting performance and an ultra-low failure rate from any cause, including radiation environments. In this section, we deal with the high-reliability specialized devices used in such circumstances and also with some of the potential new problems imported by leading-edge MOS technology.

4.9.2 Radiation-hardened computer system technology

Data-processing and control functions in spacecraft tend to employ microprocessors distributed throughout the equipment. As with memories, CMOS devices are preferable to nMOS and bipolar forms. However, several relevant differences between memories and microprocessors may also be stated. First, the radiation-induced power drain due to a few microprocessors is of less importance than that occurring in a large array of memories. Second, the application of local shielding to single processor chips is more simply achieved than for arrays of memories. The choice of microprocessor for radiation environments will rest on total-dose functional tolerance and the proneness of the system to errors induced by cosmic rays and other disturbances. 'Cold redundancy' is particularly appropriate for this type of chip.

A good example of the practical use of MOS memories and microprocessors in complex systems is ESA's On-Board Data Handling (OBDH) programme. The requirement for tolerance to radiation in this system has given rise to ESA Applied Research Programmes sponsoring special modifications of European MOS technology for radiation tolerance. All projects, space or terrestrial, have vital 'core' systems which are essential to the continued safe operation of the project. In satellite technology, such 'core' systems are the on-board computer (OBC), power conditioning system (PCS), attitude and orbit control system (AOCS), and communication systems (Tx/Rx). Because of the criticality of these 'core' systems it is worth maximizing the use of hardened technologies even though this may place a strain on the performance, mass or power budgets of equipment.

There is a major effort in progress to develop radiation-hard spacecraft OBCs using ADA software and hardened processors. These OBCs will become 'standard equipment' and consequently must have sufficient spare capacity and flexibility to serve foreseeable future needs. There currently exist hardened MIL-1750A processors (Kerr *et al.* 1991) and associated hardened bulk or SOS static memories are available to 64 or 256 kbit densities (Bion 1991). High-performance 32-bit processors are currently under development in hardened technology. MIL-1553 bus components are available in hardened form as well as components for specific protocols such as the ESA packet telemetry and telecommand standard. For the latter, 'central terminal unit' and 'remote bus interface' ASICs have been developed in SOS.

For power systems there exist hardened power MOS transistors with 1 Mrad tolerance and immune to heavy ion induced burn-out. A particularly difficult area is that of linear devices and conversion products (A–D and D–A). Progress is being made in this area (van Vonno 1991).

Hardened logic families exist or specific 'glue logic' functions can be generated as required in hardened gate array or cell-based technology. In Europe this approach has been used to develop octal interface functions in SOS. Once developed, these functions become standard 'catalogue' devices, so one gradually builds up a full range of logic functions in hardened technology. This approach is not confined to logic but is also used for more complex functions which may be perceived as a standard requirement in future projects. An example of this is an 'ion-counter' chip 8 bits wide and 24 bits deep, developed in SOS and used as a standard function in particle counting instruments (Chu and Jernberger 1991).

4.9.3 Examples of hardened technology

4.9.3.1 *Bulk silicon MOS ICs*

A hardened CMOS version of the Intel 8085 NMOS microprocessor was developed by Sandia National Laboratories. The 8OC85RH was a functional copy of the NMOS 8085 and was fully compatible with it. Total dose hardness levels of 1×10^6 rad(Si) have been recorded on tests (Kim *et al.* 1989; Sexton *et al.* 1983). Operation free of latch-up was obtained

4.10 Some advanced MOS structures: potential problems

by the use of an epitaxial substrate as the starting material in the fabrication process. The microprocessor clock could be stopped without loss of data because the memory devices, although volatile, were fully static. This was an example of design of a computer element by a laboratory sponsored by the US Departments of Defense and Energy. The Very High Speed Integrated Circuit (VHSIC) programme which followed, adopted the preferable and more effective approach of placing a radiation requirement on commercial contractors. A high radiation requirement, equivalent to the expected threat in strategic weapon systems, was imposed and has been met by several series of VHSIC devices made in bulk silicon.

4.9.3.2 *Silicon on insulator*

Dielectric isolation has been used to remove latch-up paths and reduce the volumes around a junction in which carrier generation can occur. The case for this technique in the radiation hardening of MOS logic was made in detail by Dennehy *et al.* 1969*a*. In high-rate irradiation, charge collection after a radiation pulse produces undesirable photovoltages. After the development of the silicon-on-sapphire (SOS) method in the 1960s (see Section 4.6 and Schlesier 1974), a wide range of other dielectrics were used, including separation by ion implantation (SIMOX) (Krull *et al.* 1987; Brady *et al.* 1989), zone melt recrystallization (ZMR), full isolation by porous oxidized silicon (FIPOS) (Benjamin *et al.* 1986; Tsao, 1987), epitaxial lateral overgrowth (ELO), and wafer bonding and etchback (see e.g. Stanley 1988; Fan 1989; Mitani and Goesele 1991). The thrust in the development of SOI is to remove the limitations of the SOS process mentioned earlier. The gate oxides of SOI devices are not necessarily tolerant to radiation but can be made so by the usual methods (see e.g. Leray *et al.* 1990; Bingefors *et al.* 1992). The presence of an insulator *under* the silicon channel leads to 'back-channel' conduction which is sensitive to radiation and also gives time-dependent effects (Boesch *et al.* 1990). Some further comments on SOI hardening technology are given in Chapter 14.

A hardened computer chip set, semi-custom gate arrays and cell-based design are available from European sources of SOS technology. This technology tolerates megarad total doses. Hatano *et al.* (1984) describe a clocking method for SOS which improved tolerance to total dose. The general drawback of SOS is that wafers of sapphire are not well adapted to the production process. The US government has therefore promoted research in the other SOI techniques described above.

4.10 Some advanced MOS structures: potential problems

4.10.1 Introduction

The design of the MOS device is evolving as rapidly as any electronic element invented. Since the driving force in oxide charge trapping is the field across thin insulator films, each change of geometry may change the probability of such deleterious effects. A similar argument is true for transient upsets. To illustrate the future possibilities, we give some examples of modern MOS structures now in rapid evolution. Some are designed to minimize radiation effects, others will be developed for commercial use despite the intrinsic sensitivity to radiation.

4.10.2 Lightly and moderately doped drains (LDD and MDD)

An LDD or 'shallow-drain junction structure is shown in Fig. 4.27(a). Palkuti *et al.* (1991) have shown that the lightly doped region is sensitive to charge build-up in the oxide layer. Source–drain resistance is affected by radiation as well as threshold voltage. The inset in Fig. 4.27(b) is a form of that shown in Fig. 3.6, in which we gave a simple one-dimensional treatment of the charge build-up. It shows that, with shallow–deep junctions, the charge relations in this IC structure may require the consideration of at least two dimensions, one vertical to and one parallel with the channel. Different effects are obtained from the sheet of negative charge generated at the interface and the positive charge in the bulk oxide. It should not be forgotten that charges may also be trapped in the deposited silica layer lying next to the silicon gate. The fringing fields below the gate and drain metal encourage charge separation in those layers, and this charge may also affect the lightly doped drain regions. Since LDD structures are usually of very small dimensions, the expected hot-electron injection from simple electrical stress (e.g. Yu *et al.* 1989) may call for improved prediction techniques.

4.10.3 Non-volatile memories

A polycrystalline silicon gate, set between oxide layers over the channel of an MOS device, will retain a potential imposed upon it by charge injection through the oxide. This 'electrically erasable programmable random-access memory' is used widely because it is 'non-volatile' in the sense of retaining data when the power supply is removed. The two mechanisms are photoconduction in the oxide layers and photoemission from the interfaces. Snyder *et al.* (1989) review the effect of radiation on this and the other types of non-volatile memory (e.g. silicon nitride–oxide silicon SNOS). In both cases, a high potential at the storage layer can be discharged by irradiation. The radiation tolerance value is about 0.1 megarad (1 kilogray), i.e. the the memory is 'volatile' at this dose. Based on a theoretical model, the above authors find that tolerance should be improved slightly by increasing the charge on the storage element, using thinner oxide or threshold voltage adjustment by implantation.

4.10.4 Heavy metals—disturbance of equilibrium

Metals other than aluminium and gold are being used in large-scale integrated devices as contact or track materials. Tungsten is one of the

4.10 Some advanced MOS structures: potential problems 151

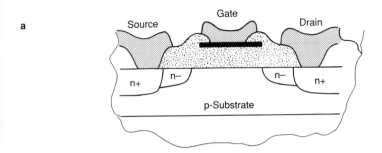

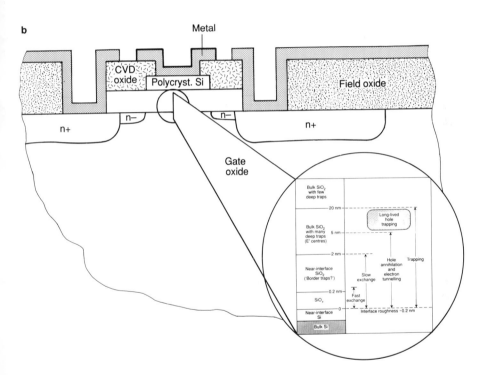

FIG. 4.27 Advanced MOS design, 'Lightly doped drain': section of a MOSFET with lightly doped drain (after Palkuti et al. 1991). (a) Cross section of actual geometry; (b) idealized geometry, showing the structure of possible charge sheets; enlarged, the charge-sheet scheme discussed in Chapter 3, which all takes place in the gate oxide region.

152 Metal-oxide semiconductor (MOS) devices

most widely used. Major commercial apparatus for deposition of tungsten by CVD and sputtering is available. In the case of fill metal for 'via' holes, as shown in Fig. 4.28, the thickness of tungsten in some cases is large enough to have a major perturbing effect (dose enhancement) on the deposition of radiation energy near the interface between the metal and the low-Z materials such as the oxide and silicon (see, for example Garth *et al.* 1980; Srour *et al.* 1984; and Section 1.4). Other metals in use are molybdenum, and tantalum. A wide range is used in screen-printed tracks.

4.10.5 Ferrite memories

An integrated CMOS-ferroelectric memory is shown in Fig. 4.29. These devices offer non-volatility at much higher dose rates than CMOS static or dynamic storage cells. The ferroelectric lead zirconate titanate films themselves are much more tolerant to total dose and neutrons than is the underlying MOS structure and exhibit advantages over the other available non-volatile technologies. The particular device shown (Benedetto *et al.* 1991) failed at a few krad. As might be expected (Section 4.6), the cause was charging in an unhardened field oxide.

4.10.6 GaAs-on-Si

Because of the difficulties of producing and handling bulk GaAs, wafers of silicon carrying thin layers of GaAs are being made (Fan 1989). Such structures will be used in conventional GaAs ICs, combinations of Si CMOS and GaAs and optoelectronic devices such as in a proposed three-dimensional optical memory (Koyanagi 1991). The diversity will provide wider scope for the radiation hardening of ICs.

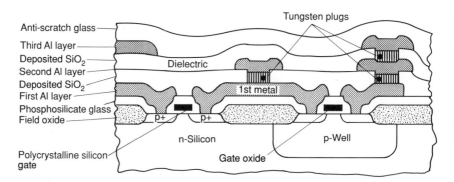

FIG. 4.28 Advanced MOS design, heavy-metal scattering: cross section of integrated circuit with multilayer metallization including tungsten vias; dose-enhancement from these metal plugs is possible.

4.10.7 High-voltage and power CMOS ICs

One of the advantages of the SOI structure is that the insulator layer can be used to separate high-voltage inputs or output stages from convention low-voltage IC logic functions. Optical isolation may be used to link the parts. High fields found in these devices may lead to larger amounts of oxide trapped charge than in low-voltage devices. Oxide charging of junction surfaces can lower breakdown voltages (Kosier *et al.* 1990). Power devices are further discussed in Chapter 7, but the structures tested by AT&T and CEA (Darwish *et al.* 1988; Desko *et al.* 1990; Dupont-Nivet *et al.* 1991) should be mentioned here as an example of the variety of possible devices (including bipolar and JFET devices) which can be isolated above dielectric films or 'tubs' of various kinds.

4.11 Conclusions

The primary problem of total-dose radiation effects in MOS devices is the unpredictable and complex charging process in the thick and thin oxide films which form an essential part of the structure. Transient upsets are more predictable but can be catastrophic, especially where pnpn effects are possible and where stored data can be lost. Memory cells will become more sensitive as the size of the memory structures becomes smaller, leading to problems for everyone in the future.

Metal-oxide semiconductor (MOS) devices are the 'workhorse' of modern electronics; the types available will become more varied and complex and

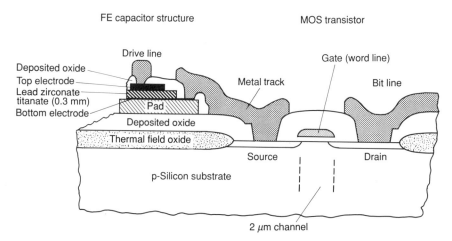

FIG. 4.29 Advanced MOS design, ferroelectric memories: cross section of a ferroelectric memory (after Benedetto *et al.* 1991); NSC ferroelectric lead zirconate titanate capacitor and a 'select' transistor. (© 1991 IEEE. Reprinted with permission.)

the sensitivity of these components to some types of radiation effect will continue to increase. This is an uncomfortable prediction since, in the research field, new recipes for 'hard' oxides and geometries with reduced sensitivity to transients (both pulsed gamma and SEU) are being developed. It is still likely because of commercial constraints; hardened recipes will be applied to only a few device types and those 'hardened' types will continue to be difficult to obtain. The suppliers of many of the more popular MOS devices, including nMOS and CMOS LSI circuits, will choose not to employ these recipes. Thus, the well-known problem of sensitivity of MOS devices is likely to remain with us for many years, changing slowly as the balance between process factors and dimensions changes. The problem for the designer of reliable equipment will remain that of understanding, predicting, and controlling the effects in the types which suit his purpose best and then conferring radiation hardness on a whole system as opposed to a single device.

To give guidelines to circuit and system designers, we have, at some length, stated the physical problems of total dose effects, proposed practicable methods of classifying and predicting the response of MOS devices from all quarters of the market place and described work proceeding in advanced research on device hardening. Given familiarity with process trends, coupled with a knowledge of the physics of radiation effects, as outlined here (we can call this 'modelling'), then predictions of damage, 'rebound' from damage, instabilities or logic upsets may be made by the designer with some confidence. Given good modelling, there is then some prospect of selecting devices and designing tolerant circuits which *make the best of the MOS technology* in an unforced manner, despite the basic sensitivity to radiation which has plagued the use of this technology in aerospace and nuclear applications. One particular cause of confusion for the designer, discussed in section 4.9, is the very great physical difference between very thin, specially treated, megarad-hard gate oxides and the oxides used in terrestrial electronics. This question is further discussed in chapters on the testing, procurement, and hardening of electronics.

References

Adams, L. and Holmes-Siedle, A. (1991). La survie de l'electronique dans l'espace. *La Recherche*, **22**, 1182–89.

Adams, L. and Holmes-Siedle, A. G. (1992). Radiation hardness assurance of space electronics. *Nuclear Instruments and Methods in Physics Research*, **A314**, 335–44.

Adams, L., Daly, E. J., Harboe-Soresen, R., Holmes-Siedle, A. G., Ward, A. K. and Bull, R. A. (1991). Measurements of SEU and total dose in geostationary orbit under normal and solar flare conditions. *IEEE Transactions on Nuclear Science*, **NS38**, 1686–92.

Anghinolfi, F. *et al.* (1990). Study of analog front-end electronics for supercollider experiments. *Proc. of the large Hadron Collider Workshop ECFA*, 90-133, Vol III. CERN, Geneva.

Aubuchon, K. G. (1971). Radiation hardening of p-MOS devices by optimization of the thermal SiO_2 gate insulator. *IEEE Transactions on Nuclear Science*, **NS18**, 117-25.

Barnes, C. E., Fleetwood, D. M., Shaw, D. C. and Winokur, P. S. (1991). Post irradiation effects (PIE) in integrated circuits. Proc. RADECS '91, Montpellier, France, Sept 9-12 1991. *IEEE Conf. Record*, Cat. No. 91 THO 400 2, **Vol. 15**, pp. 41-54.

Benedetto, J. M., *et al.* (1991). Radiation evaluation of commercial ferroelectric nonvolatile memories. *IEEE Transactions on Nuclear Science*, **NS38**, 1410-4.

Benjamin, J. D., Keen, J. M., Cullis, A. G., Innes, B. and Chew, N. G. (1986). Large area uniform silicon-on-insulator using a buried layer of oxidized porous silicon. *Applied Physics Letters*, **49**, 716-7.

Bion, T. (1991). *An 8K × 8 radiation hardened SRAM*. In Proceedings of Radiation Effects on Components and Systems, RADECS 91. Université de Montpellier, France.

Bingefors, N., Ekelöf, T., Eriksson, C., Mork, G., Paulsson, M. and Sjolund, A. (1992). Radiation hardness test with a demonstrator preamplifier circuit manufactured in silicon on sapphire (SOS). VLSI technology, University of Uppsala Preprint No. TSL/ISV 91-0056. *Nuclear Instruments and Methods*, in press.

Blamires, N. G., Totterdell, D. H. J., Holmes-Siedle, A. G. and Adams, L. (1986). pMOS Dosimeters: long-term annealing and neutron response. *IEEE Transactions on Nuclear Science*, **NS33**, 1310-7.

Boesch, H. E. and McGarrity, J. M. (1979). An electrical technique to measure the susceptibility of MOS gate insulators. *IEEE Transactions on Nuclear Science*, **NS26**, 4814.

Boesch, H. E., Jr, McGarrity, J. M. and McLean, F. B. (1978). Enhanced flatband voltage recovery in hardened thin MOS capacitors. *IEEE Transactions on Nuclear Science*, **NS25**, 1239-45.

Boesch, H. E., Jr, McLean, F. B., McGarrity, J. M. and Bailey, W. E. (1986). Saturation of threshold voltage shift in MOSFETs at high total dose. *IEEE Transactions on Nuclear Science*, **NS33**, 1191-7.

Boesch, H. E., Jr, Taylor, T. L., Hite, L. R. and Bailey, W. E. (1990). Time-dependent hole and electron trapping effects in SIMOX buried oxides. *IEEE Transactions on Nuclear Science*, **NS37**, 1982-9.

Brady, F. T., Krull, W. A. and Li, S. S. (1989). Total dose radiation effects for implanted buried oxides. *IEEE Transactions on Nuclear Science*, **NS36**, 2187-91.

Braünig, D., Gaebler, W., Fahrner, W. R. and Wagemann, H. G. (1977). *Data compilation of irradiation tested electronic components*, Report No. B-248. Hahn-Meitner Institut, Berlin.

Brown, D. B., Jenkins, W. C. and Johnston, A. H. (1989). Application of a model for treatment of time dependent effects on irradiation of microelectronic devices. *IEEE Transactions on Nuclear Science*, **NS36**, 1954-62.

Brucker, G. J., Dennehy, W. J. and Holmes-Siedle, A. G. (1965). High-energy radiation damage in silicon transistors. *IEEE Transactions on Nuclear Science*, **NS12**, 69.

Brucker, G. J., Stassinopoulos, E. G., van Gunten, O., August, L. S. and Jordan, T. M. (1982). The damage equivalence of electrons, protons and gamma rays in MOS devices. *IEEE Transactions on Nuclear Science*, **NS29**, 1966-9.

Brucker, G. J., van Gunten, O., Stassinopoulos, E. G., Shapiro, P., August, L. S. and Jordan, T. M. (1983). Recovery of damage in rad-hard devices during and after irradiation by electrons, protons, alphas and gamma rays. *IEEE Transactions on Nuclear Science*, **NS30**, 4157-61.

Chapuis, T. (1991). Test aux ions lourds du microprocesseur 68020 et du coprocesseur 68882 de chez Motorola. CNES report RA/DP/QA/CE/91-35, Toulouse, France.

Christian, S. M. (1966). Radiation tolerance of field-effect transistors. In *Field effect transistors*, (ed. J. T. Wallmark and H. Johnson) Prentice-Hall, Englewood Cliffs, NJ.

Chu, D. D. and Jernberger, A. (1991). ESA capability domain approval for CMOS/SOS ASICs. *ESA Components Conference*, Noordwijk, November 1991.

Cusick, J., Koga, R., Kolasinski, W. A. and King, C. (1985). SEU vulnerability of the Zilog Z-80 and NSC-800 microprocessors. *IEEE Transactions on Nuclear Science*, **NS32**, , 4206-11.

Danchenko, V., Stassinopoulos, E. G., Fang, P. H. and Brashears, S. S. (1980). Activation energies of thermal annealing of radiation-induced damage in n- and p-channel CMOS IC's. *IEEE Transactions on Nuclear Science*, **NS27**, 1658-76.

Danchenko, V., Fang, P. H. and Brashears, S. S. (1981). Activation energies of thermal annealing of radiation-induced damage in n- and p-channel CMOS IC's, Part II. *IEEE Transactions on Nuclear Science*, **NS28**, 4407-12.

Darwish, M. N., Dolly, M. C., Goodwin, C. A. and Titus, J. L. (1988). Radiation effects on power integrated circuits. *IEEE Transactions on Nuclear Science*, **NS35**, 1547-51.

Deal, B. E. (1980). Standardized terminology for oxide charges associated with thermally oxidized silicon. *Journal of the Electrochemical Society*, **127**, 979-81.

Dennehy, W., Brucker, G. and Holmes-Siedle, A. G. (1966). A radiation-induced instability in MOS transistors. *IEEE Transactions on Nuclear Science*, **NS13**, 273-81.

Dennehy, W. J., Holmes-Siedle, A. G. and Zaininger, K. H. (1967). Process techniques and radiation effects in MOS transistors. *IEEE Transaction on Nuclear Science*, **NS14**, 276.

Dennehy, W. J., Holmes-Siedle, A. G. and Leopold, W. (1969a). Transient radiation response of complementary-symmetry MOS integrated circuits. *IEEE Transactions on Nuclear Science*, **NS16**, 114-19.

Dennehy, W. J., Holmes-Siedle, A. G. and Zaininger, K. H. (1969b). Digital logic for radiation environments. *RCA Review*, **30**, 668-708.

Dennehy, W. J., Leopold, W., Borkan, H., and Holmes-Siedle, A. G. (1970). Radiation-hard MOS arrays. Report AFAL-TR-70-179. RCA Astro-Electronics Division, Princeton, NJ.

Desko, J. C., Jr, Darwish, M. N., Dolly, M. C., Goodwin, C. A., Dawes, W. L.,

Jr and Titus, J. L. (1990). Radiation hardening of a high voltage IC technology (BCDMOS). *IEEE Transactions on Nuclear Science*, **NS37**, 2083-8.

Dupont-Nivet, E. *et al.* A hardened technology on SOI for analog devices. Proc. RADECS '91, Montpellier, France, Sept 9-12 1991. *IEEE Conf. Record*, Cat. No. 91 THO 400 2, **Vol 15**, pp. 211-14.

Dressendorfer, P. V. (1989). *Radiation effects on MOS devices and circuits* in Ma and Dressendorfer, pp. 256-332.

Emms, C. G., Holmes-Siedle, A. G., Groombridge, I. and Bosnell, J. R. (1974). Gamma and vacuum ultraviolet irradiations of ion-implanted SiO_2 for MOS dielectrics. *IEEE Transactions on Nuclear Science*, **NS21** 159-60.

Faccio, F., Heijne, E. H. M., Jarron, P., Glaser, M., Rossi, G. and Borel, G. Study of device parameters for analog IC design in a 1.2 μm CMOS – SOI technology after 10 Mrad. Paper 0-1, IEEE NSREC '92, New Orleans.

Fan, J. C. C. (1989). New anatomies for semiconductor wafers. *IEEE Spectrum*, 2, 34.

Fleetwood, D. M. and Dressendorfer, P. V. (1987). A simple method to identify radiation and annealing biases that lead to worst case CMOS static RAM post-irradiation response. *IEEE Transactions on Nuclear Science*, **NS34**, 1408-13.

Fleetwood, D. M., Winokur, P. S., Riewe, L. C. and Pease, R. L. (1989). An improved standard total dose test for CMOS space electronics. *IEEE Transactions on Nuclear Science*, **NS36**, 1963-1970.

Fleetwood, D. M., Winokur, P. S. and Riewe, L. C. (1990). Predicting switched-bias response from steady-state irradiations. *IEEE Transactions on Nuclear Science*, **NS37**, 1806-17.

Fossum, J. G., Derbenwick, G. F. and Gregory, B. L. (1975). Design optimization of radiation-hardened CMOS integrated circuits. *IEEE Transactions on Nuclear Science*, **NS22**, 2208-13.

Freeman, R. F. A. and Holmes-Siedlé, A. G. (1978). A simple model for predicting radiation effects in MOS devices. *IEEE Transactions on Nuclear Science*, **NS25**, 1216-25.

Friedman, A. L., Waskiewicz, A. E. and Strahan, V. (1989). Post neutron performance of commercial 16K JI PROMSs. Paper PE-3, IEEE NSREC '89, Marco Island, FL.

Garth, J., Burke, E. and Woolf, S. (1980). The role of scattered radiation in the dosimetry of small devices. *IEEE Transactions on Nuclear Science*, **NS27**, 1459-64.

Grove, A. S. (1967). *Physics and technology of semiconductor devices*. Wiley, New York.

Grunthaner, F. J., Grunthaner, P. J. and Maserjian, J. (1982). Radiation-induced defects in SiO_2 as determined by XPS. *IEEE Transactions on Nuclear Science*, **NS29**, 1462-6.

Hahn-Meitner Institut (1980). *Data compilation of irradiation tested electronic components*, Report No. HMI-B353 (loose-leaf data sheets). Berlin.

Harboe-Sorensen, R., Adams, L., Daly, E. J., Sansoe, C., Mapper, D. and Sanderson, T. K. (1986). The SEU risk assessment of Z-80A, 8086 and 80C86 microprocessors intended for use in a low altitude polar orbit. *IEEE Transactions on Nuclear Science*, **NS33**, 1626-31.

Harboe-Sorensen, R. (1990). Proton and heavy ion testing of devices for analysis of SEU at low altitude. In Proceedings of the ESA Workshop on Space Environment Analysis, ESA WPP-23. ESA-ESTEC, Noordwijk, Netherlands.

Harboe-Sorensen, R., Muller, L., Daly, E. J., Schmitt, E. J. and Rombeck, F. J. (1991a). Radiation pre-screening of 4 Mbit random access memories for space application. In Proc. RADECS '91, Montpellier, France, Sept 9-12 1991. IEEE Conf. Record, Cat. No. 91 THO 400 2, Vol 15, pp. 489-504.

Harboe-Sorensen, R., Adams, L., Seran, H. and Armbruster, P. (1991b). The single event response of the Analog Devices ADSP 2100A Digital Signal Processor. Proc. RADECS '91, Montpellier, France, Sept 9-12 1991. IEEE Conf. Record, Cat. No. 91 THO 400 2, Vol 15, pp. 457-61.

Hatano, H., Sakaue, K. and Naruke, K. (1984). CMOS shift register circuits for radiation tolerant VLSI's. *IEEE Transactions on Nuclear Science*, **NS31**, 1034-8.

Heijne, E. H. *et al.* (1990). *Monolithic CMOS front-end electronics with analog pipelining*. Report No. CERN/ECP 90-13. CERN, Geneva.

Holmes-Siedle, A. G. (1982). The measurement of radiation hardness of MOS capacitors by avalanche injection of holes, US Naval Research Laboratory Contract Report No. AHS-NRL-82-1.

Holmes-Siedle, A. G. (1989). The qualification of charge coupled devices for the space environment, ESA Contract Report R1232/2. Fulmer Research, Stoke Poges, UK.

Holmes-Siedle, A. G. and Adams, L. (1983). The mechanisms of small instabilities in irradiated MOS transistors. *IEEE Transactions on Nuclear Science*, **NS30**, 413-40.

Holmes-Siedle, A. G. and Adams, L. (1986). RADFETs: a review of the use of metal-oxide-semiconductor devices as integrating dosimeters. *Radiation Physics and Chemistry*, **28**, 235-44.

Holmes-Siedle, A. G. and Freeman, R. F. A. (1978). *Radiation effects engineering handbook Report No. R*730/8. Fulmer Research Institute, Stoke Poges.

Holmes-Siedle, A. G. and Groombridge, I. (1975). Hole traps in silicon dioxide: a comparison of population by X-rays and band-gap Light. *Thin Solid Films*, **27**, 165-170.

Holmes-Siedle, A. G. and Harari, E. (1971). Optical absorption and defect structure in thin amorphous films of aluminium oxide. *Bulletin of the American Physical Society Series II*, **16**, 500.

Holmes-Siedle, A. G. and Zaininger, K. H. (1968). The physics of failure of MIS devices under ionizing radiation. *IEEE Transactions on Reliability*, **R-17**, 34-44.

Holmes-Siedle, A. G., Adams, L. and Ensell, G. (1991). MOS dosimeters — improvement of Responsivity. In Proc. RADECS '91, Montpellier, France, Sept 9-12 1991. IEEE Conf. Record, Cat. No. 91 THO 400 2, Vol 15, pp. 65-9.

Hughes, G. W. and Powell, R. J. (1976). MOS hardness characterization and its dependence upon some process and measurement variables. *IEEE Transactions on Nuclear Science*, **NS23**, 1569-72.

Hughes, H. L. and Giroux, R. A. (1964). Space radiation affects MOSFETs *Electronics*, **37**, 58-60.

Hughes, H. L. and Razouk, R. (1981). Proc. Electrochem. Society Meeting, 1981.
Hughes, R. C. (1985). Theory of response of radiation sensing field effect transistors. *Journal of Applied Physics*, **56**, 1375–9.
Jarron, P. et al. (1989). *Development of integrated CMOS circuits and pixel detectors in the CERN-LAA*. Report No. CERN-LAA-SD 89-11. CERN, Geneva.
Johlander, B. (1991). Comparative radiation test of Motorola, Texas instruments and National Semiconductor AC and ACT, *ESA-ESTEC Radiation Report RA-068*. ESA-ESTEC, Noordwijk, Netherlands.
Johnston, A. H. (1984) Super recovery of total dose damage in MOS devices. *IEEE Transactions on Nuclear Science*, **NS31**, 1427–33.
Jupina, M. A. and Lenahan, P. M. (1990). Spin dependent recombination: a 29-Si hyperfine study of radiation induced P_b centers at the Si/SiO_2 interface. *IEEE Transactions on Nuclear Science*, **NS37**, 1650–57.
Kaschmitter, J. L., Shaeffer, D. L., Colella, N. J., McKnett, J. and Coakley, P. G. (1991). Operation of commercial R3000 processors in the LEO space environment, IEEE Transactions on Nuclear Science, NS38, 1415–20.
Kerns, S. E. (1989). *Transient-ionization and single-event phenomena* in Ma and Dressendorfer, pp. 485–576.
Kerr, J. A., Garraway, A., Shaw, C. M. and Wooten, D. (1991). A radiation hard 1750 microprocessor fabricated on improved silicon on sapphire. In Proceedings of Radiation Effects on Components and Systems, RADECS '91. Université de Montpellier, France, Sept 9–12 1991. IEEE Conf. Record, Cat. No. 91 THO 400 2, Vol 15, pp. 156–8.
Kim, W. S., Mnich, T. M., Corbett, W. T., Treece R. K., Giddings, J. and Jorgensen, L. (1983). Radiation hard design principles utilized in 8085 microprocessor family. *IEEE Transactions on Nuclear Science*, **NS30**, 4229–34.
Kinnison, J. D. et al. (1991). Radiation characterization of the ADSP2100A Digital Signal Processor. *IEEE Transactions on Nuclear Science*, **NS38**, 1398–1402.
Koga, R. et al. (1990). SEU and latch up tolerant advanced CMOS technology. *IEEE Transactions on Nuclear Science*, **NS37**, 1869–75.
Koga, R. et al. (1991). On the suitability of non-hardened high density SRAMs for space applications. *IEEE Transactions on Nuclear Science*, **NS38**, 1507–13.
Kosier, S. L., Schrimpf, R. D., Cellier, F. E. and Galloway, K. F. (1990). The effects of ionizing radiation on the breakdown voltage of p-channel power MOSFETs. *IEEE Transactions on Nuclear Science*, **NS37**, 2076–82.
Koyanagi, M. (1991). 3-D optically-coupled memory. In Proceedings, VLSI'91, Edinburgh.
Krull, W. A., Buller, J. F., Rouse, G. V. and Cherne, R. D. (1987). Electrical and radiation characterization of three SOI material technologies. *IEEE Circuits and Devices Magazine*, (July), 20–5.
Lelis, A. J., Oldham, T. R., Boesch, H. E., Jr and MacLean, F. B. (1989). The nature of the trapped hole annealing process. *IEEE Transactions on Nuclear Science*, **NS36**, 1808–15.
Lelis, A. J., Oldham, T. R. and De Lancey, W. M. (1991). Response of interface traps during high-temperature anneals. *IEEE Transactions on Nuclear Science*, **NS38**, 1590–7.
Lenahan, P. M. and Dressendorfer, P. V. (1984). Hole traps and trivalent silicon

centres in metal/oxide/silicon devices *Journal of Applied Physics*, **55**, 3495-99.

Leray, J. L. *et al.* (1990). CMOS/SOI hardening at 100 Mrad (SiO$_2$). *IEEE Transactions on Nuclear Science*, **NS37**, 2013-19.

Lopez-Cotarelo, M. (1991). Study of PLD technology for use in space application, ESA-ESTEC Working Paper EWP-1614. ESA-ESTEC, Noordwijk, Netherlands.

Ma, T. P. and Dressendorfer, P. V. (ed.) (1989). *Ionizing radiation effects in MOS devices and circuits*. Wiley, New York.

Mattsson, S. (1991). Radiation assessment of complex technologies. In Proceedings of ESA Electronic Components Conference, ESA-SP-313. ESA-ESTEC, Noordwijk, Netherlands.

Maurel, J-M., Villard, L. and Adams, L. (1991). Characterization method of the total dose tolerance of a logic family: application to Texas Instruments HCMOS family. In Proc. RADECS '91, Montpellier, France, Sept 9-12 1991. IEEE Conf. Record, Cat. No. 91 THO 400 2, Vol 15, pp. 220-5.

McLean, F. B., Boesch, H. E. and McGarrity, J. M. (1976). Hole transport and recovery characteristics of SiO$_2$ gate insulators *IEEE Transactions on Nuclear Science*, **NS23**, 1506- .

McLean, F. B., Boesch, H. E., Jr. and Oldham, T. R. (1989). *Electron-hole generation, transport and trapping in SiO$_2$* in Ma & Dressendorfer, pp. 87-192.

McWhorter, P. J., Miller, S. L. and Miller, W. M. (1990). Modeling the anneal of radiation-induced trapped holes in a varying thermal environment. *IEEE Transactions on Nuclear Science*, **NS37**, 1682-9.

Meisenheimer, T. L. and Fleetwood, D. M. (1990). Effect of radiation-induced charge on 1/f noise in MOS devices. *IEEE Transactions on Nuclear Science*, **NS37**, 1696-1702.

Messenger, G. C. and Ash, M. S. (1989). *The effects of radiation on electronic systems*. Wiley, New York.

Mitani, K. and Goesele, U. (1991). Wafer bonding technology for silicon-on-insulator applications: a review. In Third Workshop on Radiation-Induced and/or Process-Related Electrically-Active Defects in Semiconductor-Insulator Systems, Research Triangle Park, NC.

Murray, J. R. (1991). A 1K shadow RAM for circumvention applications. *IEEE Transactions on Nuclear Science*, **NS38**, 1403-9.

Nicollian, E. H. and Brews, J. R. (1982). *MOS physics and technology*. Wiley, New York.

Okabe, T., Kato, M. and Katsueda, M. (1990). High-frequency annealing effects on ionizing radiation response of MOSFET. *IEEE Transactions on Nuclear Science*, **NS37**, 1670-6.

Oldham, T. R. (1984). Analysis of damage in MOS devices for several radiation environments. *IEEE Transactions on Nuclear Science*, **NS31**, 1236-41.

Oldham, T. R., McLean, F. B., Boesch, H. E. and McGarrity, J. M. (1989). An overview of radiation-induced interface traps in MOS structures. *Solid Semiconductor Science and Technology*, **4**, 986-99.

Palkuti, L., Ormond, R. D., Hu, C. and Chung, J. (1991). *IEEE Transactions on Nuclear Science*, **NS38**, 1337-41.

Poch, W. J. and Holmes-Siedle, A. G. (1969). Permanent radiation effects in complementary-symmetry MOS integrated circuits. *IEEE Transactions on Nuclear Science*, **NS16**, 227-32.

Poch, W. J. and Holmes-Siedle, A. G. (1970). Long-term effects of radiation in complementary MOS logic networks. *IEEE Transactions on Nuclear Science*, **NS17**, 33-40.

Powell, R. J. and Derbenwick, G. (1971). Vacuum ultraviolet radiation effects in SiO_2 *IEEE Transactions on Nuclear Science*, **NS18**, 99-105.

Saks, N. S. and Brown, D. B. (1990). Observation of H+ motion during interface trap formation. *IEEE Transactions on Nuclear Science*, **NS37**, 1624-31.

Saks, N. S., Brown, D. B. and Rendell, R. W. (1991). Effect of switched gate bias on radiation-induced interface trap formation. *IEEE Transactions on Nuclear Science*, **NS38**, 1130-9.

Schlesier, K. (1974). Radiation hardening of CMOS/SOS integrated circuits. *IEEE Transactions on Nuclear Science*, **NS21**, 152-8.

Schwank, J. R., Winokur, P. S., McWhorter, P.J., Sexton, F. W., Dressendorfer, P. V. and Turpin, D. C. (1984). Physical mechanisms contributing to device rebound. *IEEE Transactions on Nuclear Science*, **NS31**, 1434-40.

Schwank, J. R. and Davies, W. R. Jr, (1983). Irradiated silicon gate MOS device bias annealing. *IEEE Transactions on Nuclear Science*, **NS30**, 4100-4.

Schwank, J. R., Sexton, F. W., Fleetwood, D. M., Shaneyfelt, M. N., Hughes, K. L. and Rodgers, M. S. (1989). Strategies for lot acceptance testing using CMOS transistors and ICs. *IEEE Transactions on Nuclear Science*, **NS36**, 1971-80.

Scott, T. M. (1990). Single event test method and test results of the Intel 80386. In Proceedings of the ESA Electronic Components Conference, ESA-SP-313. ESA-ESTEC, Noordwijk, Netherlands.

Sexton, F. W. *et al.* (1983). Radiation testing of the CMOS 8085 microprocessor family. *IEEE Transactions on Nuclear Science*, **NS30**, 4235-9.

Shaneyfelt, M. R., Schwank, J. W., Fleetwood, D. M., Winokur, P. S., Hughes, K. L. and Sexton, F. W. (1990). Field dependence of the interface state density and the trapped positive charge. *IEEE Transactions on Nuclear Science*, **NS37**, 1632-40.

Shaw, D. C., Lowry, L., Barnes, C., Zakharia, M., Agarwal, S. and Rax, B. (1991). Post irradiation effects (PIE) in integrated circuits. *IEEE Transactions on Nuclear Science*, **NS38**, 1584-9.

Snyder, E. S., McWhorter, P. J., Dellin, T. A. and Sweetman, J. .D. (1989). Radiation response of EEPROM memory cells. *IEEE Transactions on Nuclear Science*, **NS36**, 2131-9.

Srour, J. R., Long, D. M., Fitzwilson, R. L., Millward, D. G. and Chadsey, W. L. (1984). *Radiation effects and dose enhancement of electronic materials*. Noyes Publications, Park Ridge, NJ.

Stanley, T. D. (1988). The state-of-the-art in SOI technology. *IEEE Transactions on Nuclear Science*, **NS35**, 1346-9.

Stanley, T. D. *et al.* (1985). The effect of operating frequency in the radiation-induced buildup of trapped holes and interface states in MOS devices. *IEEE Transactions on Nuclear Science*, **NS32**, 3982-87.

Stassinopoulos, E. G. and Brucker, G. J. (1991). *Shortcomings in ground testing, environment simulations and performance prediction*. RADECS '91, Montpellier, France, Sept 9-12 1991. IEEE Conf. Record, Cat. No. 91 THO 400 2, Vol 15, pp. 3-16.

Stassinopoulos, E. G., Brucker, G. J., van Gunten, O., Knudson, A. R. and Jordan, T. M. (1983a). Radiation effects in MOS devices; dosimetry, annealing, irradiation sequence and sources. *IEEE Transactions on Nuclear Science*, **NS30**, 1880-4.

Stassinopoulos, E. G., van Gunten, O., Brucker, G. J., Knudson, A. R. and Jordan, T. M. (1983b). The damage equivalence of electrons, protons, alphas and gamma rays in rad-hard MOS devices. *IEEE Transactions on Nuclear Science*, **NS30**, 4363-7.

Stassinopoulos, E. G., Brucker, G. J. and van Gunten, O. (1984). Total dose and dose-rate dependence of proton damage in MOS devices during and after irradiation. *IEEE Transactions on Nuclear Science*, **NS31**, 1444-7.

Stevenson, G. R. (1990). *New dose calculations for LHC detectors*. Report No. CERN 90-10, Vol III, 566. CERN, Geneva.

Sze, S. M. (1981). *Physics of semiconductor devices* 2nd ed. Wiley, New York.

Thomlinson, J., Adams, L. and Harboe-Sorensen, R. (1987). The SEU and total dose response of the INMOS Transputer. *IEEE Transactions on Nuclear Science*. **NS34**, 1803-7.

Tsao, S. S. (1987). Porous silicon techniques for SOI structures. *IEEE Circuits and Devices Magazine*, **3**, 3-7.

Vandenbroeck, J., Holmes-Siedle, A., de Keersmaecker, R., Heyns, M., Fonderie, V. and Remmerie, J. (1986). KUL Contract Report INTEL 357-86-01. Katholieke Universiteit Leuven, Belgium.

Van Vonno, N. (1991). A radiation tolerant 8-bit flash A/D converter implemented in silicon on sapphire. In Proc. RADECS '91, Montpellier, France, Sept 9-12 1991. IEEE Conf. Record, Cat. No. 91 THO 400 2, Vol 15, pp. 241-8.

Vranch, R. L., Henderson, B. and Pepper, M. (1988) Spin dependent recombination in irradiated Si/SiO_2 structures. *Applied Physics Letters*, **52**, 1161- . . .

Wallmark, J. T. and Johnson, H. (ed.) (1966). *Field effect transistors*. Prentice-Hall, Englewood Cliffs, NJ.

Walters, M. and Reisman, A. (1991). Radiation-induced neutral electron trap generation in electrically biased insulated gate field effect transistor gate insulators. *Journal of the Electrochemical Society*, **138**, 2756-62.

Williams, R. (1965), *Physical Review A*, **140**, A569.

Winokur, P. S. (1989). *Radiation-induced interface traps*, in Ma and Dressendorfer, pp. 193-255.

Winokur, P. S., Kerris, K. C. and Harper, L. (1983). Predicting CMOS inverter response in nuclear and space environments. *IEEE Transactions on Nuclear Science*, **NS30**, 4326-32.

Winokur, P. S., Schwank, J. R., McWhorter, P. J., Dressendorfer, P. V. and Turpin, D. C. (1984). Correlating the response of MOS capacitors and transistors. *IEEE Transactions on Nuclear Science*, **NS31**, 1453-60.

Winokur. P. S. *et al.* (1991). Implementing QML for radiation hardness assurance. *IEEE Transaction on Nuclear Science*, **NS37**, 1794-1805.

Yu, C-Y., Sung, J. J. and Yu, C-H. (1989). Submicrometer salicide CMOS devices with self-aligned shallow/deep junctions. *IEEE Electron Device Letters*, **10**, 487-9.

Zaininger, K. H. (1966a). Electron bombardment of MOS capacitors. *Applied Physics Letters*, **8**, 140-2.

Zaininger, K. H. (1966b). Irradiation of MIS capacitors with high-energy-electrons. *IEEE Transaction on Nuclear Science*, **NS13**, 237-47.

Zaininger, K. H. and Waxman, A. S. Radiation resistance of Al_2O_3 MOS devices. *IEEE Transactions on Electron Devices*, **ED16**, 333.

Recommended background reading

The following papers and books have been selected largely from the preceding list of references (where fuller details can be found) as a recommended reading list for students and designers wishing to establish a background in the device physics connected with MOS devices and the effects of exposure to radiation.

Adams, L. (ed.) (1994). Special edition on Space Radiation Environment and Effects. *International Journal of Radiation Physics and Chemistry*, **43**, Nr 1/2 January/February 1994.

Boesch, McGarrity, and McLean. (1978)

Braünig, D. (1989). *Wirkung hochenergetischer strahlung auf halbleiterbauelemente*. Springer Verlag, Berlin.

Brucker, Dennehy, and Holmes-Siedle. (1965).

Deal. (1980).

Dennehy, Holmes-Siedle and Zaininger. (1969).

Grove. (1967).

Holmes-Siedle and Zaininger. (1968).

Ma and Dressendorfer (eds). (1989).

Nicollian and Brews. (1982).

Sze. (1981).

5
Bipolar transistors

5.1 Introduction

Bipolar transistors consist of a pair of closely spaced p–n junctions in a single semiconductor structure. The order can be 'n–p–n' or 'p–n–p'. These devices, in both discrete and integrated form, are essential components in many electronic systems, especially in applications (such as in amplifiers), which require a high current gain or considerable 'drive' current. Radiation produces degradation of gain and an increase in leakage. The degradation of gain is well understood and derives from degradation of the transport of minority carriers across the base region. We will discuss the effects of electrons, neutrons, and gamma rays on this transport, dividing them into those occurring near the surface of the device and those that occur deeper in the material, i.e. bulk effects. In this chapter, we discuss the basic effects with emphasis on discrete silicon planar transistors. In this structure (see Fig. 5.1), the silicon substrate forms the collector region. Base and emitter regions are formed by the successive introduction of appropriate dopants by diffusion or ion implantation. The junctions are protected by a passivation layer which usually is thermally grown oxide. In general, this layer is thicker than that of MOS devices and accumulates charge when irradiated. Two books by pioneers in this field give detailed treatments of the semiconductor transport equations controlling transistor action in irradiated bipolar transistors (Larin 1968; Messenger and Ash 1986).

5.2 Effects of radiation on device function

5.2.1 Gain

The term 'gain' can refer to either of two separate parameters in a bipolar transistor. Common-base current gain (α or h_{FB}) is the ratio between collector current and emitter current and has a value rather less than unity. Common-emitter current gain (β or h_{FE}) is the ratio between collector current and base current; it will have a value, typically, of 50 to 2000. In our discussion of gain degradation, we will use the common-emitter current gain term and, for brevity, use the symbol β.

5.2.2 Degradation of gain

Gain degradation and leakage are the most striking and common effects of radiation on bipolar transistors. One cause of gain degradation is atomic

5.2 Effects of radiation on device function

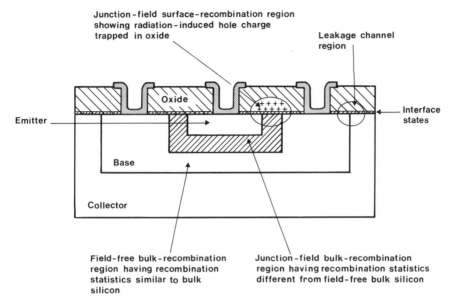

FIG. 5.1 Planar bipolar transistor, showing regions involved in radiation effects.

displacement in the bulk of a semiconductor. This bulk damage produces an increase in the number of recombination centres and therefore reduces minority-carrier lifetime. The other main cause of gain degradation is ionization in the oxide passivation layer, particularly that part of the oxide covering the emitter–base junction region (see Fig. 5.1). By a process similar to that in MOS devices (Section 4.4), charge-trapping and the generation of new interface states produces degradation of gain. The trapped surface charge and interface states cause an increase in minority-carrier surface recombination velocity, reducing gain. Under electron and proton irradiation, such surface-linked degradation often precedes minority-carrier effects.

Gain degradation is often analysed by plotting the change in reciprocal gain $\Delta(1/\beta)$ versus radiation fluence. The term $\Delta(1/\beta)$ is known as the 'gain damage figure', represented by the equation

$$\Delta(1/\beta) = 1/\beta - 1/\beta_0, \tag{5.1}$$

where β_0 and β are the gain values before and after irradiation. This is a convenient mathematical form in that the effects of bulk and surface damage are separated, as follows:

$$\Delta(1/\beta) = \Delta(1/\beta)_b + \Delta(1/\beta)_s, \tag{5.2}$$

where the suffixes b and s indicate the bulk and surface contributions respectively. However, while the bulk contribution may be reasonably well predicted from analysis of minority carrier lifetime behaviour, the surface contribution is highly dependent upon process factors (see later sections).

5.2.3 Other permanent effects

Apart from causing the degradation of gain, irradiation also causes other important effects in bipolar transistors. Increases in the junction leakage currents (e.g. collector–base reverse leakage, I_{CBO}), like the surface-linked gain degradation already mentioned, result from ionization in the surface oxide, particularly the region over the collector–base junction (see Fig. 5.1).

It is also found that heavy atomic displacement damage in the semiconductor causes an increase in the collector–emitter saturation voltage, $V_{CE(sat)}$. The rate of this increase with fluence is roughly the same as that with which gain decreases, i.e. the fluences producing a 50 per cent change in either parameter are about the same. This is because $V_{CE(sat)}$, the voltage at which both transistor junctions are just forward-biased, is an inverse function of transistor gain and is thus affected by changes in minority-carrier lifetime.

Transistors with high breakdown voltages have high resistivity (low doping) in the collector region. At very high bulk damage levels, the resistivity may be changed noticeably, causing further changes in $V_{CE(sat)}$. Such effects are of less importance in low-power transistors such as logic devices, because the switching or amplifying action is not affected by relatively high $V_{CE(sat)}$ values. On the other hand, the efficient operation of power transistors, especially switching power transistors, requires low $V_{CE(sat)}$ values. Donovan et al. (1976) indicate that failure in power transistor circuits can be caused by a 50 per cent increase in $V_{CE(sat)}$ or a hundredfold increase in I_{CEO} (a surface-controlled leakage). Table 5.1 shows the change in these and other parameters in a bipolar power transistor.

5.2.4 Transient effects

The effect of ionization on the p–n junction of a bipolar transistor may lead to transient photocurrent effects. The case of discrete bipolar transistors was discussed as an example in Chapter 3. The lowest radiation dose rates required to produce significant effects are in the region of 10^6 rad(Si) s^{-1}, well above the rates produced by either the natural space environment or typical isotope sources (see section 3.5). The response of a bipolar transistor to dose rate is usually controlled by the photocurrent generated in the collector–base junction sources. These currents may be calculated if the dose rate and junction area are known (see, for example Larin 1968). However, if a transistor is biased to operate in the active region, the flow of primary photocurrent is amplified by the current gain, h_{FE}, of the transistor (Wirth and Rogers 1964). This secondary photocurrent is linear below a dose rate of about 10^8 rad(Si) s^{-1} but, due to conductivity changes, can increase suddenly as dose rate increases above this value (again see Larin 1968 and

5.2 Effects of radiation on device function

TABLE 5.1 The effect of neutron irradiation upon three samples of a Fairchild bipolar transistor of 1970 era ($V_{CE} = 5\,V$, $V_{CB} = 30\,V$) (Snow et al. 1970)

Neutron fluence (cm^{-2})	I_m (A)	I_m/I_{mo}	β_0 at I_{mo}	β/β_0 at I_m	β at 1 A	$V_{CE(sat)}$ Pre-irrad. (V)	$V_{CE(sat)}$ Post-irrad. (V)	I_{CBO} at 30 V Pre-irrad. (nA)	I_{CBO} at 30 V Post-irrad. (nA)	Approximate[a] equivalent 1-MeV electron fluence (cm^{-2})
1×10^{14}	1.05	0.95	35.5	0.75	25.9	0.27	0.38	1.6	6.9	3×10^{16}
	0.85	0.88	33.5	0.73	21.1	0.30	0.43	0.5	6.9	
	0.90	0.83	33.7	0.72	22.7	0.29	0.42	0.1	6.0	
3×10^{14}	0.90	0.83	34.8	0.57	17.9	0.29	0.54	1.4	17.4	9×10^{16}
	1.00	0.90	34.8	0.59	20.0	0.25	0.47	0.4	15.6	
	0.90	0.89	34.3	0.59	19.1	0.25	0.49	0.7	15.3	
1×10^{15}	0.90	0.44	34.3	0.31	6.3	0.26	1.19	0.4	35.2	3×10^{17}
	0.90	0.44	33.7	0.32	6.4	0.23	1.16	0.8 A	1 A	
	0.90	0.34	34.5	0.31	6.6	0.27	1.12	0.5	35.7	
2×10^{15}	0.90	0.17	35.3	0.15	—	0.23	1.55	0.4	58.4	6×10^{17}
	0.80	0.19	37.1	0.15	—	0.25	2.41	1.1	58.9	
	0.90	0.17	34.7	0.14	—	0.25	1.86	0.4	61.7	

β_0 = pre-irradiation gain; β = post-irradiation gain; I_{mo} = current at which β_0 peaks; I_m = current at which β peaks. [a] 1 n cm^{-2} is equivalent to approximately 300 e cm^{-2} (1 MeV)

Messenger and Ash 1986). These unstable current conditions can give rise to destructive burnout of the junctions.

5.3 Bulk damage

5.3.1 General

As discussed in Chapter 3, the effect of atomic displacement in bulk silicon is a reduction in conductivity and minority-carrier lifetime. The reduction in lifetime may be represented by the following equation:

$$1/\tau - 1/\tau_0 = K_\tau \phi. \tag{5.3}$$

$K_\tau (\text{cm}^2 \text{s}^{-1})$ is known as the minority-carrier lifetime damage constant for a given type and resistivity of silicon at a given radiation energy, ϕ is the fluence, and τ_0 and τ are the initial and post-irradiation lifetime values. The value of K_τ is dependent upon material properties, e.g. the value for oxygen-free silicon is significantly less than for oxygen-rich material while phosphorous and boron dopants interact with radiation-induced defects in silicon.

This picture is complicated by various other factors which depend upon the geometry of the device, particularly that of the base region. The following are three examples of such complicating factors:

1. The effectiveness of a given recombination centre on lifetime depends upon the charge state of that centre, and this in turn is controlled by the minority-carrier equilibrium pertaining to that point. A diffused-junction device contains depletion regions near the junctions. In these regions, the electric fields are high and carriers are swept out rapidly. The material here resembles an intrinsic semiconductor, and recombination rates are very low, although the addition of new defects will certainly alter these rates. The actual fields, and hence the carrier equilibria, vary with the bias applied.

2. The nature of the production techniques commonly used for transistors. In a diffused transistor, the base and emitter are regions formed by the diffusion of dopants into the substrate silicon. Thus, the dopant concentration profile (and hence the Fermi level) in the base is far from uniform and varies by orders of magnitude. In an ion-implanted device, different dopant concentration profiles will also be found and, in addition, the implantation treatment may leave a residue of defects in the silicon.

3. When current flows in the transistor, the Fermi level is altered by the presence of excess carriers injected at the emitter–base junction and is effectively dependent upon the operating level of the transistor. Therefore, other factors being equal, the value of K_τ in a device carrying a signal

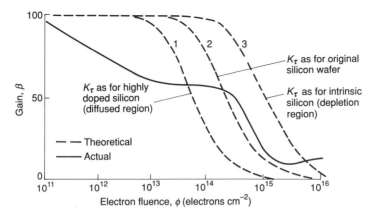

FIG. 5.2 Bipolar transistor degradation: common-emitter current gain as a function of electron fluence; the solid curve shows typical behaviour of a real device, indicating the early effects of ionization in the surface passivating layer.

current of 1 mA will be different from the value in a device carrying a current of, say, 10 A. Thus, it is a complex task to predict the K_τ value of a device from first principles.

The predicted typical relationship between gain and 1-Mev electron fluence, shown by the dashed lines in Fig. 5.2, derives from equation (5.2). A change in K_τ is expressed as a translation of the curves along the fluence axis. However, the observed gain degradation in a 'planar' (oxide-passivated) bipolar transistor does not follow this predicted course. Instead, many devices are observed to degrade in the manner shown by the solid curve in Fig. 5.2. This is a clear indication that some degrading mechanism other than minority-carrier lifetime reduction is affecting gain, and this is found to be the surface ionization effect already mentioned.

5.3.2 Influence of base width

It was noted in Chapter 3 that, for irradiated MOS transistors, the oxide thickness was a key device parameter; in bipolar devices, base width is an equally important factor and — as described below — has a significant influence upon gain degradation by bulk damage.

The Webster (1954) equation for the gain of a bipolar transistor uses the following term for relating gain β to minority carrier lifetime:

$$1/\beta = W_b^2/2D_\tau + \text{etc.} \qquad (5.4)$$

where W_b is the base width and D_τ is the minority-carrier diffusion constant. Other terms in this equation account for the contributions of surface recombination (the surface effect) and emitter efficiency. The base-width term is the controlling one for reduction in τ (i.e. bulk damage).

Assuming that the surface recombination and emitter efficiency terms remain constant, a combination of the Webster equation and (5.1) and (5.3) produces the following equation for the figure of gain damage caused by bulk damage:

$$\Delta(1/\beta) = \frac{W_b^2}{2D_\tau} \cdot K_\tau \phi. \tag{5.5}$$

This equation demonstrates the strong dependence of the gain damage figure upon base width. Thus, it is necessary to measure or calculate this term for devices under consideration. The value of W_b itself cannot be measured easily; however, the cut-off frequency f_α (the frequency at which common-base current gain α falls to one-half of its low-frequency value, say at 1 KHz) bears a close relation to W_b.

By using this dependence, Messenger and Spratt (1965) derived the following prediction for the effect of bulk damage on transistors:

$$\Delta(1/\beta) = \frac{K_\tau \phi}{2\pi f_\alpha}. \tag{5.6}$$

Here, gain degradation has been related to more easily measurable parameters. For the purpose of the prediction K_τ must be determined for at least one type of particle for any given bipolar transistor technology.

Typical figures for K_τ, measured for diffused n-p-n and p-n-p transistors under 1 MeV electron irradiation, are 1×10^{-8} and 3×10^{-8} cm^{-2} s^{-1} respectively and, for reactor neutrons, 1×10^{-6} and 3×10^{-6} cm^{-2} s^{-1} respectively.

As might be expected from solar cell investigations (see Chapter 6) (p diffusion on n-type base is more sensitive than n on p), p-n-p transistors are more sensitive than the n-p-n type by a factor of about 3 (the fluence required to yield a given degradation value is three times lower). Although the cut-off frequency f_α is included in the equation for gain prediction, an alternative parameter f_T (the gain–bandwidth product) is more easily measured and also most commonly quoted for transistors. It is the product of common-emitter gain β and the frequency of measurement at which the gain-frequency curve begins to descend.

The value of f_T is also the frequency at which β falls to unity. Table 5.2 gives examples of the relation between f_T and base width W_b.

The frequencies f_α and f_T are not exactly equal, f_T being slightly lower. The ratio between them depends on the doping profile. For ideal step junctions, $f_\alpha = \sqrt{1.22} f_T$; for diffused junctions, the ratio is 1.33. Thus, in practice, the error introduced by using f_T in the prediction of $\Delta(1/\beta)$ is not often significant and, in any case, errs on the side of safety.

5.3 Bulk damage

TABLE 5.2 Calculation of base width from base transit time or approximate gain–bandwidth product (GBP)

Approx. GBW product[a] f_T (MHz)	Base transit time[b] t_b (ns)	$\sqrt{(33.2\, t_b)}$ (s)	$33.2\, t_b$ (cm)	Base width npn[b] W_b (μm)	Base width pnp W_b (μm)
1000	0.16	5.28×10^{-9}	7.26×10^{-5}	0.73	0.63
500	0.32	1.05×10^{-8}	1.02×10^{-4}	1.02	–
200	0.80	2.63×10^{-8}	1.62×10^{-4}	1.62	–
100	1.59	5.28×10^{-8}	2.30×10^{-4}	2.29	2.0
50	3.18	10.5×10^{-8}	3.24×10^{-4}	3.24	–
20	7.95	26.3×10^{-8}	5.13×10^{-4}	5.12	–
10	15.9	52.8×10^{-8}	7.26×10^{-4}	7.26	6.3
1	1.59	5.28×10^{-6}	22.9×10^{-4}	22.9	20

[a] Ignoring t_e, t_c, etc.; GBW = gain band width

[b] Using relations: $t_b \simeq \dfrac{1}{2\pi f_T}$

$W_b = \sqrt{(2 D_n t_b)} = \sqrt{(33.2\, t_b)}$

$D_n = 16.6 \text{ cm}^2 \text{ s}^{-1}$ ($D_p = 12.5 \text{ cm}^2 \text{ s}^{-1}$)

(see Larin 1968)

It is often convenient to express transistor damage as a 'gain damage factor' (K_b). This is the gain damage figure, normalized for fluence but not for base width, f_T, etc. Thus,

$$1/\beta - 1/\beta_0 = \Delta(1/\beta) = K_b \phi. \tag{5.7}$$

Gain after a given irradiation can be calculated:

$$\beta = \frac{\beta_0}{1 + \beta_0 K_b \phi}. \tag{5.8}$$

A simple relation between K_b and K_τ is derived from eqn. (5.6):

$$K_\tau = 2\pi K_b f_T = 6.28 K_b f_T. \tag{5.9}$$

5.3.3 Influence of type and energy of radiation

The atomic displacement process which causes bulk damage is, unlike surface damage, quantitatively dependent upon the momentum of the radiation. It is useful to consider briefly the effects of energy and particle type, because by so doing, the test results of a large range of particle types and energies, including many results for reactor neutrons, will be found. Also, the particle spectrum in space is very broad.

172 Bipolar transistors

The following equations summarize four important 'damage constants' (K) which relate the relevant 'damage figure' to particle fluence.

Minority-carrier lifetime damage constant, $K(\text{cm}^2\,\text{s}^{-1})$:

$$\Delta(1/\tau) = K_\tau \phi; \tag{5.10}$$

(see also eqn (5.3)).

Gain damage constant, $K_b(\text{cm}^2)$:

$$\Delta(1/\beta) = K_b \phi; \tag{5.11}$$

(see also eqn (5.7)).

Diffusion length damage constant (for solar cells), $K_L(\text{cm}^2 \text{ per cm}^2)$:

$$\Delta(1/L^2) = K_L \phi. \tag{5.12}$$

Carrier removal rate, $\dfrac{\Delta n}{\Delta \phi}$ or $\dfrac{\Delta p}{\Delta \phi}$ (cm^{-1}):

$$n(\phi) = n_o - \frac{\Delta n}{V\phi}. \tag{5.13}$$

These constants all change with particle type and energy. This section will discuss mainly the variation of the constants for lifetime and gain damage in bipolar transistors. Figure 5.3 shows the experimental results of exposing a 2N1613 transistor to several different types of radiation. It is encouraging that the form of degradation for each type of radiation is very similar, except that $\Delta(1/\beta)$ is displaced along the flux scale (i.e. K_b is different). Thus, it may often be possible to employ test results using one particle type or energy to *predict the effects expected under irradiation by another particle type or energy.*

Figure 5.3 also shows that in all cases some slowly developing non-linear degradation takes place well before a linear growth of gain damage with fluence (indicated by a dashed line) is observed to occur. This early degradation is the 'surface effect' due to the ionizing component of the radiation.

The constant K_b can be measured from the linear portion of Fig. 5.3. K_b for reactor neutrons is seen to be over 100 times greater than for electrons. It is fortunate that low-energy protons are filtered out by normal device packaging, as otherwise the high K_b value shown for 2 MeV protons would cause a serious effect in space (see, for example, the serious proton damage to X-ray sensors discussed in Chapter 6 under the heading of CCDs). The long 'tail' of the Co-60 gamma radiation curve indicates that ionization (surface effect) is the major effect of this radiation, although bulk damage from Compton electrons is evident at very high doses (10^8 rad). Table 5.3

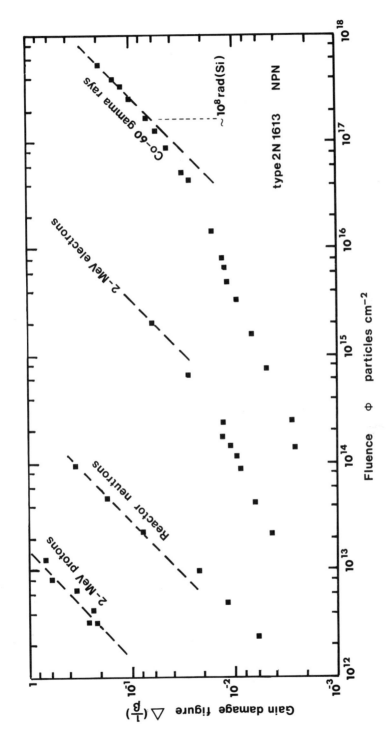

FIG. 5.3 Influence of radiation type: gain damage figure for 2N1613 devices as a function of radiation fluence (Brown 1964).

174 Bipolar transistors

TABLE 5.3 Damage constants for a number of transistors subjected to various types of irradiation (Brown, 1964)

Device type	f_T (MHz, nom)	Protons[a]		Reactor neutrons >10 keV	Electrons		Gamma rays Co-60
		2 MeV	10 MeV		2 MeV	5 MeV	

A. Transistor gain damage constant K_b (cm^2) for various types of radiation (Brown 1964; Brown and Horne 1967); collector current 2 mA[b]

2N336 npn	7	–	7.0E-15[c]	2.5E-15	4.0E-17	2.5E-17	9.0E-19
2N1132A npn	60	5.5E-14	–	–	2.5E-17	–	7.0E-19
2N1613 npn	60	7.0E-14	5.0E-15	2.5E-15	4.0E-17	7.0E-15	4.0E-19
2N2107 npn	150	–	–	–	4.5E-17	–	3.5E-19
2N2217 npn	100	–	–	8.0E-16	1.5E-17	1.7E-17	1.0E-19

B. Nominal minority-carrier lifetime damage constant K_τ (cm^2 s^{-1}) calculated from the values of K_b above[d]

2N336 npn	7	–	2.5E-7	9.0E-8	1.4E-9	9.0E-10	3.2E-11
2N1132A npn	60	1.7E-5	–	–	7.7E-9	–	2.2E-10
2N1613 npn	60	2.2E-5	1.6E-6	7.8E-7	1.2E-8	2.1E-8	1.3E-10
2N2107 npn	150	–	–	–	3.5E-9	–	2.7E-10
2N2217 npn	100	–	–	4.1E-7	7.6E-9	8.7E-9	5.1E-11

[a] Proton irradiation with cans removed
[b] The uncertainty ascribed by Brown to his K_b values is typically ±30%
[c] The notation 'E-14' indicates '×10^{-14}', etc.
[d] Calculated from (5.9) assuming f_T equal to f_α.

lists a more complete set of results by Brown (1964) and Brown and Horne (1967), giving K_b for a number of bipolar transistors and six different radiation sources. To normalize for base width, we have converted K_b to K_τ, using eqn (5.9). The K_b values of three of the n-p-n devices fall within the range given by van Lint et al. (1975) for 3 MeV electrons in 1 Ω cm silicon, namely 2 to 8 × 10^{-9} cm^{-2} s^{-1}.

It is not clear whether the variations from type to type are due to variations in resistivity in the base region, different impurities, or other dissimilarities between experiments. The data quoted by van Lint et al. (1975) are largely for the base regions of solar cells and other uniformly doped silicon. In the circumstances, the correlation is quite good.

Brucker et al. (1965) and Brucker (1967) describe a set of irradiation tests similar to those carried out by Brown. These investigators improved the experiment by adding a preliminary heavy irradiation with 125 keV electrons. This caused no displacement damage, but virtually saturated the surface damage. As a result, K_b could be determined over a wider range of fluence because the obscuring influence of surface effects had been

5.3 Bulk damage

removed. The Brucker K_b values, which we have again converted into K_τ, are shown in Table 5.4. The electron result for an n-p-n transistor at 1 mA again lies near van Lint *et al.*'s figure. Both protons and neutrons yield K_τ values higher than those of Brown's devices. Brucker also studied the effect of varying electron energy over the range 0.275 to 1 MeV and found that the energy-dependence of K_τ in transistors was similar to that found by other workers for solar cells and bulk silicon (see e.g.

TABLE 5.4 Damage constants for a number of transistors subjected to various types of irradiation with pre-irradiation to saturate surface effects: see text

Device type	f_T (MHz, nom)	Protons 16.8 MeV	Reactor neutrons >10 keV	Electrons 1 MeV	Collector current (mA)
A. Gain damage constant, K_b (cm²), for various types of radiation (Brucker et al. 1965; Brucker 1967)[a]					
2N2102 npn	100	1.5E–14[b] 8.0E–15 –	4.6E–15 2.9E–15 4.9E–15	1.9E–17 1.8E–17 4.1E–17	1 10 100
2N1132 pnp	100	1.8E–14 1.2E–14 –	4.5E–15 2.4E–15 3.4E–15	7.4E–17 6.2E–17 6.8E–17	1 10 100
B. Nominal minority-carrier lifetime damage constant, K_τ (cm² s⁻¹) calculated[c] from the values of K_b above					
2N2102 npn	100	7.7E–6 4.1E–6 –	2.4E–6 1.5E–6 2.5E–6	9.7E–9 9.2E–9 2.1E–8	1 10 100
2N1132 pnp	100	9.2E–6 6.2E–6 –	2.3E–6 1.2E–6 1.7E–6	3.8E–8 3.2E–8 3.5E–8	1 10 100

C. Damage efficiency of protons and neutrons compared with 1 MeV electrons at collector current of 1 mA (energies as above)

	$K_\tau(p, n)/K_\tau(e)$	
	Protons	Neutrons
2N2102 npn	790	240
2N1132 pnp	240	60

[a] Brucker ascribes an uncertainty to K_b values of typically ±20%
[b] The notation E–14 indicates ×10⁻¹⁴, etc.
[c] From (5.9) assuming f_T equal to f_α

TABLE 5.5 Collected values of damage constants for commercial planar silicon transistors exposed to reactor neutrons

1 Device type no.	2 Manufacturer	3 Polarity	4 Nominal cut-off frequency[a] f_T (MHz)	5 I_C value for meas. I_C (mA)	6 Exptl beta damage const.[b] K_b (10^{-15} cm² n^{-1})	7 $K_b \times f_T$ (10^{-9} cm^{-1} s^{-1})	8 Lifetime damage constant (low I_C)[c] K col. 7 × col. 2 (10^{-6} cm² n^{-1} s^{-1})
2N720A	TEX	npn	50	1	9.8	490	3.27
2N918	TEX	"	900	1	0.71	639	4.26
2N930	TEX	"	45	1	1.9	86	1.84
2N1486	RCA	"	1.2	10	110.0	132	0.88
2N1613	TEX	"	60	1	6.7	402	2.68
2N2219	TEX	"	250	1	0.99	248	1.65
2N2219	SGS	"	"	1	1.1	275	1.83
2N2219A	FAI	"	"	3	0.67	168	1.12
2N2222A	FER	"	"	1	1.6	400	2.67
2N2223A	TEX	"	50	1	6.6	330	2.20
2N3055	FER	"	20	100	44.0	880	5.87
2N3055	MUL	"	20	100	69.0	1380	9.20
2N2905	NAT	pnp	200	3	0.95	190	1.27

176 Bipolar transistors

5.3 Bulk damage

2N2905A	TEX	"	200	3	0.57	114	0.76
2N2906	TEX	"	200	3	2.1	420	2.80
2N2906	TEX	"	200	1	3.2	640	4.26
2N2907A	TEX	"	200	1	3.4	680	4.53
2N4901	TEX	"	4	100	8.7	34.8	0.23
2N4904	MOT	"	4	100	10.13	0.52	0.003
2N1132	TEX	"	90	3	4.0	360	2.40
2N2303	FAI	"	60	1	2.7	162	1.08
2N2412	TEX	"	200	1	1.7	340	2.27
2N2894	FAI	"	370	1	0.51	189	1.26
2N4028	FAI	"	150	3	2.2	330	2.20
2N5883	MOT	"	4	30	35.0	140	0.93
Cooper et al. (1979)[d]							
2N222A	—	npn	300	0.51	1.4	467	3.11
2N2907A	—	pnp	200	1	0.54	108	0.72
2N2369A	—	npn	500	10	0.20	100	0.67
2N4150	—	npn	15	1000	3.68	55	0.37
2N5666	—	npn	20	1000	5.75	115	0.77

[a] Quoted on data sheets: specified minimum, not actual

[b] $K_b = \left[\dfrac{1}{\beta} - \dfrac{1}{\beta_0} \right] \cdot \phi_n$

[c] $K = 2\pi K_b f_T$

[d] Equiv. 1 MeV neutrons quoted; $\phi_n = 2$ to 5×10^{13} cm^{-2}

178 Bipolar transistors

Chapter 6 under the heading of 'Solar Cells').

The influence of displacement defects on τ is well known to be affected by the concentration of minority carriers ('injection level') in the device. This can be explained by the Shockley–Hall–Read recombination theory. Irradiation results from mesa and wide-base power transistors at 1 MeV (Brucker 1967) and 8 MeV (Holmes-Siedle, unpublished work 1972) suggest that damage constants of the order given above can be used for these devices, despite their very different structures.

5.3.4 Irradiation results

Table 5.5 lists our analysis of a series of nuclear reactor exposures (Holmes-Siedle, unpublished work; Cooper *et al.* 1979). Listed in the first group are the maximum values of K_b for the fluence range 10^{12} to 10^{13} reactor neutrons per cm^2 ($E > 10$ keV). Where possible, we have selected an I_c value of 1 mA. K_b is the experimental value and we have calculated the 'apparent K' value from K_b as in eqn (5.9). The variations in this value are unlikely to be due to a true difference in lifetime degradation. The high values probably contain a contribution to $\Delta(1/\beta)$ from effects at the surface and in the emitter–base depletion region. The quoted value of f_T is not a measured value but the minimum value quoted in the manufacturer's data sheet. It is quite likely that some of the samples tested had a higher value than that specified. Hence, both high values of apparent K_τ and some variability are to be expected.

5.3.5 Prediction of degradation

Once a judgement has been made as to the appropriate figure for the lifetime damage constant, of a transistor, the figure can be turned into a master set of prediction curves. For example, those shown in Fig. 5.4 are for an n-p-n device, having an original gain β_0 of 100, exposed to 1 MeV electrons. These curves apply to that one value of gain only, but the trends are the same for all values of gain.

To predict the degradation of a given device numerically, the value of $\Delta(1/\beta)$ may be calculated from equations given earlier. This value of $\Delta(1/\beta)$ may then be converted by calculation to the final value of β.

Table 5.6 shows some calculations of degradation in space, using three values of electron fluence such as might be calculated for a spacecraft in a high-radiation orbit.

5.3.6 Selection principles for bipolar transistors

In general, the longest survival time of a transistor will be achieved if the following selection principles are applied:

(1) where possible, use low-gain transistors at fairly high collector current levels;

5.3 Bulk damage 179

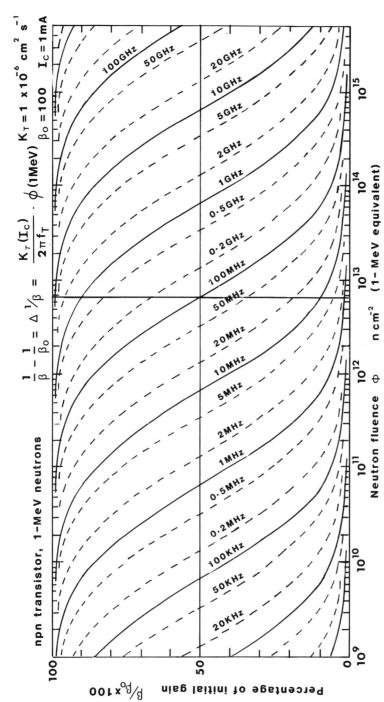

FIG. 5.4 Influence of base width: degradation of gain as a function of −1 MeV electron fluence, showing variation with alpha cutoff frequency f_α (a function of base width).

180 Bipolar transistors

TABLE 5.6 Predicted degradation of bipolar transistors in space radiation

Device type	Damage constants		Gain $(\beta)^a$ after 1 MeV electron fluence (cm^{-2}) of		
	K_T (cm^{2}s^{-1})	K_b (cm^2)	10^{13}	10^{14}	10^{15}
60 MHz transistor					
npn (e.g. 2N1613)	5.84E-9^b	1.0E-17	98.6	87.3	40.7
pnp (e.g. 2N1132)	2.27E-8	7.4E-17	93.1	57.5	11.9
200 MHz transistor					
npn (e.g. 2N2222A)	5.84E-9	5.71E-18	99.4	94.6	63.7
pnp (e.g. 2N2907A)	2.27E-8	2.22E-17	97.8	82.0	31.3
1.2 MHz power transistor					
npn (e.g. 2N1485)	4.2E-9	6.6E-16	58.8	12.5	1.5

^{a} $(1/\beta) = \dfrac{0.195 K_T}{f_\alpha} \phi = K_b \phi$

^{b} Notation as in Table 5.3

(2) when heavy-particle irradiation is involved, employ, where possible, high-frequency n–p–n devices.

5.4 Surface–linked degradation in gain

5.4.1 Introduction

In previous sections, we have described how the degradation of gain is made up of surface and bulk contributions. While the bulk damage figure $\Delta(1/\beta)_b$ can be predicted fairly well by means of the analytical techniques described, the value of the surface damage figure $\Delta(1/\beta)_s$ produced by a given radiation dose is much less easily predicted for a given transistor design. The latter value is highly dependent upon the surface preparation used, which usually consists of the growth of a passivating oxide in steam. Each manufacturer tends to use a different process and, even when meeting ultra-high reliability specifications, is liable to vary the process significantly from time to time.

Figure 5.5 shows the gain degradation exhibited by a batch of transistors from one day's production, as a function of dose during gamma irradiation, which produces mainly surface effects at the doses shown. Units with identical functional specifications, but from different manufacturers, may vary even more widely after irradiation. For these effects, the 'surface damage figure', $\Delta(1/\beta)_s$, was found to be a useful tool in that it provides a routine method of recording and characterizing the surface effect. An example of

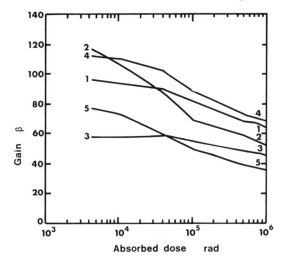

FIG. 5.5 Variability of surface-effect gain as a function of deposited dose in n–p–n bipolar transistor, type 2N2102; the variability of gain degradation due to the surface ionization effect is shown by the results for a batch from one day's production, irradiated by Co–60, with 10 mA collector current (Poch and Holmes-Siedle 1968).

the variation of surface damage figure with the collector current I_c used for measurement (roughly equivalent to emitter injection level) is shown in Fig. 5.6. A reasonably close linear relationship between I_c and damage figure is demonstrated.

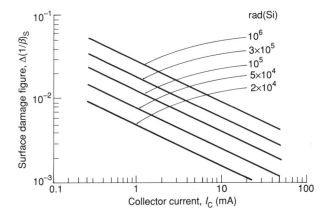

FIG. 5.6 Specification of surface damage figure in a bipolar transistor: form of an engineering specification for 'worst case' as a function of collector current in 2N916 transistors of a given manufacturer, at a 99% probability of keeping below the indicated degradation level (Poch and Holmes-Siedle 1968).

All transistors exhibit lower gain values at low current than at high current. This is because, at low currents, the flow of minority carriers is greater near the surface. The recombination of carriers at the surface removes carriers more effectively at low currents. It appears that this surface gain factor is always present and is simply increased by irradiation, an unexpected result since interface states are surface recombination sites and these are increased by irradiation. The change in gain can thus be interpreted as a change in the surface recombination term in the well-known Webster equation for the bipolar transistor. Stanley and Martin (1978) obtained a large amount of data on the surface effect in bipolar transistors as part of the hardening programme for the Voyager project. $\Delta(1/\beta)$ versus collector current after irradiation by electrons of energy near 2.5 MeV was measured and found to vary from type to type by as much as 100 times.

5.4.2 Statistical prediction of surface damage

The radiation-induced loss of gain for a batch of devices analysed statistically, using $\Delta(1/\beta)_s$ as a measure of damage, has been found to have a frequency distribution fitting the normal Gaussian curve (Poch and Holmes-Siedle 1968). For example, a set of 32 2N2102-type planar n-p-n transistors was irradiated to a dose level of 10^6 rad (10^4 Gy) gamma radiation. Measuring β and β_0 at a collector current of 0.7 mA, the mean value of the damage figure was 22×10^{-3} and the standard deviation 7.4×10^{-3}.

If the assumption of normal distribution is valid and this 2N2102 sample can be considered typical of the entire population from that production line, then the probability that the anticipated value of the ionization damage figure for the production conditions holding at the time will not exceed a specified maximum at a specified radiation dose can be calculated. Values of $\Delta(1/\beta)_s$ determined in this manner can be treated as a 'worst case' upper limit of anticipated transistor gain degradation. It is therefore feasible to predict degradation in the form of a 'worst expected surface damage figure' plotted against collector current. This form of prediction may be a useful method of specifying surface effects in design documents.

5.4.3 Collector–base leakage current

Increase in collector–base leakage current I_{CBO} is usually due to the formation of a surface channel. Although slight increases in the I_R term for the collector–base junction are produced by a reduction in the minority-carrier lifetime, the values are usually a minor component of the reverse leakage. By contrast, a charge build-up in the oxide layer over the junction can produce a surface channel which conducts strongly. As a result, I_{CBO} values which, for small-signal planar transistors, are usually 10^{-9} A or thereabouts, may increase tenfold for a dose in the 10^4 rad (10^2 Gy) range. Onset as a function of dose is often quite sudden and probably not amenable to

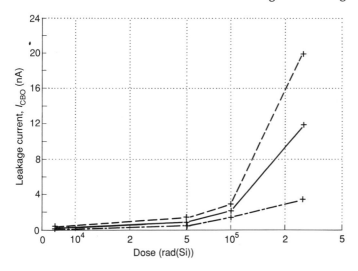

FIG. 5.7 Example of radiation-induced leakage current in a bipolar transistor, showing variability of response within a batch of five specimens (Braünig et al. 1980. Reprinted with permission).

statistical treatment. Figure 5.7 shows a case of sudden onset and scattered results in a batch of five specimens.

5.4.4 The 'maverick' device

Unfortunately, one of the characteristics of bipolar transistor performance is the occurrence of the 'maverick' or anomalous device. Such a device may be discovered in an otherwise normal batch, all processed in an essentially similar manner. It is anomalous in that it exhibits a radiation-induced gain degradation much more severe than the 'worst expected case' based on a Gaussian distribution as described earlier. While most of the devices follow a well-defined band of surface damage figure, the anomalously sensitive unit may show a degradation level which is higher by a factor of 50 or more. To design all circuits and associated shielding to tolerate this abnormal sensitivity and wide range of degradation characteristics would impose severe penalties with regard to size, weight, power drain, and complexity. On the other hand, the occurrence of such a degree of degradation in a particularly vital component would be catastrophic to equipment performance. The existence of a 'maverick' population could easily be missed if small-sample tests were performed and, therefore, preselection test programmes should — if possible — specify large statistical samples of devices.

5.4.5 Annealing of surface effects

The thermal annealing of the surface effects in bipolar transistors follows the same trends as that in MOS devices. Interface states should anneal out at temperatures in the 100°C to 200°C range and trapped charge should be removed between 150°C and 300°C. Some relaxation may occur even at room temperature. Many, but not all, bipolar devices can be annealed by baking although quite a large amount of damage persists in some types (Brucker 1967; Poch and Holmes-Siedle 1968). Thermal annealing as a basic preselection technique is not widely used.

The irradiate–anneal (IRAN) preselection procedure involves testing the entire quantity of any device proposed for use. On the basis of test results and design criteria, in the form of specified allowable degradation, the acceptable devices are retained (those that degraded within acceptable limits) and the unacceptable ones set aside for other, less critical, uses. Normally, the original (pre-irradiation) electrical characteristics can be restored, through an annealing process, to the samples selected for use without unacceptable loss of reliability. It must then be assumed that any subsequent in-flight irradiation to the same dose levels will cause the devices to degrade to approximately the same extent as during the test in the simulated environment. Following this procedure, the engineer has the added advantage of knowing in advance exactly what the degradation will be. Experiments by JPL to further evaluate the feasibility of this technique showed that IRAN preselection has limitations. JPL experience (Stanley and Martin 1978) uncovered a number of cases where the degradation of devices subjected to re-irradiation was not the same as that of the original IRAN irradiation. It would appear that the technique works only for certain bipolar devices. We are not aware of any programmes since Voyager using the IRAN technique.

5.4.6 Slow thermal annealing of bulk damage

Radiation-induced displacement defects (bulk damage) in silicon do not anneal easily, because long-lived vacancies and interstitials created by the radiation are usually complexed with an impurity atom (oxygen or dopant). The defects which concern us most are those that are stable at room temperature, e.g. the 'A' centre (complex of a vacancy with oxygen) and the centres designated 'E' (phosphorus vacancy complex), 'J' (di-vacancy) and 'K' (divacancy–oxygen complex). These centres are completely stable at temperatures below 200°C, but the damage often anneals between 200°C and 450°C completely. It is not easy to diagnose the damage effect in a commercial device. However, some information on the defects involved in transistor damage has been obtained by isochronal annealing, in which devices are heated for the same time at several evenly spaced temperature steps (Brucker 1967).

Messenger and Ash (1992) discuss the rapid annealing which is observed when a bipolar transistor is exposed to a pulse of neutrons at room temperature. Gregory and Sander (1970), discuss the same rapid effects, and also the slow annealing which occurs when the irradiated devices are then heated (isochronal and isothermal anneal).

5.4.7 Saturation voltage

The well-known 'knee' in a plot of collector current I_c versus collector-emitter voltage V_{CE} (with I_B constant) for a bipolar transistor (see Larin 1968) occurs at a low value of V_{CE} (typically 0.1 to 1 V). 'Saturation' occurs when V_{CE} falls to the same order of voltage as the forward voltage on the base–emitter junction V_{CE}. As V_{CE} falls, the collector–base junction, which must be biased in the reverse direction in order to produce high gain, becomes forward-biased. The transistor is then said to be in 'saturation'. As a result the gain falls dramatically, producing the 'knee' in the I_c and V_{CE} curves. For certain types of application (saturated switches, power transistors), it is necessary for transistors to operate in saturation. An increase in the saturation voltage $V_{CE(sat)}$ is usually harmful. By decreasing gain and increasing the resistivity of the silicon, particle irradiation can increase the $V_{CE(sat)}$ value. For measurement purposes, $V_{CE(sat)}$ is usually defined as the V_{CE} value required to produce a given value of I_B/I_c ('forced gain')

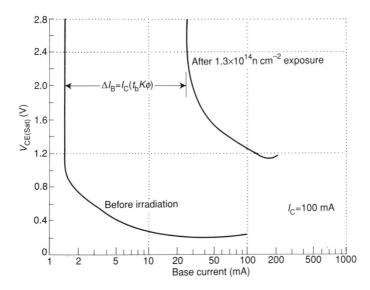

FIG. 5.8 Change in saturation voltage of a silicon n–p–n medium-power bipolar transistor, type 2N1613, under neutron–gamma irradiation at 1.3×10^{14} n cm^{-2} (after Larin 1968).

near the 'knee' described above. Figure 5.8 shows the changes induced in the saturation region by neutron irradiation (Larin 1968). Owing to the effects described earlier, gain has fallen, but increases in silicon resistivity have also affected the V_{CE} values required for a given ratio of I_B to I_c in saturation. For example, at $I_B = 50$ mA (forced gain = 2), the required V_{CE} value has been changed from 0.2 to 1.5 V by an exposure of 1.3×10^{14} n cm^{-2} (reactor neutrons).

An increase in the values of $V_{CE(sat)}$ under particle irradiation is important for 'saturation bipolar logic' devices such as the TTL series. In logic devices, when the pull-down transistor is turned hard 'on', the value of the voltage drop across the device is low and equal to $V_{CE(sat)}$. If the silicon forming those junctions increases in bulk resistivity, then that voltage drop will increase. This effect in turn causes a change in logic output and moreover produces higher power dissipation in the silicon.

In power transistors, changes of the above sort are also serious because, in 'high-voltage types', silicon of low resistivity is employed in the collector so that low values of breakdown are avoided. When the initial doping levels are low, a given particle fluence will alter the resistivity of the silicon more radically than that of a heavily doped material. Thus, in high-voltage power transistors, the increase in $V_{CE(sat)}$ proceeds more rapidly under neutron exposure than in 'low-voltage' amplifying or 'fast', logic devices possessing heavily doped collectors. For example, a fluence of 10^{12} n cm^{-2} may cause a 'high-voltage' device to undergo 100 per cent change in $V_{CE(sat)}$, while a 'low-voltage' device may undergo only a few per cent change.

5.5 Long-lived radiation effects in bipolar integrated circuits

Most of the mechanisms described so far have been for discrete bipolar devices. The degradation problem is often not so severe in integrated circuits, probably because oxides are of higher quality, collector currents are often high (greater than 1 mA) and the circuits are inherently tolerant of mild loss of gain. Degradation of bipolar integrated circuits results from radiation effects of a similar nature to those in discrete devices. However, because of circuit constraints, different device parameters may need to be employed for testing.

5.5.1 Digital ICs

The comment made earlier that degradation in digital integrated circuits is often less severe than in discrete devices is particularly applicable to bipolar integrated circuits. This is because the 'drive' currents are often high (usually in the mA range) and the logic intrinsically tolerant. Although I²L devices in particular are designed to operate at lower currents and logic voltages, commercial bipolar devices are normally much less affected by a given ionizing dose than their equivalents in the MOS field.

5.5 Long-lived radiation effects in bipolar integrated circuits

Emitter-coupled-logic (ECL) and transistor-transistor-logic (TTL) are generally tolerant to doses up to 10^6 rads. Special processing, in particular of the oxides, can improve this tolerance to make 'ultra-hard' logic.

Integrated-injection-logic (I^2L) is a form of bipolar transistor logic in which several transistors are 'merged' or fed by the same emitter, as in Fig. 5.9. This design offers a speed–power product comparable to other bipolar technologies, except that speed is a function of the bias point chosen. When a low-speed low power dissipation operating point is chosen, I^2L can be a serious rival to CMOS as a logic element for LSI. This technology is therefore of some concern because the very low collector currents used can make devices very sensitive to surface degradation. Stanley *et al.* (1977) showed that even a highly complex I^2L device, a microprocessor, would survive to well over 10^6 rad; Raymond and Pease (1977) found significant differences from manufacturer to manufacturer and these results are discussed below. While generally following the trends for bipolar devices, the details serve as a good example of the behaviour of integrated circuit transistors of small geometry and so are discussed in some detail.

I^2L devices should be less effected by ionization affects than MOS devices (that is, if given the same degree of care in control of surface oxide properties). On the other hand, because some wide-base lateral PNP transistors are included in the logic network, sensitivity of I^2L to displacement damage (see Chapter 3), will be higher than for other bipolar processes such as TTL. Raymond *et al.* (1975) and Pease *et al.* (1975) performed pulsed neutron and gamma testing of I^2L from a number of different manufacturers and their results are sumarized in Figs 5.10 and 5.11.

Berndt (1988), in his survey of bipolar hardening techniques, noted that, apart from the lateral devices, modern bipolar technologies have sufficiently thin base regions that gain is relatively unaffected by radiation.

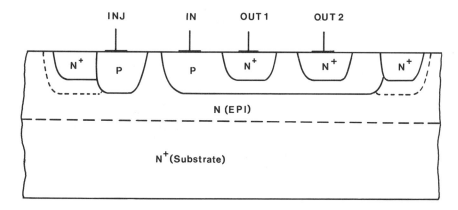

FIG. 5.9 Integrated injection logic: the structure of a typical element.

188 Bipolar transistors

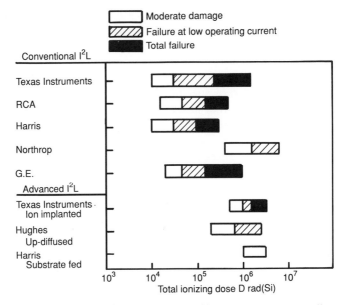

FIG. 5.10 Integrated injection logic: summary of long-term ionization effects induced by gamma irradiation in devices from various sources (Raymond and Pease 1977 © IEEE 1977. Reprinted with permission).

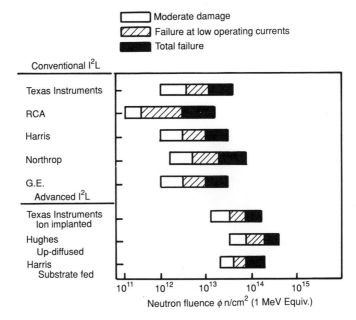

FIG. 5.11 Integrated injection logic: summary of long-term displacement effects induced by neutron/gamma irradiation (Raymond and Pease 1977 © IEEE 1977. Reprinted with permission).

5.5 Long-lived radiation effects in bipolar integrated circuits

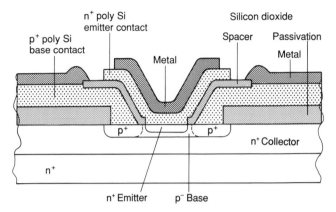

FIG. 5.12 Polycrystalline silicon (poly-Si) emitter transistor: cross-section of a typical annular, integrated design (after Enlow et al. 1991 © 1991 IEEE. Reprinted with permission).

5.5.2 Emitter-base surface effects in integrated transistors

Integrated bipolar transistors must be small in order for the ICs to compete in performance with corresponding MOS ICs. They also make use of heavily-doped polycrystalline silicon layers, a contact material often referred to briefly as 'polysilicon'. One geometry contrasting strongly with the 'discrete' geometries discussed earlier is the 'polysilicon-emitter transistor' shown in Fig. 5.12. Enlow et al. (1991) describe devices of this kind from several sources. A typical annular ('donut') design has an emitter diameter of only 11 μm. The passivation of the emitter-base junction is often thermally-grown silicon dioxide, as in the planar, discrete devices described earlier. Response to ionizing radiation is similar to the latter class of devices. Gamma-ray experiments (Enlow et al. 1991) showed severe degradation in β at about 10^5 rad (10^3 Gy), and time-dependent behaviour which is different from MOS circuits (see section 4.6.1).

5.5.3 Analog ICs

Bipolar analog (also known as 'linear') integrated circuit technology tends to be somewhat neglected in radiation studies and yet this technology provides us with the vital 'front-end' electronics without which there would be no signals to be processed by the digital electronics.

Van Vonno (1988) listed some typical bipolar analog functions as;

- fast operational amplifiers.
- fast sample and hold amplifiers and peak detectors.
- analog to digital converters of high accuracy.
- analog to digital converters of medium accuracy and high speed.

190 Bipolar transistors

These functions may present radiation effects problems in 'nucleonic' applications (accelerators, fission, fusion etc.) which frequently involve exposure to severe radiation environments.

Since most analog devices are in the form of high gain amplifiers, often working with very low input currents, it is not surprising that radiation-induced gain and leakage effects, described earlier, should strongly affect their operation. The primary parameters of interest include offset voltage V_{os}, input offset current I_{os}, and input bias current I_{IN} or I_B. These are, in general, highly sensitive to radiation. The strong influence of processing on radiation sensitivity was demonstrated by Palkuti *et al.* (1976) in a survey of operational amplifiers of the popular LM108 type. Extensive lot sampling over different dates of manufacture showed that about 30 per cent of the wafer lots tested were extra-sensitive while some lots of excellent radiation tolerance could be found. A detailed failure analysis, using local SEM irradiation was used to identify the most vulnerable transistors in the circuit. They were found in the high gain second stage amplifier. Low power operational amplifiers are extremely sensitive to ionizing radiation. Even in the tens of kilorad range, changes in DC parameters beyond specification limits occur in a significant number of devices. Comparators can also be sensitive to ionizing radiation. Failure modes include an increase in input offset current and input bias current, reduction in drive capability and latching of the output to the positive supply. Voltage regulator and reference devices are relatively unaffected by radiation. Price and Stanley (1975) have published the results of tests at JPL on operational amplifiers, comparators and voltage regulators using electron and Co-60 radiation. Gauthier and Dantas (1985) surveyed rad-hard analog to digital converters and found failure levels ranging from 30 krad (I^2L) to greater than 1 Mrad (bipolar). Allman (1992) performed a wide ranging survey of analog to digital converters including bipolar and mixed bipolar-CMOS (BiCMOS) technologies. He found that tolerance to gamma rays ranged from 5 to 50 krad.

The design of radiation-tolerant PNP transistors is one of the major problems in analog bipolar circuit development. These are already of poorer performance than NPN transistors and are significantly affected by radiation due to their 'lateral' structure, with sensitive field oxide covering the base region. Van Vonno (1988) discusses the hardening of analog circuits by means of new designs including the use of high-impedance JFET devices, differential amplifiers and base current cancellation.

5.5.4 Isolation technology

The radiation response of both digital and analog bipolar integrated circuits is strongly influenced by the choice of isolation technology. In the past, junction isolation has been employed and this is still used for many bipolar designs. Junction isolation is relatively tolerant to radiation but may lead to excessive leakage currents at a dose of the order of 1 Mrad. For 'ultra hard'

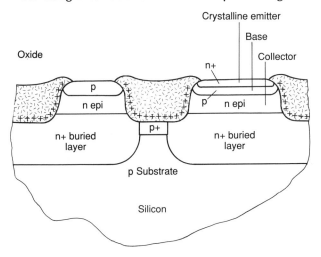

FIG. 5.13 Integrated bipolar transistor with recessed field oxide and walled crystalline emitter. The periphery of the junction between the p-type base and n+ emitter is in contact with the oxide wall, leading to surface degradation effects under radiation (Pease et al. 1983 © IEEE. Reprinted with permission).

designs, dielectric isolation or silicon on insulator (SOI) technologies are used. Modern bipolar technologies make use of oxide isolation, either recessed field oxide or trench isolation. These two forms of isolation are shown in Figs 5.13 and 5.14. The structures bear some resemblance to MOS structures shown in Chapter 4, herein lies a possible reason for the comparatively high radiation sensitivity of many oxide-isolated designs.

Oxide isolation reduces junction capacitance and allows much greater

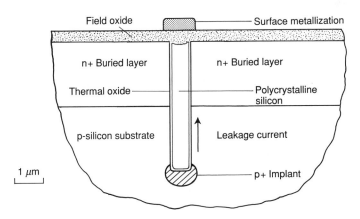

FIG. 5.14 Trench isolation structure: cross-section of an isolation trench, which acts as a parasitic MOSFET (Enlow et al. 1989 © 1989 IEEE. Reprinted with permission).

192 Bipolar transistors

TABLE 5.7 Summary of total dose failure levels in recessed-oxide digital bipolar microcircuits. (Pease et al. 1983)

MANUFACTURER	PART TYPE	TECHNOLOGY	FEATURES	MINIMUM THRESHOLD FAILURE LEVEL (krad(Si)	FAILURE MODE/ PARAMETER FAILURE
ADVANCED MICRO DEVICES (AMD)	2901C (4 BIT SLICE)	IMOX	SHALLOW OXIDES, WALLED EMITTER	80–100	C-E LEAKAGE IN OUTPUT BUFFER
	29823 (9 BIT REGISTER FILE)	IMOX	SHALLOW OXIDES, WALLED EMITTER	100–200	C-E LEAKAGE and FUNCTIONAL FAILURE
	29116 (16 BIT PROCESSOR)	IMOX	SHALLOW OXIDES, WALLED EMITTER	150	C-E LEAKAGE and FUNCTIONAL FAILURE
	93L422 (256 × 4 RAM)	IMOX	SHALLOW OXIDES, WALLED EMITTER	<20	C-E LEAKAGE and FUNCTIONAL FAILURE
FAIRCHILD	FAST LOGIC F00/F181 (QUAD 2-IN NAND/ 4 BIT ALU)	ISOPLANAR II SCHOTTKY	NO INPUT DONUTS	20–50	BURIED LAYER TO BURIED LAYER LEAKAGE
	F20/F373 (DUAL 4-IN NAND/ OCTAL LATCH)	ISOPLANAR II SCHOTTKY	INPUT DONUTS	<10	B.L. TO B.L. LEAKAGE
	F373	ISOPLANAR II SCHOTTKY	EXPERIMENTAL	100–200	B.L. TO B.L. LEAKAGE
	MEMORIES 93422A (256 × 4 RAM)	ISOPLANAR I	NESTED EMITTER, NO DONUTS	<20	B.L. TO B.L. LEAKAGE/V_{OH} AND BIT ERRORS
	93422	ISOPLANAR Z	FIRST PROCESS LOT, ENGINEERING SAMPLES	1000	NO PARAMETRIC OR FUNCTIONAL FAILURES

5.5 Long-lived radiation effects in bipolar integrated circuits

Manufacturer	Device	Process	Structure	Dose (krad)	Effect
FAIRCHILD	93Z511 (16 K PROM)	ISOPLANAR Z	WALLED EMITTER, NO DONUTS	20–50	B.L. TO B.L. LEAKAGE BIT ERRORS, "WINDOW" OBSERVED
	93Z511	ISOPLANAR Z	WALLED EMITTER, NO DONUTS	1000	NO PARAMETRIC OR FUNCTIONAL FAILURES
	MICROPROCESSOR 9445 (16 BIT μ PROCESSOR)	ISOPLANAR I²L	WALLED EMITTER ON I/O, NO DONUTS	50–70	FUNCTIONAL
MOTOROLA	74F00 (QUAD 2-IN NAND)	MOSAIC	NESTED EMITTER, INPUT DONUTS	10	B.L. TO B.L. LEAKAGE
NATIONAL	74ALS00/74ALS20	ALS	WALLED EMITTERS, INPUT/OUTPUT DONUTS	15	B.L. TO B.L. LEAKAGE
SIGNETICS	74F00/74F374 (QUAD 2-IN NAND/ OCTAL FLIP-FLOP)	SIMILAR TO ISOPLANAR II SCHOTTKY	WALLED EMITTERS, INPUT DONUTS OUTPUT DONUTS	<10	B.L. TO B.L. LEAKAGE
TEXAS INSTRUMENTS	ALS LOGIC ALS20 (DUAL 4-IN NAND)	ALS 1	NESTED EMITTER, INPUT DONUTS	20	B.L. TO B.L. LEAKAGE
	ALS161 (BINARY COUNTER)	ALS 1	NESTED EMITTER INPUT DONUTS	50	B.L. TO B.L. LEAKAGE

TABLE 5.7 contd

MANUFACTURER	PART TYPE	TECHNOLOGY	FEATURES	MINIMUM THRESHOLD FAILURE LEVEL (krad(Si)	FAILURE MODE/ PARAMETER FAILURE
	ALS138 (1 OF 8 DECODER)	ALS 1.5	WALLED EMITTER	100–200	ASSUMED C-E LEAKAGE
	ALS654/ALS574 (OCTAL TRANCEIVER/ OCTAL FLIP-FLOP)	ALS 1	NESTED EMITTER, EXPERIMENTAL PROCESS VARIATIONS	15 TO 1000	B.L. TO B.L. LEAKAGE
	SBP9989 (16 BIT μ PROCESSOR MICROPROCESSOR	NON-ISOLATED I²L	NESTED "EMITTER," n + SUBSTRATE	40 >1	SIDEWALL CURRENT and FUNCTIONAL FAILURE

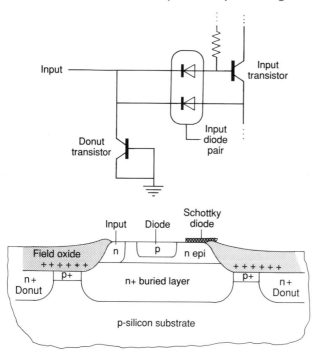

FIG. 5.15 'Donut' structure with input protection diodes (Pease et al. 1983. © 1983 IEEE. Reprinted with permission).

packing densities and higher speeds to be achieved. Pease *et al.* (1983) surveyed a wide range of recessed oxide technologies from different manufacturers (see Table 5.7), and found failure levels ranging from less than 10 krad to greater than 1 Mrad. Table 5.7 refers to technology as it existed in 1983 but the author confirms that its conclusions are still largely valid in the 1990s. The 'walled emitter' is shown in Fig. 5.13. In the 'nested emitter' technology the emitter diffusion is totally surrounded by the base. 'Schottky logic' (LS and ALS) has integrated schottky clamping diodes to limit the logic swing and improve speed. The 'donut', Fig. 5.15, is an annular buried guard ring surrounding the input protection diodes.

The most frequent failure mechanism found in Pease's study was increased leakage due to the inversion of the silicon as caused by charge buildup in the isolation oxide. This field effect behaviour has been described in Chapter 4. Enlow *et al.* (1989) studied isolation trenches lined with oxide and then filled with polycrystalline silicon. Again the predominant failure mechanism was inversion of the silicon due to field effects from the isolating oxide.

5.6 Transient upsets in bipolar integrated circuits

The effects of pulses of ionization in junction devices was discussed in Chapter 3 and Section 5.2. The basic behaviour of photocurrents in logic circuits was further discussed in Chapter 4. Being current-sensitive devices, bipolar integrated circuits (IC) can be extremely vulnerable to upset or latch-up induced by such currents. Much of the early work on radiation hardening of ICs was an attempt to reduce the effects of bipolar ICs of the currents induced by gamma-ray pulses. Most bipolar ICs have a large isolation junction which not only generates photocurrent but supplies the fourth junction necessary for 'latch-up' (see Sections 3.6.3 and 4.6). Junction-isolated ICs can exhibit latch-up as a response to a gamma dose rate as low as 10^6 rad(Si).sec^{-1} (see, for example Messenger and Ash 1989). For bipolar IC technologies dielectric isolation (see section 4.9) is an expensive way of removing the latch-up problem and increasing 'hardness' by an order of magnitude. This improvement by itself is not sufficient to meet the needs of all military equipment (see Chapter 15). Messenger and Ash quote Gover and Rose (1984) on the spread of failure values for various transistor-transistor logic IC technologies exposed to short and long flash X-ray pulses (all fail below 10^9 rad.sec^{-1}). Research on reducing junction responses and designing more tolerant circuits has produced devices which do not upset at 10^9 rad(Si).sec^{-1}. A 'burnout' effect caused by a single

TABLE 5.8 Summary of surface and bulk radiation effects in bipolar transistors

Type of effect	Phenomenon	Deleterious effects	Important radiation types	Damage units
Surface effects	Ionization in oxides	Changes in surface properties	Space Gamma radiation Electron beams	rad(Si)
Bulk effects	Atomic displacement	Changes in current-carrier properties (bulk damage)	Nuclear reactors Nuclear weapons Particle beams	equivalent 1 MeV electrons or neutrons
Transient effects	Photocurrent generation	No long-lived effects unless latch-up or burn-out levels are reached	Pulsed radiation	not applicable
	Rapid annealing	Short periods with low device performance	Pulsed radiation	see "Bulk Effects"

energetic ion in bipolar transistors has been reported by Titus *et al.* (1991) (see also Section 3.7.3).

5.7 Summary

In this chapter we have described the long-lived effects of radiation on bipolar transistors and how these effects may be separated into surface and bulk mechanisms as shown in Table 5.7. We have discussed how these effects influence the radiation response of bipolar integrated circuits. We have shown how new trends in isolation technology bring MOS failure mechanisms into the field of bipolar devices. As a consequence, radiation sensitivity increases dure to various forms of surface effect, including leakage. As in the case of MOS technology, the physical analysis of the responses of the insulator is a profitable approach to a method of systematic prediction. Bipolar transistors are normally more tolerant of gamma radiation than MOS transistors but improvements in the types with recessed oxide are needed, and have been achieved by process changes. Because junction areas are often large, transient effects in bipolar transistors are large, relative to MOS devices, and unsophisticated designs of integrated circuit may exhibit non-destructive logic upset at dose rates as low as 10^6 rad(Si). sec^{-1}. Small, modern bipolar transistors are often very tolerant to neutrons or other sources of bulk damage, but are sensitive to single-event upsets.

References

Allman, M. (1992). *Analogue to digital converters for space application.* ESTEC Working Paper EWP 1648. ESA-ESTEC, Noorwijk. Netherlands.

Berndt, D. F. (1988). Designing hardened bipolar logic. *Proceedings of IEEE*, **76(11)**, 1490–6.

Braünig, D., *et al.* (1977). *Data compilation of irradiation tested components. Report No. B-248.* Hahn–Meitner Institut, Berlin.

Brown, R. R. (1964). Proton and electron permanent damage in silicon semiconductor devices, Boeing Report D2-90570.

Brown, R. R. and Horne, W. E. (1967). Space radiation equivalence for effects on bipolar transistors, NASA CR-814. US Dept of Commerce, Washington, DC.

Brucker, G. J. (1967). Correlation of radiation damage in silicon transistors bombarded by electrons, protons and neutrons. In Proceedings of Symposium on Radiation Effects in Semiconductor Components, Toulouse.

Brucker, G. J., Dennehy, W. J. and Holmes-Siedle, A. G. (1965). High-energy radiation damage in silicon transistors. *IEEE Transactions on Nuclear Science*, **NS12**, 69–75.

Cooper, M. S., Retzler, J. D. and Messenger, G. C. (1979). Combined neutron and thermal effects on bipolar transistor gain. *IEEE Transactions on Nuclear Science*, **NS26(6)**, 4758–62.

Donovan, R. F. Hauser, J. R. and Simons, M. (1976). A survey of the vulnerability of contemporary semiconductor components to nuclear radiation, AFAL-TR-74-61. US Air Force Avionics Laboratory, Dayton, OH.

Enlow, E. W., Pease, R. L., Combs, E. W. and Platteter, D. G. (1989). Total dose induced hole trapping in trench oxides. *IEEE Transactions on Nuclear Science*, **NS36(6)**, 2415–22.

Enlow, E. W., Pease, R. L., Combs, W., Schrimpf, R. D. and Nowlis, R. N. (1991). Response of advanced bipolar processes to ionizing radiation. *IEEE Transactions on Nuclear Science*, **NS-38**, 1342–51.

Gauthier, M. K. and Dantas, A. R. V. (1985). Radiation hard analog to digital converters for space and strategic applications. *JPL Report 85-84*. Jet Propulsion Laboratories, Pasadena, US.

Gover, J. E. and Rose, M. A. (1984). IEEE NSREC Short Course, Colo. Springs, Sections 1, 2.

Gregory, B. L. and Sander, H. H. (1970). Transient annealing of defects in irradiated silicon devices. *Proc. IEEE*, **58(9)**, 1328–41.

Larin, F. (1968). *Radiation effects in semiconductors*. Wiley, London.

Messenger, G. C. and Ash, M. S. (1986). *The effects of radiation on electronic systems*. Von Nostrand Reinhold, New York.

Messenger, G. and Spratt, J. (1965). Displacement damage in silicon and germanium transistors. *IEEE Transactions on Nuclear Science*, **NS12**, 53–74.

Palkuti, L. J., Sivo, L. L. and Gregor, R. B. (1976). *IEEE Transactions on Nuclear Science*, **NS23(6)**, 1756.

Pease, R. L., Galloway, K. F. and Stehlin, R. A. (1975). Radiation damage to integrated injection logic cells. *IEEE*, **NS-22**, 2600–4.

Pease, R. L., Turfler, R. M., Platteter, D. G., Emily, D. and Blice, R. (1983). Total dose effects in recessed oxide digital bipolar microcircuits. *IEEE Transactions on Nuclear Science*, **NS30(6)**, 4216–23.

Poch, W. J. and Holmes-Siedle, A. G. (1968). A prediction and selection system for radiation effects in planar transistors. *IEEE Transactions on Nuclear Science*, **NS-15(6)**, 213–9.

Price, W. E. and Stanley, A. G. (1975). *IEEE Transactions on Nuclear Science*, **NS22(6)**, 2669–74.

Raymond, J. P., Wong, T. Y. and Schuegraf, K. K. Radiation effects on bipolar integrated injection logic. *IEEE Transactions on Nuclear Science*, **NS-222**, 2605–10.

Raymond, J. P. and Pease, R. L. (1977). *IEEE Transactions on Nuclear Science*, **NS24(6)**, 2327–36.

Snow, E. H., Albus, H. P., Yu, A. Y. L., Hurlson, R. E. and Tremere, D. A. (1970). Study of radiation effects on novel semiconductor devices. *USAF Contract Report No. AFCRL-70-0586*. Fairchild Research and Development, Palo Alto, CA.

Stanley, A. G. and Martin, K. E. (1978). Radiation damage in silicon transistors. *Radiation Physics and Chemistry*, **12**, 133–42.

Stanley, A. G., Mallen, W. and Springer, P. (1977). *IEEE Transactions on Nuclear Science*, **NS24(4)**, 1977–8.

Titus, J. L., Johnson, G. H., Schrimpf, R. D. and Galloway, K. (1991). Single event burnout of bipolar junction transistors. *IEEE Transactions on Nuclear Science*, **NS-38**, 1315-22.

Van Lint, V. A., Gigas, G. and Barengoltz, J. (1975). *IEEE Transactions on Nuclear Science*, **NS22**, 2663-8.

Van Vonno, N. (1988). Designing hardened bipolar analog ICs. *Proceedings of the IEEE*, **76**, 1496-1501.

Webster, W. M. (1954). On the variation of junction transistor amplification factor with emitter current. *Proccedings of the Institute of Radio Engineers*, **42**, 914-20.

Wirth, J. L. and Rogers, S. C. (1964). The transient response of transistors and diodes to ionizing radiation. *IEEE Transactions on Nuclear Science*, **NS-11**, 24-38.

6
Diodes, solar cells, and optoelectronics

6.1 Introduction

In this chapter, we combine our discussion of diodes and optoelectronics because the two series of devices are heavily intertwined in device history and function. Solid-state diodes take many forms. Some are 'optical' (i.e. are designed to react to light or to emit light) and others work in the dark. Before junction devices were invented, an optical vacuum diode, working by photoemission, was called an 'electric eye'. Optoelectronics comprise vacuum and solid-state devices which may be made either to emit or to sense light. The light often bears signal information. Frequently, an emitter and a sensor are used together as a transmitter and a receiver of information, for example an optical signal processor. Transmitter and receiver may be separated by an optical medium such as a light guide or a window, or by free space. Optical fibres and passive optical media are dealt with in Chapter 8, but we will mention here the details of some modern optical signal processing devices in which the transmission medium can be considered as a separate system component. For example, thin-film waveguide, couplers, and interferometers can be discrete or part of a completely integrated optical device (IOD). Optical gyroscopes contain integrated mirrors and beam-splitters. Other optoelectronic devices are sensor elements in cameras; these have high sensitivity to light. High-energy radiation may act both as a source of damage and of interference.

Optoelectronic devices in which the active elements are semiconductors are frequently sensitive to radiation because the absorption or generation of light in a solid medium is influenced by the defect structure of that medium. For example, light-emitting diodes operate by the recombination of excess carriers via impurity centres in a III–V compound. Particle-induced damage centres can interfere with this process. The detection of light in silicon usually operates by the collection of minority carriers generated by the light. Both bulk and surface radiation damage can interfere with this collection process. One case discussed below, the charge-coupled device (CCD), is an example of the multiple effects which radiation can have on a highly sophisticated optoelectronic structure.

In optoelectronic devices, a transparent window often forms part of the encapsulation around a sensor, so that optical media may be an intrinsic part of optoelectronics. This part of the science is well developed, since the first studies of radiation damage were on radiation-induced coloration in transparent (i.e. wide-band-gap, insulating) media. In general, vacuum tubes and other devices with glass envelopes are highly tolerant to radiation, but, unless this tolerance is mandatory, they are now rarely used, owing to their bulk and their requirements for high electrical power and voltage. However, miniature vacuum diodes and triodes are now being made (see Chapters 9 and 15).

Major test programmes have been carried out for space and military projects in the USA and Europe. Crouzet (1985/1980) performed a study for ESA in 1985. Barnes (1984) carried out a major review of radiation effects in optical devices. Friebele (1991) reviewed the subject and several major conferences have been held on the radiation hardening of optoelectronics (Evans 1982; Greenwell 1984/1986; Levy and Friebele 1985; Gillespie and Greenwell 1987; Bruce 1991), though with a strong emphasis on optical-fibre links.

6.2 Diodes: general

6.2.1 Introduction

The semiconductor diode performs a large variety of electronic functions. These include rectification or 'blocking', switching, photocurrent generation, light emission, and Zener breakdown at an electronic barrier, most commonly a diffused p-n junction. Other barriers include epitaxial and implanted junctions, the hetero junction, and the Schottky barrier. Materials include silicon, germanium, and all the compound semiconductors. p-n junctions from subelements of all integrated circuits (e.g. source-drain junctions of MOS devices), but this section will discuss mainly the discrete silicon p-n junction diode.

Unless heavy-particle irradiation is involved, rectifying action is not affected seriously by radiation; optical diodes, however, may be seriously affected (Eisen and Wenger 1982; R. Martin et al. 1983).

For transient radiation, Messenger and Ash (1986) noted that diode response is one of primary photocurrent and is dependent on diode area. Low-power diodes generate in the range $10^{10} - 10^{-13}$ A rad^{-1} s^{-1} and larger-power diodes in the range $10^{-8} - 10^{-9}$ A rad^{-1} s^{-1}.

6.2.2 Mechanisms

The changes in minority-carrier lifetime, τ, and resistivity produced in silicon by bulk radiation damage are reflected in the response of p-n junction devices to radiation. In a p-n junction, both forward and reverse I-V characteristics contain '$1/\tau$' lifetime terms. For forward current,

202 Diodes, solar cells, and optoelectronics

$$I_F \propto \frac{\sqrt{\mu}}{\tau}, \tag{6.1}$$

where I_F is the forward current for a given voltage and μ is the minority-carrier mobility. The reverse current is composed of diffusion and generation current. Both currents are increased if τ is reduced:

$$I_{gen} \propto 1/\tau.$$

The reverse diffusion current which flows when a diode is illuminated also contains a lifetime term:

$$I_{diff} \propto 1/\tau.$$

Thus, we can estimate the effects which a given amount of lifetime damage will produce on currents passing through diodes under bias, although, in the case of reverse leakage current, the magnitude of surface leakage is usually greater than either of the above terms and less predictable. Carrier removal will of course also increase resistance and hence the voltage drop produced by a current flowing across the base region. The principles described above apply also to the metal–semiconductor junction (Schottky barrier) used in some nuclear diodes, photodiodes and integrated circuits.

'Surface effects' on several device structures (e.g. rectifier diodes and transistors) are described in Chapter 5. The term embraces a wide range of effects which vary with the treatment applied to p–n junctions where they meet a surface. In oxide-passivated ('planar') diodes, it is predictable that surface leakage will be increased by irradiation. Normally, the magnitude of surface effects will be determined by the ionizing dose received.

6.3 Solar cells

6.3.1 General

A solar cell is simply a p–n junction of large area (e.g. $5 \, \text{cm}^2$) which is specially constructed to collect the maximum amount of light from the sun and convert the photogenerated carriers into current and power. In order to admit the light to the active region, the diodes commonly consist of a very shallow junction of heavily n-type material on a thicker region of more lightly doped p-type material. This 'n-on-p' or 'n/p' silicon solar cell is the form used to power most spacecraft flying in the 1990s, although 'p-on-n' was used originally and cells made from polycrystalline silicon, III, II–V and II–VI compounds are tried from time to time and are in experimental use on terrestrial power units. Because some light-generated carriers have to diffuse to the junction from large depths, cell efficiency is a function of the minority-carrier diffusion length and, by the same token, is strongly affected by radiation-induced defects.

6.3.2 Diffusion-length degradation and equivalent fluences

As described in Tada *et al.* (1982), a major indicator of solar cell efficiency is the short-circuit current, I_{SC}. This parameter is strongly related to diffusion length, L:

$$I_{SC} = A.\ln(L) + B. \tag{6.2}$$

Typical figures for a 10 Ω cm n/p solar cell in tungsten light are $L = 100\ \mu\text{m}$, $I_{SC} = 30\ \text{mA cm}^{-2}$ under illumination by a solar simulator at 135 mW cm^{-2}. Figure 6.1(a) and (b) show that, when the same solar cell type is irradiated by various particles, the two parameters degrade monotonically and in very similar fashion.

Shockley–Read–Hall recombination theory predicts that, when few excess carriers are present, diffusion lengths and minority-carrier lifetimes in silicon should change according to

$$1/\tau - 1/\tau_0 = K_\tau \phi, \tag{6.3}$$

$$1/L - 1/L_0 = K_L \phi, \tag{6.4}$$

where τ is the minority-carrier lifetime after irradiation, τ_0 the initial minority-carrier lifetime, K_τ the minority-carrier lifetime damage constant (dependent on material, type of particle, injection level, and temperature), L diffusion length after irradiation, L_0 the initial diffusion length, K_L the minority-carrier diffusion-length damage constant (dependent on material, type of particle, injection level, and temperature), and ϕ the particle fluence.

The level of excess carriers affects the damage constant. This was found in irradiations with neutrons by Curtis and Germano (1967) and in other measurements quoted by Tada *et al.* (1982). This measurement implies that degradation due to heavy particles will be worse for a solar cell operating a low light levels. Diffusion length in neutron-irradiated solar cells measured in sunlight is about twice that found at low light levels (Stofel *et al.* 1969); this comparison appears in Fig. 6.1(a), and the reasons are discussed in Chapter 3.

6.3.3 Background

Except in highly technical handbooks (e.g. Tada *et al.* 1982; Cooley and Janda 1963), there are few accounts available of the history of radiation effects on solar cells. The subject nevertheless has a fascinating history, and the events can be traced through the proceedings of the annual Photovoltaic Specialists' meetings (see e.g. Cuevas *et al.* 1990) and the European Space Power conferences (see e.g. Crabb 1991).

Interest in the effect of radiation on the solar cell began early in the study of power from diodes. This was not directly because of use in space (Bell Labs invented the device for remote telephone boxes) but because large-area

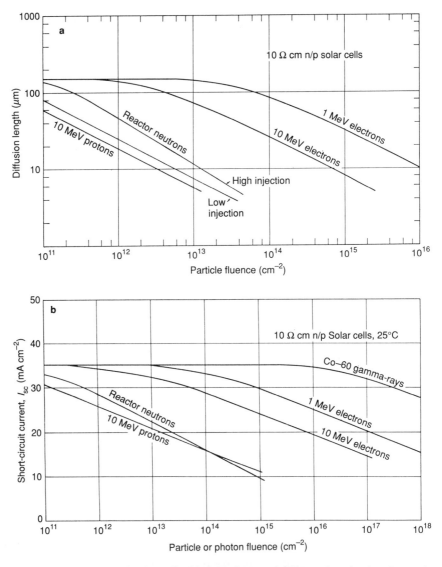

FIG. 6.1 Degradation of solar cells: (a) degradation of diffusion length of 10 Ω cm n/p solar cells versus fluence for protons, neutrons, and electrons; (b) the same for short-circuit current under 135 mW cm^{-2} solar simulator illumination, with gamma-ray results added (Tada et al. 1982. Reprinted with permission).

silicon diodes were tried as converters in isotope-powered microbatteries (P. Rappaport, unpublished work). The beta rays emitted by the chosen isotopes caused degradation of the p-on-n cells used, and experiments were started using the 1 MeV van de Graaff accelerators at Bell Labs (Rosenzweig, Gummel and Smits, 1963) and RCA Laboratories (Loferski and Rappaport 1958).

This was one of the first instances of a serious practical need to understand the damage in bulk silicon. The chances of success were propitious because the structure is so simple and the base region, where the damage was most effective, was not seriously affected by the device processing (a one-step diffusion). Much of the important physics of silicon defects has been investigated using this uncomplicated structure with its high sensitivity to bulk radiation damage and low sensitivity to surface effects.

The first solar cells flown in space were of the p-on-n type. These degraded rapidly in the van Allen belts owing to the degradation of minority-carrier lifetime in the n-type base region. It was realized that lifetime in p-type silicon was not degraded so rapidly. RCA then developed an n-on-p cell which had approximately ten times the endurance in space, and these were a major commercial success. The reason for the difference was the large cross section in recombination for the oxygen-vacancy centre in n-type silicon. Owing to the different Fermi level and charge states in p-type material, the oxygen centre hardly figures at all in the diffusion-length degradation. Degradation in the n/p solar cell, while improved, still did not meet all of the needs for long spacecraft missions. RCA, based on Vavilov's work (1962/1963) and indications found in lithium-drifted detectors, began the study of lithium-doped silicon. As hoped, lithium, being a mobile impurity, was found to move to the radiation-induced defect. Diffusion length, degraded by an exposure to electrons, was seen to recover by a large percentage in a few hours (Wysocki 1966). The efficiency of p/n solar cells doped with lithium, an n-type dopant, recovered similarly. The mechanisms were worked out by others (see e.g. Brucker *et al.* 1968). The method did not in fact replace the n/p solar cell, owing to the cost of lithium diffusion and worries about the long-term stability of the lithium–defect complex. Subsequent development of solar power for space lay mainly in improving the n-on-p silicon technology, the power/weight ratio, and orbital lifetime of large solar cell arrays with further exploration recently of III–V and II–VI semiconductors (Flood 1992).

6.3.4 Predicting the degradation of solar cell arrays

The cells in solar cell arrays in spacecraft are protected by thin cover glasses. The resistivity of the silicon base region may be 1 to 10 Ω cm. Degradation-versus-time predictions in a given orbit are made by a computer program in which all these factors are adjusted (Debruyn and Jensen 1983). Refinements of these schemes were used to predict and manage the high levels of

206 Diodes, solar cells, and optoelectronics

degradation suffered by the Hipparcos spacecraft when launched into a low-apogee elliptical orbit (Crabb et al. 1991).

The typical degradation of silicon cell current, with a slope of about 14 per cent per decade, is shown to hold for many different particles, as shown in in Fig. 6.1(b). Particle differences, mechanisms and new technology to 1982 are reviewed in Tada et al. (1982), which includes a description of computer programs for predicting damage in a given type of cell. More recent reviews are given by Weinberg (1990) and Flood (1992).

GaAs and InP solar cells for space application are the subject of intensive research. They offer higher efficiency than silicon and have higher radiation hardness when cover glasses are used (Loo et al. 1990, Weinberg 1991).

6.3.5 Equivalent fluences

It is often possible to 'simulate' the effect of irradiation of one particle by using another, nore easily available, type of particle. This is discussed more fully under bipolar transistors (Chapter 5). However, much of the research into this principle was performed usinq solar cells as test vehicles (Tada et al. 1982). See also the discussion in Section 3.2.2.4.

Table 6.1 shows the relative effects of various particles on solar cells. in the diodes in question, a given fluence of reactor neutrons is 'equivalent to' a larger fluence of 10 MeV electrons and a still larger fluence of 1 MeV electrons. In the case shown in Table 6.1, 1 reactor n cm^{-2} is equivalent to 2000 normally incident 1 MeV e cm^{-2}. The ability to convert in this way is useful when comparing test results.

The principle of equivalent fluences applies to most solar cells because, given this structure, atomic displacement effects ('bulk damage') are the dominant degradation mechanisms. The usefulness of test data in predicting

TABLE 6.1 Critical particle fluxes, ϕ_{crit}, required to produce 25 and 50 per cent degradation of short-circuit current under illumination of 10Ω cm n-on-p silicon solar cells at zero air mass[a]

	ϕ_{crit}(25%)	Damage ratio K_r(particle)	ϕ_{crit}(50%)	Damage ratio
		K_r(1 MeV e)		
10 MeV protons	8.0×10^{12}	6250	1.8×10^{13}	6666
Reactor neutrons	2.5×10^{12}	2000	6.0×10^{13}	2000
10 MeV electrons	3.0×10^{14}	16.7	8.0×10^{15}	15
1 MeV electrons	5.0×10^{15}	1.0	1.2×10^{17}	1.0
Co-60 gamma rays	1.8×10^{18}[b]	0.003	–	–

[a] Values taken from Tada et al. (1982)
[b] This fluence yields a dose of approx. 10^9 rad(Si) (10^7 Gy(Si))

performance will be influenced by (a) the 'purity' of the response of the device (in particular, surface effects can interfere with the accuracy of the bulk damage factors, see e.g. Sections 5.4 and 6.4) and (b) the need for test fluences to match as regards the amount of bulk damage deposited in the structure; for example a test in a reactor to a fluence of 10^{13} n cm^{-2} is, according to Table 6.1, equivalent in diodes to a 1 MeV electron fluence of 2×10^{16} e cm^{-2}. This is a much larger degree of damage than is accumulated in typical space missions. On the other hand, the dependence of damage on the energy of electrons and protons can be accommodated quite well in predicting damage in certain devices, see e.g Section 6.12.3 and Van Lint *et al.* (1980). The latter authors say: 'The most common need is to establish an upper limit for the damage produced by one particle type using experimental data obtained from another particle type'. They stress that calculation, although an approximate procedure, is a useful one.

6.4 Low-power rectifier diodes

Low-power rectifier diodes are usually of a planar structure and exhibit low leakage before irradiation. Thus, small radiation-induced alterations in surface charge may produce noticeable effects on the measured leakage even though such leakage does not often become a serious functional hazard. A typical result of electron beam irradiation, carried out by JPL on a sample group of GE 1N4148 signal diodes, is given by Stanley *et al.* (1976) and Price *et al.* (1981/1982). After a fluence of 10^{13}e cm^{-2} of energy 2.2 MeV, an ionizing dose of approximately 300 krad(Si), the mean reverse leakage current at 15 V increased from a fraction of a nanoampere to about 1 nA. However, the device did not exceed the specified value of 25 nA. Only a very small change in the forward characteristic was observed. The authors of the JPL test report commented that diodes, when exposed to such doses as those used, exhibit 'inherent radiation hardness' and that consequently not many tests on diodes were performed. The above is borne out by test results from other European programmes. However, Wagemann *et al.* (1973) also discuss the occurrence of 'mavericks', i.e. exceptionally sensitive diodes.

One US manufacturer advertises a radiation-characterized form of diode, type 1N5430. This diode has a guaranteed performance after a fast neutron dosage of 10^{14} n cm^{-2} ($E_n > 10$ keV); isotope gamma rays of the order of 10^5 rad can be assumed to accompany the neutrons). Typically, the reverse current at 50 V increases from 20 to 22 nA at 10^{14}n cm^{-2} and 32 nA at 10^{15}n cm^{-2}. Minimum and maximum limits of certain post-irradiation parameters are given in the data sheets.

Generally, low-power silicon rectifiers exposed to a fluence of 3×10^{12} reactor n cm^{-2} show a small increase in forward voltage drop of the order of 5 per cent. Such a change can normally be neglected. On the other hand, in some tests (Eisen and Wenger 1982), changes in reverse leakage were found to be as large as 500 per cent if the initial leakage was low, e.g. in the 10 nA range, the post-irradiation values being in the region of 100 nA. In this case, the effect would be produced by the ionization accompanying the reactor neutron fluxes used for testing.

It is to be expected that reverse leakage currents will be both noticeable and more severe when a diode is under a high reverse voltage or when low leakage currents may affect the circuit as, for example, in photomultiplier circuitry.

6.5 High-power rectifier diodes

Larin (1968) gives a calculation of the reduction of majority carrier density and minority-carrier lifetime in the very wide (100 μm) base region of an n^+-p-p^+ power diode after irradiation to a neutron fluence of about 10^{14} n cm^{-2}. Forward voltage drop at high current (100 mA) is degraded from about 0.7 to 2.5 V – a serious degradation. In high-power diodes, an increase in the forward voltage drop may be of greater significance than in low-power devices because power dissipation will increase noticeably. High diffusion length is required to maintain a low voltage drop. Eisen and Wenger (1982) report that an increase in forward voltage drop of 100 per cent at 1 A was observed in some power rectifiers exposed to 3×10^{12} n cm^{-2} (reactor). It is probable that designs which exhibit a much smaller change can be made if base-region doping is increased.

6.6 Zener diodes and diodes in avalanche breakdown

At reactor neutron exposures in the 10^{12} n cm^{-2} range, Zener breakdown voltages do not change more than 10 mV, well within the limits of the usual commercial devices. Braünig *et al.* (1981) show a similar change from a fluence of 10^{13} cm^{-2} of 2.2 MeV electrons. Shedd *et al.* (1971) give a model for the effect of radiation pulses in the region of 10^9 rad (Si). sec^{-1} on neutron-tolerant voltage reference diodes in the avalanche condition. The transient current is reduced with respect to normal photocurrent in reverse bias because a junction in avalanche is in a condition of low impedance.

6.7 Microwave diodes

Messenger and Ash (1986) note that microwave diodes (p-i-n, Impatt, Trapatt and Gunn) are inherently 'hard' to both total dose and neutron irradiation but damage in radar phase shifters was reported by Brucker *et al.* (1978).

6.8 Light-detecting devices

6.8.1 Photodiodes

In the photovoltaic mode, the photodiode operates on the same general principle as the solar cell. The photovoltaic mode is useful when the signal amplitude is required to reproduce the variations in light intensity. The speed of response, however, is limited by the time taken for carriers in the base to diffuse to the junction.

In the photoconductive mode, only the carriers created within the electric field of the junction (depletion region) are collected, which increases the speed of response. In addition, this removes some of the dependence of the performance on minority-carrier lifetime and thus reduces the effect of particle damage. Some effects will still occur, because leakage and series resistance are also increased by neutron bombardment and affect the light-generated signal.

In reactor exposures of photodiodes (Holmes-Siedle, unpublished data) neutron fluence of about 3×10^{12} n cm^{-2} (1 MeV) reduced the photocurrent, in the photovoltaic mode at 'short circuit', of type TIL 78 and other devices. At this fluence, the photo current was reduced by about 90 per cent. In another device, the Ferranti MS7B, operated at open circuit (roughly equivalent to photoconductive mode), the signal decreased by 32 per cent for the same order of neutron fluence. The changes of responsivity, dark current, series resistance, and linearity all showed smooth increases (steady damage constants) against equivalent 1 MeV neutron fluence (Korde *et al.* 1989).

Barnes and co-workers (Barnes 1984) developed special photodiodes of

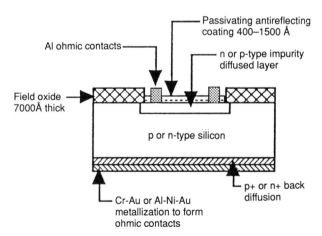

FIG. 6.2 Structure of a photodiode (Korde *et al.* 1989. © 1989 IEEE. Reprinted with permission).

210 Diodes, solar cells, and optoelectronics

various III–V compunds which were optimized to respond as little as possible to ionizing radiation. By the use of direct-bandgap semiconductors (e.g. GaAlAs, InGaAsP) as opposed to indirect (Si, Ge), the sensor layers could be made thinner, thus absorbing less ionizing radiation per light photon absorbed. Furthermore, the collection of minority carriers from volumes outside the depletion region was reduced by several methods (Wiczer *et al.* 1982). A wide collection of commercial and developmental photodiodes absorbing in the range 0.7 to 1.6 μm were tested in a pulsed X-ray beam at the same time.

It has been found possible to produce radiation-tolerant silicon photodiodes for electron-beam-scanned systems (Brucker and Cope 1971) and far-UV detection (Korde and Canfield 1989). The junction passivating oxide is protected from the beam by scanning from the rear and by adding a protective gold film. Figure 6.2 shows the structure of a photodiode and Fig. 6.3 shows typical degradation of the forward *I–V* characteristic of this device under neutron irradiation.

6.8.2 Detector diodes for high-energy physics

Nuclear scientists have developed diodes for nuclear particle detection which can withstand high fields at the junction without high leakage (Knoll 1989).

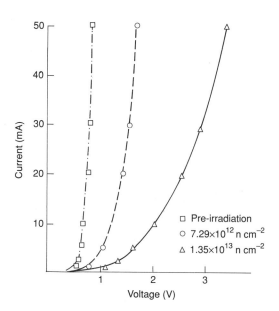

FIG. 6.3 Degradation of a photodiode: effect of reactor neutrons on the forward *I–V* characteristic of a photodiode on 5000 Ω cm n-type silicon (UDT RD-100). □, Pre-irradiation; ○, 7.29×10^{12} n cm^{-2}; △, 1.35×10^{13} n cm^{-2} (Korde *et al.* 1989. © 1989 IEEE. Reprinted with permission).

In early work, Schottky barriers were preferred because the 'dead region' was smaller than that obtained with junction technology. Kraner and others have developed shallow implanted junctions having very high base resistivities (e.g. p^+-n-n^+ structures) and very low leakage. Such devices are however very sensitive to the introduction of defects (Kraner *et al.* 1983, 1989; Lemeilleur *et al.* 1991; Hall 1992). Li and Kraner (1991, 1992) irradiated several types of p-n-diode having base resistivities varying from 10 to 30 000 Ω cm. The fluences of Pu-Be fast neutrons, which have an average energy of 4.5 MeV, ranged from 0.5 to 5×10^{11} n cm^{-2}. Leakage currents grew linearly with fluence. The leakage per unit fluence varied strongly for resistivity values above 2000 Ω cm, typically increasing fivefold. No changes in donor concentration were observed, that is, no type inversion occurred.

The parameter controlling leakage is the generation lifetime, which governs the rate of emission of holes in the depletion region. For accelerator experiments being designed, such as the US Superconducting Supercollider and European Large Hadron Collider (Heijne 1983; Groom 1989), the expected neutron fluences at the detectors can be as high as 10^{14} n cm^{-2} (1 MeV equivalent); the lifetime damage effectiveness of a Pu-Be neutron from the test source used above is about 1.7 times that of a 1 MeV neutron. Kraner *et al.* (1989) note that, at this fluence value, the room-temperature diode leakage currents could go as high as 7 mA cm^{-2}, which would probably make room-temperature operation impossible. Contacts could change their character and the resistivity of the base could alter radically (see for example Konozenko *et al.* 1973). Because incidental impurities, especially oxygen, play such a large part in such devices, research is needed to understand the changes taking place in nuclear physics detector devices.

6.9 Phototransistors

A phototransistor consists of an npn or pnp structure designed such that the base region is efficiently exposed to a light beam. The minority carriers produced constitute the signal normally supplied electrically by the base contact and, when the normal V_{CE} bias is supplied, collector current flows at a value proportional to the light intensity. The 'base current' is of course amplified by h_{FE} (see Chapter 5) and neutron damage will accordingly affect the responsivity of phototransistors. The gain of a phototransistor is in fact directly proportional to the minority-carrier lifetime in the base region (Stanley 1970). Under neutrons, light-activated relays (thyristors, etc.) should behave in the same general way as phototransistors.

The degradation of the output current of an LS600 phototransistor after exposure to 2×10^{12} reactor cm^{-2} was 80 per cent (Holmes-Siedle, unpublished work). Hardened phototransistors are reported (Matzen *et al.* 1991) which degraded from a gain of 225 to 140 after irradiation with 1×10^{13} n cm^{-2} or 1 Mrad(Si). Given the strong dependence of gain on minority-carrier lifetime, it is unlikely that phototransistors of greatly improved tolerance to radiation can be prepared, and this form of light sensor should be dispensed with when the predicted degradation cannot be tolerated. Since a phototransistor can be considered as a photovoltaic diode with a built-in amplifier, it is reasonable to assume that most signal applications, normally dealt with by the use of phototransistors, can be realized using a photodiode followed by a radiation-tolerant amplifying device (junction FET, high-frequency transistor). Photosensitive field-effect transistors are obtainable, but no radiation test data are available at present.

6.10 Light-emitting diodes (LEDs) and lasers

6.10.1 General

When a p–n junction in a III–V semiconductor is forward-biased, the carriers pass into and across the junction, and recombine near the junction. Given suitable defect centres, the recombination is accompanied by light of a photon energy slightly less than the band-gap energy of the semiconductor. The efficiency of conversion of electrical power into photon events is directly proportional to the minority-carrier lifetime in the active regions of the diode. if the semiconductor chip is cut so as to provide a resonant cavity, then the light emitted will be coherent and the device is termed a solid-state laser. III–V lasers have found more advanced uses than the LED, especially in directed energy applications, optical information storage, and communication (Barnes 1984).

Particles can affect the light output efficiency of emitters in two ways:

(1) as for transistors, neutron damage reduces the minority-carrier lifetime in the active regions;

(2) as particles produce new defects (recombination centres), the radiation-induced defects compete for carriers with the pre-existing luminescent defects, and this may further reduce the light output efficiency.

6.10.2 LEDs

The importance of the two damage effects will vary quite strongly with the impurities used, and it is not surprising therefore that some designs of light-emitting structure are found to be more affected by neutron irradiation than others. The more sophisticated epitaxial structures (often multilayer ones) are often strongly affected by a neutron exposure of 10^{12} n cm^{-2}. The non-epitaxial zinc-diffused gallium arsenide diodes are less strongly affected, while Si-doped GaAs diodes are damaged in the same way as the epitaxial type. Barnes (1977) has studied neutron/gamma effects, and Stanley (1970) tested LEDs using high-energy electrons. JPL has also exposed LEDs to 2.5 MeV electrons (K. Martin *et al.* 1985). Thomson and Janssens (1978) showed that there is significant injection annealing of LEDs which may balance the radiation degradation. Judicious choice of forward current and duty cycle could extend the life of LEDs in the space environment.

Barry and co-workers have employed an ingenious method of measuring the minority-carrier lifetime in light−emitters including GaAs and SiC LEDs (Barry *et al.* 1990, 1991) and thereby determined threshold energies of electron damage. Griffin *et al.* (1991) obtained neutron damage constants for GaAs.

6.10.3 III–V lasers

III–V lasers are fabricated by epitaxy of many layers of varying ternary or quaternary alloy composition. High-power semiconductor lasers have been examined for the effect of neutron damage by Sandia National Laboratories (Carson and Chow 1989). Threshold current density increased after exposure to 10^{13} cm^{-2} pulsed reactor neutrons (1 MeV equivalent). The authors also explain the mechanisms which can lead to lowering of the threshold for self-damaging thermal effects, based on increased recombination velocity in the surface facets.

6.11 Optocouplers

Optocouplers (opto-isolators) are devices which employ a light beam to achieve electrical isolation between a signal input and the rest of an electronic circuit. The usual types consist of a light-emitting diode facing a photodiode or phototransistor chip across a small thickness of an optical medium such as a polymer, glass or, in some cases, air. A new series of 'photologic' consists of an optical-fibre input and an electrical output, usually a logic voltage. The response of these devices to radiation damage is simply a combination of the degradation of the component parts. Thus, for example, it is not surprising that isolators employing phototransistors are more sensitive than those employing photodiodes (see earlier parts of this chapter).

A degradation of about 70 per cent in the current transfer ratio (I_{in} vs. I_{out}) was found after exposure of a phototransistor (type TIXL 101) to a neutron fluence of 2×10^{12} reactor n cm^{-2}. Because of surface effects, 10^5 rad gamma rays also produced degradation. Brucker (1978) estimated the threshold of damage to opto-isolators to be in the region of 5×10^{10} n cm^{-2} (14 MeV) for the photodiode types and of 10^{10} for phototransistor types. He derived these estimates from data by Barnes (1977) on light-emitters (threshold of damage 10^{11} n cm^{-2} (14 MeV)) and some of his own data on p–i–n diodes, showing some degradation at 10^{11} n cm^{-2} (14 MeV), by the use of a factor of two for relative damage effect. Brucker (1978) thus gives a figure of 5×10^{10} n cm^{-2} (14 MeV) for opto-isolators, but the above explanation shows that this is very much a worst case, i.e. the fluence at which the first detectable damage is observed in quite sensitive devices. The data on the TIXL 101 phototransistor show that, for some types

214 Diodes, solar cells, and optoelectronics

of opto-isolators, circuits which will survive much higher levels than 5×10^{10} n cm^{-2} could in fact be designed even with commercial devices.

6.12 Charge-coupled devices (CCDs)

6.12.1 General

An extensive treatment of CCDs is given below. This class of devices is a very good example of the interaction of the disorder caused by high-energy particles with a device designed for very precise sensing, for example the detection of signals equivalent to a fraction of a photoelectron (Janesick *et al.* 1991). As explained by Burt (1978) and shown in Fig. 6.4, a series of MOS field plates is used to shift or store charges in the semiconductor layer. A row of such plates can be clocked so as to move data ones and zeros along a row. The logic is the same as a dynamic MOS shift register but the structure is much simpler. Charges generated by light or particles are easily stored and shifted. Such optical CCDs are analogue sensors but the output is readily

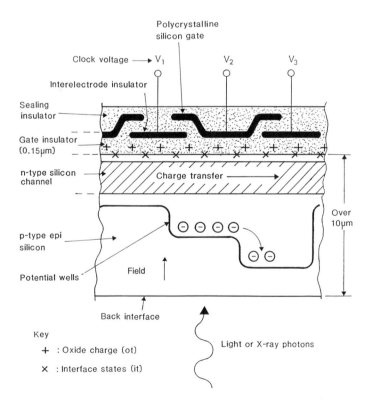

FIG. 6.4 Cross section of a back-illuminated charge-coupled device, showing polycrystalline silicon electrode structure, buried n-type channel, potential wells, and charged states near the Si–SiO$_2$ interface.

6.12 Charge-coupled devices (CCDs)

covertible to digital form. The uses of CCD imagers include miniature closed-circuit TV cameras suitable for robot guidance, key sensors in spacecraft control and stabilization systems (e.g. star trackers), and particle or electron beam imagers in accelerators (see e.g. Watts 1988; Andersen *et al.* 1991). Large CCD imaging chips containing 4096 × 4096 picture elements are being designed for the focal plane of astronomical cameras, including X-ray spectrometers. This application demands exceptionally high charge-transfer efficiency (CTE). The above scientific requirements have motivated research and test efforts aimed at understanding and reducing radiation effects in CCDs (Janesick *et al.* 1989, 1991; Dale and Marshall 1991; Holland *et al.* 1991; Hopkinson, 1991). A small charge-injection device (CID) has been used by several authors as a useful tool for bulk damage studies (Srour *et al.* 1986; Marshall *et al.* 1990); the CID is similar to the CCD and many models are common to the two technologies.

While CCDs and CIDs are metal-oxide semiconductor (MOS) structures, they contain many untypical device features such as a photosensitive depletion region and, sometimes, a thinned epitaxial substrate. MOS switching devices and CCDs are made by similar but not identical processing methods, so that the basic physical effects observed in the insulators are common to both types. However, radiation effects in the silicon sometimes, untypically, dominate the damage picture, as explained by Holmes-Siedle (1989).

Some differences between two classes are as follows. MOS switching devices are mainly used in digital logic, while the CCD imager is an analogue optical sensor. Digital devices tend to operate on a low, single rail voltage. CCDs employ at least ten independent control voltages, applied to a complex array of electrodes. The clocking of digital devices is always rapid; with CCDs, slow scan is common. Finally, unlike the switching device, the CCD is much more sensitive to changes in the minority-carrier transport and charge trapping capability of defects in the silicon; for penetrating photons, the carriers travel from relatively large depths in the silicon.

The parameters used in the slow-scan, ultra-low-light mode used for space science make this CCD mode particularly vulnerable to irradiation. For example, when detecting X-ray photon events, it is desirable to have performance levels as follows:

(1) ability to resolve the charge due to one electron in a well;
(2) charge-transfer inefficiency to be maintained very low — in the region of 10^{-6} electrons lost per pixel transfer

Leakages are particularly important because:

(1) CCDs are sensitive to small 'dark currents';
(2) the slow-scan mode multiplies the effect of dark current.

As Burt and co-authors (1978) explain, early designs of CCD exhibited

216 Diodes, solar cells, and optoelectronics

poor charge-transfer efficiency because of signal loss to interface state trapping. The burying of the channel by the use of implantation was adopted specifically to remove the potential well containing the signal charge from the influence of the interface states (they possibly extend a few atomic distances into the silicon).

Figure 6.5 shows a layered array of CCD chips on special ceramic substrates designed at the Rutherford-Appleton Laboratory and Brunel University, UK (see e.g. Watts 1988). The array is arranged to surround the beams colliding at the vertex of the Stanford Linear Accelerator Collider (SLAC) and detect the tracks of particles created. The array will not receive direct bombardment by the SLAC particle beam but has already accumulated kilorad levels of secondary radiation during the experiment.

6.12.2 Ionization effects in CCD structures

6.12.2.1 *General*

Many imaging CCDs have exhibited high sensitivity to ionization, with typical failure levels in the range 3 to 10 krad. Some work has been performed on radiation hardening, and hardened arrays are made for military applications. The most noticeable consequence of ionization is an increase in dark current owing to interface state generation (Saks 1980; Debusschere 1984; Janesick *et al.* 1988; Hopkinson and Chlebek 1989; Boudenot and Augier 1991). This is not the only reason for device failure, however. If high CTE is required, then displacement damage due to particles can be the dominant cause of failure. This important problem will be discussed in detail below.

FIG. 6.5 Charge-coupled devices as nuclear detectors: a layered array of CCD chips on special ceramic substrates designed at the Rutherford–Appleton Laboratory and Brunel University, UK (see e.g. Watts 1988). The array is arranged to detect the tracks of particles created at the vertex of the Stanford Linear Accelerator Collider (SLAC). Reproduced courtesy of S. Watts and the Rutherford–Appleton Laboratory, Chilton, Oxfordshire, UK.

6.12 Charge-coupled devices (CCDs)

Using 150 kV X-rays, Co-60, Sr-90, and protons to 50 MeV, it has been found that ionization effects are independent of the type of radiation and only a function of total dose (Debusschere 1984; Roy *et al.* 1989; Holland *et al.* 1990). Some initial studies showed that imaging CCDs can be made more tolerant to ionization using the well-established MOS hardening techniques (Saks 1980; Debusschere 1984).

CCDs can, surprisingly, withstand ionization effects from a direct scanning electron beam up to 10^5 electrons per pixel, given proper design. Stearns and Wiedwald (1988) tested the electron-beam imaging performance of several CCDs (RCA SID501, TI4849 and TK512M), using an SEM electron beam in the energy range 0.3–30 keV, impinging on the back, i.e. the surface not containing the electrodes. Imaging was successful and no severe ionization effects were reported (no exposure doses are given). Clark and Lowe (1991) exposed the front side of a TI CCD to 25 krad (250 Gy) of 8 keV X-rays and various exposures of soft X-rays from a synchrotron beam, with no degradation of its imaging power. Ravel *et al.* (1991) have demonstrated that CCDs can be used in the place of diode arrays in vacuum tubes known as electron bombarded semiconductor imagers which employ electron beams of energy about 5 keV and are both 'sensitive and robust' despite the absorption of large doses from Si K_α X-rays in the oxide layer. Allinson *et al.* (1991) characterized interface damage in CCDs which interfered with the recording of X-ray diffraction patterns; they also developed simple routine methods of annealing this damage between measurements.

6.12.2.2 *Oxide trapped charge (ot)*

Radiation-induced oxide trapped charge resides in the 'bulk' of the oxide (i.e. out of electrical communication with the silicon–silicon dioxide interface). In Deal's notation (Deal 1980), shifts due to this charge receive the subscript 'ot'. CCD insulators are often made of two layers to reduce pinhole defects, but this model probably serves adequately. In the above model, the change in flatband or threshold voltage induced by oxide trapped charge at low doses can be expressed as follows (Freeman and Holmes-Siedle 1978):

$$\Delta V_{ot} = R.A.D, \quad (6.5)$$

where R contains constants of the device, A is the probability of charge trapping, and D is the dose. Although the model is designed for cases where voltage is applied to the gates during irradiation, we can also quote constants for the unbiased case. The mechanism of the bias effect in a grown oxide is straightforward. An applied field in the oxide separates the carriers induced by the radiation in the insulator (see Chapter 4).

To minimize pinholes, CCD devices may employ insulator–insulator sandwich structures under the gate electrode. The new interface introduces a new site for charge traps. The bias-dependence of sandwich insulators is

218 Diodes, solar cells, and optoelectronics

usually different from that of single oxides (see e.g. Poch and Holmes-Siedle 1968).

If the potential diagram of a buried n-channel CCD is studied, it will be noted that the field in the oxide is equivalent to the negative irradiation bias condition (V_1-; see Chapter 4). This is because a higher DC positive bias is applied to the channel than occurs at any time in the clock voltages applied to the gates. This alleviates radiation effects in some insulators.

Irradiation with a square-wave field on CCD gates has been tested (Holland *et al.* 1991). For calculating magnitudes, it appears sufficient to assume that the field is on the whole of the time.

6.12.2.3 *Interface trapped charge (it)*

Effect on surface potential. In Deal's notation (Deal 1980), charge trapped in these states receives the subscript 'it'. Depending on the sign of the interface charge, ΔV_{it} may add to or subtract from the effect of ΔV_{ot}. Freeman and Holmes-Siedle (1978) and Holmes-Siedle and Adams (1983) give simple expressions and data for the buildup of interface states in MOS transistors as a function of dose. Killiany *et al.* (1974), Debusschere (1984), and Boudenot and Augier (1991) give a description of the effect of radiation-induced interface states on CCDS. A fuller discussion is found in Chapter 4.

Effect on surface generation. In CCDs used to integrate light or X-ray signals over long periods of time, sources of dark current are a significant source of signal degradation (Hopkinson and Chlebek 1989; Boudenot and Augier 1991). The interface states are recombination–generation centres. Surface states in n-channel CCDS, given the right conditions, will give rise to many electrons per second at room temperature. Radiation will of course increase these currents by creating new states (Saks 1980; Janesick *et al.* 1991).

6.12.2.4 *MOS transistor elements in CCDs*

The effect of positive oxide charge on MOS transistors in general is explained in Chapter 4. Increasing dose leads to increasing negative shifts in the I–V characteristic. CCDs with n-type charge-transfer channels employ n^+ junctions for the FETs and diodes employed. Positive charge build-up in the insulator causes the threshold voltages, V_T, to shift in the negative direction. If the FET is designed to be in the 'OFF' state when V_T is zero, this condition may be disturbed after irradiation (the VTNZ condition; see chapter 4). This condition could disturb 'reset' and other clocking operations used in the output network. However, it appears that in many CCD designs, moderate shifts V_T can be accommodated by adjusting the power supply voltages (Holland *et al.* 1990). Amplifier FETs used in the CCD output exhibit an increase in noise under ionizing radiation, the amount depending on device design (Roy *et al.* 1989; Holland *et al.* 1991).

6.12.2.5 Charge-transfer channel

The effect of insulator charge on the charge transfer operation does not consist of degradation in efficiency but is a shift of potentials, which may or may not affect operation. This is illustrated in Fig. 6.6. The lines drawn below the gates represent the channel potential ϕ_s versus lateral distance (x) (Roy et al. 1989). The potentials before irradiation are represented by solid horizontal lines and, after irradiation, by dashed lines. The calculated value of the potential is about 10 V higher than the voltage applied to each gate. Thus, for the gate shown as R02, the surface potential ϕ_s is

$$\phi_s = V_G + 10 = 22 \text{ V}$$

with +12 V clock voltage as shown by the solid line.

Radiation-induced positive charge in the gate insulator can be represented by an additional voltage on the gate. This gives an increase, V_{rad}, in the band-bending due to the field produced by the charge sheet. Thus, as shown by the dashed line,

$$\phi_s = V_{rad} + V_G + 10 = 32 \text{ V}$$

(+12 V clock voltage and +10 V from radiation-induced charge). This increase in ϕ_s has no effect on the storage and transport of charge from gate to gate, so long as they are all affected similarly. However, at the end of the row, the signal will not be efficiently transferred to the output diode if the potential of the final well is not higher than that of the conduction band in the n+ diffusion. The former is controlled by the output gate, having a bias voltage V_{OG}; when clocking the signal out, the diode potential is set by the reset drain voltage V_{RD}, shown here as +17 V. The setting represented by the solid lines gives efficient transfer of signal charge from the output gate ($\phi_s = 12$ V) to the output diode ($\phi_s = 17$ V). The setting represented by the dotted line gives inefficient or zero transfer of charge. In other words,

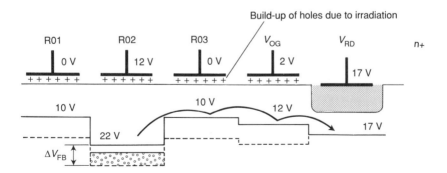

FIG. 6.6 Potentials in a charge-coupled device (CCD): potential values for the wells near the output gate before and after high-energy irradiation (after Roy et al. 1989).

220 Diodes, solar cells, and optoelectronics

irradiation has interfered with charge transfer to the output diode. However, this can be rectified if the values of V_{OG} and V_{RD} are adjusted to restore the previous conditions.

6.12.2.6 *Voltage shifts*
We will now discuss the magnitudes of charging and voltage shifts expected, using the simple formula (eqn (6.5)) explained earlier:

$$\Delta V_T = R.A.D.$$

It was also noted above that the amount of charge trapped below a CCD gate electrode depends on the value and length of time for which a voltage is applied during irradiation, and possibly after irradiation.

For the simple case of a grown SiO_2 insulator of thickness 140 nm, the figures for the model developed by Freeman and Holmes-Siedle (1978) are:

V_I *positive*:

soft oxide $A = 1.0$ $R.A = 500\,mV\,krad^{-1}$ ($50\,mV\,Gy^{-1}$)

hard $A = 0.01$ $R.A = 5\,mV\,krad^{-1}$ ($0.5\,mV\,Gy^{-1}$)

V_I *zero*:

soft   – $R.A = 100\,mV\,krad^{-1}$ ($10\,mV\,Gy^{-1}$)

hard   – $R.A = 1\,mV\,krad^{-1}$ ($0.1\,mV\,Gy^{-1}$)

Cooling the CCD causes a slowing down of trapping processes in the oxide, tending to 'freeze-in' the hole charge. The effect this has on cooled CCDs is discussed in Chapter 4.

6.12.3 Displacement damage in CCD structures

6.12.3.1 *General*
Bombarding particles which impart large amounts of kinetic energy to the silicon atoms can produce imperfections in the silicon lattice of a wide variety of types, collectively called 'displacement damage' (van Lint *et al.* 1980). In the case of heavy particles encountered in space vehicles and in ion implantation, the centres may form clusters. Most of these defect centres are intensely active as minority-carrier recombination–generation centres or majority-carrier removal centres. For CCDs, the carrier collection and transfer processes are profoundly affected by these defects. The issues connected with space proton damage in CCDs are reviewed in Chapter 3.

Two points are worth noting here:

1. Cluster defects produced by neutrons have been implicated in near-catastrophic damage effects occurring in CCDs on vehicles powered by radioisotope generators (Janesick *et al.* 1991).

6.12 Charge-coupled devices (CCDs)

2. Large and complex defects can be left in silicon wafers by the implantation process, even though the high-temperature heat treatment given after implantation is intended to eliminate these defects completely. This type of disturbance of the silicon in the buried channel should not be ignored when it is realized how sensitive the CCD mode of operation is to small perturbations in the silicon. The defects present before irradiation are often referred to as 'bulk states' (McNutt and Meyer 1981; Janesick *et al.* 1991).

The electrical changes due to displacement damage in silicon include:

(1) increase in recombination and generation of minority carriers, leading to decreased minority carrier lifetime and diffusion length;
(2) carrier removal, leading to increased resistivity and also to the temporary removal of carriers, followed by re-emission.

One of the most important parameters is the CTE or CTI of the devices. The changes in other parameters are noted; for example dark current spikes are created if the temperature is above $-50°C$.

The impact of displacement defects is different in various regions of the CCD, such as the channel region, depletion region, substrate–epitaxial layer boundary, substrate, etc. The effect of a gradual build-up is usually gradual degradation rather than sudden catastrophic failure, but the effects are shown below to be serious.

Displacement damage creates vacant sites in the silicon lattice (hence 'bulk damage'). In phosphorus–implanted buried channels, complexes of vacancies with the dopant phosphorus (an E-centre) are the most effective in increasing CTI (Janesick *et al.* 1991). These defects act as generation and trapping centres. The generation centres also produce local 'hot spots' of dark current, which when isolated to single pixels appear as dark current spikes above the more uniform dark current background arising from the interface states.

While the generation of these defects in CCDs has been studied in the past (see e.g. van Lint *et al.* 1980), the impact of bulk defects in a modern generation of scientific CCD imagers is particularly important, owing to the increased sensitivity and quality of the new devices (Holland *et al.* 1990). For example, the effect of a single bulk trap is proportionately greater in the X-ray CCD, where the charge count is determined to a few electrons, than in the early devices, which possessed much poorer as-processed CTE values.

In recent series of tests on CCDs in the USA and UK, devices were irradiated with protons from van de Graaff generators or cyclotrons (Janesick *et al.* 1991; Holland *et al.* 1991; Dale *et al.* 1989, 1990). An energy of 10 MeV was chosen as the prime value for investigating broad trends and differences between device technologies, being judged particularly suitable from studies of the proton spectrum expected in the Earth's inner proton belts. Ancillary

222 Diodes, solar cells, and optoelectronics

experiments were done at 1.5 MeV, and at 6.5 MeV and 210 MeV at other facilities (Holland *et al.* 1990).

Since the main application was the use of CCDs as imaging spectrometers, a new test programme was designed to test the key property, CTE or CTI, in conditions as close as possible to space use (Holland *et al.* 1991). To achieve this, chips were irradiated at −90°C, under power. The devices were tested with the aid of a weak Fe-55 source mounted in the cryostat which produces manganese X-rays of energy 5.9 keV. The X-ray data were recorded and archived on site.

Proton damage as a function of temperature was seen to be an important parameter, since the release time constants of any proton-induced traps change with temperature. The resulting data indicate the type of traps which are generated by the damage and also indicate the best operating temperature after irradiation. 1990 tests on a device possessing a charge confinement structure had indicated that the CTE of the devices could be 'hardened' to radiation damage by an amount proportional to the degree of charge confinement (Holland *et al.* 1990). Further devices were irradiated to confirm this result, using two charge confinement techniques. The first (reported in the 1990 results) is referred to as a narrow buried channel, where the buried channel implant in the parallel direction is reduced from a width of 17 μm down to 5 μm. The second technique, referred to as a notch (Janesick *et al.* 1991) or supplementary buried channel involves depositing an additional implant strip 5 μm wide along the centre of the standard 17 μm buried channel. These devices had the supplementary channel in the serial register. For small signals, as is the case in X-ray spectroscopy, the transferred charge concentrates in the central 5 μm and thereby encounters fewer traps when being clocked to the output node.

6.12.3.2 *CTI growth with bulk damage: a model*

The system engineer needs a comparative measure of the effectiveness of the array of equipment hardening methods which are avialable. The first prediction tool needed is a model of CTI growth versus proton fluence at some standard particle energy. We present a model below which allows quantitative predictions of the effect of changing design parameters, including device technology and temperature. From experiments using 10 MeV protons, Holland *et al.* (1991) derived a prediction model, which is shown in Fig. 6.7. The curve has the following form:

$$\Delta(\text{CTI}) = 2 \times 10^{-13} \Phi_{10}, \tag{6.6}$$

where Φ_{10} is the fluence of normally incident 10 MeV protons used. We can now extend this model to other proton energies, temperatures, and clocking rates, and possibly other CCD types. If we introduce factors based on the experiments described, then the prediction equation becomes

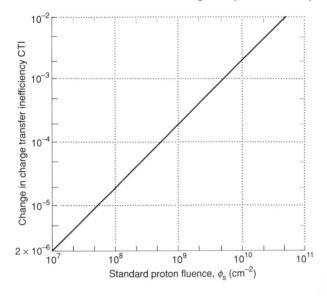

FIG. 6.7 Prediction of charge-transfer inefficiency: model for the growth, vs. equivalent 10 MeV fluence, of charge-transfer efficiency of the EEV P86000 (CCD-02) charge-coupled device, measured without warm-up after bombardment at −90°C.

$$\Delta(\text{CTI}) = 2 \times 10^{-13} \sum (\Phi_E \cdot F_{\text{NIEL}}) \cdot f_T \cdot \frac{L_{\text{pix}} \cdot W_{\text{chan}}}{22 \times 17} \cdot \frac{f_{t(\text{clock})}}{t_{\text{release}}}. \quad (6.7)$$

The damage constant is the slope of Fig. 6.7, explained above. The term $\Phi_E \cdot F_{\text{NIEL}}$ is derived by converting the proton energy spectrum into a number of finite energy bins and assigning to each bin an average factor, F_{NIEL} for the non-ionizing energy loss relative to 10 MeV protons. The result is a fluence value of 'Damage-equivalent normally incident 10 MeV protons'. The temperature factor f_T, relative to −90°C can be calculated from eqn (6.6), at least over the stated range. Experimental evidence suggests that charge loss scales directly with the area of the transfer channel, so that the terms $L_{\text{pix}}/22$ and $W_{\text{chan}}/17$ perform a simple normalization to the geometry of the device in question. The reduction of W in the narrow buried channel samples, described above, confirms our assumption.

It has been demonstrated that the CTI may be affected by the clocking rate (Janesick *et al.* 1991). This mechanism works by having a minority of the signal packets undergo charge trapping and then clocking out the rest before the traps have released their charge. In the case of the X-ray CCD for astronomical applications, the majority of sources observed will be weak, producing only a few X-rays (a few thousand electrons) per image against a totally dark background. In this regime, the traps will frequently be unoccupied and it is believed that minor increases in the clocking speed will

224 Diodes, solar cells, and optoelectronics

not strongly improve the global CTE. The results of CTI in the serial register suggests that there may be a weak dependence on clocking rate. A more general result is produced if we include the trap release time in the equation.

6.12.3.3 *Other CTI growth models*
We discuss here the comparative merits of the above model with that developed by Janesick *et al.* (1991). The model also assumes linearity in growth of CTI with proton fluence and scaling with pixel area. Temperature and clock rate are not modelled, but the dependence of charge loss on defect numbers is explicitly modelled. Janesick and co-workers use the TRIM program to predict the number of displacements produced in the pixel volume. The actual charge loss proves to be far lower than that predicted, because vacancy annihilation is not accounted for. They thus introduce a radiation trap inefficiency factor, R, to account for this. Thus, while we introduce an empirical damage constant, 2×10^{-13}, Janesick and co-workers employ a theoretical damage constant D and a normalizing factor R. The results of the two models for different device types at energies near 10 MeV agree quite well, indicating that a general model for predicting CCD responses to particles may be possible in the near future.

The shape of the energy-dependence curve as the proton stopping range approaches the depth of the inactive 'front shielding' is still a matter of some debate. Janesick employs a 'Proton Transfer Curve' which is specific for a given 'dead layer' at the surface of the CCD (Janesick *et al.* 1991). This assigns zero damage to a proton of energy less than about 0.2 MeV (this cut-off is given by TRIM when modelling a monoenergetic, normally incident proton). Our model is based on the assumption that, for a CCD shielded behind an absorber equivalent to a sphere of several millimetres of aluminium, the presence or absence of a few micrometres of silicon or oxide at the surface cannot possibly affect the response to a broad external proton spectrum which, after transport, passes into the active region. We therefore use the dependence of non-ionizing energy loss (NIEL) on proton energy. Thus, we recommend the assignment of a finite damage factor to protons of all energies which turn up in the active microvolume.

6.12.3.4 *Goals for bulk damage limitation*
Studies on a spacecraft mission dominated by protons have set preliminary goals for a mission environment for scientific CCDs of not nore than 10^9cm^{-2} of damage-equivalent normally incident 10 MeV protons unless other means (say cooling or special technology) can be found to alleviate the inefficiency introduced by this environment (Holland *et al.* 1991).

6.12.4 Conclusions on CCD degradation

The mechanisms which threaten CCD action under a radiation environment are:

(1) oxide charging and the resultant field effects;
(2) interface states and the resultant dark currents;
(3) silicon atom displacement damage;
(4) interelectrode leakages.

Charge-coupled devices are a complex form of metal-oxide semiconductor structure with analogue functions which are highly sensitive to disturbance by environmental effects. Space radiation and isotope radiation will affect sensors such as scientific X-ray telescopes, star and beacon trackers, and video cameras. The magnitudes of these effects are significant but manageable. Although the effects are manageable, we emphasize that there is a finite risk of their getting out of hand unless extreme care is taken during the design and qualification of sensors based on CCDs. After considering research such as that described above, many space agencies have decided to adopt CCDs as primary sensors, a decision which is adventurous but correct considering the benefits gained from the technology.

6.13 Electro-optic crystals

Inorganic and organic materials possessing certain types of crystal symmetry may demonstrate the electro-optic effect. Examples of this effect are birefringence, doubling of laser light frequency, light-valve, piezoelectric and pyroelectric action, and other ferroelectric effects. Little study has been performed of space radiation effects on these materials as a class. Apart from a predictable darkening of some crystals due to colour centres or radiolysis, no known pattern exists in the radiation effects to be found in the megarad range. Test data, especially for quartz, will be found in many compilations (see Chapter 8). $LiNbO_3$, a widely used optical material, has been tested with gamma rays and showed little effect on power handling up to a dose of 1 megarad (Grabmayr 1990). Tests of implanted light guides in $LiNbO_3$ are described by Roeske *et al.* (1987)

6.14 The new optics

The computer laboratories of the world are now investigating a wide range of uses for optics in electronics. In designing very fast computers for data handling, there are several advantages in using beams of light in the place of metal tracks to interconnect the component parts of computers. Out of this requirement has grown a series of inventions which go to make up 'optical computing' (Jahns and Huang 1989).

Many of the semiconducting optoelectronic devices concerned are made from gallium arsenide and related compounds. These are generally insensitive to radiation damage but may show a response to high-energy radiation at a high dose rate (responding to the radiation in the same way as to light). This situation can be avoided in space and so this class of devices is unlikely to be 'soft'. By contrast, some of the transparent dielectrics used in optics, including various forms of doped silica, may be strongly affected by radiation damage.

Germania-doped silica is much more strongly affected than pure silica, as discussed in Section 8.4. This could lead to a major loss of ability to transmit a light signal. Similar degradation affects many normally transparent dielectrics which contain 'doping' or natural impurities (Friebele 1991). The foreign atoms form 'colour centres' which absorb light (see Chapter 8). Natural quartz, for example, becomes purple under irradiation (Friebele 1991) because of colour centres derived from the aluminium and iron impurities.

Some of the plans for optical computers involve the mounting of the optical switching and memory elements on a quartz slab. The technical term for this structure is 'planar integration of free-space optical components' (Jahns and Huang 1989). The light will pass many times through this slab, giving a long optical path and hence the opportunity for light loss. If such a computer is to be used in a radiation environment, then losses from colour centres will have to be foreseen and avoided by the correct choice of material for the slab and reduction of the total optical path for a given signal.

6.15 Vacuum devices and extreme environments

Vacuum tubes, such as TV tubes, photomultipliers, 'vacfets', and photocells are intrinsically tolerant to radiation (see Chapter 15) The is because the parts frequently comprise only pieces of metal, spot-welded together and evacuated within an inert 'bottle'. The mechanisms of light generation and sensing in vacuum devices are much less strongly affected by radiation than in the solid state. However, the transparent envelope may be darkened in the window region. It is also likely that the lenses and accompanying circuitry are affected by radiation and, in most of the 'hardened TV cameras' on offer, it is the latter parts which have been modified for radiation tolerance.

Where imaging is required in very intense environments, say for visual inspection within a reactor, accelerator or radiographic system, it may be possible to avoid exposing lenses, sensors or circuits, either by the use of radiation-tolerant fibre bundles or scanning laser beams. In radiology and radiotherapy, it is desirable to intensify the shadowgraphs of the item imaged, and this leads to damage of the intensifier system unless the layout is specially designed (Holmes-Siedle and Stewart, unpublished reports).

Vacuum display devices are unlikely to play a major part in equipment exposed to space or nuclear radiation, but cathode ray tubes carry their own radiation damage problems with them. The electron beam may damage the target, burning-in images, and the bremsstrahlung produced may darken the glass screen or damage other local parts.

6.16 Conclusions

In this chapter we have discussed many different types of diode and some other structures as well. Silicon diodes, by virtue of their simple structure and potential for large area, are used in both the optoelectronic mode and in the dark. In the latter case, only the rectifying characteristic of the p-n junction is used. In this mode, the degradation of performance induced by irradiation is generally much less than that of transistors. Only at high irradiation levels do these diodes develop leakages or undue forward resistance when irradiated under demanding stress conditions. On the other hand, in the optical mode, solid-state light-converting devices react badly to particle damage — another example of the general tendency of high-energy radiation to create disorder, this time in a structure which is frequently dependent on high crystalline order for the transport of carriers over relatively long distances.

The discussion in this chapter shows that optoelectronic devices in a hostile environment are subject to multiple effects and that particles or photons may tend to dislocate a highly tuned, high-technology system. It is clear that the technology continues to develop dramatically because of the great effectiveness of optoelectronic devices (for example the charge-coupled device in science) and present new problems because of the precise adjustments demanded of the materials technology. No one optoelectronic system is preferable for radiation environments, although vacuum devices will generaly be more tolerant. However, while the latter will serve for very 'hot' environments, solid-state devices will be needed where miniature scale is needed. For example, the CCD will replace the TV tube in cameras entering reactor environments. The use of fibre-optic links is expanding and is favoured for industrial data links; optical computing (optical links on the micro-scale) will expand in the future. Some general cautions include the following: when considering normally inert transparent elements under radiation, the engineer must remember that they luminesce as well as darken, especially in the blue; light sensors will always be extra-sensitive to transient radiation effects, treating the radiation-induced carriers as extra light signals; and solid-state light-sensing devices often react badly to particle damage because of interference with the transport of light-generated carriers.

References

Allinson, N. M., Allsopp, W. E., Magorrian, B. G. and Quayle, A. (1991). Effects of soft X-ray irradiation on solid state imagers. *Nuclear Instruments and Methods in Physics Research*, **A310**, 267–72.

Andersen, J. N. *et al.* (1991). Photoemission spectroscopy at MAX-Lab. *Synchrotron Radiation News*, **4**, 15.

Barnes, C. E. (1977). Development of efficient, radiation-insensitive GaAs: Zn Leds. *IEEE Transactions on Nuclear Science*, NS24, 2309–14.

Barnes, C. E. (1984). Radiation effects in optoelectronic devices, Sandia Report No. SAND84-0771. Sandia National Laboratories, Albuquerque, NM.

Barry, A. L., Maxseiner, R., Wojcik, R., Briere, M. A. and Braunig, D. (1990). An improved displacement damage monitor. *IEEE Transactions on Nuclear Science*, NS37, 1726–31.

Barry, A. L., Lehmann, B., Fritsch, D. and Braünig, (1991). Energy dependence of electron damage and displacement threshold energy in 6H silicon carbide. *IEEE Transactions on Nuclear Scicence*, NS38, 1111–5.

Boudenot, J. C. and Augier, P. (1991). Total dose effects on charge coupled devices (CCD) reverse annealing phenomenal. In Proceedings of the ESA Components Conference, ESA SP-313.

Braünig, D., Gaebler, W., Fahrner, W. R. and Wagemann, H. G. (1977). *GfW Handbook for data compilation of irradiation tested electronic components*, Report No. HMI-B248. Hahn-Meitner Institut, Berlin.

Bruce, A. J. (ed.) (1991). Proceedings of the Symposium on Solid State Optical Materials. American Ceramic Society, Westerville, OH.

Brucker, G. J. (1978). TFTR diagnostics engineering report no. 3, Report No. PH-I-004. Princeton Plasma Physics Laboratory.

Brucker, G. J. and Cope, A. D. (1971). Radiation sensitivity of silicon imaging sensors on missions to the outer planets, Final Report, JPL Contract 953106. RCA Astro Electronics, Princeton, NJ.

Brucker, G. J., Faith, T. J. and Holmes-Siedle, A. G. (1968). Capacitance measurements and diffusion constants in lithium containing solar cells. *IEEE Transactions on Nuclear Science*, NS15, 61.

Brucker, G. J., Rosen, A. and Schwarzmann, A. (1978). Neutron damage in p-i-n diode phase shifters for radar arrays. *IEEE Transactions on Nuclear Science*, NS25, 1528–33.

Burt, D. (1978). Fabrication technology for charge-coupled devices. In *Charge coupled devices*, (ed. M. J. Howes and D. V. Morgan). Wiley, New York. 81–102.

Carson, R. F. and Chow, W. W. (1989). Neutron effects in high power Ga As laser diodes. *IEEE Transactions on Nuclear Science*, NS 36, 2076–82.

Clarke and Lowe (1991). Real-time X-ray studies using CCDs. *Synchrotron Radiation News*, 4(6), 24–9.

Cooley, W. C. and Janda, R. J. (1963). *Handbook of space radiation effects on solar cells*, NASA Publication No. SP-3003. NASA, Washington, DC.

Crabb, R. L. *et al.* (1991). Power system performance prediction for the Hipparcos spacecraft. In European Space Power, Conference, ESA SP-320, European Space Agency, Noordwijk.

Crouzet, S. A. (1985). Evaluation de composants optoelectroniques destinés aux applications spatiales, ESA Contract Report CR(X) 1442; Report under Contract 5799/84.

Cuevas, A., Sinton, R. A. and Swanson, R. M. (1990). Point- and planar-junction solar cells for concentration applications: fabrication, performance and stability. In Proceedings of the 21st IEEE Photovoltaic Specialists' Conference, Kissimimee, FL, pp. 327–32. IEEE Press, New York.

Curtis, O. L. and Germano, C. A. (1967). Injection-level studies in neutron irradiated silicon. *IEEE Transactions on Nuclear Science*, **NS14**, 68–77.

Dale, C. J. and Marshall, P. W. (1991). Displacement damage in Si imagers for space applications. In SPIE/SPSE Electronic Imaging Science and Technology Conference, *Charge coupled devices and solid state optical sensors II*, San Jose, CA, SPIE Proceedings Vol. 144, 770-87 SPIE, Bellingham, WA.

Dale, C. J., Marshall, P. W., Burke, E. A., Summers, G. P. and Bender, G. E. (1989). The Generation lifetime damage factor and its variance in Si. *IEEE Transactions on Nuclear Science*, **NS36**, 1872–81.

Dale, C. J., Marshall, P. W. and Burke, E. A. (1990). Particle-induced spatial dark current fluctuations in focal plane arrays. *IEEE Transactions on Nuclear Science*, **NS37**, 1784-91.

Deal, B. E. (1980). Standarized terminology for oxide charges associated with thermally oxidized silicon. *IEEE Transactions on Electron Devices*, **ED27**, 606–7; *Journal of the Electrochemical Society*, **127**, 979–81.

Debruyn, J. D. and Jensen, L. H. (1983). *The UNIFLUX system*, ESTEC EWP-1309. ESA, Noordwijk, Netherlands.

Debusschere, I. (1984). Radiation evaluation of the ESA 512 imaging technology, ESTEC Contract 5170/82/NL/MS. Leuven University, Belgium.

Eisen, H. and Wenger, C. (1982). *Radiation effects on semiconductor devices—data summary*, Report HDL-DS-82-1. Harry Diamond Laboratories, Adelphi, MD.

Evans, G. A. (ed.) (1982). *Laser and laser systems reliability*, SPIE Vol. 328. SPIE, Bellingham, WA.

Flood, D. J. (1992). Radiation damage resistance requirements for advanced space solar cell development. *Proceedings of the international workshop on radiation effects in semiconductor devices for space application*, Takasaki, Japan, (Japanese Atomic Energy Research Institute).

Freeman, R. F. A. and Holmes-Siedle, A. G. (1978). A simple model for predicting radiation effects in MOS devices. *IEEE Transactions on Nuclear Science*, **NS25**, (6), 1216–25.

Friebele, E. J. (1991). Radiation effects. In *Glass science and technology*, (ed. D. R. Uhlmann and N. J. Kreidl). Academic Press, Boston.

Gillespie, C. H. and Greenwell, R. H. (ed.) (1987). *Optical techniques for sensing and measurement in hostile environments*, SPIE Vol. 787. SPIE, Bellingham, WA.

Grabmayr, G. (1990). On the test of optical power handling capability and radiation hardness of $LiNbO_3$., ESTEC Contract Report 7885/88/NL/PB(SC).

Greenwell, R. H. (ed.) (1984/1986). *Fiber optics in adverse environments*, SPIE Vols 506 and 721. SPIE, Bellingham, WA.

Griffin, P. J., Kelly, J. G., Luera, T. L. and Lazo, M. S. Neutron damage equivalence in GaAs LEDs, *IEEE Transactions on Nuclear Science*, **NS38**, 1216-25.

Groom, D. E. (1989). Radiation levels in detectors at the SSC. In Proceedings Vol. I of ECFA Study Week, Barcelona, CERN Reports 89-10, 96 and 103. CERN, Geneva.

Hall, G. (1992). Modern charged particle detectors. *Contemporary Physics,* **33**, 1-14.

Heijne, E. (1983). Radiation damage: experience with silicon detectors in high-energy particles beams at CERN. In *Miniaturization of high-energy physics detectors* (ed. A. Steanini). Plenum Press, New York and CERN Report 83-06.

Holland, A., Abbey, A. and McCarthy, K. (1990) Proton damage effects in EEV charge coupled devices. *Proceedings of SPIE*, **1344**, Paper 35.

Holland, A., Holmes-Siedle, A. G., Johlander, B. and Adams, L. (1991). Techniques for minimizing space proton damage in scientific charge-coupled devices. *IEEE Transactions on Nuclear Science*, NS38, 1663-70.

Holmes-Siedle, A. G. and Adams, L. (1983). The mechanism of small instablities in irradiated MOS transistors. *IEEE Transactions on Nuclear Science*, NS30, (6), 4135-40.

Holmes-Siedle, A. G. (1989). Radiation problems in the use of charge-coupled devices in space, Fulmer Reports Nos R1232/1 and 2. Fulmer Research Ltd., Stoke Poges, UK.

Hopkinson, G. R. and Chlebek, C. (1989). Proton damage events in an EEV imager. *IEEE Transactions on Nuclear Science*, NS36, 1865-71.

Hopkinson, G. R. (1991). Radiation testing of thomson-CSF CCDs for the SILEX programmer, Final Report, ESA Contract No 7787/88/NL/DG.

Jahns, J. and Huang, A. (1989). Planar integration of free-space optical components. *Applied Optics*, **28**, 1602-5.

Janesick, J. R., Elliott, T. and Pool, F. (1988). Radiation damage in scientific charge coupled devices. *IEEE Transactions on Nuclear Science*, NS36, 572-8.

Janesick, J., Soli, G. and Elliott, T. (1991). The effects of proton damage on charge-coupled devices. In SPIE/SPSE Electronic Imaging Science and Technology Conference: *Charge coupled devices and solid state optical sensors II*, San Jose, CA, SPIE Proceedings, Vol. 1147, 87-108.

Killiany, J. M., Baker, W. D., Saks, N. S. and Barbe, D. F. (1974). Effects of ionizing radiation on charge-Coupled device structures. *IEEE Transactions on Nuclear Science*, NS21, 193-200.

Knoll, G. F. (1989). *Radiation detection and measurement*. Wiley, New York.

Konozenko, I. D., Galushka, A. P., Starchik, M. I., Levchuk, I. V. and Khivrich, V. I. (1973). Investigation of radiation damage in p-type Si with hydrogen impurity and high purity n-type Si. In J. E. *Radiation damage and defects in semiconductors*, (ed. J. E. Whitehouse), IOP Conference Series No. 16, pp. 289-94. Institute of Physics, London.

Korde, R. and Canfield, L. R. (1989a). Silicon photodiodes with stable, near theoretical quantum efficiency in the soft X-ray region. *Proceedings of the SPIE*, **1140**.

Korde, R., Ojha, A., Braasch, R. and English, T. C. (1989b). The effect of neutron irradiation on silicon photodiodes. *IEEE Transactions on Nuclear Science*, NS36, 2169-75.

Kraner, H. W., Ludlam, T., Kraus, D. and Renardy, J. (1983). Radiation damage in silicon surface barrier detectors. In *Miniaturization of high-energy physics detectors*, (ed. A. Steanini). (Plenum Press, New York).

Kraner, H. W., Li, Z. and Posnecker, K. U. (1989). Fast neutron damage in silicon detectors. *Nuclear Instruments and Methods in Physics*, **A279**, 266-71.

Larin, F. (1968). *Radiation effects in semiconductor devices*. Wiley, New York.

Lemeilleur, F., Glaser, M., Heijne, E. H. M., Jarron, P. and Occelli, E. (1991).

Neutron-induced radiation damage in silicon detectors. In IEEE Nuclear Science Symposium and Medical Imaging Conference, Santa Fe, NM.
Levy, P. W. and Friebele, E. J. (ed.) (1985). *Radiation effects in optical materials*, SPIE Vol. 541. SPIE, Bellingham, WA.
Li, Z. and Kraner, H. W. (1991). Studies of frequency dependent C–V characteristics of neutron irradiated p + n silicon detectors. *IEEE Transactions on Nuclear Science*, **NS38(2)**, 244–50.
Li, Z. and Kraner, H. W. (1992). Studies of dependence of changes in silicon detector electrical properties caused by fast neutron radiation on the oxidation thermal process.
Loferski, J. J. and Rappaport, P. (1958). Radiation damage in Ge and Si detected by carrier lifetime changes: damage thresholds. *Physics Review*, **111**, 432.
Loo, R. Y., Kamth, G. S. and Sheng, S. L. (1990). Radiation damage and annealing in GaAs solar cells. *IEEE Transactions on Electron Devices*, **37**, 485–97.
Marshall, P. W., Dale, C. J. and Burke, E. A. (1990). Proton-induced displacement damage distributions and extremes in silicon microvolumes. *IEEE Transactions on Nuclear Science*, **NS37**, 1776–83.
Martin, K., Gauthier, M., Coss, J. R., Dantas, A. R. V. and Price, W. E. (1985). *Total dose radiation effects data for semiconductor devices*, JPL Publ. No. 85-43, Vol. 1. NASA–JPL, Pasadena, CA.
Matzen, W. T., Hawthorne, R. A. and Kilian, W. T. (1991). Radiation hardened phototransistor. *IEEE Transactions on Nuclear Science*, **NS38**, 1323–8.
McNutt, M. J. and Meyer, W. E. (1981). Bulk impurity charge trapping in buried channel charge coupled devices. *Journal of the Electrochemical Society*, **128**, 892–6.
Messenger, G. C. and Ash, M. S. (1986). *The effects of radiation on electronic systems*. Van Nostrand Reinhold, New York.
Poch, W. J. and Holmes-Siedle, A. (1968) Permanent radiation effects in complementary-symmetry MOS integrated circuits. *IEEE Transactions on Nuclear Science*, **NS16**, 227.
Price, W. E., Martin, K. E., Nichols, D. K., Gauthier, M. K. and Brown, S. F. (1981/1982). *Total dose radiation effects data for semiconductor devices*, Report No 81-66. Jet Propulsion Laboratory, Pasadena, CA.
Ravel, M. K. and Reinheimen, A. L. (1991). Backside-thinned CCDs for keV electron detection. *SPIE Proceedings*, **1447**, 109–22.
Roeske, F., Jander, D. R., Lancaster, G. D., Lowry, M. E., McWright, G. M., Peterson, R. T. and Tindall, W. E. (1987). Preliminary radiation hardness testing of LiNbO$_3$ Ti optical directional coupler modulators operating at 810 nm. In *Optical techniques for sensing and measurement in hostile environments*, (eds. C. H. Gillespie and R. H. Greenwell), SPIE Vol. 787, SPIE, Bellingham, WA.
Rosenzweig, W., Gummel, H. K. and Smits, F. M. (1963). Solar cell degradation under 1-MeV electron bombardment. *Bell System Technical Journal*, **42**, 399–414.
Roy, T., Watts, S. J. and Wright, D. (1989). Radiation damage effects on imaging charge coupled devices. *Nuclear Instruments and Methods in Physics Research*, **A275**, 545–57.

Saks, N. S. (1980). A new technique for hardening CCD imagers by suppression of interface state generation. *IEEE Transactions on Nuclear Science*, **NS27**, 1727–34.

Shedd, W., Buchanan, B. and Dolan, R. (1971). Transient radiation effects in silicon diodes rear and in avalanche breakdown. *IEEE Transactions on Nuclear Science*, **NS 18**, 304–9.

Srour, J. R., Hartmann, R. A. and Kitazaki, K. S. (1986). Permanent damage produced by single proton interactions in silicon devices. *IEEE Transactions on Nuclear Science*, **NS33**, 1597–1604.

Stanley, A. G. (1970). *IEEE Transactions on Nuclear Science*, **NS17**, 239.

Stanley, A. G., Martin, K. F. and Douglas, S. (1976). *Radiation design criteria handbook*, Tech Memo 33-763. Jet Propulsion Laboratory, Pasadena, CA.

Stearns, D. G. and Wiedwald, J. D. (1988). The response of charge coupled devices to direct electron bombardment. Submitted to *Reviews of Scientific Instruments*.

Stofel, E. J., Stewart, T. B. and Ornelas, J. R. (1969). Neutron damage to silicon solar cells. *IEEE Transactions on Nuclear Science*, **NS16(5)**, 97–105.

Tada, H. Y., Carter, J. R., Anspaugh, B. E. and Downing, R. G. (1982). *Solar cell radiation handbook*, (3rd ed), NASA–JPL Publication 82-69. TRW Systems Group and Jet Propulsion Laboratory, Pasadena, CA.

Thomson, I. and Janssens, G. (1978). Radiation sensitivity and recovery characteristics of 1 MeV electron-irradiated optocouplers and GaAs LEDs, ESTEC Internal Report MISC 078.

Van Lint, V. A. J., Flanagan, T. M., Leadon, R. E., Naber, J. A. and Rogers, V. C. (1980). *Mechanisms of radiation effects in electronic materials*, Vol. I. Wiley, New York. (no Vol. II published to date)

Vavilov, V. S. *et al.* (1962/1963) *Fizika Tverdogo Tela*, **4**, 3373; *Journal of the Physical Society of Japan*, **18**, 136 (quoted in Vavilov, V. S. and Ukhin, N. A. (1977). *Radiation effects in semiconductors*. Plenum Press, New York).

Wagemann, H. G., Spencker, A., Braünig, D., Roncin, J. C. and Pelous, G. (1973). Radiation test and preselection of maverick diodes for the project Symphonie satellite. *Microelectronics and Reliability*, **12**, 467–72

Watts, S. J. (1988). CCD vertex detectors. *Nuclear Instruments and Methods*, **A265**, 99–104.

Wiczer, J. J., Dawson, L. R., Osbourn, G. C. and Barnes, C. E. (1982). Permanent damage effects in Si and AlGaAs/GaAs photodiodes. *IEEE Transactions on Nuclear Science*, **NS29**, 1539–44.

Wysocki, J. T. (1966). Lithium-doped radiation resistant silicon solar cells. *IEEE Transactions on Nuclear Science*, **NS13**, 168.

Walters, R. J. *et al.* (1991). Space radiation effects in InP solar cells. *IEEE Transactions on Nuclear Science*, **NS38**, 1153–9.

Weinberg, I. (1990). Radiation damage mechanisms in silicon and gallium arsenide solar cells. In *Current topics in photovoltaics*, Vol. 3, (Coults, T. and Meakin, J., (eds.)). Academic Press, London.

Weinberg, I. (1991). Radiation damage in InP solar cells. *Solar Cells*, **31**, 331.

7
Power devices

7.1 General

The power subsystems of large space equipment, radiation-generating equipment, and nuclear power sources are frequently required to handle high currents and voltages above the kilowatt range. In addition, the other subsystems require local regulation by, amongst other devices, power transistors, thyristors, and large rectifiers. Silicon power devices of the type described operate on the same general principles as junction devices discussed earlier, but their construction is often different: the chip is larger and may even be in the form of a 'pellet' having the full diameter of the source ingot. Epitaxial growth is used more extensively in power devices, while special doping (e.g. neutron transmutation) and unusual geometry are required to achieve a high junction breakdown voltage, low 'ON' resistance and good heat removal. In general, locally low levels of doping are required for high-voltage devices. In radiation-effects engineering terms, this is significant because carrier removal due to particle irradiation will be more noticeable. Also, in order to accommodate large depletion regions, large values of base width are required in the design of power transistors and thyristors. This can impart unusually high sensitivity to gain-degradation (see Chapter 5).

In some respects, the power MOS device is also structurally different from its low-power relatives (see, for example, VMOS and HEXFET construction). Nevertheless, the response of the gate oxide layer – probably the dominant effect in space and gamma radiation – can still be predicted by the methods described earlier.

7.2 Bipolar power transistors

The degradation of gain in transistors under particle degradation has been discussed earlier, and details of commercial power transistors were given. It can be seen that unless these devices are carefully selected, they are at risk from moderate quantities of electrons, protons, alpha-particles and neutrons, by virtue of the bulk damage mechanism. For example, neutron-induced damage becomes serious below a fluence value of $10^{12}\,\text{n}\,\text{cm}^{-2}$ (1 MeV). To understand the generally high radiation sensitivity of power devices, it is necessary to outline the theory of transistor breakdown and its influence on base width. The avalanche breakdown voltage of the collector–base junction of a silicon transistor, BV_{CBO}, is highest when the collector region is very lightly doped. To a first approximation,

$$BV_{CBO} = \frac{2 \times 10^{17}}{N_{coll}}, \qquad (7.1)$$

where N_{coll} is the doping concentration in units per cm^3 in the collector region. Thus, for a breakdown voltage greater than 100 V, a doping level of less than 2.5×10^{15} cm^{-3} is necessary. This is equivalent to a resistivity higher than $2\,\Omega$ cm.

Now the practical limit of collector-to-emitter voltage in transistor operation is in fact lower than this value. The reason is that two collective effects occur in the npn or pnp structures, the 'punch-through' and the 'open-base' effects. At high collector-base voltages, the depletion region of the collector–base junction extends completely through the base region. If this occurs, the collector-emitter voltage is, in a sense, 'shorted' because the p-type base region, once depleted, no longer rectifies. A large current, I_{CE}, flows between emitter and collector. This punch-through conduction is at its highest when the base contact is 'open'. The base region is then floating (i.e. capable of taking up the potentials due solely to I_{CE}) and the net result is that I_{CE} (in this case termed I_{CEO}) is increased further by the amplifying action of the n-p-n structure. As a result, the 'breakdown value of V_{CE} with base open' (BV_{CEO}) is lower than BV_{CBO} with a ratio

$$\frac{BV_{CBO} \cdot (h_{FE})}{BV_{CEO}}, \qquad (7.2)$$

where h_{FE} is the forward gain of the transistor.

To achieve a high value of V_{CE} before punch-through occurs, a wide base is required. Other factors being equal,

$$BV_{PT} \propto W_B,$$

where BV_{PT} is the punch-through value of V_{CE}, and W_B is the base width.

The above rationale leads to the conclusion that, for the design of high-voltage power transistors (devices in which the value of the rated V_{CEO} is high), we require two physical features which lead to high sensitivity to neutron irradiation, namely:

(1) a low doping level for the collector;
(2) a high base width.

In accordance with eqn (7.2), the initial value of h_{FE} must be kept as low as possible. Thus, only a small 'margin' for neutron-induced degradation will exist. Typical values for the rated parameters of a power transistor are given in Table 7.1.

TABLE 7.1 Typical parameters of two types of n-p-n bipolar power transistor

Device type	V_{CEO}(max) (V)	h_{FE}(max) (I_c/I_b)	f_T(min) (MHz)	Power (max) (W)	Process specification
1	400	40	50	10	P 48
2	100	120	2	120	P 4A

Source: National Semiconductor (USA) data sheets

To give an example of particle effects at neutron fluences of 2×10^{12} and 7×10^{11} n cm^{-2}, a transistor of type 1 would lose respectively about 50 and 10 per cent of its gain, while a specimen of type 2 would lose respectively 99 and 90 per cent. This illustrates how voltage and power specifications alter the 'hardness' obtainable from bipolar power transistors.

Despite the large junction areas, it is unlikely that transient, ionization-induced leakage currents will be of any importance at the expected dose rates in space. A typical response to ionizing dose rate is 5×10^{-9} A rad s^{-1}. Similarly, at the high values of 'drive' used, the impact of surface effects on gain and leakage (which are induced by ionization and are important for low-power transistors — see Chapter 5) is unlikely to add significantly to the large particle effects described above.

In the Europe and the USA, radiation-hardened power transistors have been developed. Epitaxial collector regions are carefully adjusted for doping level and minimum acceptable widths (to minimize change in V_{CE}(sat) and base regions are adjusted to the minimum acceptable width (to minimize gain degradation). An example of the data for some advanced power transistors capable of operating reasonably well after a neutron fluence of 10^{14} n cm^{-2} has been given earlier. Other comparative neutron tests have been performed in the 10^{12} to 10^{13} n cm^{-2} range on single diffused and planar-epitaxial power transistors.

7.3 Thyristors (silicon-controlled rectifiers)

To perform triggering and latching functions, the four-layer thyristor structure requires high values of gain in the overlapping n-p-n and p-n-p structures of which it is composed. Similar arguments indicate the need for:

(1) high base widths and resistivity values, as described for power transistors;
(2) a high doping uniformity to avoid the local triggering which is often induced if local regions of low resistivity are present;
(3) low minority-carrier lifetime values in certain junction regions to avoid charge storage effects which induce slow turn-off. (Brotherton *et al.* 1982)

236 Power devices

Optically triggered thyristors are becoming widely used. In these devices, some severe radiation-induced effects may be anticipated, arising in both the optical and the semiconductor media. For reactor neutron fluences higher than 10^{12} n cm^{-2} (1 MeV), it has been found that the characteristics of switching may change radically in existing commercial thyristors. The gate current at the triggering point may increase over 10 times and also the 'holding current', while the forward voltage drop in the 'ON' condition may double owing to resistivity effects. Bosnell and Widdows (unpublished data 1976) found that, for certain commercial thyristor types, triggering parameters degrade suddenly as the neutron dose is increased. Few problems appear until a fluence of 2×10^{12} cm^{-2} has been passed, but thereafter, parameters change at a more than linear rate.

7.4 Power MOSFETs

7.4.1 Parameter changes under radiation

While MOS transistors for logic switching are being reduced to outline dimensions smaller than 10 μm on a side for VLSI chip technology, MOS transistors for power control are being enlarged to sizes up to 10 mm on a side. However, the power MOS transistor actually comprises many thousands of vertical MOS structures connected in parallel (typically, over 50 000). Hence, the technology is similar to that used for integrated circuits. The gate oxide technology is similar to MOS integrated circuits and, hence, the same rules for total-dose hardness apply.

The response of power MOS devices to radiation is mainly characterized by:

(1) threshold voltage shift;
(2) transconductance degradation;
(3) reduction in breakdown voltage;
(4) burn-out induced by transients.

As for other MOS devices, the shift of threshold voltage, V_T, is negative and shows a significant bias dependence for both n-channel and p-channel devices (Seehra and Slusark 1982). A typical failure dose for 'grounded gate' condition may be 100 krad (1 kGy) and for DC conditions (positive or negative) 15 krad (150 Gy).

Transconductance degradation occurs as a result of mobility reduction due to interface state generation. However, significant changes in transconductance occur only at comparatively high doses, above 1 Mrad (10 kGy) (Roper and Lewis 1983).

Drain–source breakdown voltage (BV_{DS}) shows significant reduction as a function of total dose in certain technologies. The mechanism is believed to be charge trapping in the field oxide and generation of interface traps

7.4 Power MOSFETs

at the field oxide–silicon interface (Blackburn *et al.* 1983). The trapped charges alter the potential at the surface of drain and source junctions and, hence, they become part of the junction termination and influence the breakdown voltage. The reduction of breakdown voltage is a strong function of the drain–source voltage applied during irradiation. Typical changes in BV_{DS} noted during irradiation of a 500 V n-channel device are -40 V at 1 Mrad for zero bias and -140 V at 50 krad (500 Gy) for 300 V bias. The behaviour at high drain-source voltages (>200 V) during irradiation shows a sharp decrease in BV_{DS} at low voltages and then an increase at about 150 krad (1500 Gy), followed by a small, gradual decrease. This implies that, even in the rather low total-dose environment of space, the reduction in breakdown voltage may be significant and must be considered in conjunction with normal derating requirements.

7.4.2 Radiation-tolerant power MOS circuits

Scope exists for radiation tolerance to be introduced as part of the circuit design procedure when using power MOS. Bearing in mind the earlier discussion of V_T and BV_{DS} bias-dependence, it is clear that circuit designs should minimize both the magnitude and the application time of bias voltages. For example, 'overdrive' (i.e. the use of a negative gate voltage to turn off an n-channel transistor) should be applied with caution because the enhanced radiation-sensitivity resulting from the use of gate bias may dominate. Inverter circuits should be tested as complete circuits, as it is found that a combination of dynamic switching and self-heating of transistors may lead to a higher tolerance to radiation than simple assessments would indicate.

7.4.3 Transient and heavy-ion-induced burn-out

Transient induced burn-out of power MOS devices is of significant concern in military and space technologies. There has been significant research in this topic in recent years and at least two manufacturers are now offering radiation hard devices which are also immune to burn-out.

The mechanism bears some resemblance to latch-up in integrated circuits (see Chapters 3 and 14) and results in destructive 'burn-out' of the device (Oberg and Wert 1987). The vertical structure of an n-channel power MOS introduces a parasitic bipolar transistor structure which is normally in the 'OFF' condition when the transistor is 'OFF'. Charge injection from a transient gamma pulse or a heavy-ion strike may introduce sufficient charge to turn 'ON' the parasitic structure. The results in heavy conduction, leading to destructive 'burn-out'.

The ion LET required for 'burn-out' decreases as V_{DS} increases and, in general, immunity from heavy-ion-induced 'burn-out' requires the use of a V_{DS} of less than 50 per cent of the maximum rated value.

7.5 Static induction transistor (SIT)

The static induction transistor, which bears a close resemblance to the JFET (Chapter 9), has been proposed as a radiation-hard power-switching device (Hanes *et al.* 1988). The SIT is a 'buried gate' FET, the gate being a p-type grid structure buried in an n^- base which is in a lightly doped epitaxial layer grown on an n^+ substrate. One disadvantage of the SIT is that it is a 'normally on' device which requires a new approach to power control circuitry. A 350 V 100 A SIT has been shown to be radiation-hard to 10^8 rad and 10^{14} n cm^{-2} (see also Section 9.1). An advantage of the SIT is that the manufacturing process is compatible with that proposed for 'smart power' devices (see next section).

7.6 'Smart power' devices

Significant research is being directed to the development of integrated 'smart power' devices. Such devices are intended to 'sense' the demand of a given load and adjust the operation of power conditioning and switching devices accordingly. Detection of overload conditions, current and thermal, is incorporated and protection mechanisms are automatically brought into operation. This leads to a requirement to integrate high-voltage–high-current power transistors with precision linear control circuitry and low-voltage control logic. The radiation response of 'smart power' devices tends to be dominated by the control circuitry, which includes comparators, amplifiers, and control logic. Hanes *et al.* (1988) proposed diode-transistor logic employing junction FETs as control circuitry in a radiation-hard 'smart power' device. Darwish *et al.* (1988) reported on a dielectrically isolated, bipolar–CMOS–diffused MOS circuit with 200 V output drive tolerant to 30 krad total dose, 2×10^{14} n cm^{-2} and 10^9 rad (Si) s^{-1} (10^7 Gy (Si) s^{-1}).

7.7 Conclusions

Of the many classes of semiconductor devices used in space equipment and other lightweight electronics, silicon power devices are among those few whose high sensitivity to damage from specific particles must be given consideration. Calculations given in earlier chapters show that, because of the high base-width values of such devices, electron and neutron fluences as low as 10^{14} and 10^{11} cm^{-2} respectively may cause significant degradation in bipolar junction devices. The behaviour of the power MOS device resembles that of its corresponding smaller relation, the MOS switching device, with the additional problem of 'burn-out' induced by transients.

Significant progress is being made in the development of radiation tolerant 'smart power' devices, which will certainly find wide application in the future.

References

Blackburn, D. L., Benedetto, J. M. and Galloway, K. F. (1983). The effect of ionizing radiation on the breakdown voltage of power MOSFETs. *IEEE Transactions on Nuclear Science*, **NS30**, 4116-21.

Brotherton, S. D. and Bradley, P. (1982). Defect production and lifetime control in electron and gamma irradiated siliconn. *Journal of Applied Physics*, **53**, 5720.

Darwish, M. N., Dolly, M. C., Goodwin, C. A. and Titus, J. L. (1988). Radiation effects on power integrated circuits. *IEEE Transactions on Nuclear Science*, **NS35**, 1547-51.

Hanes, M. H., Bartko, J., Hwang, J-M., Rai-Choudhury, P. and Leslie, S. G. (1988). Radiation-hard static induction transistor. *IEEE Transactions on Nuclear Science*, **NS35**, 1475-9.

Oberg, D. L. and Wert, L. J. (1987). First non destructive measurements of power MOSFET single event burn out cross sections. *IEEE Transactions on Nuclear Science*, **NS34**, 1736-41.

Roper, G. B. and Lowis, R. (1983). Development of a radiation-hard n-channel power MOSFET. *IEEE Transactions on Nuclear Science*, **NS30**, 4110-5.

Seehra, S. and Slusark, W. J. (1982). The effect of operating conditions on the radiation resistance of VDMOS power FETs. *IEEE Transactions on Nuclear Science*, **NS29**, 1559-64.

8
Optical media

8.1 General

Transparent materials have large values of E_G, the forbidden gap. This is because, if E_G is greater than 3.1 eV, then visible light, which has photon energies less than 3.1 eV, cannot excite carriers and hence the material does not absorb visible light. However, when exposed to radiation, transparent materials may darken. This happens if the radiation produces new energy levels in the gap into which carriers can be excited. It is now likely that visible light can be absorbed. A classic example is the production of 'colour centres' in alkali chloride crystals by X-rays or particles (Schulman and Compton 1963; Agullo-Lopez, 1989). Figure 8.1 shows some examples of colour centres. Nearly all multicomponent silicate glasses, as used in optics, produce an absorption in the blue end of the spectrum when exposed to radiation (Lell 1962). The radiation-induced loss at 400 nm is usually ten times higher than at 800 nm and there is even higher loss in the UV, including some absorption maxima characteristic of the structure of the defect which is trapping charge. The result for most silica glasses is a red or yellow colour, depending on the exact spectral shape of the absorption. Another oxide series which has yielded important knowledge of radiation-induced coloration is the alkaline earth oxide series, for example, magnesium oxide (Crawford and Slifkin 1972, Henderson *et al.* 1977).

In this chapter, we discuss the physics, chemistry, and practical problems associated with windows, lenses, optical coatings, and optical fibres. Fibres are a special case, as they have extraordinarily long optical paths and the materials are manufactured by very special methods. For the other, more conventional components, the optical path lengths are short, ranging from a few micrometres (coatings) to a few centimetres (lenses), and the materials are usually manufactured by melting.

Unless otherwise stated, we will assume that we are dealing with pieces of glass or other optical medium in the thickness range 1 mm to 1 cm and that we are concerned with performance losses greater than about 20 per cent, such as may be evidenced by visible darkening under irradiation. Translating this loss into other absorption units, the range of concern comprises values greater than the following, roughly equivalent values of loss in different units:

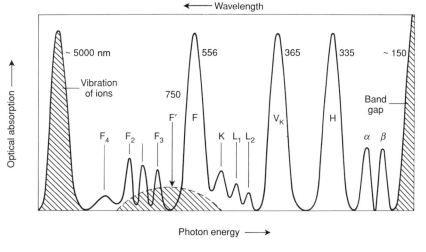
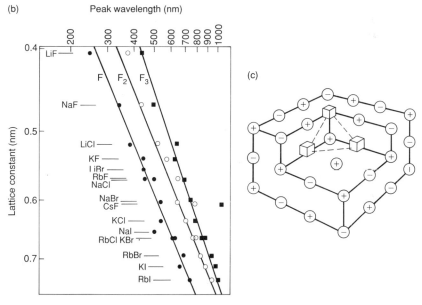

FIG. 8.1 Optical absorption bands in alkali halide crystals: (a) the spectrum; a perfect crystal would have no absorption peaks between the shaded pair (band-to-band and vibrational absorptions). As a result, the materials are transparent to the eye. On irradiation, the most obvious peak in the visible, the F centre, is produced and the crystal exhibits a 'pure' colouration, often aesthetically pleasing. KCl plates turn a beautiful purple; the peak wavelength values for the KCl bands are shown in nm. The UV bands include the absorption due to trapped holes. (b) the Mollwo–Ivey Law shows that the wavelengths of the F peaks can be predicted for members of the halide series. (c) the structure of the F3 colour centre: the cubes represent halogen atom vacancies, a single one, when negatively charged, constitutes an F-centre (after Agullo–Lopez et al. 1989. Reprinted by permission of Academic Press, Orlando, Florida).

absorption : 20 per cent
absorption : 1 dB
loss rate in 1 mm : 1 dB mm^{-1} or 10^6 dB km^{-1}
optical density : 0.1
absorption coefficient : 0.2303 cm^{-1} or 2303 μm^{-1}
transmission : 90% (assuming 100% initially)

If the above loss was caused in a thickness of 1 mm by a dose of 1000 rad (10 Gy), then the specific radiation-induced loss would be expressed as 10^{-3} dB mm^{-1} rad^{-1} or 10^{-1} dB mm^{-1} Gy^{-1}.

8.2 Window materials

8.2.1 General

The effects of displacement and ionization in optical media were discussed in Chapter 3. Metal halides are used only in specialized applications, but their reaction to radiation is of great physical interest, because of their simple cubic structure, the availability of large single crystals, easily prepared (like rock salt), and the well-understood defects produced by the displacement of the anion (halogen). It happens that the halogen in halides is much more easily displaced than the oxygen in oxides. Surprisingly, even a photon of relatively low energy such as an X-ray can displace bromine or chlorine in potassium chloride. Thus, the science of radiation-induced defects started on 'colour centres' in these simple compounds (see Fig. 8.1). However, because of the lack of a technological application for them, the information is used only as a model for the technological materials.

8.2.2 Colour centres in halides and oxides

The original academic work on colour centres concerned alkali halides. Some colour centres have been used in generating laser beams. Recently, the field has been extended to cover transparent oxides, such as silica and magnesium oxide and their glassy derivatives. A common radiation-induced defect in glasses is the production of a yellow or 'smoky' coloration. This is produced by the 'tail' of the UV absorption peak which intersects the visible region in the blue. A colour centre associated with aluminium impurity atoms is thought to be the main defect in 'smoky' quartz (O'Brien 1955; Mitchell and Paige 1956). Other workers (Friebele et al. 1978) found a similar state of affairs in their spectral measurements for phosphorus-doped optical fibres.

Transparent materials used in technology are often polycrystalline or amorphous. As a result, defects exist in the material as processed, and irradiation merely serves to excite electrons into these defects. The latter is the case for most optical glasses, many of which become deeply coloured on exposure to several kilorads of gamma rays or particles. The effect is put to a useful purpose in some types of dosimeter, which use either the absorption or the associated luminescence of glasses or polycrystals to measure ionizing dose.

The nomenclature for simple colour centres is as follows. Anion vacancy centres are known as F-centres, which may exist in charged, uncharged, or aggregated forms, which bear appropriate subscripts. Cation vacancies are known as V-centres, of which there is also a variety (see Fig. 8.1).

Sapphire (Al_2O_3) is regarded as a very radiation-tolerant window material. The colour centres are well-researched (Crawford 1983; Levy 1961; Pells 1984) and have their optical absorption peaks mainly in the UV region.

8.2.3 Silicate glasses

The effect of composition on radiation-induced absorption in multicomponent glasses is reviewed by Lell *et al.* (1966) and Stroud *et al.* (1965). Figure 8.2 shows absorption curves from Lell *et al.* (1966) which demonstrate the variety of absorptions to be expected from optical media and the effect of alkali. The examples shown are for special silicate glasses, but the principles also hold for optical glasses, although absorption is often lower in the latter. In Fig. 8.3 and Table 8.1, Evans and Sigel (1975) have compared the radiation-induced losses in a variety of useful optical glasses. The relevant values at 800 nm vary between approximately the same sensitivity as that of the glasses shown in Fig. 8.2 (e.g. a loss for lead flint glass of $1.3\,dB\,km^{-1}\,rad^{-1}$) and a sensitivity 300 times lower (zinc crown at $0.04\,dB\,km^{-1}\,rad^{-1}$).

The following gives an example of the influence of the glass composition on the structure of 'colour centres' produced in optical glasses. A space television camera used for weather observation was evaluated for performance in the weather satellite orbit (Holmes-Siedle, unpublished). A 10-component camera–lens system, irradiated to 10^5 rad (10^3 Gy), suffered pronounced loss of transmission after irradiation. There was some recovery up to 50 per cent of original performance over a few months. Some lens elements looked grey, but the most common colour was straw yellow.

In the design of optical systems tolerant to radiation, the implication is that light in the blue or UV region will be more difficult to handle than red or near-IR. For ultra hard optics, pure sapphire and synthetic fused silica have been shown to be extremely insensitive to radiation-induced losses in the visible region (see for example Cooley and Janda 1963).

Some optical glasses having improved tolerance to radiation have been produced by Corning, Schott, and Pilkington. These are based on the effect of cerium oxide on the radiation-induced absorptions in the UV and visible.

Figure 8.2 shows the absorption spectra of pure and doped silica material irradiated with ionizing radiation. Whereas there are two optical absorption peaks in pure fused silica, the alkali-doped samples exhibit three such peaks. The intensities of the latter are larger by at least a factor of ten, having a specific radiation-induced loss greater than $1\,dB\,km^{-1}\,rad^{-1}$ in the UV

244 Optical media

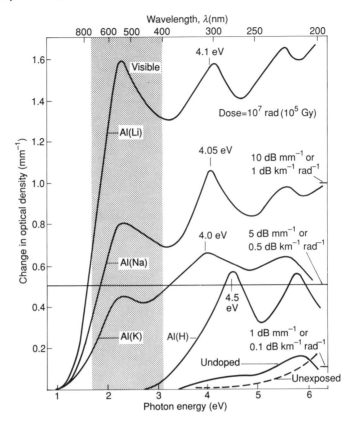

FIG. 8.2 Typical radiation-induced loss spectra in fused silica doped with aluminium and alkali metals, from 200 to 900 nm; approximate specific radiation-induced loss values for a dose of 10^7 rad (10^5 Gy) are indicated (after Lell 1962. Reproduced with permission).

region, as against 0.2 dB km^{-1} rad^{-1} for pure silica, and a value of about 0.8 at 600 nm as against 0.1 dB km^{-1} rad^{-1} for pure silica. Aluminium doping, also shown in Fig. 8.2, shows the type of effect which the addition of other intentional network-forming dopants can produce. The contrast between alkali and aluminium illustrates a characteristic difference between network-disrupting agents (such as alkali) and network-formers (Al, B, P). Each disrupting alkali atom produces a non-bridging Si–O group which can then trap a radiation-generated free hole before it recombines. These same trapped hole species appear to be present in vapour-deposited silica fibre and other heat-treated silicas, probably because thermal treatments can break Si–O–Si bonds and hence create non-bridging oxygens. The hole is trapped in one of the p-orbitals of the oxygen atom.

The absorptions occurring in multicomponent optical glasses are usually

8.2 Window materials

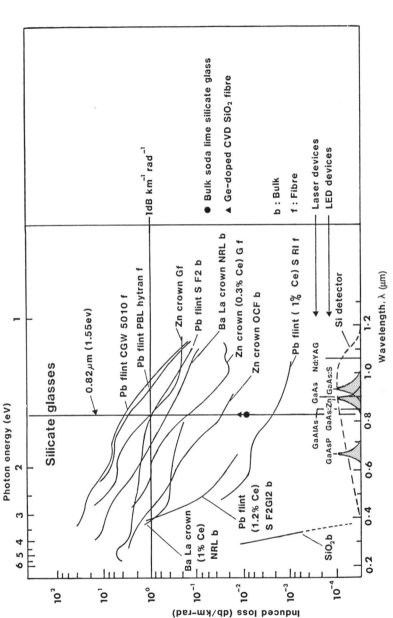

FIG. 8.3 Spectra of radiation-induced losses in bulk silicate glasses and fibres shortly after room-temperature irradiation at doses ranging from 2×10^3 to 5×10^5 rad(Si) at 7×10^2 rad(Si) s^{-1}. Curves from Evans and Sigel (1975); data points from Evans and Sigel (1974), as given in Table 8.1. (© 1974 and 1975 IEEE. Reprinted with permission).

TABLE 8.1 Radiation-induced loss per unit dose in selected bulk glasses and fibres (after Sigel and Evans (1974))

Source	Code	Type	Core material	Form of glass	Response[a] (dB km^{-1} rad(Si)$^{-1}$) $\lambda = 0.8\ \mu m$	$\lambda = 0.9\ \mu m$	$\lambda = 1.05\ \mu$
Corning	(CGW)	5010	Pb Flint	Fibre	5.4	2.5	0.50
Pilkington	(PBL)	Hytran	Pb Flint	Fibre	4.5	1.9	0.50
Galileo	(G)	0001AA	Zn Crown	Fibre	1.5	0.49	0.25
Schott	(S)	F2	Pb Flint	Bulk	1.3	0.69	0.21
Dividing line for values above and below 1 dB km^{-1} at $\lambda = 0.8\ \mu m$							
NRL		GL2382	BaLa Crown	Bulk	0.65	0.35	0.16
Galileo	(G)	0001AB	Zn(0.3% Ce) crown	Fibre	0.27	0.0062	0.0026
NRL[b]	—	GL2364	BaLa (1% Ce) Crown	Bulk	0.21	<0.18	—
Owens Corning	(OCF)	X-4147A	Zn crown	Bulk	0.040	0.020	<0.016
Schott[c]	(S)	F2G12	Pb (1.2% Ce) flint	Bulk	<0.01	—	—
Schott	(S)	R1	Pb(~1% Ce)	Fibre	0.0031	0.0015	0.0010
Corning	—	—	SiO$_2$(Ti)	Fibre	8×10^{-1}	—	—
Corning	—	—	SiO$_2$(Ge)	Fibre	1.4×10^{-2}	—	—
NRL	—	—	Soda–lime silicate	Bulk	1×10^{-2}	—	—
NRL	—	—	Suprasil I	Bulk	$<1 \times 10^{-5}$	—	—

[a] Except where mentioned, readings made 1 hour after γ-irradiation; 1 dB km^{-1} rad (Si)$^{-1}$ = 10^2 dB km^{-1} Gy (Si)$^{-1}$
[b] 30 min after γ-irradiation [c] 9 min after γ-irradiation

orders of magnitude stronger than those which occur in the pure or doped silicas discussed above. This is understandable when it is realized that such glasses contain relatively large percentages of alkali or other network-disrupting oxides (Na_2O, CaO, BaO, etc.). Each atom of the metal can support a non-bridging oxygen ion of the type which can trap holes. Thus, soda glass is visibly affected at a few thousand kilorads, while the effect of a few hundred rads on lead-oxide glass fibre is large enough for a fraction of a metre to provide the sensing element of a personnel dosimeter. Multicomponent glasses developed for fibre cores may also change colour (Holmes-Siedle 1982). Based on US results, it is expected that, if the exposure dose is in the kilorad range, the use of multicomponent glasses will not be possible even for short runs of fibre.

8.2.4 Particle-induced defects

Studies which have contributed much to our understanding of bulk damage in silica were performed by Arnold (1973), using electrons and ions. Studies by Levy (1960) have confirmed that reactor neutrons produce similar defect centres. Early work suffered from the fact that the silica used was often relatively impure. A significant rate of generation of new displacements will only occur in silica samples operating in intense particle beans (e.g. in nuclear equipment). For example, Primak (1975) has described the heavy irradiation of fused silica bars in a fusion-neutron simulation source and calculated the relative defect production rates for neutrons in the fusion energy range. In these circumstances, radiation causes an increase in the density of silica. Rädlein (1991) discusses similar phenomena expected in silica-based telescope materials (e.g. Zerodur) in the space radiation environment. Ionization of the order of 10^7 rad (10^5 Gy) in the surface region (see Appendix E) can cause surface compaction of the order of 10^{-4}. A square-root dependence on dose is expected. A telescope in geostationary orbit risks deformations of the order of one wavelength.

8.3 Coatings

The small optical path length in a typical optical coating (of the order of micrometres) prevents the development of serious radiation-induced losses in the visible region. The optical properties of films and the effects of radiation on film-forming glasses are reviewed by Wong and Angell (1976). It is unlikely that changes of refractive index, shrinking or swelling will affect inorganic optical coating performance under exposure to nuclear and space environments.

8.4 Optical light guides

8.4.1 Introduction

The transmission of information by conversion into the modulations of a light beam is a useful technique for an electrically noisy environment and for broad-band communication. RF and magnetic fields have no effect on light beams. For ruggedness, reliability, and high data rates over long distances, fibres made from silica glasses are the most highly developed form of optical link. The effect of radiation on silica glasses has been intensively studied and it is possible to say at the outset that, although the effect of radiation on silica materials is considerable, it is often tolerable if good control of the purity of the materials is exercised (see e.g. Friebele 1985; Lyons, 1985).

8.4.2 Sources of radiation-sensitivity in silica and glasses

In this discussion, we will take the common case of optical fibre materials composed of a glassy silica network (O–Si–O–Si structures) in which varying degrees of imperfection are introduced by bond strain and atoms which disrupt the network. Unlike the well-known case of the alkali halides, ionization does not give rise to the displacement of any atoms and hence no new 'colour centres' are formed. This is borne out in the case of pyrolytic fused silica, which can be irradiated to a very high dose (10^8 rad) without visible coloration appearing. Synthetic *crystalline* silica (quartz) undergoes no visible coloration, but spectroscopy in the UV region reveals a radiation-induced absorption peak. This is probably due to impurities remaining from the flux used for crystal growth. Natural quartz is very impure; when irradiated to a dose of 10^5 rad, strong 'smoky' brown coloration is produced. As we will see later, pyrolytic silica, when doped heavily or even when subjected to strains during fibre-drawing, produces colorations which, in the 10^6 rad (10^4 Gy) range, are visible to the naked eye and may be significant at doses as low as 10^3 rad (10 Gy).

8.4.3 Prediction models for optical fibre loss versus dose

8.4.3.1 *Fundamentals*

As mentioned earlier, the fundamentals of the production of colour centres in silica-containing materials have been worked out in great detail by a number of investigators, for example Mitchell and Paige (1956) Weeks and Nelson (1960) and Weeks and Lell (1964) at Oak Ridge, and Griscom (1985). These investigations form a body of work which provides a base for models of the effect of radiation on the optical properties of fibres. A proposed prediction model (Holmes-Siedle, 1980) is outlined below.

For the deterioration of optical performance, we shall employ the specific loss, L_0 in dB km^{-1} rad^{-1}. The range over which this term should be used must be limited because of the known saturation of radiation-induced loss at high doses (10^5 to 10^6 rad) and the known transient effects such as the

8.4 Optical light guides

decay of loss with time after a pulse of radiation. However, as a means of conveying orders of magnitude in sensitivity — as in the present discussion — the term is useful.

Radiation-induced absorptions in fibres in the 400 to 1500 nm range tend to lie in the range 10^{-2} to $10\,\text{dB km}^{-1}\,\text{rad}^{-1}$. Figure 8.4 shows the simple arithmetical relations between transmission, optical density, and loss (loss in dB is found by multiplying optical density by 10; absorption coefficient is a scientific unit which can be easily related to oscillator strength). The optical path length of the irradiated sample bears a linear relation to the loss expressed in the usual engineering term, dB per km. We can thus extrapolate simply from short lengths to kilometre distances.

Finally, if radiation-induced loss increases linearly with dose, we can express the characteristic radiation-sensitivity of a fibre in terms of 'specific radiation-induced loss' in $\text{dB km}^{-1}\,\text{rad}^{-1}$ at a given wavelength. This should be a characteristic of the material, not altered by sample thickness or type of ionizing radiation; it constitutes a *useful performance parameter* for a fibre system. $1\,\text{dB km}^{-1}\,\text{rad}^{-1}$ equals $100\,\text{dB km}^{-1}\,\text{Gy}^{-1}$.

8.4.3.2 Simple mathematical model

It is reasonable to expect the build-up in the number of occupied colour-centre defects in (and, hence, the optical absorption of) irradiated silica to be proportional to dose if the fraction occupied is small. The plot or 'growth curve' of detectable absorption against dose could thus have the form of a straight line over several decades before the fraction occupied becomes large. At this point, the curve will first become sublinear and finally saturate (become flat). An analysis of the data of Friebele *et al.* (1978) shows that the loss-versus-dose curves do follow this scheme. It appears that the low-dose, linear regions of the curves in a variety of different silica materials have slope lying between 10^{-2} and $1\,\text{dB km}^{-1}\,\text{rad}^{-1}$ for the wavelength 820 nm.

Thus, as a model for fibre degradation, we can postulate upper and lower limits as shown in Figure 8.4. This corresponds to an equation:

$$L = L_0 D \quad (L \ll L_{\text{sat}}) \tag{8.1}$$

where L is the loss and D the dose. The slope L_0, representing the 'specific radiation-induced loss', is constant for a given material and, at a light wavelength of $0.82\,\mu\text{m}$ has the values 10^0 and 10^{-2} for the upper and lower bounds respectively. L_{sat} is the saturation value of loss.

We expect that L_0 is a function of defect concentrations N_1, N_2, etc., the peak absorption wavelength, λ_{peak}, the wavelength of observation, λ_{obs}, the oscillator strength of the defect, S, and the capture cross section of the defect for charge-carriers, σ. If f, g, etc. are functions of the shape of the absorption band (i.e. of the broadening and 'tailing') and the separation between λ_{obs} and λ_{peak}, then the form might be:

250 Optical media

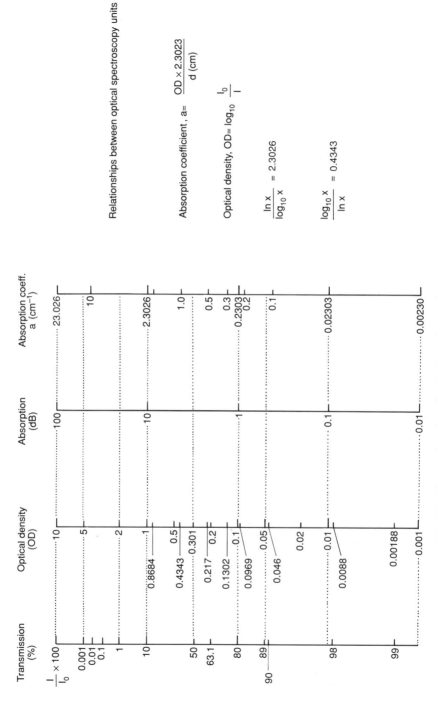

FIG. 8.4 Conversion chart for finding relations between optical absorption units.

8.4 Optical light guides

$$L_0 \propto S_1\sigma_1 N_1 f(\lambda) + S_2\sigma_2 N_2 g(\lambda) + \ldots \text{etc.} \qquad (8.2)$$

CVD silica used in fibres is deposited rapidly in layers containing varied amounts of dopant, and subjected to several transient heat treatments and then to collapsing and drawing stresses. Therefore, it may not behave as simply as these models imply. For example, the experimental growth curves given by Friebele (1978) show signs of at least two saturating processes. Such effects can be handled mathematically by providing in the equations for the filling of several centres. The question of the creation of new centres by particles can be accommodated by making L_0 a function of particle flux. Experiments by Mattern *et al.* (1975) suggest that neutron displacement effects are of small significance, even in intense reactor environments.

8.4.4 Vapour-deposited fibre technology

Figure 8.5 shows a graph which summarizes a set of radiation tests on communications fibres of the more advanced type. The fibres were all made by the successive deposition of silicon dioxide layers on a former or 'bait'. The constituents used are of the very highest purity, but the refractive indices of different regions are manipulated by the addition of dopants to the stream of silicon tetrachloride which is the source of the silica. The common dopants are gases containing germanium, phosphorus, and fluorine (Fig. 8.5 indicates impurities Ge, P, F in brackets). Some forms have as their core section an ultrapure undoped silica, very similar to Suprasil. The two solid lines marked 'worst case expected' (A) and 'best case expected' (B) form approximate boundaries for all of the data of the above type which was included in a survey we performed (Holmes-Siedle 1980). It will be seen that, initially, the curves rise in a linear fashion and, in case B, with a slope of 10^{-1} dB km^{-1} rad^{-1}. Upper and lower limits (lines A and B respectively) were found during our survey of radiation-induced loss in all-silica optical fibres. Some examples of experimental results (dotted lines) are given.

Germania-doped quartz provides a striking example of coloration under radiation. Figure 8.6 shows a cross section of a preform rod from which optical fibres are drawn. The active region has been built up by depositing layers of silica on the inside of a tube of natural fused quartz. The central circle is a mixture of germania (germanium dioxide) and silica (silicon dioxide). Although both of these starting materials are very pure, and separately show little response to radiation, the mixture shows a red coloration when exposed to gamma rays. Holmes-Siedle (1982) explained the visual appearance by reference to two light absorption peaks, one caused by germanium and another by phosphorus. The natural quartz shows a purple coloration but only where it has been in contact with the silica deposit. Friebele *et al.* (1991) have correlated the radiation responses of single-mode fibres with the fabrication methods used.

252 Optical media

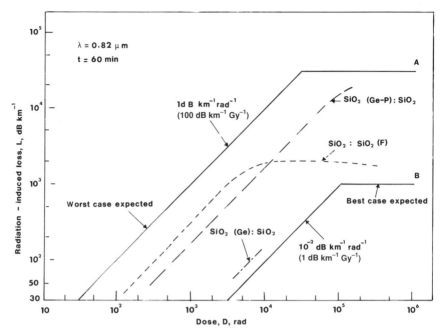

FIG. 8.5 Loss vs. dose: results of radiation tests on communications fibres, showing upper (A) and lower (B) limits (symbols in brackets represent doping impurities—see text, Section 8.4.4).

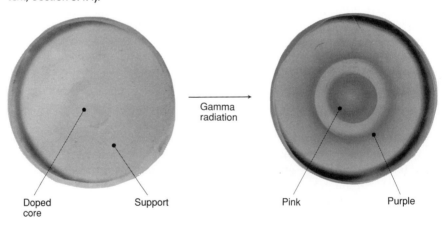

FIG. 8.6 Visible coloration in germanium-doped silica: photograph of an irradiated optical-fibre preform, of which the central portion (about 2 mm diameter) is germanium-doped CVD silica; (left) unirradiated control, (right) irradiated to 200 krad (2 kGy). 10 mm length photographed in diffused light about 2 months after irradiation. The clear inner annulus is undoped silica and the darker outer annulus is natural quartz which is coloured purple near its inner boundary. The irradiated core is pink in colour and shows growth rings from the CVD deposition process (Holmes-Siedle 1982).

8.4 Optical light guides 253

8.4.5 Fibres drawn from Suprasil rods

Plastic-coated silica (PCS) fibres are quite different from the above in several respects. Fabrication is less sophisticated and the fibres produced are not so rugged under the conditions used for splicing or interconnection. A silica rod (of 'bulk' – outer-surface vapour-deposited (OVPO) – type, made for many years for other purposes such as optical blanks, out of Suprasil, Amersil, Spectrosil, Corning 7940 or one of the other long-used synthetic silicas) is drawn in an oxy–hydrogen flame and then immediately passed through a bath of coating polymer. These PCS fibres are quite low in cost but, with this method, it is difficult to achieve numerical apertures less than 0.15. Low numerical apertures are required for high-rate communication. However, diagnostic instruments and local area networks may not require high data-rate transmission and could employ PCS fibres.

Data for Suprasil by Mattern *et al.* (1975) are plotted in Fig. 8.7. Curve B is carried over from Fig. 8.5. Data for bulk Suprasil fall below this line. As might be expected, there is an opportunity for impurities to be diffused during the drawing procedure into the fibre or stresses introduced. The radiation-induced loss in silica is known to be associated with the presence of impurities (alkalis, hydrogen, etc.). It is therefore not surprising that the one point recorded for BTL fibre, made from Suprasil, lies above that for the bulk Suprasil rod and near to curve B, i.e. the loss is worse because defects have been introduced by processing. In a fibre made from pure Suprasil, coated with plastic (PCS), by Galileo Corp., which was tested by Friebele *et al.* (1978), curve E, the loss is already 5 dB km^{-1} at a dose of 10^2 rad. As can be seen from the diagram, the build-up curve peaks at 10^4 rad and then reduces again. Another curve, F, has a similar shape, but shows a lower loss at all points. The data show a difference of three orders of magnitude in the radiation-induced loss to be found in ultrapure bulk silica at the same dose level, as is illustrated by the response of bulk Suprasil SS1, Suprasil SSWF and Corning 7940 at 10^8 rad (10^6 Gy). Even nominally pure silica fibres should therefore be carefully selected and tested. The effort devoted to the selection of fibres will be influenced by the loss which can be tolerated, the lengths which are in highly exposed locations, and the total lengths used.

8.4.6 Fibre luminescence

Mattern *et al.* (1975) measured the radiation-induced luminescence energy delivered to a photosensor connected to a fibre during a pulse of X-rays. The dose rate was over 10^{10} rad s^{-1}. The constant calculated for the light yield from Corning type B silica fibre at 800 nm was 2×10^{-14} mJ cm^{-1} nm^{-1} rad^{-1}.

We will assume the following values for a fibre cable under exposure:

254 Optical media

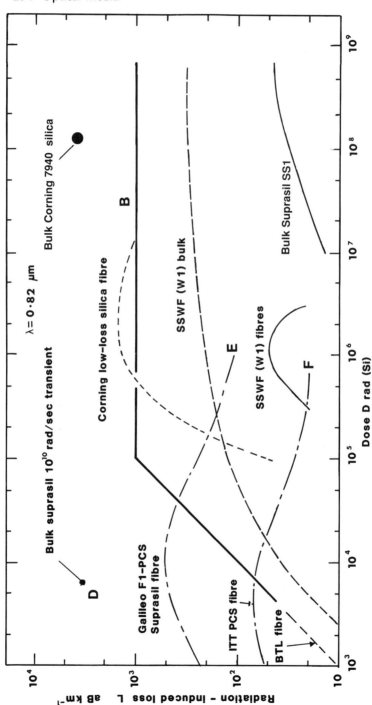

FIG. 8.7 Radiation-induced loss of ultrapure silica in bulk and fibre forms as a function of radiation dose. Curve B is the lower limit, found by A. Holmes–Siedle in an earlier survey of all-silica (non-plastic) fibres. Apart from point D, all other results are for gamma rays at low dose rate (about 100 rad sec^{-1}). Dotted lines and open circle—from Mattern et al. 1974; chain-dotted lines and curve E—Evans and Sigel, 1978; curve F—Sigel et al. 1979; filled circle—Evans and Sigel, 1975

dose rate : 1 rad s^{-1} (10^{-2} Gy s^{-1})
length of cable : 10 cm
bandwidth of sensor : 300 nm
responsivity of photodiode : 1 A W^{-1}

The calculated output of a photosensor under these conditions is 6×10^{-11} A. This is unlikely to produce any problems in the form of 'background' interference in normal electrical or light signals, which would normally generate far higher current values. on the other hand, Mattern mentions that other materials may yield higher efficiencies. However, Golob *et al.* (1977) found that the luminescent efficiency of five modern fibre types was even lower than the above. Both neutron and X-ray pulses in the $10^2 - 10^4$ rad range produced results indicating luminescence yields in the range 10^{-17} to 3×10^{-15} mJ cm^{-1} nm^{-1} rad^{-1}. Thus, the above value of 2×10^{-14} mJ cm^{-1} nm^{-1} rad^{-1} (2×10^{-12} mJ cm^{-1} nm^{-1} Gy^{-1}) is the highest so far found. The luminescence spectrum in one instance is shown in Fig. 8.8. The light was of Cerenkov type (Knoll 1989).

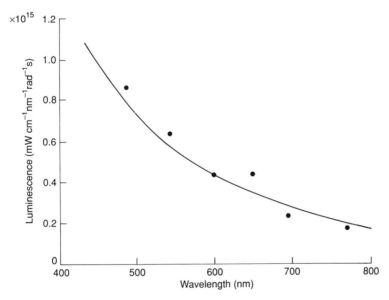

FIG. 8.8 Dose rate effect luminescence induced by pulsed electron irradiation in a step-index fibre, as a function of emission wavelength; $\theta = 40°$. The curve shows the $1/\lambda^3$ Cerenkov theoretical relation; the points are experimental data. (After Golob *et al.* 1977. Reprinted with permission).

256 Optical media

8.4.7 Polymeric optical fibres

For short data links and for some imaging bundles of fibres, polymeric fibres (PMMA or polystyrene) are used. The effect of radiation on some single data links has been studied by West (1986). Bursts of radiation led to transient absorptions which rapidly recovered. These appeared to be connected with components of the plastic other than the basic polymer chains. Study of the relative degradation of glass and polystyrene rods used for fibre drawing (Holmes-Siedle, unpublished work) showed that the optical properties of the polystyrene had a tolerance to radiation which was of the same order as or better than that exhibited by the glass. This has economic implications in the design of imaging systems for certain radiation environments, since the polymeric form of fibre bundle is less costly to manufacture.

8.5 Scintillators

In high-energy physics, one of the most versatile forms of radiation sensor is the scintillation counter. The light pulses emitted by the scintillator bear information on the energy and number of particles and the geometry and light decay can be varied to give further information. The scintillator volume is often quite large so that the light has to be transmitted efficiently to the output point (e.g. the surface of a block or the end of a fibre). Radiation damage can reduce both the light transmission and the quantum efficiency of these important optical media which may be liquid, solid, organic, or inorganic. in high-energy physics experiments, scintillators may be close to intense sources of radiation. Schönbacher (1987) and Ilie et al. (1991) collected and tabulated a large amount of data from European institutions on this topic. The most common way of expressing damage is reduction of the pulse height as emitted by a given counter setup. For example, the pulse height from crystals of caesium iodide of dimensions 3 cm × 3 cm × 10 cm was reduced by a dose of 1000 rads (10 Gy) of Co-60 gamma rays to a normalized value of about 0.5 (Kobayashi and Sukuragi 1987).

The tolerance of organic scintillators appears in general to be higher than for inorganic crystals and glasses, Schönbacher's review suggests that many organics are still operable at 10^6 rad (10^4 Gy) (see also Marini et al. 1985). Bezuglii et al. (1965) exposed solid scintillating polymers to 4 megarads (40 kGy); Schönbacher places the results in the following order of falling radiation tolerance : polyvinylxylene, polyvinyltoluene, and polystyrene, the largest losses being about 50 per cent in both light yield and optical transmission. Polyvinyltoluene and acrylic scintillators were also tested at CERN (see Section 10.3).

Since high-energy physics experiments may run for years, there is a good chance that darkening, caused early in the run, will have faded considerably by the end (see e.g. Sirois and Wigmans 1985). Fading is a well-known phenomenon in alkali halides, in which colour centres anneal at room temperature. A lesser degree of fading is expected in organics since the colorations produced by irradiation will not necessarily 'anneal', often being stable organic radiolytic compounds (see Chapter 10). However, if a liquid organic scintillator were used, it would be possible to 'harden' the system by circulating or exchanging the damaged parts of the medium.

8.6 Conclusions

In the thickness range 1–10 mm, many window materials and lenses show significant losses in the dose range 10^4 to 10^6 rad (10^2 to 10^4 Gy). Thus, optics exposed to space and nuclear radiation should require special engineering and should receive high priority in testing. As much as 50 per cent recovery of loss may be expected with time after irradiation, particularly if the temperature is raised. Some 'hardened' optical glasses are available commercially.

The theory of radiation-induced trapped charges, which cause the losses in transmission in optical materials, has benefited from the extensive research done on defect production in alkali halides and the coloured states of these defects. However, in engineering materials such as silica glasses, the defects which cause the colours in transparent materials are not usually formed by the radiation but exist in the material 'as manufactured'. The reduction of sensitivity to radiation in such materials can be effected by eliminating or modifying these defects.

As the optical paths in fibre-optic systems may be very long, radiation-induced loss effects due to space or reactor radiation may be quite severe. Multicomponent glass fibres will show greater loss under radiation than ultrapure synthetic silica fibres. It is possible for polymeric fibres to perform as well as, or better than, glasses. The prediction of loss versus time and luminescence effects is possible, if samples of the fibre technology concerned are tested under radiation.

References

Agullo-Lopez, F., Catlow, C. R. A. and Townsend, P. D. (1989). *Point defects in materials*. Academic Press, London.

Arnold, G. (1973). *IEEE Transactions on Nuclear Science*, **NS20**, 220-8.

Bezuglii, V. D. and Nagornaya, L. L. (1965). Effect of radiation on the stability of plastic scintillators. *Journal of Nuclear Engineering*, (Transl. from Russian) **19** A/B, 490.

Cooley, W. C. and Janda, R. J. (1963). *Handbook of radiation effects in solar cell power systems*, NASA SP-3003. NASA-JPL, Pasadena, CA.

Crawford, J. H. (1983). *Semicond. Insul.* **5**, 599.

Crawford, J. H., Jr. and Slifkin, L. M. (ed.) (1975). *Point defects in solids, Vol I general and Ionic Crystals*. Plenum Press, New York.

258 Optical media

Evans, B. D. and Sigel, G. H., Jr. (1975). *IEEE Transactions on Nuclear Science*, **NS22**, 2462-7.

Evans, B. D. and Sigel G. H., Jr. (1984). Permanent and transient radiation induced losses in optical fibers. *IEEE Transactions on Nuclear Science*, **NS21**, 113-1.

Friebele, E. J. (1991). Correlation of single-mode fiber radiation response and fabrication parameters. *Applied Optics*, **30**, 1944-57.

Friebele, E. J., Sigel, G. H., Jr. and Gingerich, M. E. (1978). Radiation response of fiber optic waveguides in the 0.4 to 1.7 μm region. *IEEE Transactions on Nuclear Science*, **NS25**, 1261-6.

Friebele, E. J. et al. (1985). Overview of radiation effects in fiber optics, in Levy et al. 70-88.

Golob, J. E., Lyons, P. B. and Looney, L. D. (1977). Transient radiation effects in low loss optical waveguides. *IEEE Transactions on Nuclear Science*, **NS-24**, 2164-8.

Griscom, D. L. (1985). Nature of defects and defect generation in optical glasses, in Levy et al., 38-9.

Henderson, B. and Wertz, J. E. (1977). *Defects in the alkaline earth oxides.* Taylor & Francis, London.

Holmes-Siedle, A. G. (1980). Radiation effects on the joint European Torus: guidelines for preliminary design, Fulmer Report No. R857/2. Fulmer Research Institute, Stoke Poges, UK.

Holmes-Siedle, A. G. (1982). Irradiation of optical fibre blanks as a test of impurity content and radiation sensitivity. *IEEE Transactions on Nuclear Science*, **NS29**, 1405-9.

Ilie, D., Ilie, S., and Schönbacher, H. (1991). *Review of studies on radiation damage to scintillating materials used in high-energy physics experiments.* Report No. CERN-TIS-CFM /IR /91-18, European Organization for Nuclear Research (CERN), Geneva.

Kobayashi, M. and Sukuragi, S. (1987). Radiation damage of CsI(Tl) crystals above 10^3 rad. *Nuclear Instruments and Methods*, **A254**, 275.

Lell, E. (1962). Radiation effects in doped fused silica. *Physics and Chemistry of Glasses*, **3**, 84-94.

Lell, E., Kreidl, N. J. and Hensler, J. R. (1966). Radiation effects in quartz, silica and glasses, ed Burke, J. E. *Progress in Ceramic Science,* **4**, 1-93. Pergamon Press, Oxford.

Levy, P. and Friebele, E. J. (ed.) (1985). Critical Reviews of Technology: optical materials in radiation environments. SPIE volume 541 (SPIE Bellingham, WA).

Lyons, P. B. (1985). *Fiber optics in transient radiation fields*, in Levy et al., 89-96.

Marini, G. et al. (1985). *Radiation damage to organic scintillator materials.* CERN Report No. 85-08. European Organization for Nuclear Research (CERN), Geneva.

Mattern, P. L. Watkins, L. M., Skoog, C. D. and Barsis, E. H. (1975). Absorption induced in optical wave guides by pulsed electrons as a function of temperature, low dose rate gamma and beta rays and 14 MeV neutrons. *IEEE Transactions on Nuclear Science*, **NS22**, 2468-74.

Mitchell, E. W. and Paige, E. G. (1956). *Philosophical Magazine*, **1**, 1085.

O'Brien, M. C. (1955). *Proceedings of the Royal Society (London)*, **A231**, 404.

Pells, G. P. (1984). Ceramic materials for reactor application. *Journal of Nuclear Materials*, **122-3**, 1338.

Primak, W. (1975). *The compacted states of vitreous silica*. Gordon and Breach, New York.

Rädlein, E. (1991). *Simulation der wechselwirkung von weltraumstrahlung mit glas und glaskeramik*. Dissertation for Dr. Ing. Technischen Universität, Clausthal, Germany.

Schönbacher, H. (1987). *Review of radiation damage studies in scintillating materials used in high energy physics experiments*. TIS Commission Report, TIS-RP/201. European Organization for Nuclear Research (CERN), Geneva.

Schulman, J. H. and Compton, W. D. (1963). *Color centers in solids*. Pergamon Press, Oxford.

Sirois, Y. and Wigmans, R. (1985). Effect of long term low level exposure to radiation as observed in acrylic scintillators. *Nuclear Instruments and Methods*, **A240**, 262.

Stroud, J. S., Schreurs, J. W. H. and Tucker, R. F. (1965). In *7th International Congress on Glass*, pp. 1-42.18.

Weeks, R. A. and Nelson, C. M. (1960). *Journal of the American Ceramic Society*, **43**, 399-404.

Weeks, R. A. and Lell, E. (1964). *Journal of Applied Physics*, **35**, 1932-8.

West, R. H. (1986). Radiation induced losses in pure silica core fibres. *SPIE Proceedings*, **721**, 50-6.

Wong, J. and Angell, C. A. (1976). *Glass: structure by spectroscopy*. Dekker, Basel.

9
Other components

This chapter includes descriptions of electronic and mechanical components which do not merit a chapter to themselves. Other hardware requires mention because of particularly high tolerance or intolerance of exposure to radiation. A more systematic analysis of radiation responses of structures which are definable by material was given in section 3.8.

9.1 Junction field–effect transistors

9.1.1 Introduction

The field-effect transistor (FET) has many of the characteristics required to withstand heavy radiation damage. Amplification is produced by the transport of majority carriers through a channel which is quite heavily doped. Unlike the MOSFET, described in Chapter 4, current flow is remote from the surface of the semiconductor. Thus, both the major effects which degrade bipolar and MOS transistors in a radiation environment (ionization and bulk minority-carrier recombination) are absent. Commercial junction FET devices have indeed proved in practice to be tolerant of the effects of heavy ionization and bulk damage, and some work has also been done in the USA to produce special silicon JFET devices which are even more tolerant of neutrons (Shedd *et al*. 1969). Despite these advantages, the use of the JFET in 'hardened' equipment has not been extensive. For space environments, where ionization dominates bulk damage, the advantages are not as great as in nuclear environments. However, in the field of instrumentation amplifiers, the properties of the Si JFET as a low-noise, high-impedance input device have already been appreciated in commercial use and are attractive for some uses in space. The static induction transistor principle is essentially a grid-controlled vertical FET and its performance under neutrons appears hopeful. The gallium arsenide FET technology is now increasingly used in microwave circuits and high-speed analogue and digital switching applications. The modulation FET employing an AlGaAs layer can achieve higher speeds than MESFETs.

9.1.2 Mechanisms of degradation of FETs

As expected, gamma and particle irradiations to a total dose of 10^6 rad (10^4 Gy) cause only minor degradation in Si JFETs (Rudie 1972; Newall

9.1 Junction field–effect transistors

et al. 1981; Zuleeg *et al.* 1982). Normally, transconductance, pinch-off voltage, 'On' resistance and other channel parameters are practically unchanged. Owing to oxide charge build-up, surface leakage may manifest itself as an increase in gate to source leakage but this appears to remain in the nanoampere range even under the highest doses used. This leakage can be troublesome owing to the high input impedance presented by a Si JFET (Messenger and Ash 1986). Allen *et al.* (1984) found that leakage induced by gamma rays could be a significant problem in circuits using JFETs in a low-leakage application. They identified the cause of the leakage to be an inverted layer under the source and drain metallization oxide, and proposed that the problem could be minimized by the use of a highly doped guard band around the channel. Transient effects are characterized by a primary photocurrent, amplified as a secondary photocurrent. This secondary photocurrent is observable at a lower dose rate than in bipolar transistors. Reactor neutron irradiation causes minor degradation at 10^{13} n cm^{-2}, but severe degradation at 10^{15} n cm^{-2}.

Notthoff (1971) found that matched pairs of discrete commercial Si JFET devices, which are suitable as input devices for a differential amplifier, operated well after exposure to a reactor neutron fluence of about 10^{14} n cm^{-2}. Compared with bipolar equivalents, silicon integrated operational amplifiers with JFET structures as input devices may therefore be preferred for very high radiation environments. Other radiation-hardened circuits which have been built using Si JFETs include multiplexers and ladder networks for data acquisition circuits. An application which suggests itself for future space use is as the amplifying element for a highly exposed photodetector, say in the parts of a data link which must survive for a long time in a robot exposed to reactor or RTG radiation. A 'photo JFET' would consist essentially of a photodiode with built-in amplification by the JFET mechanism. No test data are available for this type of device. However, if radiation problems associated with amplification or light detection arise, the possibilities of JFET devices should always be investigated.

The silicon static induction transistor is a large power device which operates on the FET principle but is very different from the small planar structures described so far (see Section 7.5). A grid of metal fingers, embedded in the semiconductor, acts in a manner analogous to the vacuum triode. Hanes *et al.* (1988) found that the device principle is very tolerant of bulk damage. TRIGA reactor neutrons caused changes in resistivity but a fluence of 10^{14} cm^{-2} did not cause failure. A fluence of 10^{15} cm^{-2} changed an original channel resistivity of 30 ohm-cm to 6000 ohm-cm, which prevented good 'on' performance. A long anneal at 350°C was required to remove the neutron damage.

The introduction rates for defects in III–V compounds by high-energy particles (electrons, protons, and ions) has been described by Zuleeg *et al.* (1982), and Knudson *et al.* (1985). The results have been applied to the prediction of resistivity changes in FET structures. FETs with Schottky-barrier gate structures are known as metal–semiconductor FETs (MESFETs) and the operation of these and other FETs is described by Sze (1981). The mobility of III-V semiconductors leads to good high-frequency operation. GaAs FETs are used routinely as low-noise input devices for microwave amplifiers, while very fast logic devices are under development. Several investigations (Zuleeg et al., 1978, 1982) have shown that neutron irradiation of GaAs FETs has the effects predicted. From studies of bulk properties on III-V compounds, for example, the reduction of conductivity can be predicted. As for silicon JFETs, the resulting device degradation only becomes apparent in the range 10^{14} to 10^{15} n cm^{-2} (1 MeV) (Buehler 1968).

Janousek et al. (1988) studied neutron effects in GaAs JFETs and MESFETs and found negligible degradation up to 6×10^{13} n cm^{-2}, but significant degradation at 2×10^{15} n cm^{-2}. They also found that ion-implanted devices were more suceptible to damage than epitaxially grown material.

Modulation-doped AlGaAs-GaAs heterostructure MODFETs were studied by Krantz et al. (1988), who found that the neutron degradation was dominated by the GaAs material parameters only. There was a threshold voltage change of about 30 per cent for a fluence of 10^{15} n cm^{-2}.

GaAs MESFETs and resistors were studied by Galashan and Bland (1990) in connection with direct-coupled FET logic (DCFL) integrated circuits. They confirmed that ion-implanted devices were more sensitive than epitaxial or bulk devices. V_T shifts were small up to 4×10^{14} n cm^{-2} and the shift was proportional to the energy of the ion implant. The extrapolated failure level for DCFL inverters was 1.8×10^{15} n cm^{-2} and 52.4 MeV protons were found to be more damaging than the equivalent neutron dose as normally calculated.

9.2 Transducers

9.2.1 General

Electronic transducers can be described as devices which perform measurement. Space vehicles, remote scientific equipment and other forms of automated apparatus working in a radiation field are likely to require transducers. Transducer technology covers a very large variety of types and it is therefore difficult to predict the effects of radiation.

Optical transducers have been dealt with in Chapter 6. Further important categories include:

(1) mechanical sensors (strain gauges, displacement sensors, pressure gauges);

(2) temperature sensors (thermistors, thermocouples);

(3) particle sensors;
(4) magnetic sensors (including Hall devices).

The requirements for sensitivity are very variable but, clearly, radiation effects will have more impact on applications requiring high sensitivity. Radiation-induced degradation is likely to cause inaccuracy, and the associated high-gain instrumentation amplifiers and oscillators may be more vulnerable. Radiation-induced noise may be important at low levels of signal.

The science of transducers is growing and new designs appear frequently. Therefore, in project studies, each new design with its associated electronic preamplifiers and signal conditioning circuitry should be carefully examined for sensitivity to radiation before it is adopted for operation in radiation environments.

9.1.2 Previous transducer studies

Studies of the impact of space radiation on transducers are rare. However, in the nuclear field, Brucker (1977, 1979) wrote useful reports on his survey of piece parts used in diagnostic equipment on nuclear reactors. Their contents are mainly based on a search of the literature and are followed by recommendations for choices wherever two device types are available. As shown in Tables 9.1 and 9.2, the sensors appear to fall into two classes, those which are useless at $10^{14}\,\text{n cm}^{-2}$ (14 MeV) and those which still operate

TABLE 9.1 Radiation damage thresholds for diagnostic sensors (Brucker 1977)

Detector	Application	Threshold fluence (n cm^{-2}) (1 MeV equivt)	Threshold dose (rad(Si))
HgCdTe	Photovoltaic	5×10^{11}	10^6
PbSnTe	Photovoltaic	5×10^{11}	10^6
GaAs	Photovoltaic	10^{12}	10^6
Si(Li) or Ge(Li)[a]	Reverse-biased diode	10^{10}	3×10^5
Surface barrier diode	Reverse-biased diode	10^{14}	10^7
Pyroelectric	Temperature change	10^{14}	10^7
Schottky diode[b]	Reverse-biased diode	10^{14} to 10^{15}	5×10^5 to 10^8
Crystal Ge (Cu- or Hg-doped)	Photoconductive	5×10^{13}	10^6
NaI	Scintillator	10^{14}	10^5

[a] Both these detectors can be annealed back to their initial state by high temperature.
[b] Depending on construction details, the minimum or maximum value applies.

with little or no degradation. In the former group are opto-isolators, charge-coupled devices (CCDs) (see Chapter 6 for further comment), and a piezoelectric pressure transducer employing $BaTiO_3$ used in the shear mode. In the 'resistant' class fall ZnS scintillators, InSb devices for sensing heat radiation, silicon thermistors, and lead zirconate titanate pressure transducers used in the compression mode. At high particle fluences, germanium devices are not advisable.

Tables 9.3, 9.4, and 9.5 give some test data and basic information on the 'corruption' of the signals from transducers by transient, radiation-induced ionization or leakage effects. Two radiation sensors shown in Table 9.3, HgCdTe and a pyroelectric detector, are the most sensitive to noise. The

TABLE 9.2 Threshold damage levels: permanent damage in parts for diagnostics (Brucker 1977)

Detector or device[a]	Ionization (rad)	Bulk damage (n cm^{-2}) (1 MeV equivt)
Optical isolator (PD)	10^6	5×10^{10}
Optical isolator (PT)	5×10^4	10^{10}
Zinc sulphide	–	10^{14}
Indium antimonide (doped 10^{15} atoms cm^{-3})	10^8	10^{14}
Germanium bolometer (doped 10^{15} atoms cm^{-3})	10^7	2×10^{13}
Germanium (PV) (gallium-doped)	3×10^6	10^{12}
Germanium (PC) (gallium-doped)	6×10^7	5×10^{13}
Diamond	–	10^{14}
CCD (surface-channel) (dark-current failure)	10^4	10^{11}
CCD (buried-channel) (dark-current failure)	10^4	5×10^{10}
Silicon thermistor (boron-doped)	–	10^{14}
RCA memory CDP 1821	5×10^3	10^{15}
Harris memory HMI 6508	10^3	10^{15}
RCA memory CD 4061	10^5	10^{15}
NMOS memory—4 Kilobit	10^3	10^{15}
Barium titanate	9.5×10^6	7.6×10^{10}
Lead zirconate titanate	4×10^{10}	3.6×10^{18}
Quartz crystal (X-ray diffraction effects)	–	10^{19}

[a] PC: photoconductive; PD: photodiode; PT: phototransistor; PV: photovoltaic. Channeltron: temporary fatigue commences after 10^{10} counts; gain can be restored by a clean-up treatment.

TABLE 9.3 Detector radiation noise thresholds (Brucker 1977)

Material or detector	Threshold flux or dose rate (rad s^{-1})
HgCdTe	10^2
Fibre optics	10^{10}
Pyroelectric	3×10^{-3}

TABLE 9.4 Threshold data upset levels (Brucker 1977)

Detector or device	Dose rate (rad s^{-1})	Neutron rate (n cm^{-2} s^{-1})
Optical isolator	10^4	–
CCD (scramble of data in a register)	10^5	–
RCA memory CDP 1821	6×10^{10}	–
Harris memory HMI 6508	8×10^7	–
Intersil memory IM 6508	8×10^7	–
RCA memory CD 4061	10^8	–
Kilobit nMOS-memory	5×10^8	–
Barium zirconate titanate	–	2.1×10^4
Lead zirconate titanate	–	1.2×10^{12}

TABLE 9.5 Carrier generation rates and currents for exposure to a unit dose rate, one rad per second or 10^{-2} Gy per second (Brucker 1977)

Detector or device	Generation (carriers per cm^3 s)	Current (A cm^{-3})
Silicon	4×10^{13}	6.4×10^{-6}
Germanium	10^{14}	1.6×10^{-5}
InSb	4×10^{14}	6.4×10^{-5}
Diamond	10^{13}	1.6×10^{-6}
Channeltron	6.4×10^{-5} e cm^{-2} s^{-1}	10^{-13} A

signal amplitude would have to be known for a proper noise evaluation to be made. Table 9.5 records a calculation of current generation, and Table 9.6 notes some measurements/calculations by the Princeton Fusion Experiment Group which may prove useful in defining background interference in X-ray detectors, including photomultipliers.

Two careful studies of space radiation effects in photovoltaic devices have some relevance to spaceborne sensors (Cooley and Janda 1963; Tada *et al.* 1982). These studies explain damage in solar cells and give results for transparent materials.

266 Other components

TABLE 9.6 The neutron and gamma ray responses of some detectors (Brucker 1977)

Detector[a]	Material	Radiation	Signal	Units
SPMT	ZnS	Reactor n	54	$\dfrac{\text{eV cm}^2}{\text{n mg}}$
IC	Argon	Reactor n	8	$\dfrac{\text{eV cm}^2}{\text{n mg}}$
PMT	Glass	Reactor n	2.4	Electrons per n[b]
PMT	Quartz	Reactor n	0.28	Electrons per n[b]
PMT	Glass	0.662 MeV	6×10^{-3}	Electrons per photon[b]
IC	Argon	2.5 MeV n	2.8	$\dfrac{\text{eV cm}^2}{\text{n mg}}$

[a] IC: ionization chamber; PMT: photomultiplier; SPMT: scintillator optically coupled to PMT
[b] Photocathode electrons

9.3 Temperature sensors

Thermocouples are insensitive to typical space radiation levels and can be used even inside nuclear reactors (say, fluences of 10^{20} n cm^{-2}) as long as the wire insulation associated with them (e.g. glass and ceramic tube) does not become leaky (Ricketts 1972). The same general statements apply to platinum resistance thermometers.

Thermistors are composed of semiconductors but, again, the wire insulation is usually the most sensitive element and most devices will withstand megarad doses (Ricketts 1972). Modern fibre-optic temperature sensors will, of course, exhibit the optical darkening discussed in Chapter 8.

9.4 Magnetics

Magnetic components are frequently used for non-volatile data storage e.g. magnetic tape, cores, discs, bubble memories, and plated wire memories. All these components are generally considered to be very radiation-tolerant.

Non-volatile random-access memories using a layer of ferroelectric capacitors as the memory array, fabricated in conjunction with conventional silicon IC techniques, are available. Messenger and Ash (1986) found this technology to be tolerant to 10^7 rad (10^5 Gy) and 10^{14} n cm^{-2}. Radiation effects would be expected to be dominated by the driving, sensing and amplifying semiconductor devices associated with the memory elements.

Hall-effect sensors are small 'Hall bars' made of ferrite (e.g. type SBV 566). When a 'control current' of about 50 mA is passed through the bar and a magnetic field, B, is applied perpendicular to the current, a Hall voltage, V_H, of a few millivolts appears across the voltage output arms at right angles to the current flow and field direction, and is proportional to current and field. An operational amplifier with feedback — frequently placed in the same package as the Hall bar — is used to amplify V_H to a few volts. Diverse transistors (e.g. 2N2222) can be added to act as relays or to give a TTL logic voltage.

The transport properties of ferrite (mobility, etc.) will not be greatly affected by neutrons in the 10^{15} n cm^{-2} (1 MeV) range. Thus, the site of any degradation suffered is more likely to be in amplifier circuits. The latter is essential for readout. The Hall bar can be connected remotely to the amplifier chip by means of a cable but, of course, the low value of the original Hall voltage means that, as the leads become longer, so is the Hall-effect sensitivity reduced by noise, voltage drop, thermal EMFs, etc.

9.5 Superconductors

Superconductors possessing high critical temperatures (T_c) are of interest for space and military uses as RF filters and cavities, high-sensitivity low-noise detectors, thin-film tunnel junctions, low-loss interconnects, and various passive functions. Maisch *et al.* (1987) performed irradiations with 55.7 MeV electrons and 63 MeV protons on plasma-spray-deposited $YBa_2Cu_3O_7$. They found a fluence-dependent decrease in T_c and an increase in room temperature resistance, and concluded that the technology was usable in space with only moderate shielding. Electron and proton radiation damage has also been studied in $YBa_2Cu_3O_7$ by Chrisey *et al.* (1988). Two preparation techniques were employed, plasma spray and laser evaporation. Both types showed an increase in room temperature resistivity. A reduction in transition temperature was seen only in the laser-evaporated films. Buskirk *et al.* (1988) studied single-phase $GdBa_2Cu_3O_6$ and $YBa_2Cu_3O_6$ using 63.5 MeV electrons. They found no change in the transition temperature and a decrease in room temperature resistivity (in contrast to the previous reference). The results were consistent with gamma testing by other workers. It is suggested that the change in room temperature resistivity is an ionization effect and that changes in T_c are explained using the non-ionizing energy loss (NIEL).

Josephson junctions are an application of superconductors which can perform switching in tens of picoseconds with microwatt power dissipation. Two superconducting electrodes are separated by an insulator thin enough to allow electron tunnelling. The switch has zero resistance at a current less than the critical current I_c and switches to a high-resistance state when the current is greater than I_c. Magno *et al.* (1982) studied alpha-induced upsets and concluded that the ionization created by the alpha track caused a local increase in current above I_c which then caused current-crowding in other areas. This would indicate that Josephson junctions will almost certainly be sensitive to single-event upsets in the space environment.

268 Other components

9.6 Mechanical sensors

9.6.1 General

Transducers of mechanical movement include sensors for displacement, pressure, acceleration, vibration, sound, etc. The effect of radiation depends, of course, on the principle used to measure the movement. The earliest electrical types were moving coils. Later, strain gauges made of metal and silicon were developed. More recently, use has been made of optical fibres and piezoelectric effects in ceramics and polymers, and of silicon microstructures for the manufacture of accelerometers, pressure gauges, tactile sensors, oscillators, etc.

In coil solenoid sensor structures, the effects of radiation will centre on the degradation of the coil insulation and the mechanical linkages (e.g. the polymeric parts). In optical types, darkening of glasses will be most important. Metallic and ceramic strain gauges are unlikely to be affected. The piezoelectric polymer film, PVF_2, has been shown to survive doses of more than 10^7 rad (10^5 Gy) without loss of performance, but degrades at higher doses with evolution of HF. Note that PVF_2 appears much less sensitive to radiation than PTFE.

9.6.2 Silicon micromechanisms

As the mechanical properties of silicon are not affected by even large doses of electrons, protons, neutrons or gamma rays, it is probable that radiation-tolerant silicon micromechanical devices can be built. However, these often incorporate diffused junctions or even integrated transistors which, as discussed elsewhere, must of course be considered in the usual way. Discrete strain gauges, capacitive pressure sensors and accelerometers, however, should not be severely affected by typical space radiation levels.

9.7 Miscellaneous electronic components

Devices included in this section are those that either have a relatively high tolerance to radiation or cannot logically be categorized under any other section. Examples of the first category are non-semiconductor components such as resistors, inductors, and electron tubes. Table 9.7 shows a list of components and materials known to be sensitive to radiation. Information on reactor tests on miscellaneous components can be found in a number of handbooks on nuclear hardening (see e.g. Rudie 1972; Ricketts 1972; and references therein) and the data bank compilations listed in Chapter 13. Tables in section 3.8 systematically survey damage mechanisms in a large range of electronic device structures.

9.7 Miscellaneous electronic components

TABLE 9.7 Materials and devices with generally poor radiation tolerance

Semiconductors	Paints
Optical lenses	Reflective coatings
Optical fibres	Wood, paper, string, cloth
Optical windows	Thin insulators
(e.g. encoder plates)	Photosensitive materials
Elastomers	Gas sensors
(e.g. plastic bellows)	Liquid-ion sensors
Plastic bearings	Surface-active reagents
Lubricants	Piezoelectric transducers
Adhesives	Micropositioners
Hydraulic fluids	

9.7.1 Capacitors

Capacitors consist of large areas of conductor separated by a thin insulator. It is usually desirable for the insulator to exhibit very low leakage ($R_{insulation} > 10^{13}\,\Omega$). Total dose permanently degrades the insulation resistance of organic insulators, but barely at all in the case of air or ceramic. If the dose rate is significant, the insulator can exhibit 'transient photoconductivity' (see Section 10.7). In solid capacitors, some charge becomes trapped during irradiation and may be released slowly some time later to produce a small, long-term conductivity after irradiation has ceased.

9.7.1.1 Total-dose effects

The electrical effects of total dose become severe in capacitors until a dose of about 10^7 rad is reached. Plastic and paper capacitors are the most likely to exhibit radiation-induced changes in leakage and dielectric loss. Glass and ceramic capacitors may be unaffected up to 10^{10} rad (10^8 Gy). For reactor irradiation, damage thresholds are in the region of 10^{14} to 10^{16} n cm^{-2} ($E > 10$ keV) (Ricketts 1972) and it is likely that the source of damage in this case is the accompanying gamma-ray dose.

Tantalum capacitors show both radiation-induced conductivity and stored-charge variation (Smith *et al.* 1977; Srour and McGarrity 1988). Unbiased exposure results in voltage build-up as well as a change in dielectric conductivity. The voltage increases rapidly with dose and is opposite in polarity to the 'built-in' voltage. Saturation occurs when the two voltages are equal. The radiation response is affected by the extent of previous reverse-bias applications. The radiation response is reduced by using capacitors in 'back-to-back' (series) or 'inverted parallel' (opposite polarity electrodes connected together) configurations. Tantalum foil capacitors have a reduced radiation response, owing to the internal 'back-to-back' construction.

9.7.1.2 *Dose-rate effects*

Electrolytic capacitors are subject to an unusual photovoltaic type of leakage (Rudie 1972) but, in fact, leak less than plastics and paper technologies (Ricketts 1972). Natural radiation dose rates in space are normally not high enough to yield significant radiation-induced conductivity (RIC) in capacitor dielectrics. The order of magnitude for RIC is given in Chapter 10.

9.7.2 Resistors and conductors

Discrete resistors have been irradiated at very high intensities in nuclear reactors and flash X-ray machines. Most of the changes observed are due to the breakdown of encapsulating media such as silicone potting compounds. Only mild resistance changes take place at a fluence of 10^{13} fast neutrons in carbon resistors, which are the most sensitive class. The dose rates required to give transient changes are of the order of 10^{10} rad s^{-1} (10^8 Gy s^{-1}). Precision wire-wound resistors are many orders less sensitive (Ricketts 1972). Diffused silicon resistors increase in resistance at high fluences owing to carrier removal and mobility degradation.

Conduction in metals is not affected by gamma rays or space particles. In reactor irradiation, changes are observed in the ultra-high fluence range (10^{20} n cm^{-2} and above) owing to alterations in crystal lattice structures (Dienes and Vineyard 1957).

9.7.3 Quartz crystals

Srour and McGarrity (1988) note that ionizing radiation causes permanent shifts in frequency and changes the Q of a crystal. The major factors determining radiation response are the type of material, type of cut, operating frequency, and mode of operation. Impurity-related defect complexes in the quartz are an important cause of radiation-induced changes. Changes in local electronic bonding cause changes in the elastic constant, resulting in frequency shifts. Migration of alkali ions results in long-term delayed conductivity which degrades the Q value.

The piezoelectric properties of quartz and most other single crystals are not catastrophically affected by neutrons, even at 10^{18} n cm^{-2}. However, oscillator frequencies must often be accurate to extremely close limits (say 1 part in 10^6), and neutron damage to the lattice in the range 10^{12} n cm^{-2} and gamma doses in the 10^5 rad (10^3 Gy) range can produce permanent changes in oscillation frequency of about 1 part in 10^7. Thus, ultra-high-precision oscillators may experience inconvenient drifts, but predictable drift of this sort can probably be adjusted by recalibration.

'Swept' quartz is less susceptible to radiation effects. 'Sweeping' involves applying an electric field along the z-axis at elevated temperature for an extended period of time. If this is performed in air, hydrogen ions or holes replace swept-out alkali ions. If sweeping is performed in vacuum or an inert atmosphere, the resultant crystal is even more radiation-tolerant (Srour and McGarrity 1988). High-grade synthetic quartz, suitable for oscillators, and having a reproducible response to radiation is now available commercially. King (1974), in a review of the effect of radiation on 5 MHz, fifth-overtone resonators, notes a negative change of 50 ppm in frequency for 10^6 rad in natural quartz and one-fifth of this amount in Z-growth electronic-grade synthetic quartz, in the positive direction. Neutron damage produces positive shifts at rates which vary from 0.56×10^{-15} to 3×10^{-15} ppm per n cm^{-2} for unpurified synthetic quartz.

Norton *et al.* (1984) performed extensive evaluation of quartz resonators for space application. They found typical frequency shifts in the range of 10^{-10} per rad (10^{-8} per Gy) for swept quartz. The frequency shifts were less as the radiation dose accumulated, and they suggested that a 'preconditioning' of 20–50 krad (0.2 to 0.5 kGy) would be beneficial. They found that radiation response was highly lot-dependent but reproducible within a given lot. Proton- and gamma-induced frequency shifts were similar, as might be expected for an ionization effect.

A different form of resonator construction, known as electrodeless or BVA, was studied by Suter *et al.* (1988). They found an improvement of between 10 and 100 times compared with conventional adhered electrode, even though the BVA resonator used unswept quartz. Preconditioning with 20 krad (200 Gy) reduced the radiation-sensitivity, as for adhered electrode resonators. It is suggested that the improvement in radiation peformance may be connected with the reduction in mechanical stress in the electrodeless construction.

9.7.4 Vacuum tubes

Despite the large power drain and size of vacuum tubes, their use in very high neutron/gamma environments has proved to be useful. For amplifying tubes (valves), no known degradation mechanism (except outgassing in some cases) provides an upper fluence limit to operation. Imaging devices, of course, suffer optical-path and possibly photosensor degradation, but vidicons are used in reactors and have been tested for use in space missions.

Miniature vacuum tubes are the subject of significant research effort. These are produced using silicon micromachining techniques to produce fine needle-shaped field-emission electron emitters or thermionic cathodes; the 'anode' is also silicon and the whole structure is fabricated in one piece of silicon using submicrometre techniques. Many thousands of such 'tubes' can be fabricated in a square millimetre or so. Flat display panels have been demonstrated using this technology. No radiation data have been reported on this technology but it could be expected to be radiation-hard.

9.7.5 Semiconductor microwave devices

Both silicon and III–V semiconductors are used for microwave signal generation, switching, and amplification. Amplifiers are dealt with in our

272 Other components

discussion of transistors. The other two microwave device types employ the action of majority carriers and fall into a different, lower class of radiation-sensitivity.

The main effects of radiation on the above class of device are:

(1) reduction of majority-carrier concentration by bulk displacement damage;
(2) transient increase in majority carriers produced by a burst of radiation.

These effects have not been found to produce significant problems in geostationary orbit. The effects of intense neutron and flash X-ray irradiation on Gunn diodes and silicon p-i-n switches have been investigated by Berg and Dropkin (1970, 1971) and Chaffin (1971).

Borrego *et al.* (1978) tested a wide range of 2-4 GHz GaAs MESFETs and found no change in performance below 10^7 rad (10^5 Gy). For neutron irradiation significant changes were seen in DC parameters and LF noise at fluences between 5×10^{13} and 8×10^{14} n cm^{-2}.

9.7.6 Miscellaneous hardware

This term includes connectors, cables, gaskets, O-rings, switches, and other items mentioned in Table 9.7. In these categories of hardware, an approach to prediction is to examine the properties of the component materials used in their construction as in Chapter 3 for metals, semiconductors, and other inorganics, and in Chapter 10 for organics. Normally, the mechanical properties of structural plastics show the onset of damage in the range 10^7 to 10^8 rad (10^5 to 10^6 Gy) . Elastomers are more sensitive than rigid plastics, owing to the larger cross-linking changes in the former. Leakage effects in connectors and cables are dealt with by calculation of radiation-induced conductivity. The humble wood, paper, string, and cloth should not be forgotten. When made of the radiolytically sensitive cellulose, they disintegrate after exposure of fractions of a megarad (tens of kGy) of ionizing radiation. Cellulosic plastics are similarly sensitive. Except for reactor vessel environments (see Chapter 2), we can regard metals as being ultra-tolerant of displacement and ionization effects.

References

Allen, D. J., Coppage, F. N., Hash, G. L., Holck, D. K. and Wrobel, T. F. (1984). Gamma induced leakage in junction field effect transistors. *IEEE Transactions on Nuclear Science*, **31**, 1487-91.

Buehler, M. G. (1968). Design curves for predicting fast-neutron induced resistivity changes in silicon. *Proc. IEEE*, Oct., 741.

Berg, N. and Dropkin, H. (1970). Neutron displacement effects in epitaxial Gunn diodes. *IEEE Transactions on Nuclear Science*, **NS17**, 233-8.

Berg, N. and Dropkin, H. (1971). The effect of ionizing radiation on Gunn diode oscillators. *IEEE Transactions on Nuclear Science*, **NS18**, 295–303.

Borrego, J. M., Gutmann, R. J., Moghe, S. B. and Chudzicki, M. J. (1978). Radiation effects on GaAs MESFETs. *IEEE Transactions on Nuclear Science*, **NS25**, 1436–43.

Brucker, G. J. (1977). TFTR diagnostic engineering report no. 2, Report No. PH-I-001. Princeton Physics Laboratory, NJ.

Brucker, G. J. (1979). TFTR diagnostic engineering report no. 3, Report No. PH-I-004, Princeton Physics Laboratory, NJ.

Buskirk, F. R., Neighbours, J. R., Maruyama, X. K., Sweigard, E. L. Dries, L. J. Huang, C. Y. and Junga, F. A. (1988). Radiation effects in bulk samples of high Tc superconductors $YBa_2Cu_3O_6$ and $GdBa_2Cu_3O_6$. *IEEE Transactions on Nuclear Science*, **NS35**, 1486–90.

Chaffin, R. J. (1971). Permanent damage in PIN microwave diode switches. *IEEE Transactions on Nuclear Science*, **NS18**, 429–35.

Chrisey, D. B., Maisch, W. G., Summers, G. P., Knudson, A. R. and Burke, E. A. (1988). The influence of radiation damage on the superconducting properties of thin film $YBa_2Cu_3O_7$. *IEEE Transactions on Nuclear Science*, **NS35**, 1456–60.

Cooley, W. C. and Janda, J. R. (1963). *Handbook of radiation effects in solar cell power systems*, NASA SP-3003, Appendix A. NASA–JPL, Pasadena, CA.

Dienes, G. J. and Vineyard, G. H. (1957). *Radiation effects in solids*. Interscience, New York.

Galashan, A. F. and Bland, S. W. (1990). Neutron and proton irradiation of shallow channel GaAs direct-coupled field effect transistor logic devices and circuits. *Journal of Applied Physics*, **67**, 173–9.

Hanes, M. H., Bartko, J., Chang, J-M., Rai-Choudhury, P. and Leslie, S. G. (1988). Radiation-hard static induction transistor. *IEEE Transactions on Nuclear Science*, **NS35**, 1475–9.

Janousek, B. K., Yamada, W. E. and Bloss, W. L. (1988). Neutron radiation effects in GaAs junction field effect transistors. *IEEE Transactions on Nuclear Science*, **NS36**, 1480–6.

King, J. C. (1974). Hardening quartz resonators to ionizing radiation—a review, Report No. SAND 74-0136. Sandia Laboratories, Albuquerque, NM.

Knoll, G. F. (1989). Radiation detection and measurement. Wiley, New York.

Knudson, A. R., Campbell, A. B., Stapor, W. J., Shapiro, P., Mueller, G. P. and Zuleeg, R. (1985). Radiation damage effects of electrons and H, He, O, Ce and Cu ions on GaAs JFETs. IEEE Transactions on Nuclear Science, NS-32, 4388–95.

Krantz, R. J., Bloss, W. L. and O'Loughlin, M. J. (1988). High energy neutron irradiation effects in GaAs modulation-doped field effect transistors (MODFETs): threshold voltage. *IEEE Transactions on Nuclear Science*, **NS35**, 1438–43.

Magno, R., Nisenoff, M., Shelby, R., Kidd, J. and Campbell, A. B. (1982). Alpha induced upsets in Josephson junctions. *IEEE Transactions on Nuclear Science*, **NS29**, 2090–4.

Maisch, W. G. et al. (1987). Radiation effects in high-T_c superconductors for space application. *IEEE Transactions on Nuclear Science*, **NS34**, 1782–5.

Martin, K. E., Gauthier, M. K., Coss, J. R., Dantas, R. V. and Price, W. E. (1985). *Total dose radiation effects data for semiconductor devices*, JPL Publication 85-43. NASA-JPL, Pasadena, CA.

Messenger, G. C. and Ash, M. S. (1986). *The effects of radiation on electronic systems*. Van Nostrand Reinhold, New York.

Newall, D. M., Ho, P. T., Menick, R. L. and Pelose, J. R. (1981). Total dose hardness of microwave GaAs field-effect transistors. *IEEE Transactions on Nuclear Science*, **NS28**, 4403–6.

Norton, J. R., Cloeren, J. M. and Suter, J. J. (1984). Results from gamma ray and proton beam radiation testing of quartz resonators. *IEEE Transactions on Nuclear Science*, **NS31**, 1230–5.

Nothoff, J. K. (1971). Radiation responses of matched silicon junction field-effect transistors. *IEEE Transactions on Nuclear Science*, **NS18**, 397–403.

Ricketts, L. W. (1972). *Fundamentals of nuclear hardening of electronic requipment*. Wiley, New York.

Rudie, N. J. (1976). *Princilples and techniques of radiation hardening*. Western Periodicals, North Hollywood, CA.

Shedd, W., Buchanan, B. and Dolan R. (1969). The response of silicon JFETs to neutrons. *IEEE Transactions on Nuclear Science*, **NS16**, 87–95.

Smith, L. J., Apodaca, L., DeMartino, V. R. and Trew, J. W. (1977). Charge loss and recovery characteristics of irradiated tantalum capacitors. *IEEE Transactions on Nuclear Science*, **NS25**, 2230–5.

Srour, J. R. and McGarrity, J. M. (1988). Radiation effects on microelectronics in space. *Proceedings of the IEEE*, **76**, 1443–69.

Suter, J. J., Maurer, R. H. and Kinnison, J. D. (1988). The susceptibility of electrodeless quartz crystal BVA resonators to proton ionization effects. *IEEE Transactions on Nuclear Science*, **NS35**, 1451–5.

Sze, S. M. (1981). *Physics of semiconductor devices*. (2nd edn.). Wiley-Interscience, New York.

Tada, H. Y., Carter, J. R., Auspaugh, B. E. and Downing, R. G. (1982). Solar cell radiation handbook, (3rd ed.). JPL Publication 82-69. TRW Systems Group and Jet Propulsion Laboratory, Pasadena, CA.

Zuleeg, R. and Lehovec, K. (1978). Neutron degradation of ion-implanted and uniformly doped enhancement mode GaAs JFETS. *IEEE Transactions on Nuclear Science*, **NS25**, 1444–9.

Zuleeg, R., Nothoff, J. K. and Troeger, G. L. (1982). Ionising radiation response of GaAs JFET's and DCFL circuits. *IEEE Transactions on Nuclear Science*, **NS29**, 1656–61.

10
Polymers and other organics

10.1 Introduction

Organic compounds are sensitive to radiation by virtue of the irreversible chemical processes which take place when covalent bonds, in this case C-C and C-H bonds, are excited or ionized by irradiation. Frequently, the bonds are ruptured and the reactive fragments then form new compounds, a process known as radiolysis. We have not met this process in our discussion of electronic devices and optics because the inorganic materials which largely make up such devices do not undergo decomposition in this way. However, organics in the form of plastics now play an indispensable role in conventional product engineering, examples being moulded structures, packets, fibre-reinforced boards, insulating sheathing, insulating and hydraulic oils, and so on. In more advanced engineering, organics, again mainly polymeric, are also extremely diverse, examples being photoresists, protective, encapsulating and passivating layers, inert binders for pigments and photographic media, and active media for sensing light, force or temperature. Further, there is a huge volume and variety of non-polymeric, often finely divided organic solids which come into our lives as foods, drugs, papers, dyes, explosives, and myriad natural products. Finally, organic liquids play a part in most of the above fields, including drinks, solvents, synthetic intermediates, lubricants, and insulating oils. An unexpected proportion of this diverse list may be exposed to radiation in industrial processes. For example, plant material may be used as a filler in radiation reprocessing, foods are irradiated as a preservation measure, and a toxic material such as dioxin vapour may be irradiated to destroy it. The diversity of the above list of materials makes it clear that, while all are likely to have some reactions in common (e.g. generation of hydrogen), it is a large task to list and categorize accurately the changes in physical parameters or to develop broad prediction schemes in the same way as for silicon devices. On the other hand, much research has been reported around the world by radiation chemists and workers in fields such as nuclear engineering, sterilization methods, and materials processing.

The interpretation of effects in extended structures such as macromolecules (polymers, proteins, etc.) is particularly complex. Many surveys have been written (see Section 10.3.1). Our main object in this chapter is to classify the main forms of organic degradation (or occasionally improvement) under irradiation and summarize some of the more important examples and problems met with polymers in engineering and science. We give a brief description of permanent radiation effects in different classes of polymer and discuss also radiation-induced photocurrents in cables, connectors, and capacitors. Biological effects are briefly discussed in Chapter 2. For electronic systems within spacecraft enclosures, the integrated doses are usually well below the threshold for serious permanent degradation in electrical and mechanical properties, which is a level of 10^7 rad (10^5 Gy) or more.

10.2 Radiolytic reactions

In organic materials under irradiation, a chain of reactions always occurs. The chain may be very complex but always starts with very rapid electronic phenomena (Sangster 1970) which include the conversion of the incident particle or photon into many particles of much lower energy. This is followed by the generation of reactive, short-lived intermediate compounds, usually free radicals (Sangster 1989). This is why the mechanisms of organic reactions under irradiation are often investigated by 'pulse radiolysis' techniques using light or pulsed particle accelerators (Baxendale and Busi 1982). This is also why the effects of irradiation with different particles in organics are qualitatively similar, since they have in common the early 'ionization' stage. The degree of reaction can usually be predicted if the energy deposited in rad(C) (or Gy(C)) is calculated, not forgetting the distribution of dose with depth and the influence of the ambient atmosphere and neighbouring surfaces and species (Turner *et al.* 1988; Egusa, 1991; Sasuga *et al.* 1991). The product of the reactions of the intermediates mentioned above is often a mixture of compounds, as described below.

The standard term for expressing the radiation-chemical yield of such reactions is G, the number of molecules or particles produced per 100 eV of radiation energy deposited. One reference work (Milinchuk and Tupikov 1989) gives extensive lists of the yields and rate constants of well-characterized reactions in non-polymeric organics such as butane or sodium acetate. Sodium acetate produces hydrogen ($G = 0.55$) and methane ($G = 0.71$). Butane produces hydrogen ($G = 4.3$), methane ($G = 3.9$), ethane ($G = 0.48$), and several other hydrocarbons.

An effect common to most organic compounds is the formation of radiolytic hydrogen gas from ruptured C–H bonds. The radiolytic hydrogen yield, $G(H_2)$ provides one of the most fundamental and straightforward measurements of the sensitivity of an organic material to radiation. For example, van de Voorde (1970b) correlates the radiation tolerance of epoxy polymers by comparing their $G(H_2)$ values and their degree of mechanical degradation under gamma rays. G-values vary from 0.1 to 0.01 for epoxy polymers but are greater than 1.0 for nylon and poly(methyl methacrylate).

When an organic molecule is bombarded by radiation, the reactive intermediates formed will interact, but if there are other species in the media around them, they will also react with those agents. Two prime examples of 'other species' are oxygen and water. Both are important in determining the final products, and oxidative degradation from exposure in air (dissolved oxygen) is a serious damage mechanism for polymers. Organics are attacked by radiation-induced OH radicals in water (Ali and Clay 1979; Klein and Schuler 1978); some biological radiation damage occurs by this mechanism (Chatterjee et al. 1986; Turner et al. 1988). The rates of reaction with free radicals or molecules in a solid organic may be slow, leading to 'late effects' and dose-rate effects (Clough et al. 1984; Mayer 1987). Two major industrial uses of radiation—the improvement of materials by radiation processing and the sterilization of foodstuffs and surgical equipment—produce some desirable and some unwanted radiation-chemical reactions in organics. These are discussed in major conferences (see e.g. Silverman 1979).

10.3 Radiation tolerance of polymers and organic compounds according to application

10.3.1 General

Many surveys have been written on the effects of radiation on polymers (see for example Bolt and Carroll 1963; Chapiro 1962; Charlesby 1960; Dole 1972, 1973; Hanks and Hamman 1969; Milinchuk and Tupikov 1989; Swallow 1973). Pinner (1958) amalgamated data, generated mainly by Oak Ridge National Laboratories, into a 'league table' for the utility of polymers under irradiation entitled 'Resistance of commercial organic polymers to ionizing radiation'. This form of chart has since been used widely to summarize the work of many groups. Pinner's chart had the form:

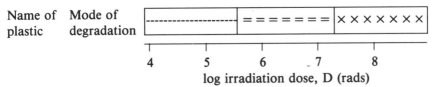

where the three bar symbols indicated progressive damage to a mechanical parameter such as ultimate elongation. The first symbol (----------) indicated 10 per cent change, and the second (= = = =) 90 per cent and the third, destruction. The table was used with acknowledgement by Swallow (1973) and has often mistakenly been attributed to Swallow by later authors, who have repeated the format with little alteration and added some new data, say for filled as well as unfilled polymers. The numerical values for degradation have been replaced by codings for damage or 'utility' such as 'incipient', 'mild to moderate', 'limited use', etc. A more accurate mode of expressing degradation is of course a plot of the growth of damage versus dose, as shown later for polypropylene, or a 'radiation index' (RI) such as the logarithm of the dose in Gy at which elongation at break, E, a well-defined mechanical property, is halved (IEC 1977). The RI for a polyolefin cable sheath tested by Schönbacher et al. (1989) was 5.5 (i.e. E was halved at a dose of $10^{5.5}$ Gy ($10^{7.5}$ rad)).

A league chart was issued by van de Voorde and Restat (1972 and Appendix F). Our tabular analysis of this chart, given in Appendix F, shows that only a few types of plastic are affected at 10^7 rad (10^5 Gy), but few are not severely affected at 10^8 rad (10^6 Gy). As a reservation, we should note that data on mechanical properties are usually the main basis of such league charts. Fillers can reduce radiation effects (see for example Pinner 1958; and optical or electrical changes can usually be detected at much lower doses than mechanical damage effects). Therefore the league charts should be used with great caution in engineering work or, as Pinner (1958) says: 'Quantitative extrapolation must be circumspect'.

10.3.2 Polymers in electronics

Polymers are used very widely as subelements of electronic piece parts and circuits, for example:

(1) epoxy and polyurethane resins in circuit boards as adhesives and potting or encapsulation compounds;
(2) Mylar, polycarbonate and polypropylene in capacitors;
(3) varnishes and epoxy compounds for coil insulation and magnet cores;
(4) silicones and polyimides in many specialized electronics applications;
(5) PVC, PTFE and sulphones in sockets and connectors and on wires.

In radiation environments with ionization doses of less than 10^7 rad (10^5 Gy) it is unlikely that any of the above will exhibit serious degradation, except PTFE and polypropylene. When exposed in air or oxygen, degradation begins to occur below this dose value. Egusa (1991) discusses some special degradation effects in glass-reinforced polymers (see section 10.3.4).

10.3.3 Remote handling

Remote handling in nuclear plants and space involves microelectronically controlled handling devices working very close to, or surrounded by, radiation sources. Whereas the electronics can sometimes be removed from the worst radiation flux, some parts cannot. This includes front optical elements, actuators, and preamplifiers. Emergency equipment may have to be fully immersed in the radiation field. It has been stated that robots used in the clearing of radioactive debris from the Chernobyl disaster failed owing to radiation effects on the electronics. Such limitations may imply that polymeric parts are not the limiting factor in remote control equipment. However, some plastics parts, such as grippers, may be in highly exposed positions (10^7 rad h^{-1}, 10^5 Gy hr^{-1}) in the machinery of a nuclear or radiation processing plant (Holmes-Siedle 1985). Clearly, total doses can exceed

10.3 Radiation tolerance of polymers and organics

10^8 rad (10^6 Gy) in a short time, meaning that few polymers will not be severely affected (see Section 10.3.1).

10.3.4 Accelerator parts

As described in Chapter 2, accelerator magnets and electronics are exposed to secondary photons and neutrons. Egusa (1991) has shown that, in polymer composites with carbon or glass fibres, as used in cryogenic magnets and circuit boards, the scattering of neutrons leads to an unexpectedly high damage rate from these particles in the epoxy composites supporting coils.

In large accelerators such as the 450 GeV proton machine at CERN, Geneva, there are large numbers of magnets, up to 1000 in the installation. Materials for the magnet, including polymeric sheathing, are chosen with minimum engineering tolerances because space is limited. Tests for candidate polymers were taken to doses of 10^9 rad (10^7 Gy) (Phillips 1972, 1981; Schönbacher *et al.* 1979 a, b, 1989). Low electrical conductivity must be maintained at all costs, while mechanical integrity is important if the coils are subject to mechanical shifting. Fluorescent organics, used as scintillators are used in many ways in the high-energy physics measurements carried out around the collision area of the accelerator experiment (see next section).

10.3.5 Optical fibres, windows and scintillators

The problem of darkening of optical media in general is dealt with in Chapter 8. For some applications, polymeric fibres are competitive with glass or silica. It has been shown that even unpurified clear polymers such as polystyrene do not undergo serious optical loss at doses in the megarad range (Holmes-Siedle, unpublished work); this is considerably better than some glasses. Radiation-induced optical loss could become a problem in opto-isolators or light sensors/emitters which employ plastic windows or lenses for isolation or encapsulation. Changes in the absorption spectra of polymeric transparencies are unlikely to be as predictable as for inorganic materials. The effects of ionizing radiation on fluorescent organic materials is discussed in Section 8.5. The darkening of polyvinyltoluene and polyacrylic scintillators is described in the CERN compilation of *essais de radioresistance* (Beynel *et al.* 1982).

The active ingredients of organic scintillators are highly fluorescent compounds such as chloroanthracene (see e.g. Belcher *et al.* 1957). Over and above light losses brought about by the darkening of the medium, the fluorescence of the organic molecule will be changed or destroyed if radiation breaks it up. For example, if chlorine were split from the chloroanthracene molecule, the remaining anthracene would still fluoresce, but at a different wavelength and efficiency. Scission of the carbon–carbon rings would destroy the fluorescence. There is thus scope for increasing the tolerance of scintillators to radiation by classical radiochemical techniques.

10.3.6 Lubricants

Lubrication by organic oils is dependent on the geometry and size of the lubricating molecules. Thus, it is not surprising that radiation, when it accelerates molecular scission, cross-linking and oxidation, should also hasten the degradation of lubricating properties. Oxidative degradation is also caused by heat, and it is found that oils suitable for high temperatures are also normally more resistant to radiation (van de Voorde 1970a). Antioxidants and bases which deal with oxidation and its products will normally improve radiation-tolerance. Other additives may be used to reduce radiolysis of the lubricant molecule.

10.3.7 Bombardment of coatings in space

The surfaces of spacecraft are bombarded by photons, electrons, protons, and alpha particles. The numbers and energies of these particles are such that, in a skin layer of the order of 0.1 mm, chemical disruption, coloration and electrostatic charging effects are produced. The effects will be pronounced in polymeric insulators, and special research is done to ensure that thermal blankets and paints will withstand these effects over the required mission period.

10.4 Radiation processing

10.4.1 Sterilization of products

Irradiation of packaged surgical devices in cobalt-60 cells containing megacuries (PBq) is replacing the use of ethylene oxide gas to sterilize before packaging. Gamma rays are preferable so that large cartons can be irradiated; for electron beams, packages have to be thin. Traditionally, the dose used has been 2.5 Mrad (25 kGy) (this is 'overkill' even for radiation-tolerant bacteria). On testing polymers to 5 Mrad (50 kGy), users found strong coloration and sometimes embrittlement (Mayer 1987). Polycarbonate was not affected mechanically, but the medical devices assumed a yellow coloration unacceptable to users. PVC turned deep orange. Manufacturers experimented with stabilizers until, as Mayer says, 'Today [1987], it is almost as difficult to find a PVC tubing which will turn yellow as it was to find one that would not in 1980'. Other unacceptable effects in medical applications are the odour of decomposition products, which may also be extractable in a solvent. Dosimetry using colour changes in organic materials is used as a control in sterilization processing and dose monitoring in accelerator chambers (see section 1.5).

10.4.2 Irradiation of foods

The irradiation of food varies with the product and the length of time it must keep. Sprouting in vegetables can be inhibited by 10 krad (100 Gy). Doses up to 0.5 Mrad (5 kGy) are used to lengthen the shelf-life of packaged foods, especially seafood (see, for example, annual reports of Bhabha Atomic Research Centre, Trombay, India). Doses up to 1 Mrad (10 kGy) have been used to improve the cooking properties and digestibility of beans; this treatment has the beneficial effect of reducing 'flatulence factors'—undigested saccharides which normally pass to the lower intestine.

10.4.3 Radiation curing of plastics

Silverman (1979) gives an impressive list of the number of commercial components which are fabricated with the help of radiation to cure the polymers, which include elastomers and coatings including those on frying pans. Several hundred electron-beam and gamma machines are in action around the world. Curing doses vary but are likely to be in the region of 10^7 rad (10^5 Gy), so that damage to parts of the product not being cured must be considered.

10.4.4 Resists

Photoresist technology began with the use of poly(methyl methacrylate) layers, which react to radiation or UV light. Chain scission in the polymer enhances the dissolution rate and allows 'development', in which the exposed part can be removed by a solvent or 'developer' (Hiraoka 1977). Greater sensitivity and resolution can be obtained from 'negative' resists, in which radiation causes very efficient cross-linking groups (for example a diazidophenyl compound) to be activated and join up the main polymer chains (for example polyisoprene). In the exposed regions, the polymer hardens (Reichmanis 1989). The key to sensitivity is the strong tendency of an azide compound to decompose under irradiation to nitrogen and a free radical:

$$R-N_3 \rightarrow R-N: + N_2.$$

Using electron and proton beams, patterns with line-widths less than 0.5 μm can be defined.

10.5 Long-lived degradation in polymers

10.5.1 Relative sensitivity

There are many books and handbooks on the long-lived effects of radiation on the engineering and chemical properties of commercial and pure polymers. Detectable degradation of the mechanical, optical and electrical properties of polymers usually appears at a dose value lower than 100 Mrad (10^6 Gy). In some cases, doses less than this can actually *improve* toughness. Doses in the 3 Mrad range are actually used for the routine sterilization of surgical plastics. An exception to this is the case of fluoropolymers, some of which show catastrophic mechanical degradation if exposed in air at

TABLE 10.1 Course of reaction in polymers exposed to radiation—behaviour of the carbon backbone

Cross-linking of chains		Degradation (scission) of chain
Polyethylene	Epoxy resins	Polyisobutylene
Polypropylene	Silicones	Polytetrafluorethylene (Teflon)
Polystyrene	Poly(ethylene terephthalate)	Polymethacrylates (Perspex, Lucite)
Poly(vinyl chloride)	Most elastomers	Cellulose & derivatives
		Butyl rubber

megarad doses. In most polymers, long-lived electrical leakage and increased dielectric loss are likely to appear well before mechanical degradation occurs. Different forms of radiation normally produce the same qualitative effects in polymers, namely the rupture of chemical bonds, the evolution of gaseous molecules, destruction of crystallinity, and change of colour. All of these are ionization effects. Not unexpectedly, the quantitative effect for different forms of radiation can be predicted by calculating the dose. In polymers, the rupture of chemical bonds usually results in one of two sequels: cross-linking or scission, possibly complicated by attack on the intermediates by an outside species such as air or water.

Table 10.1 lists the polymer types most likely to undergo cross-linking or scission, as the case may be. As the radiation dose increases, polymers which tend to cross-link show an increase in strength and toughness; however, they ultimately become brittle. Those tending to scission will become steadily weaker and may eventually liquefy. In both groups, hydrogen gas and corrosive gases may be evolved and colorations are produced owing to complex side-reactions. As may be imagined from the very large variety of compositions, engineering plastics differ greatly from one to another in radiation-tolerance. The threshold dose for radiation damage varies by at least three orders of magnitude. Because the demands made on plastics in different applications are so diverse, it is often difficult to predict degradation effects except by considering individual cases and by radiation testing (IEC 1977). Most of the 'league tables' compiled are based on the measurement of tensile properties (see Appendix F); Kapton, polystyrene, phenyl-silicones, polyethylene, Mylar and epoxy are usually near the top of any such tables, while nylon, poly(vinylchloride) most fluoropolymers and all elastomers are usually near the bottom.

10.5.2 Effects of additives and fillers

The long-lived effects of radiation on plastics can be strongly affected by the chemical additives used, especially by the quantities of antioxidant and plasticizer present. In both cases, the additives delay the onset of mechanical and other effects caused by oxidation, chain scission and embrittlement. Therefore, to increase radiation-tolerance in plastics, we can specify plastics which are rich in these additives. Fillers and pigments may sometimes alter the radiation-induced behaviour of the polymeric matrix (Pinner 1958; Egusa 1991).

10.5.3 Combined effects of stress (fields, vacuum and temperature) and ageing with irradiation

All plastics age on storage owing to the evaporation or exhaustion of antioxidants and plasticizers. Hostile environments may accelerate radiation-induced effects. For example, if plasticizers vaporize when a polymer is irradiated *in vacuo*, then radiation-induced embrittlement can be said to be accelerated by the use of vacuum; however, since oxygen is gradually removed from a polymer in a vacuum, oxidative scission may be greatly reduced. Increased temperatures will encourage oxidative attack and outgassing. Exposure to an electrical field or light is also likely to accelerate these effects. PVC cable material is more severely affected by radiation if irradiated slowly at increased temperature (Clough *et al.* 1984).

10.6 Radiation–tolerance of plastics according to technology

10.6.1 General

Engineering polymers are frequently grouped according to the manufacturing processes used and the function required. Broad process classifications are thermoplastics, thermosets, and elastomers. Functional classifications include structural, insulating, coating, adhesive, and optical. Other specialist classes include 'active' polymers such as resists, piezoelectric films, charge storage layers (electrets), and light-sensitive or colour-change media. Demands on stability (mechanical, electrical, and optical) vary very widely in these materials. 'League tables' of generic plastics names can be misleading unless the properties measured and the tolerances allowed are specified. For example, Appendix F suggests that few plastics show damage in vacuum at doses below 10^6 rad (10^4 Gy). On the contrary, long polymeric optical fibres may be rendered useless by a dose of 10^4 rad (10^2 Gy). In the following section we make comments on plastics by technology. Systematic listings of test data can be found in van de Voorde (1970a, b), Dauphin (1977), Schönbacher and Stolarz-Izycka (1979a, b), Maier and Stolarz (1983), Beynel *et al.* (1982); Bouquet (1986); Campbell (1981), Wilski (1987), and Phillips (1981). The latest series of test reports by Schönbacher *et al.* (1989) expresses mechanical property degradation according to the recent IEC standard (1977), including a radiation index (RI). Hammoud *et al.* (1987) Laghari *et al.* (1990) and van de Voorde (1970a, b) compare degradation

284 Polymers and other organics

of electrical and mechanical properties, showing that these do not necessarily reach failure at the same dose.

10.6.2 Thermoplastics

10.6.2.1 *Structural plastics*

A very large variety of parts of machines and electronic equipment are made by the moulding of thermoplastics. Radiation-tolerance estimates related to the mechanical performance of about 35 different types of polymer are listed by Schönbacher and Stolarz-Izycka (1979b). These indicate that very few thermosetting plastics show damage effects below a dose of 1 Mrad (10^4 Gy). Polymers which exhibit damage at this dose include the cellulose group, polypropylene, polyformaldehyde (Delrin), and PTFE. In the case of PTFE, the results quoted are for irradiation in air. In vacuum, the tolerance of PTFE is increased ten-fold. The most tolerant thermoplastics include polyimides (Kapton), Mylars and polystyrene. In all halogenated polymers (PTFE, PVdF, PVC), the possibility of release of halogen acid vapour must be considered. Sulphur dioxide is released from polysulphone plastics. Detailed listings of gaseous releases can be found in Milinchuk and Tupikov (1989). Tensile and breaking strength and elasticity are usually the properties of most interest in this group.

10.6.2.2 *Plastics films as dielectrics and coatings*

Thin thermoplastic films (usually extruded or solution-cast) are often used at the limit of their mechanical or electrical strength in capacitors, varnishes or wrappings. Polyester (Mylar) and polyimide (Kapton) films are found to be very stable under irradiation. Little degradation is found at doses of 10^9 rad 10^7 Gy). Polyethylene film develops an odour at about 10^7 rads (10^5 Gy) and loses some mechanical strength at 10^8 rad. Polypropylene is worse in this respects and this case illustrates two important variables in testing and prediction of polymer performance: Fig. 10.1 is a plot of results from Hammoud *et al.* (1987) and Wilski (1987). Wilski found that the change of elongation-at-break is strongly affected by the dose rate. The explanation for this is that, at the high rate, oxygen in the polymer is rapidly consumed by reactive intermediates. At low rates, oxygen concentration is maintained by diffusion of oxygen from the surface. Hammoud *et al.* compared electrical and mechanical properties of the same film material and found that electrical properties survived well after the material had become brittle. Transient conductivity is an important parameter in selecting dielectrics for capacitors and is discussed in a later section. Gas evolution, including acid vapours, is significant at dose values near 10^8 rad (10^6 Gy).

10.6.3 Thermosetting plastics

Thermosetting plastics are usually formed by the mixture of a fluid prepolymer such an epoxy compound with an agent which initiates polymeriza-

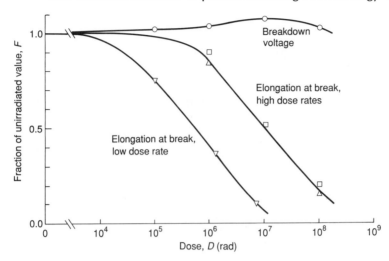

FIG. 10.1 Change of mechanical and electrical properties of polypropylene under irradiation: △, ▽, monofilaments under gamma-rays at 445 and 20 000 rad h^{-1} (Wilski 1987); □, ○, 25.4 μm isotactic film under electron beam at >18 rad h^{-1} (Hammoud et al. 1987).

tion (a hardener such as an amine). This produces a highly cross-linked polymer. Composite structural elements, PCBs (printed-circuit boards), and adhesives are made from this class of polymer. The manufactured formulas are very varied and, therefore, response to radiation is diverse. However, because of the high original degree of cross-linking, the radiation-tolerance is expected to be, and often is, very good. Some epoxy adhesives lose only about 10 per cent of their tensile shear strength at a dose of 10^9 rad (10^7 Gy) in air, (van de Voorde 1970b). Epoxy glass laminates also show good performance at this dose level.

10.6.4 Elastomers

The need to maintain elasticity is of course the critical property in this group. The most stable elastomers are polyurethanes and phenylsilicones (usable to well above 10^8 rad (10^6 Gy). Nitrile, styrene–butadiene and natural rubbers can also be used up to 10^8 rad (10^6 Gy). Butyl rubber liquefies and neoprene evolves HCl at similar dose levels. Most proprietary polyurethane foam rubbers can be used as electrical encapsulant materials at a dose level of 10^9 rads (10^7 Gy) in vacuum at temperatures between −85°C and +250°C. Silicone and polysulphide sealants are probably less tolerant to radiation.

10.7 Radiation–induced conductivity in insulators

The exposure of solids to ionizing radiation produces current-carriers in the form of electrons and holes. If the original conductivity is small, then the presence of carriers produces an observable increase in the background conductivity of the material σ_B. The increases follow the rules of normal conduction and we can thus calculate a 'radiation-induced conductivity' (RIC) value σ_R (see for example Wintle, 1966; Thatcher *et al.* 1969; van Lint *et al.* 1980). Ideally, this will increase instantaneously to a steady level when exposure starts, and decay when it stops. If σ_R is large and σ_B is small, we can use the simple formula:

$$R_R = \frac{L}{\sigma_R - A}, \qquad (10.1)$$

where L is the sample length, A is the sample area and R_R is the resistance of the insulator during irradiation (assumed to be very much greater than the background resistance). We can say that a new resistor R_R appears in parallel with resistor R_B which represents the small background conductivity. The expected magnitude of this new resistor R_R can be found from the curve in Fig. 10.2. A typical radiation-induced response of a polymer is given by the curve:

$$\sigma_R = 10^{-18} \cdot \dot{d} \qquad (10.2)$$

where $\dot{d}$ is the dose rate in rad(c). min^{-1} and σ_R is in Ω^{-1}cm^{-1}. The measured results for many polymers lie about the curve, and the linear formula may be used for approximate predictions of RIC. The photocurrent I_R is of course calculated from R by Ohm's law. For some polymers, the dependence of σ_R on $\dot{d}$ is however not linear, e.g. the points given in Fig. 10.2 for polyethylene show R to be proportional to $\dot{d}^{0.7}$.

In Fig. 10.2, some representative dose-rate values for space vehicles, fusion reactors, and gamma pulses from nuclear bursts are marked on the bottom scale. The upper limits for the radiation-induced conductivity values for these three conditions are respectively 10^{-18}, 10^{-13} and $10^{-6}\Omega^{-1}$cm^{-1}. We can see that, for a 1 cm cube of polymer, these represent R_R resistors of values 10^{18}, 10^{13} and $10^6\Omega$ respectively. As regards the practical impact, we can see that probably only in the last case would this change of resistance produce any noticeable reduction in the performance of a polymer as an insulator. This is because the insulation resistance R_B will be typically of the order of $10^{13}\Omega$. Adding a resistor of the same, or higher, value in parallel will not produce a serious disturbance of the insulator's function. The linear formula given above gives us a simple engineering approach to apply to leakage questions for space systems or ground installations. However, Fig. 10.2 shows us that it may not give the 'worst case' RIC value for low dose

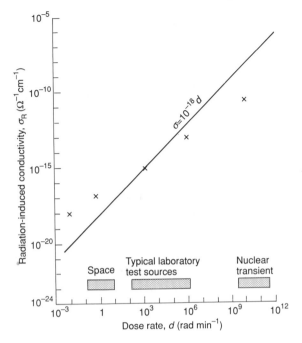

Fig. 10.2 Polymer leakage: prediction curve for radiation-induced conductivity in many polymers; the crosses show an exception, polyethylene; ranges of dose rates for various environments are shown.

rates. Kurtz *et al.* (1983) describe a method for reducing RIC in dielectrics by adding an electron-accepting organic compound to the commercial polymer film. Sullivan *et al.* (1971) describe a method for routinely measuring the conductivity of plastic films with time after a pulse of 10^{13} rad sec^{-1} (10^{11} Gy sec^{-1}).

10.8 The space environment

The integrated doses accumulated within spacecraft enclosures during the normal mission time are not usually high enough to affect the operation of equipment. However, some exceptions may be cited:

(1) in space, polymers will be directly exposed to bombardment at the surface of the spacecraft and hence receive a high doses, along with the effects of atomic oxygen;

(2) The optical properties of some polymers may degrade at doses as low as 10^4 rad (10^2 Gy), especially in optical fibres (long optical paths);

(3) polymeric transducers (e.g. photoconductors scintillators or pyroelectric crystals) may be damaged easily by radiation;

(4) remote manipulators on spacecraft will have to work in especially severe

environments, e.g. when changing the isotope charge in a power generator.

(5) The release of reactive gases may corrode nearby metal films.

10.9 Conclusions

In organic materials under irradiation, a chain of reactions, in which oxygen and moisture from the environment may be incorporated, starts with rapid electronic phenomena, followed by the generation of reactive, short-lived intermediate compounds and finally a complex mixture of chemical products. Radiation effects in polymers include permanent degradation and transient conductivity. Mechanical, electrical, and optical properties may be changed at doses above 1 Mrad (10^4 Gy), but only a few polymers degrade severely at a dose below 10^7 rad (10^5 Gy). Heavy surface bombardment in space may have severe effects on plastics sheeting or coatings. The commonly published bar charts refer mainly to effects on mechanical properties, not to electrical or optical effects, which may be more severe (see Appendix F). In engineering terms, designers should recognize that polymeric materials in special high-performance uses may be subject to radiation-induced degradation and should scrutinize plastics items intended for specialist applications in the same manner as for semiconductor parts.

References

Ali, S. M. and Clay, P. G. (1979). Gamma radiolysis of cellulose acetate. *Journal of Applied Polymer Science*, **23**, 2893–7.

Baxendale, J. H. and Busi, F. (ed.) (1982). *The study of fast processes and transient species by electron pulse radiolysis*. Reidel, Dordrecht.

Belchek, E. H. and Geilinger, J. E. (1957). Improved scintillating media for radiation dosimetry. *British J. Radiology*, **30**, 103.

Beynel, P., Maier, P. and Schönbacher, H. (1982). Compilation of radiation damage test data, Part III: Materials used around high-energy accelerators, Report No. CERN 82-10. CERN, Geneva.

Bolt, R. O. and Carroll, J. G. (1963). *Radiation effects in organic materials*. Academic Press, New York.

Bouquet, F. L. (1986). *Radiation effects on electronics*. Systems Co., Panorama City, CA.

Campbell, F. J. (1981). Radiation damage in organic materials. *Radiation Physics and Chemistry*, **18**, 115–25.

Chapiro, A. (1962). *Radiation chemistry of polymer systems*. Wiley, New York.

Charlesby, A. (1960). *Atomic radiation and polymers*. Pergamon Press, Oxford.

Chatterjee, A., Koehl, P. and Magee, J. L. (1986). Theoretical consideration of the chemical pathways for radiation-induced strand breaks [in DNA]. *Advances in Space Research*, **6**, 97–105.

Clough, R. L. *et al.* (1984). Long-term radiation effects on commercial cable insulating materials. *Nuclear Safety*, **25**, 238.

Dauphin, J. (1977). Guidelines for the simulation of the degradation of materials to particulate radiation in space, ESA PSS-34/QRM-07T. European Space Agency,

References

Dole, M. (1972/1973) *The radiation chemistry of macromolecules*, Vols I, II. Academic Press, New York.
Egusa, S. (1991). Effects of neutrons and gamma rays on polymer matrix composites as low temperature materials. *Radiation Physics and Chemistry*, **37**, 135–40.
Hammoud, A. N., Laghari, J. A. and Krishnakumar, B. (1987). Electron radiation effects on the electrical and mechanical properties of polypropylene. *IEEE Transactions on Nuclear Science*, **NS34**, 1822–6.
Hanks, C. L. and Hamman, D. J. (1969). The effects of radiation on electrical insulating materials, Report No. REIC-46. Radiation Effects Information Center, Battelle Memorial Institute, Columbus, OH.
Hiraoka, H. (1977). Radiation chemistry of poly(methacrylates). *IBM Journal of Research and Development*, **21**(2), 121–30.
Holmes-Siedle, A. G. (1985). Radiation effects on robotic manipulators, EEC Project Report. Taylor Hitec, Chorley, UK.
IEC (International Electrotechnical Commission) (1977). Guide for determining the effects of ionizing radiation on insulating materials, IEC Standard 544 (in 4 parts) Geneva. [This specification includes a standard format for presenting test data.]
Klein, G. W. and Schuler, R. H. (1978). Oxidation of benzene by radiolytically produced OH radicals. *Radiation Physics and Chemistry*, **11**, 167–71.
Kurtz, S. R., Arnold, C. and Hughes, R. C (1983). The development of a radiation hardened polymer dielectric by chemical doping. *IEEE Transactions on Nuclear Science*, **NS30**, 4077–80.
Laghari, J. R. and Hammoud, A. N. (1990). A brief survey of radiation effets on polymer dielectrics. *IEEE Transactions on Nuclear Science*, **NS37**, 1076–83.
Maier, P. and Stolarz, A. (1983). Long-term radiation effects on commercial cable-insulating materials irradiated at CERN, Report CERN 83-08. CERN, Geneva.
Mayer, B. (1987). Recent developments in radiation-sterilizable plastics. In Proceedings of ANTEC '87, pp. 1190–2. Society of Plastics Engineers, Brookfield Court, CT.
Milinchuk, V. K. and Tupikov, V. I. (1989). *Organic radiation chemistry handbook*, (trans.T. J. Kemp) Ellis Horwood, Chichester.
Phillips, D. C. (1978). The effect of radiation on electrical insulators in fusion reactors. *Report No* AERE-8923. UK Atomic Energy Authority, Harwell.
Phillips, D. C. *et al.* (1981). The selection and properties of epoxide resins used for the insulation of magnet systems in radiation environments. *Report No. CERN* 81-05. CERN, Geneva.
Pinner, S. H. (1958). Irradiation of polymerisation products. *Reports on Progress in Applied Chemistry*, **43**, 463–76.
Reichmanis, E. (1989). Radiation chemistry of polymers for electronic applications. In *The effects of radiation on high-technology polymers*. American Chemical Society, Washington, DC.
Sangster, D. F. (1970). *Principles of radiation chemistry*. Arnold, London.
Sangster, D. F. (1989). In *The effects of radiation on high-technology polymers*. American Chemical Society, Washington, DC.
Sasuga, T., Kawanishi, S., Niishi, M., Seguchi, T. and Kohno, I. (1991). Effects of ion irradiation on the mechanical properties of several polymers. *Radiation Physics and Chemistry*, **37**, 135–40.

Schönbacher, H. and Stolarz-Izycka, A. (1979a). Compilation of radiation damage test data, Part I: Cable insulating materials, Report No. CERN 79-04. CERN, Geneva.

Schönbacher, H. and Stolarz-Izycka, A. (1979b). Compilation of radiation damage test data, Part II: Thermosetting and thermoplastic resins, Report No. CERN 79-08. CERN, Geneva.

Schönbacher, H. and Tavlet, M. (1989). Compilation of radiation damage test data Part I, 2nd edn. halogen-free cable-insulating materials. *Report No. CERN*, 89-12. CERN, Geneva. [Note: this edition includes, for the first time, data in the format recommended by the International Electrotechnical Commission, IEC standard 544].

Silverman, J. (1979). Current status of radiation processing. *Radiation Physics and Chemistry*, **14**, 17-21.

Sullivan, W. H. and Ewing, R. L. (1971). A method for the routine measurement of dielectric photoconductivity. *IEEE Transactions on Nuclear Science*, **NS18**, 310-7.

Swallow, A. J. (1973). *Radiation chemistry*. Longman, London.

Thatcher, R. K. and Kalinowski, J. J. (1969). *TREE handbook*, Report No. DASA 1420, Battelle memorial Institute, Columbus, OH.

Turner, J. E. *et al.* (1988). Studies to link the basic radiation physics and chemistry of liquid water. *Radiation Physics and Chemistry*, **32**, 503-10.

Van de Voorde, M. H. (1970a). Effects of radiation on materials and components, Report No. CERN 70-5. CERN, Geneva.

Van de Voorde, M. H. (1970b). Action des radiations ionisantes sur les resines epoxydes, Report No. CERN 70-10. CERN, Geneva.

Van de Voorde, M. H. and Restat, C. (1972). Selection guide to organic materials for nuclear engineering, Report No. CERN 72-7. CERN, Geneva.

Van Lint, V. A. J., Flanagan, T. M., Leadon, R. E., Naber, J. A. and Rogers, V. C. (1980). *Mechanisms of radiation effects in electronic materials*. Wiley-Interscience, New York. pp. 205-69.

Wilski, H. (1987). The radiation-induced degradation of polymers. *Radiation Physics and Chemistry*, **29**, 1-14.

Wintle, H. J. (1966). Radiation-induced conductivity in insulators. *Radiology*, **33**, 706-7.

Further reading

Clegg, D. W. and Collyer, A. A. (1991). *Irradiation effects on polymers*. Elsevier Applied Science, London. [An up-to-date book on radiation effects in polymers escaped our notice until recently. A collection of articles by authorities, edited by Clegg and Collyer, gives detailed coverage of recent technical and scientific advances. An example of a recent advance is an explanation of the high radiation tolerance of an engineering polymer, ULTEM, a polyetherimide, which has advantages over Kapton, a polyimide. Contributors include Clough, Phillips, Burnay, and Dole.]

11
The interaction of radiation with shielding materials

11.1 Introduction

The understanding of radiation effects in systems depends in part on a good understanding of the fundamental interactions between particles and photons and the materials in question. These interactions affect both shielding calculations and predictions of local effects in device structures. The local effects have mainly been discussed elsewhere in relation to specific device structures. The interaction of radiation in shields is a broad subject well covered in standard texts such as Price *et al*. (1957), Profio (1979), and Chilton *et al*. (1984). Other useful discussions of interactions are given in Evans (1955), Hubbell (1969), Knoll (1989), and Delaney (1992). In this chapter we discuss selected aspects of radiation shielding in general and also discuss two portions of interaction theory not widely covered: first, the interactions having a practical impact on the stopping of space radiation in spacecraft, especially applicable to analysing radiation effects in unmanned spacecraft, and second, some interactions important for soft X-rays.

Previous chapters have described the radiation environment to which Earth-orbiting electronic devices may be exposed, and have indicated how these devices react to that exposure. It will now be useful to describe the physical principles which govern the transport of energy from the incoming radiation flux to the device materials. Of fundamental importance in this respect is the estimation of the attenuating effect of other materials protecting the device from the external radiation environment. This protection may be provided fortuitously by spacecraft structural members or covers ('built-in absorbers'), or may have to be added specifically for protection ('add-on absorbers').

This chapter describes some of the physical processes involved in radiation transport through matter, its attenuation and scattering, and generation of secondary radiation. Radiation attenuation data for the important spacecraft trajectories through the Earth's trapped radiation environment are given in graphical form. These data should allow the manual evaluation of radiation doses in simple absorber geometries.

Subsequent chapters will discuss ways of making the best use of shielding material at minimum cost and weight penalty, and describe computerized

292 The interaction of radiation with shielding materials

methods which deal with the complexity of some radiation-dose calculations for spacecraft.

11.2 Particle radiation transport and range

11.2.1 General

Energetic particles passing through material can undergo a variety of interactions leading to energy loss, scattering and/or the generation of secondary particles. in Chapter 2 it was seen that the principal components of the Earth's radiation environment are energetic electrons and protons. Interactions involving these particles are summarized below. The reader is referred to standard texts mentioned earlier for a more detailed description.

11.2.2 Range

The range of a charged particle in a material is the thickness of material penetrated before the particle loses all its energy. 'Practical range' (R_p) is the most probable range of a particle of a given incident energy. Some electrons, in particular, clearly penetrate thicknesses greater than R_p and are often termed 'straggling electrons'. The term 'maximum range' takes this into account; the distinction will be illustrated more clearly in the subsequent discussion of transmission coefficients.

Range may be expressed either in units of depth (length) or, more commonly, as the product of depth and density (g cm^{-2}). This unit is equivalent to mass per unit area and used frequently in radiation studies as a measure of absorber thickness (it may be termed 'shield thickness' in some diagrams). For convenience, we shall henceforward refer to this unit as 'mass thickness'. Some values for electron and proton ranges in aluminium are tabulated in Appendix C.

The range of a charged particle at a given incident particle energy, when expressed in g cm^{-2}, is closely similar for all materials. There are second-order effects dependent on excitation potential and atomic number, but — at the energies likely to be encountered in the space environment — the spread in range is no more than a factor of two. Plots of the practical ranges of electrons and protons in a number of materials are shown in Fig. 11.1. The data of both particles in Si is taken from Berger and Seltzer (1964, 1966) and from Barkas and Berger (1964). Ranges in aluminium are given by Linnenbom (1962), and proton ranges in aluminium are also given by Cooley and Janda (1963). Data on various particle ranges in materials, including germanium and liquid propane, is also given by Berger and Seltzer (1982). Range data are tabulated in Ziegler *et al.* (1985) and are available in PC software form from Ziegler as 'TRIM' (Ziegler 1991).

Unlike photons, neutrons are slowed down by the *nuclei* of atoms. Therefore, more so than for photons, the process varies with the individual

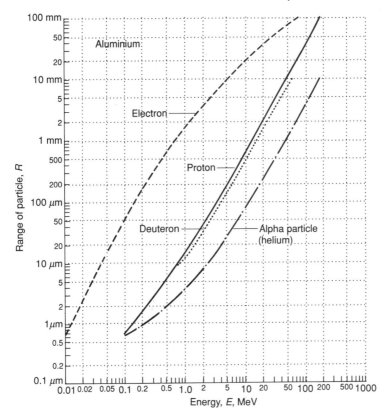

FIG. 11.1 Particle range in materials: practical ranges of electrons, protons and alpha particles in aluminium as a function of incident energy. For numerical values, see Appendix, tables B6 to B8.

tendencies of nuclei to capture or be excited by the neutrons. For water, the slopes of the attenuation of neutrons and photons from an isotope are both exponential with depth but the slopes are steeper for all energies of neutron (see e.g. Chilton *et al.* 1984).

11.3 Transport of electrons

When we are considering transport and damage processes, electrons can be regarded as rather 'light' particles which interact with material mainly at the atomic (i.e. not the subatomic) level, producing excitation and ionization while losing momentum and being scattered in the process. Because of their low mass relative to nuclei, electrons are readily scattered through large angles, and complete 'back-scattering' of electrons from materials can be significant, especially in high-Z materials. The deceleration of the

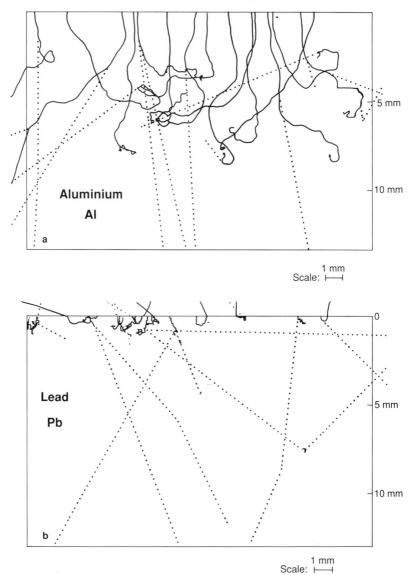

FIG. 11.2 Trajectories of 5 MeV electrons in (a) aluminium and (b) lead, computed with the GEANT Monte Carlo code; electrons are injected normally from above. Dotted lines indicate the paths of bremsstrahlung photons induced by the electrons. Note the heavy backscatter from lead (ESA 1989). (ESA with permission.)

electrons in the strong electric fields of the atom results in the generation of energetic photons, a process known as 'bremsstrahlung' (braking radiation). Figure 11.2 illustrates electron–bremsstrahlung behaviour in matter, showing trajectories in aluminium and in lead. These trajectories were computed by Daly (in ESA 1989) using the CERN Monte Carlo code GEANT (see Chapter 12 for a discussion of computational methods). The figure shows clearly the tortuous nature of electron motion and the production of penetrating bremsstrahlung. Per unit thickness, lead is clearly a better absorber of electrons than aluminium but, if units of mass thickness (distance × density) are used for calculating the range, then there are only small differences between materials of widely different atomic numbers (Knoll 1989). We can see, however, that lead produces greater backscatter.

11.3.1 Transmission coefficients for electrons

We have already indicated that the transmission of electrons through material is a highly complex process. They do not travel in continuous straight lines, but follow highly scattered paths. The degree of scattering depends, for example, upon the material, the incident particle energy, and the angle of incidence. A number of sophisticated analytical treatments have been developed to estimate the dose deposited and the residual flux after transmission of electrons through material of given 'stopping power'. Among the best known are the various Monte Carlo methods, whereby an electron track through a material is divided into a large number of 'steps' very similar to the real case. In a typical treatment of an aluminium medium, as described by Berger and Seltzer (1968), the step size is chosen so that electron energy decreases on average by a factor of 2^{-8} per step.

When a stream of 'monoenergetic' electrons passes through a material, the energy and particle flux are both reduced. The emergent energy spectrum is broadened and somewhat asymmetric; however, a clearly dominant peak or 'most probable energy' is preserved, and its value will be less than the incident energy.

The Number transmission coefficient (NTC) is the ratio between the total emergent particle fluence N_e and the total incident particle fluence, N_i:

$$\text{NTC} = N_e/N_i \qquad (11.1)$$

Values of NTC for the transmission of omnidirectional electrons of energy 0–6 MeV through plane aluminium shielding of various thicknesses are tabulated by Berger and Seltzer (1968).

Figure 11.3 shows a graph of values of NTC as a function of aluminium absorber thickness (actually mass thickness in g cm^{-2}) for various incident energies. A feature of each curve is the 'tail', an effect of 'straggling electrons' which penetrate further than might otherwise be expected. The tails of the curves cut the thickness axis at the 'maximum range' for a particular energy. The broken lines, which neglect the straggling effect, cut the axis

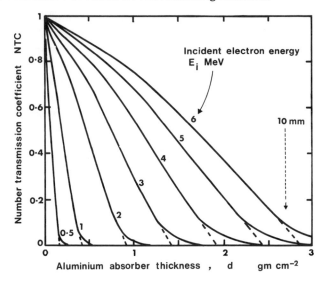

FIG. 11.3 Electron transmission: number transmission coefficient (NTC) of omnidirectional electrons through a plane aluminium absorber, as a function of thickness.

at the 'practical range'. The practical ranges indicated in Fig. 11.3 agree well with those given for aluminium by, for example, Linnenbom (1962).

11.3.2 Stopping power

For electrons of energy less than 5 MeV, typical of the Earth's trapped radiation belts, almost all energy loss on passing through material is by interaction mechanisms that result in ionization of the material, i.e. the creation of electron–hole pairs with little momentum transfer to the atoms. The rate of loss of energy E with distance traversed, known as the 'stopping power' of the material, is given by the following equation (Berger and Seltzer 1982):

$$-\frac{dE}{dx} = \frac{2e^4 z^2 N_A Z}{mv^2 A} B_e, \qquad (11.2)$$

where N_A/A is the number of atoms of atomic number Z per unit volume ($N_A \simeq 6 \times 10^{23}$ atoms per mole), A is the atomic mass, z and v are the charge number (1 for an electron) and velocity of the incident particle respectively, e and m are the charge and mass of an electron respectively, and x is the 'path length' or distance measured along the track of an electron.

B_e is known as the 'stopping number' of the material; it is a function of particle energy, but rises only slowly with E. Therefore, dE/dx at first falls rapidly with increasing E (and hence v), being dominated by the $1/v^2$-dependence. It reaches a minimum as v approaches a limit at the speed of light. At higher energies (corresponding to a relativistic increase in electron mass), dE/dx rises slowly with the now dominant B-dependence, since v is now limited.

The minimum stopping power for electrons occurs at energies in the range 1–2 MeV; electrons at this energy are said to be 'minimum ionizing'. The amount of data available on stopping power for various materials is considerable (see for example Barkas and Berger 1964; Berger and Seltzer 1964, 1966; Ziegler 1991). Note that, normally, stopping power is quoted in units of energy lost per unit 'mass thickness' measured along the particle path in MeV g^{-1} cm^2.

In addition to energy loss by collision, there is a further contribution to stopping power due to radiation loss (i.e. bremsstrahlung generation). At energies below 1 MeV, this is extremely small when compared with collision loss as a mechanism for stopping electrons. It is a rising function of energy, but does not dominate over collision loss except at energies well over 10 MeV. For the typical near-Earth electron spectrum, only a very small fraction of the energy passes into bremsstrahlung, but the latter radiation is so much more penetrating that it emerges as a significant remaining 'background' when all electrons have been stopped (see Appendix E).

11.3.3 Internal spectrum

The energy spectrum of the particle radiation emerging from the inner surface of an absorber will clearly be a degraded form of the incident external spectrum. Particles of lower energy, having ranges less than the shield thickness, may be completely stopped, while at higher energies there will be a reduction in flux.

Figure 11.4 shows a flat incident isotropic electron spectrum after passing through an aluminium slab. This behaviour was computed with the ITS/Tiger Monte Carlo code by Daly (ESA 1989). The change in the electron spectrum is shown in 11.4(a), while 11.4(b) shows how the emerging spectrum depends on direction. Note that the fluxes shown are normalized to unit incident current and are differential in energy. On average, it takes four electrons distributed isotropically to produce unit current, two of which are moving away from the medium. In addition to this, 20 groups are used, each with a width of 0.25 MeV. Therefore, the unshielded flux is 0.4 electrons per MeV per unit incident current (20 × 0.4 × 0.25 MeV = 2 electrons). Figure 11.4(a) shows that with small amounts of shielding, the higher-energy part of the spectrum is degraded but that this is accompanied by an enhancement of the low-energy portion of the spectrum, because of secondary electrons. As the shielding increases, both low- and high-energy electrons are attenuated and lose energy, producing a rounding of the spectrum. Figure 11.4(b) shows that a significant fraction of the electron flux emerges at a relatively large angle to the slab normal.

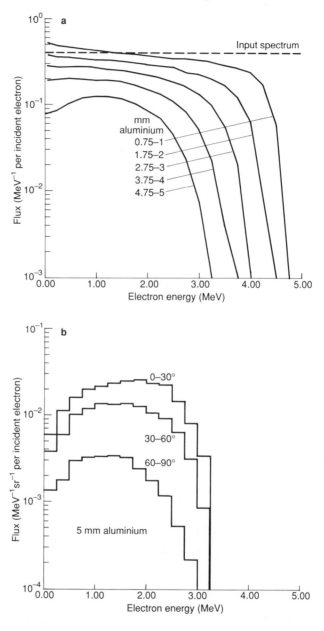

FIG. 11.4 Variation of electron spectra in a 5 mm slab of aluminium the input electrons are isotropic, with a uniform spectrum up to 5 MeV (broken line in (a)). (a) Spectra at various depths; the fluxes are differential in energy and normalized to unit incident current. (b) Transmitted spectra in three ranges of angle to the slab normal; these fluxes are differential in both energy and solid angle (ESA 1989). (ESA with permission.)

11.4 Transport of protons and other heavy particles

11.4.1 Interactions

Energetic protons and ions, being heavier particles, are not subject to the high degree of scattering experienced by electrons and normally follow virtually straight paths. It is relatively easy to compute their slowing and energy deposition in materials, and they have well-defined ranges.

There is a small, but not negligible, probability of protons and other ions interacting with atomic nuclei, causing fragmentation of the nucleus or emission of secondary neutrons and protons. Such fragments and secondaries are hazards which cannot be ignored in some circumstances, such as in manned missions, single-event processes or heavily shielded detectors. Secondary neutrons produced in spacecraft materials, or even the upper atmosphere, can again interact 'inelastically' with nuclei, producing gamma-rays which may interfere with gamma-ray telescopes.

11.4.2 Energy loss and attenuation

The rate of ion energy loss is, as for electrons, defined by a 'stopping power' formula:

$$-\frac{dE}{dX} = \frac{4e^4 z^2 N_A Z}{mv^2 A} B_i \tag{11.3}$$

where z is the ion charge number, Z the material atomic number, B_i the ion stopping number, N_A the Avogadro constant (6×10^{23}), and A the atomic mass number. Clearly, fully ionized energetic ions such as cosmic rays deposit energy very rapidly in a material, and this gives rise to single-event phenomena (SEU, latch-up and radiobiological). This energy loss expression can be readily evaluated to give particle ranges, residual energies and energy deposit, the latter relating to dose and upset. See e.g. Table B7.

11.5 Electromagnetic radiation: bremsstrahlung, X- and gamma-rays

11.5.1 General

In spacecraft engineering and operations, energetic electromagnetic radiation (photons) can be encountered in a number of forms:

(1) bremsstrahlung radiation produced by the slowing of energetic electrons in the atomic electric fields of a material;
(2) X-radiation produced by electron-beam excitation of atomic transitions;
(3) nuclear emissions, Cerenkov radiation, etc.

The interaction of electromagnetic radiation with matter is thus a topic of some importance in predicting or testing the response of devices to radiation,

300 The interaction of radiation with shielding materials

particularly where device electrode materials of high atomic number, such as gold and molybdenum, lie in close relation to the active region of the device. A discussion of the relevant general features of interactions is given here; discussions of specific problems such as dose enhancement and attenuation by package material are given in Sections 4.10 and 15.4.

Since bremsstrahlung, X-rays and other electromagnetic radiation are different manifestations of the same type of radiation, they are absorbed according to the same laws, i.e. laws of photon interaction. However, whereas bremsstrahlung is usually 'white', i.e. spread over a broad spectrum, gamma-rays have well-defined 'lines', i.e. energy peaks, corresponding to atomic and nuclear energy states.

11.5.2 Bremsstrahlung

Photons produced by the bremsstrahlung mechanism are a significant problem in spacecraft shielding because the photons are generally much more penetrating than the primary electrons themselves. The production of bremsstrahlung is higher in materials of high atomic number, Z, and is proportional to the square of Z. Bremsstrahlung attenuation depends strongly on energy; photoelectric absorption usually dominates at energies below 0.1 MeV, Compton scattering at energies around 1 MeV, and pair-production at high energy, above 10 MeV. These processes all result in the production of further electrons.

Figure 11.5(a) shows the bremsstrahlung spectra resulting from the flat electron spectrum shown in fig. 11.4(a). These were also produced with the ITS/TIGER Monte Carlo code by Daly (ESA 1989). Clearly there is a bias towards the production of low-energy photons and while the absorber thickness is less than the electron range, the continuing production of bremsstrahlung leads to an increasing photon flux. Figure 11.5(b) shows that a significant fraction of the transmitted bremsstrahlung photon flux is at a large angle to the slab normal.

11.5.3 Other electromagnetic radiation

In passing through matter, energetic charged particles give rise to Cerenkov radiation. This is a result of particles travelling faster than the speed of light in the medium; the result is the emission of radiation which may interfere with an optical system's detectors. Nuclear interactions can also give rise to gamma-radiation which can be troublesome, as mentioned in the next subsection. Finally, gamma-radiation is emitted from radioisotope thermoelectric generators used in deep-space interplanetary missions.

11.5 Electromagnetic radiation

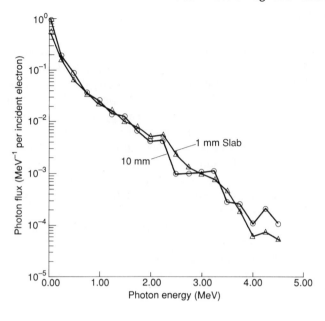

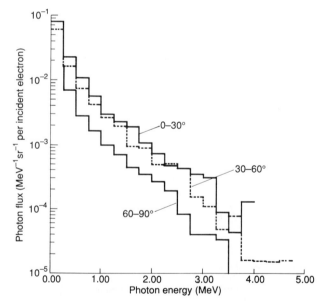

FIG. 11.5 Bremsstrahlung spectra in an aluminium slab: electron spectrum of Fig. 11.4(a). (a) Spectra at 1 and 10 mm; the fluxes are differential in energy and normalized to unit incident current. (b) Transmitted spectra in three ranges of angle to the slab normal; fluxes are differential in both energy and solid angle (ESA 1989). (ESA with permission.)

302 The interaction of radiation with shielding materials

11.5.4 Production and attenuation of electromagnetic radiation

11.5.4.1 *Production*

Figure 11.6 (Wyard 1952) shows the spectrum of bremsstrahlung X-ray photons generated when a 1 MeV electron beam strikes a considerable thickness of material. As described later, this broad spectrum results from a complex process, but it will be seen that the peak of the spectrum emitted is at a photon energy of about half the energy of the impinging particle. Because it is broad, the emission is sometimes called 'White Radiation'. The efficiency of Bremsstrahlung generation is strongly dependent upon the material; a heavy element will generate bremsstrahlung much more effectively than a lighter one. This efficiency is roughly in proportion to the atomic number. This dependence is not followed so simply, however, when dealing with the attenuation of bremsstrahlung, as described below.

When electrons bombard a target, the characteristic emission lines of the target material are superimposed on the 'white radiation' described above. Table 11.1 gives some wavelengths for common target materials. For most materials, the K lines will be of greatest interest to us. However,

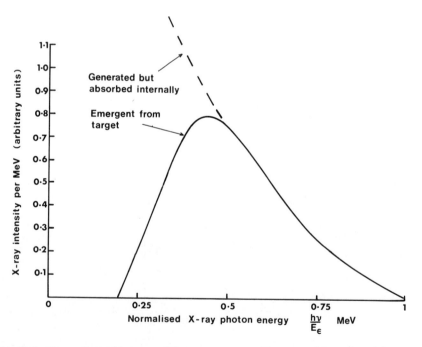

FIG. 11.6 Generation of bremsstrahlung: spectrum of bremsstrahlung X-ray photons generated when electrons are slowed down in a material. For a 1 MeV electron, the energy scale can be read in MeV. The intensity expressed in rads integrated over the whole emergent spectrum will then be of the order of 10^{-12} rad(Si) per unit incident flux (10^{-14} Gy(Si) per unit incident flux). (From Wyard 1952).

TABLE 11.1 Typical characteristic X-ray emission lines for elements of importance in spacecraft and devices

Element	Atomic no., Z	Highest K line (keV)	Highest L line (keV)
Be	5	0.11	–
C	6	0.28	–
O	8	0.53	–
Al	13	1.56	0.07
Si	14	1.84	0.09
Ti	22	4.96	0.46
Cr	24	5.99	0.65
Mn	25	6.54	0.64
Fe	26	7.11	0.79
Co	27	7.71	0.79
Ni	28	8.33	0.94
Cu	29	8.98	1.02
Mo	42	20.00	2.51
Ta	73	67.42	11.67
W	74	69.48	12.10
Pt	78	78.34	13.56
Au	79	80.66	14.78
Tl	81	85.45	15.33
Pb	82	88.06	15.84
U	92	115.39	21.66

for tantalum, tungsten, gold, platinum, and lead, the L lines, which have photon energies in the 7–10 keV region, may also be of significance, especially in laboratory testing practices. Figure 11.7 shows the photon spectra for several values of bombarding electron energy and a tungsten target, calculated for a commercial radiographic X-ray machine using rules given by Johns and Cunningham (1969). The K emission appears as a complex set of lines. It can be seen that a copper filter greatly modifies or 'hardens' the spectrum and reduces the influence of the characteristic lines.

Figure 11.8 shows another spectrum for a low-energy X-ray tube operated at a potential of 25 kV. Here, the tungsten L lines are seen and, because of the low acceleration potential, the K lines are not excited. Both types of X-ray tube are used for device testing (see Chapter 13).

11.5.4.2 Attenuation

An important feature of electromagnetic radiation in the 5–1000 keV range (X-rays) is the strong difference in attenuation from material to material

304 The interaction of radiation with shielding materials

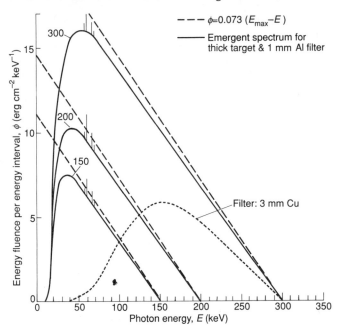

FIG. 11.7 Photon energy spectra calculated for a commercial tungsten-target radiographic X-ray machine using graphs by Johns and Cunningham (1969). The potentials of the bombarding electron beams are constant at 150, 200 and 300 V. Filtration by 3 mm copper is shown in one example. The characteristic K lines of tungsten at 57.592, 59.310, 67.236 and 69.089 keV are not to scale. The broken lines represent the theoretical formula for energy fluence, the solid and dotted lines are the fluences emerging through filters of 1 mm Al and 3 mm Cu respectively.

(in contrast to electron attenuation, see section 11.3) and the very wide range of attenuation per unit thickness over that energy span (Attix 1989; Price *et al.* 1957; Profio 1979).

In ideal circumstances, known as 'narrow geometry', X-rays are attenuated according to Lambert's law:

$$I/I_0 = e^{-\mu\rho d}, \qquad (11.4)$$

where ρ is the density of the stopping material and d is the thickness; ρd is thus a normalized 'mass per unit area' in $g\,cm^{-2}$ (or $kg\,m^{-2}$) and indicates roughly the stopping power for a given structure; μ is the attenuation coefficient for the material and is a total of the energy absorption and scattering powers of the constituent atoms. This coefficient is derived from the sum of all the interactions of a photon with a material (photoelectric

11.5 Electromagnetic radiation

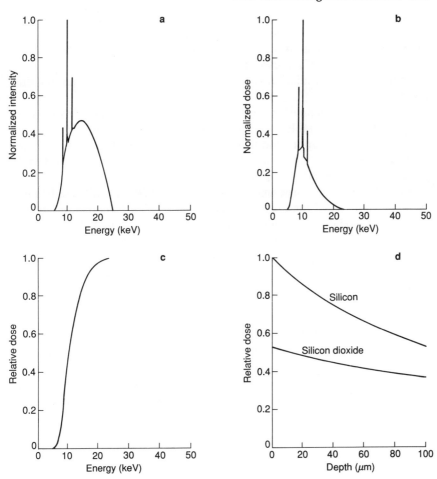

Fig. 11.8 X-ray energy spectra and doses deposited for a low-voltage machine with 150 μm Al filter: (a) energy spectrum; (b) dose spectrum; (c) integrated dose; (d) depth–dose profile. Reproduced by permission of ARACOR, Sunnyvale, CA.

absorption, Compton scattering, etc.) and is often expressed in the form of a cross section (a probability) of these interactions:

$$\sigma_{tot} = \sigma_1 + \sigma_2 \ldots, \tag{11.5}$$

where σ_{tot} is the total cross section (probability) per atom and σ_1, σ_2, etc. are the probabilities for particular physical processes.

The relation of attenuation coefficient to the above is:

$$\mu_{tot} = 6.02252 \times 10^{23} \sigma_{tot}/(\text{atomic mass}), \tag{11.6}$$

where σ_{tot} is in barns per atom, ρ is in g cm^{-3}, and μ_{tot} is in cm^{-1}.

The attenuation coefficient is sometimes given as the 'mass attenuation coefficient' (μ/ρ) in units of cm^{2}g^{-1}. For dosimetric calculations (Attix 1989), another form of μ is used, namely the 'energy absorption coefficient', which includes only radiation energy absorbed and ignores attenuation due to scattering.

A list of mass attenuation coefficients for high-energy photons is given in Appendix C. The table shows that, in the region of 1 MeV, the dependence of absorption upon atomic number is quite weak. Thus, for example, 1 g cm^{-2} of lead (0.1 mm) will produce 7 per cent attenuation of 1 MeV photons, while the same 'mass thickness' of aluminium (3.7 mm) will produce 6 per cent. Only large thicknesses of shield material will significantly attenuate photons of these energies produced by Van Allen belt electrons.

We mentioned that all materials are roughly equivalent in the efficiency of stopping X-rays in the megavolt range. Unfortunately, this statement ceases to hold at lower energies. At 100 keV, for instance, very great differences in absorption exist between, say, steel, plastics, and biological materials. These differences are also tabulated in Appendix C. They greatly complicate testing with X-rays and make all dosimetry more complex. For example, the attenuation of X-rays at 5 keV is typically 10 000 times higher than at 1 MeV (see also section 11.5.5.3). This large factor implies that radiation testing procedures are less complex when high energies of photon are used, because package attenuation and scattering are less important. Tabulations and plots of the absorption coefficients versus energy are given in Veigele *et al.* (1971). Commercial computer databases containing these values are also available.

11.5.4.3 *'Build-up'*

In calculating bremsstrahlung doses in spacecraft, we must remember that the 'narrow geometry' required for the exact use of Lambert's law does not hold. Scattering effects, sometimes called 'build-up', increase the number of transmitted photons by a factor of about 2 at all depths greater than about 5 mm. Another effect of scattering is also referred to as 'build-up': in gamma-radiation testing, we add 'build-up material' around a sample to promote Compton scattering equilibrium (Johns *et al.* 1969).

11.5.5 Soft X-rays and vacuum ultraviolet (VUV): generation and special effects of long-wavelength X-rays

11.5.5.1. *Introduction*

While X-rays with energies of 5 keV and above have been used for many years, sources and uses of X-ray photons with lower energies have only recently become the subject of intensive research (Michette 1989). The same may also be said of the higher energy range of vacuum UV light, 0.1 to 100 eV. Owing to the short wavelelengths of these X-ray and VUV photons, they have uses in the lithography of very fine patterns, in microscopy and

11.5 Electromagnetic radiation

astronomy. Because of their varied stopping power, they require interesting filters and must be handled in vacuum (Samson 1967); because they can be focused, they are of stronger interest in science than are hard X-rays (Koch 1983).

In this section we will describe some of the machines used to generate soft X-rays and VUV photons and describe some forms of damage caused by them.

11.5.5.2 *Machines and optics*

Tables 11.2 and 11.3 (Michette 1989) show some sources of soft X-rays, which include plasmas, synchrotron accelerators, and simple electron-beam tubes. Haelbich *et al.* (1983) have described the set-up by IBM of a beam tube for soft X-ray lithography on the VUV storage ring of the National Synchrotron Light Source at Brookhaven National Laboratory, USA. IBM moved on to a a small commercial synchrotron source built in England (Wilson 1990). Electrons of energy 700 MeV are generated in a superconducting magnet chamber; the synchrotron radiation power in the 1 keV region is adequate for semiconductor wafer-resist exposures at economic production rates.

11.5.5.3 *Absorption constants*

Figure 11.9 shows the absorption and scattering properties of photons from 100 to 10^6 eV, which include strong changes at K, L and M absorption edges and are strongly affected by the atomic weight of the material. Of great interest in biological research are the so-called water and protein

TABLE 11.2 Sources of soft X-rays (Michette 1989)

Source	Wavelength range[a] λ (nm)	Comments
Electron synchrotron	~0.2 to 200+	Continuum with no lines
Laser plasma on:		
C,K,Ar,Cl target	~2–5	Line emission
Al, Fe target	~6–20	Mainly line
Yb, Sm target	~4–60	Mainly continuum with weak lines
Electron beam on		
C-containing insulator	~2–25	Mainly continuum with weak lines
Z pinch gas puff	~1	Lines characteristic of gas superimposed on continuum

a. $\dfrac{1.2398541}{\lambda(nm)} = E(photon)\ (keV)$

308 The interaction of radiation with shielding materials

TABLE 11.3 Comparison of soft X-ray sources (Michette 1989)

Source	Relative intensity (photons sec^{-1} mrad^{-2} in 1% bandwidth)	
	Peak	Average
	(synchrotron bending magnet = 1)	
Synchrotron		
Bending magnet	1	1
Undulator	$\sim 10^3$	$\sim 10^3$
Solid-target		
Carbon K	$\leq 5 \times 10^{-12}$	$\leq 5 \times 10^{-11}$
Copper L	$\leq 10^{-10}$	$\leq 10^{-9}$
Aluminium K	$\leq 10^{-9}$	$\leq 10^{-8}$
Laser plasma		
High power[a]	~ 10	$\sim 2 \times 10^{-11}$
Low power[b]	$\sim 10^{-2}$	$\sim 2 \times 10^{-10}$
Electron beam[c]	$\sim 10^{-4}$	$\sim 2 \times 10^{-11}$
Gas puff[d]	~ 1	$\sim 2 \times 10^{-8}$

[a] Laser energy ~100 J, pulse repetition rate ~10^{-3} Hz.
[b] Laser energy ~1 J, pulse repetition rate ~10 Hz.
[c] Repetition rate ~100 Hz.
[d] Repetition rate ~20 Hz.

windows at about 200 eV (2 nm) (Robinson 1989). The absorption lines due to silicon are evident in the diagram of the quantum efficiency of a charge coupled device, shown in Fig. 11.10 (Acton *et al.* 1991).

11.5.5.4 Effects
For most materials, the radiation damage by soft X-rays and VUV photons is similar to that from hard X-rays, being due to ionization caused by electron excitation. The effect of soft X-rays in silicon devices has been investigated by space astronomy workers, since exposures to test sources (e.g. Fe-55 photons) and to focused images in space may be sufficient to disturb the insulator layers in scientific silicon detectors (see Chapter 6). More basic studies have been done to extend knowledge of damage effects in oxide films to the soft X-ray range see, for example Dozier *et al.* 1987; Williams *et al.* 1988). Radiation effects in silicon diodes for detection of soft X-rays are described by Korde *et al.* (1989). The junction effects can be severe and have to be prevented by a thin shield of gold deposited over the junction. Dosimetry of soft X-rays can be done using radiochromic films (Haelbich *et al.* 1983).

Soft X-rays give unique images of biological materials (Robinson 1989; Jones and Gordon 1989). It is possible to produce microscopic images of

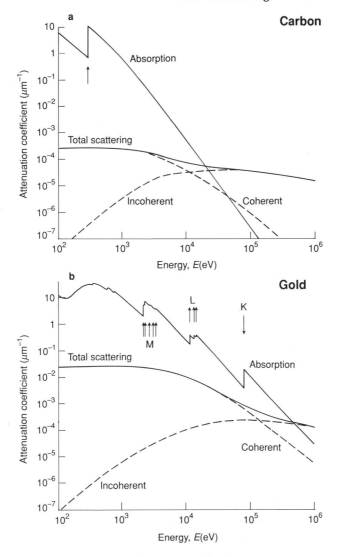

FIG. 11.9 Linear attenuation coefficients for scattering and absorption of photons in (a) amorphous carbon and (b) gold, as a function of energy; K, L, and M are X-ray absorption edges, multiple in the case of gold (after Michette 1989).

cell structures of size about 30 nm using photons of energies around 200 eV. This method compares well with electron microscopy because it is much easier to work with wet, live samples of cells. Apart from those practical advantages, water stops soft X-rays less strongly than protein.

310 The interaction of radiation with shielding materials

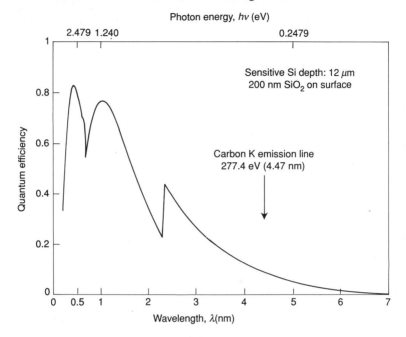

Fig. 11.10 Quantum efficiency of a silicon detector in the soft X-ray region; the calculation, by Acton et al. (1991), was for a silicon charge-coupled device, front-illuminated.

Unfortunately, however, the minimum soft X-ray exposure of the cell causes serious radiation damage which kills and may disrupt the cell. The samples for contrast microscopy are held between two thin plastic foils in the wet state and X-rays are passed through, giving a true focused image using Fresnel lenses. The damage due to this method has been compared with that due to transmission electron microscopy (Goltz 1991; Robinson 1989).

Quantifying and then minimizing biological radiation effects has received much attention. Exposing a cell element to 1000 soft X-ray photons (of energy, say, 200 eV) deposits about 10^9 rad (10^7 Gy) in that region. This is perhaps a million times the lethal dose for a cell, and little can be done to rectify the situation. The current effort is rather to prevent disruption of the structure. Using pulsed X-ray sources may achieve this.

11.6 Radiation attenuation by shielding; deposition of dose in targets

11.6.1 Dose versus depth

Ionization induced by the various particles described above, on reaching

11.6 Radiation attenuation by shielding; deposition

a device buried in an absorber, results in damage as described in Chapter 3. Dose, the energy deposited per unit mass of material, is the basic parameter for evaluating ionization-induced damage. The dose deposited in a 'target' depends on the particle type and on the energy which it retains after passing though any surrounding absorber. The dose is computed from the product of the particle fluence at a particular point and the restricted stopping power of the particle. The qualifier 'restricted' means that only the energy loss which results in *locally* deposited energy is considered; the generation of secondaries which travel some distance before depositing their energy are excluded (Berger and Seltzer 1982).

The evaluation of biological effects of radiation is much more complex and involves the use of an empirical 'dose equivalent' factor (see section 1.3). There is some uncertainty attached to the use of the concept of dose for heavy ions, which have unique local effects in materials or living tissues. There is a range of early and late biological effects which vary with dose rates (Bücker and Facius 1988). Displacement is discussed in Section 11.7.

11.6.2 Shielding: relation between space radiation flux and deposited dose

In performing evaluations of space radiation environments, it is often most convenient to consider a silicon target shielded by aluminium, since this relates closely to typical electronic devices in spacecraft and is the basis for much testing work and calculation. The effects of shielding on space radiation are:

(1) reduction of fluxes of primary particles;

(2) creation and subsequent absorption of secondary radiation;

(3) changes in the energy spectra.

Therefore the dose in a target depends on the amount of shielding in a way that is not simple to calculate.

For engineering purposes, a 'dose–depth' curve is often used to represent the space radiation environment, its modification by shielding, and the resulting dose in components. Usually the individual components of the environment (protons, electrons, bremsstrahlung, etc.) and the sum of the components are plotted as a function of the thickness of shielding material, usually assumed to be a sphere or slab of aluminium. This removes all information about particle energies or even particle type and is usually an integration over orbits or complete missions.

This subsection provides data from which the dose in silicon can be calculated for various thicknesses of aluminium shielding. Computer programs for such purposes are readily available and we recommend that readers who have a continuing need for such calculations obtain copies of these programs. Chapter 12 contains details on programs and how to obtain them.

We will now consider how to calculate the dose which is produced behind a given thickness of aluminium shielding by electrons, electron-induced bremsstrahlung, and protons resulting from the broad spectra of electrons and protons found in space. Seltzer (1979) used the ETRAN Monte Carlo computer code to simulate the propagation of electrons and photons (bremsstrahlung), in order to compute the dose, D, generated at different depth values, x, in planar, semi-infinite (one-dimensional) aluminium shielding for a range of incident *isotropic, monoenergetic* electron energies.

The $D(x, E)$ data set thus created allows an arbitrary input spectrum $f(E)$ to be folded with $D(x, E)$ to yield a sum dose at depth x due to the spectrum. This is normally done by the SHIELDOSE computer program, but can be done manually. In addition to electrons and bremsstrahlung, Seltzer's data set includes doses from protons, computed by means of the straight-ahead, continuous-slowing-down approximation. The data set is organized in normalized form, with the depth scale normalized with range, and the doses energy-, range- and current-normalized and made dimensionless. For ease of use, these are converted to dose (Si) per unit incident isotropic (4π) electron or proton flux, as shown in Figs. 11.11–11.13. There is a pair of graphs for each of the important radiation types: electrons, electron-induced bremsstrahlung, and protons. For each type, the graphs (a) give the doses as functions of incident particle energy for various shielding depths, while the graphs (b) give the doses as functions of shield depth for various incident particle energies.

In performing an evaluation of the expected radiation dose on a space mission, the shielding geometry influences the result. It should be recognized that the doses provided in these figures are for a *planar shield* where, clearly, radiation comes principally from one side; paths which are not normal to the face encounter increased amounts of shielding. The widespread use of planar and slab geometries arises from the efficiency with which Monte Carlo analyses can be performed with them. A less optimistic basis for the evaluation of shielding effects is to assume *spherical* shielding. When the dose point is located at the centre, minimum shielding is encountered in all directions. This is preferable for initial evaluation, except where there is good justification for the assumption of planar shielding, for example when considering surface materials. If solid-angle sectoring is employed for analysis of complex geometries, doses based on spherical shielding should be used.

The method makes it possible to calculate the relationship between ionization dose and the depth of absorber (for the case of silicon shielded by aluminium) for any spacecraft orbit where the integrated flux vs. energy spectrum is known. The calculation is performed by combining the particle spectrum with the relevant dose transmission curves (Figs 11.11 to 11.13). In principle, the procedure described is identical for electrons, electron-induced bremsstrahlung, and protons.

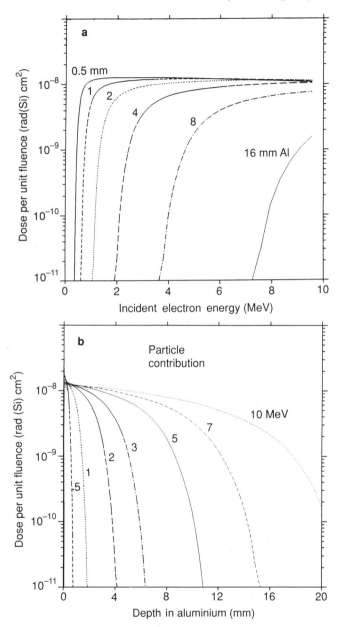

FIG. 11.11 Electron dose deposition in planar semi-infinite aluminium shielding: (a) as a function of incident electron energy for different shield depths; (b) as a function of shield depth for different incident electron energies. Dose is for silicon and is normalized to unit incident isotropic flux. (Daly 1988). (ESA with permission.)

314 The interaction of radiation with shielding materials

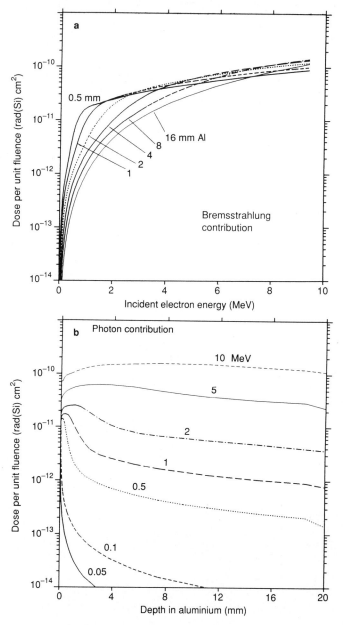

FIG. 11.12 Bremsstrahlung dose deposition; as in Fig. 11.11 (Daly 1989). (ESA with permission.)

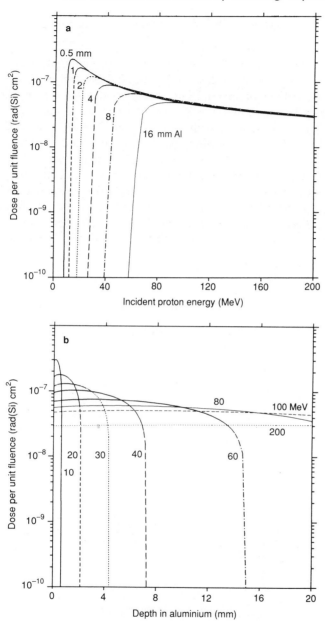

FIG. 11.13 Proton dose deposition; as in Fig. 11.11 (Daly 1989). (ESA with permission.)

316 The interaction of radiation with shielding materials

11.7 Atomic displacement damage versus depth

The degradation of the energy and number of particles as they proceed through a slab of absorber has been explained in detail in preceding sections dealing with the ionization dose versus depth relations for absorbers surrounding electronic equipment. Curves analogous to those for dose transmission may be required for a limited range of electronic devices which are also sensitive to displacement damage (see Chapter 3) especially solar cells and charge coupled devices (see Chapter 6). Thus, in the same way as before, generalized damage power versus energy curves may be constructed so that damage versus depth curves can be calculated for a given set of particle spectra. W. L. Brown *et al.* (1963) constructed such curves ('BGR' curves) for silicon on the same general basis as the dose transmission curves; see Figs 11.14 and 11.15). They are useful in, for instance, studies of solar cells. In more recent work (see Section 3.2) it has been proposed that

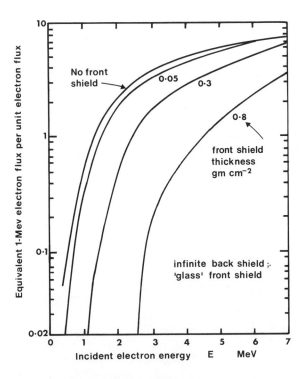

FIG. 11.14 Electron damage transmission: 'BGR' curves of damage-equivalent 1 MeV electron flux as a function of energy of a monoenergetic isotropic flux of electrons incident upon n-on-p solar cells with different front shieldings (typical solar-cell cover glass); back shield assumed infinitely thick. For uniform all-round shielding, multiply the vertical scale by 2. (After Brown *et al.* (1963). Reproduced by permission.)

11.7 Atomic displacement damage versus depth

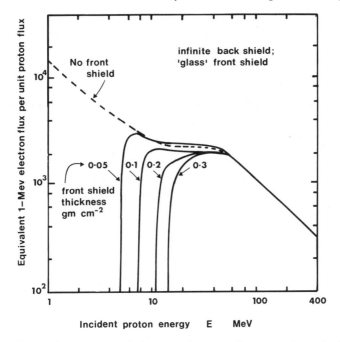

FIG. 11.15 Proton damage transmission; as in Fig. 11.14 (Brown et al. 1963). The shape of this curve has been revised in the light of new data (see Section 3.2). Reproduced by permission of Bell Telephone Laboratories.

the 'no front shield' curve in Fig. 11.15 should follow a simpler, monotonic form connected with the theoretical non-ionizing energy loss (NIEL) (see Table 3.1). For a complete description of solar cell degradation behind shields and the concept of damage equivalence for predicting such degradation, the reader is referred to Tada *et al.* (1982). The latter work gives, for a variety of solar cells, the degradation to be expected in cell output power, open-circuit voltage and short-circuit current as a result of given fluences of 'equivalent 1 MeV electrons'. Such calculations can also guide the internal design of space vehicles.

Some power transistors have base regions wide enough to exhibit such gain degradation in space that their use in the power units of a spacecraft is usually avoided. On the other hand, high-frequency transistors in 5-year circular equatorial orbits will not experience sufficient exposure for serious damage to occur. This holds for transistors of frequency cutoff about 100 MHz. 'Serious damage' here may be taken to represent a decrease of 10 per cent in gain. For dealing with such calculations, the 'BGR' damage curves will suffice so long as the vertical scale is multiplied by 2 to allow for non-infinite back-shielding. For the more sensitive charge-coupled device more sophisticated programmes have been developed (see Chapter 6). It should be noted that, for typical fast-switching transistors in space, the 'surface effect' due to ionization will be more severe than the atomic displacement effect.

11.8 Influence of material type on radiation stopping

11.8.1 Deposition of dose

The procedure for deriving the dose–depth curves described in the preceding sections has been applied to the typical case of a silicon device protected by an aluminium absorber. To estimate the doses deposited in materials other than silicon, the appropriate 'stopping power' data should strictly be used as the basis of the dose per unit fluence calculation. In practice, however, stopping power (when expressed in 'mass thickness' units) does not vary greatly from material to material.

Figures 11.16 and 11.17 show electron and proton stopping powers for various materials, normalized, for convenience, to the equivalent thickness of aluminium. Stopping power values for silica, for instance, are not easily available, but it is reasonable to assume that they are no more than a few per cent greater than those for silicon. Clearly, it would be essential to make the appropriate corrections when calculating the dose in materials such as water, human tissue, polyethylene or heavy materials such as gold, where stopping power values are further removed from those of silicon or aluminium.

11.8.2 Shielding materials

When calculating radiation doses behind different absorbing or shield materials, it has often been the practice to 'convert' all materials to equivalent aluminium thickness by making the appropriate density correction. The use of a curve of dose versus depth in aluminium for all materials thus requires the assumption that the transmission coefficients for a given 'mass thickness' are the same for any material. This may lead to considerable error, especially in the case of electron doses.

For orbits where electrons are the main source of ionization (e.g. the geostationary orbit), we recommend that detailed system designs treat spacecraft as a 'multilayered structure', that is, as a series of slabs of different materials. Methods based on treating all materials as aluminium equivalents are useful, but — for these orbits — are adequate only for preliminary feasibility studies. The need for the multilayer approach arises from the strong degree of scattering of electrons by matter and the strong dependence of this effect on atomic number. As a result, a slab of material of *high* atomic number transmits fewer electrons than one of *low* atomic number with the same 'mass thickness'. Figure 11.18 shows this effect for several different materials (Mar 1966). The electrons are scattered out of the transmitted

11.8 Influence of material type on radiation stopping

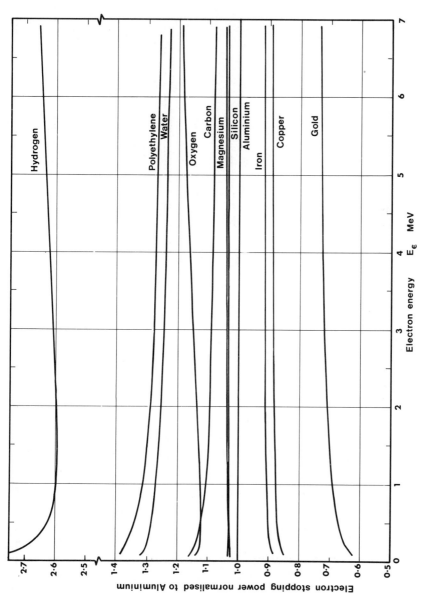

Fig. 11.16 Electron-stopping power for various materials, normalized against aluminium, as a function of incident energy.

320 The interaction of radiation with shielding materials

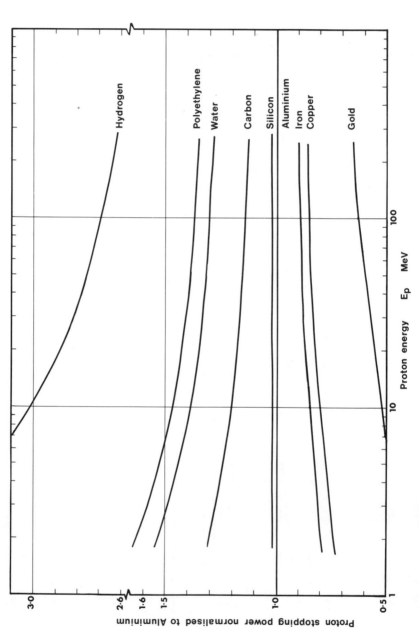

FIG. 11.17 Proton-stopping power for various materials, normalized against aluminium, as a function of incident energy.

11.8 Influence of material type on radiation stopping

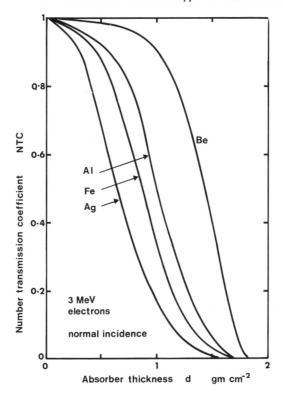

FIG. 11.18 Electron transmission for various materials: number transmission coefficient for 3 MeV electrons at normal incidence as a function of absorber thickness (Mar 1966).

beam, many, in fact, in a backward direction. Figure 11.19, based on data from Wright and Trump (1962) shows experimental results for the backscattering of megavolt electrons from thick targets. It will be seen that lead is 5 to 10 times more efficient at back-scattering than aluminium. It is not surprising, therefore, that the opposite, but far weaker, dependence of electron 'stopping power' on atomic weight is swamped by the scattering effects.

As an illustration of the effect of high-Z shielding material on backscattering and X-ray generation, Fig. 11.20 shows dose–depth curves (for electron and bremsstrahlung doses) for equivalent mass-thicknesses of aluminium and lead absorbers in the geostationary orbit environment. The curve for aluminium is derived from the old AEI-7 model and has been shown in an earlier figure; the curve for lead should be regarded as a provisional estimate. It can be seen that, for shield thicknesses less than 1 g cm^{-2}, lead is a more efficient absorber. However, this degree of protection will

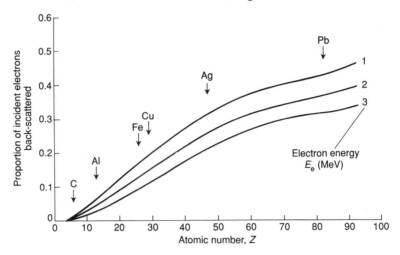

FIG. 11.19 Electron back scattering: fractional back scattering as a function of atomic number of absorber material (from Wright and Trump 1962).

be satisfactory only for devices that tolerate total doses of greater than 10^4 rad(Si) (10^2 Gy (Si)). For more sensitive devices requiring greater protection, the bremsstrahlung generated by lead is likely to provide an irreducible and unacceptable background. The same strong dependence on atomic weight does not hold for protons, on account of their greater mass. It must be emphasized that Fig. 11.20 is intended only to demonstrate a trend and should not be used directly in generating design rules. It represents a simplified case of a uniform absorber of a single material.

If we consider the case of a composite absorber consisting of, say, lead and aluminium layers, then the composite dose–depth relation would clearly be expected to lie somewhere between the extremes of lead and aluminium, and would depend upon the proportion of the overall mass-thickness contributed by each material. It must be remembered, however, that most of the bremsstrahlung is generated in the outermost part of the absorber where the lower energy electrons are stopped. Therefore, the effect of adding lead to the outside of aluminium shielding would be to limit the minimum dose to something close to the lead bremsstrahlung level. If, on the other hand, lead was the innermost component of the composite, the two advantages of the better electron shielding of lead and the lower bremsstrahlung generation rate for aluminium could, in principle, be combined. In practice, a large amount of absorber will often consist of epoxy–glass laminate circuit boards (see Appendix C). For these, transmission coefficients lie between the values for carbon and aluminium.

Using Monte Carlo calculation methods, Mar (1966) has made an

11.8 Influence of material type on radiation stopping

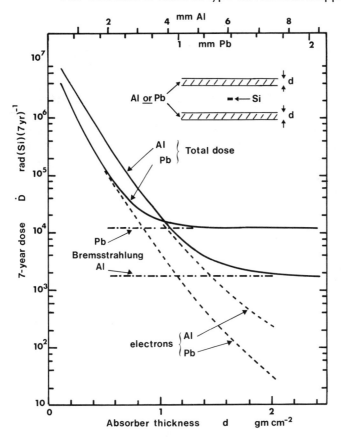

FIG. 11.20 Dose–depth curves (using AEI-7 model) for the geostationary orbit, showing the effect of different absorber materials. The curve for lead is a provisional estimate.

important study of electron transmission as a function of atomic number and derived complex expressions for number transmission coefficient and transmitted flux–energy distribution for monoenergetic incident electrons.

$$\text{NTC}(E, Z, X) = \exp\left[-\left[\frac{0.634\,EZ^{0.23}}{X^{0.848}}\right]^{-k}\right], \quad (11.9)$$

where $k = 7(Z - 3.25)^{-0.24}$, and E, Z and X are incident electron energy, atomic number, and shield thickness, respectively. The transmitted differential flux distribution is represented by the expression

$$\phi(E_e, Z, X)\,dE = A\left[\exp[-b(E - E_e)]\right]dE, \quad (11.10)$$

where E is the emergent electron energy, A is a constant, and $b = (X/E)^{-1.46}(1.53 - 0.0147\,Z)/E$.

A boundary condition, $\phi\,dE = 0$ for $E_e > E_{max}$, applies to this formula, where E_{max} is the peak or 'most probable' transmitted energy. This expression reflects therefore the finite width of the emergent energy spectrum after transmission of a monoenergetic incident flux (see Section 11.3). Mar's approximation in this case is to consider the transmitted spectrum as a vertical edge (E_{max}) with an exponentially decaying distribution at energy less than E_{max}. This is proper, except that, as shield thickness increases, the transmitted energy spectrum becomes broader and the use of a vertical edge is less appropriate.

11.8.3 Routine calculation of particle transmission

It has been demonstrated that because of heavy scattering effects, electron transmission in different materials is extremely variable. To merely use 'stopping powers' in estimating shielding effects can lead to errors; number transmission coefficients (NTC) must also be employed. Calculating these from first principles requires Monte Carlo procedures too long for routine work, so it is fortunate that Mar's formula provides a shorter route. The calculation of bremsstrahlung is highly complex and best performed by computer.

11.9 Conclusions

This chapter has described some of the physical principles involved when radiation interacts with materials. We have indicated the practical methods required to make adequate estimates of the dose deposited in a sensitive device material which is protected from the direct action of the space radiation environment by external absorber materials. We have demonstrated the complexity of the physical processes and noted the approximations which may be made. We should note particularly that some approximations which may be justifiable in preliminary calculations, when order of magnitude is the main question, should be removed in later, detailed calculations where serious compromises between the weight and life of a vehicle are required. When the engineering decisions in question demand accuracy outside the scope of the manual methods described and repetitive calculations are involved, then computer methods are greatly superior. These are described in Chapter 12.

References

Acton, L., Morrison, M., Janesick, J. and Elliott, T. (1991). Radiation concerns for the Solar-A soft X-ray telescope. *SPIE* Vol 1447, *Charge-Coupled Devices and Solid Slate Optical Sensors II*, 123–39.

References

Attix, F. H., (1989). *Introduction to radiological physics and radiation dosimetry*. Wiley, New York.

Barkas, W. H. and Berger, M. J. (1964). *Tables of energy losses and ranges of heavy charged particles*, NASA Publication 1113, pp. 103-7. National Academy of Sciences, National Research Council, Washington, DC.

Berger, M. J. and S. M. Seltzer, S. M. (1964). *Tables of energy losses and ranges of electrons and positrons*, NASA Publication 1113, pp. 205-69. National Academy of Sciences, National Research Council, Washington, DC.

Berger, M. J. and Seltzer, S. M. (1966). *Additional stopping power and range tables for protons, mesons, and electrons*, NASA SP-3036. Office of Technology Utilization, Washington, DC.

Berger, M. J. and Seltzer, S. M. (1968). *Penetration of electrons and associated bremsstrahlung through aluminum targets*, NASA SP-169, pp. 285-322. Office of Technology Utilization, Washington, DC.

Berger, M. J. and Seltzer, S. M. (1982). *Stopping powers and ranges of electrons and positrons*, (2nd edn), NBSIR 82-2550A. National Bureau of Standards, Washington. DC.

Brown, W. L., Gabbe, J. D. and Rosenzweig, W. (1963). Results of the Telstar radiation experiments. *Bell System Technical Journal*, **42**, 1505-59.

Bücker, H. and Facius, R. (1988). Radiation problems in manned spaceflight with a view towards the space station. *Acta Astronautica*, **17**, 243.

Chilton, A. B., Shultis. J. K. and Faw. R. E. (1984). *Principles of radiation shielding*. Prentice Hall, Englewood Cliffs, NJ.

Cooley, W. C. and Janda, R. J. (1963). *Handbook of space radiation effects on solar cell power systems*, NASA SP-3003. US Dept of Commerce, Washington, DC.

Daly, E. J. (1989). *The radiation environment; the interaction of radiation with materials; computer methods*. ESA (1989).

Delaney, C. F. G. and Finch, E. C. (1992). *Radiation detectors*, Oxford University Press.

Dozier, C. M., Fleetwood, D. M., Brown, D. B. and Winokur, P. S. (1987). An evaluation of low-energy X-ray and cobalt-60 iradiations of MOS transistors. *IEEE Transactions on Nuclear Science*, **NS34**, 1535-9.

ESA (1989). *Radiation design handbook*. ESA Document PSS-01-609. ESTEC, Noordwijk.

Evans, R. D. (1955). *The atomic nucleus*. McGraw-Hill, New York.

Goltz, P. (1991). Calculations on radiation dosages of biological materials in phase contrast and amplitude contrast X-ray microscopy. In *X-ray Microscopy III*. Springer, Berlin.

Haelbich, R. P., Silverman, J. P., Grobman, W. D., Maldonado, J. R. and Warlaumont, J. M. (1983). Design and performance of an X-ray lithography beam line at a storage ring. *Journal of Vacuum Science and Technology*, **B1**, 1262-5.

Hubbell, J. H. (1969). Photon cross-section attenuation coefficients and energy-absorption coefficients from 10 keV to 100 GeV, National Bureau of Standards Report NRSDS-NBS 29. US Government Printing office, Washington, DC.

Johns, H. E. and Cunningham, J. R. (1969). *The physics of radiology*. Thomas, Springfield, IL.

Jones, K. W. and Gordon, B. M. (1989). Trace element determinations with synchrotron-induced X-ray emission. *Analytical Chemistry*, **61**, 341A–358A.

Knoll, G. F. (1989). *Radiation detection and measurement.* Wiley, New York.

Koch, E. E. (ed). (1983). *Handbook on synchrotron radiation*, Vol. 1. North-Holland, Amsterdam.

Korde, R. and Canfield, L. R. (1989). Silicon photodiodes with stable, near theoretical quantum efficiency in the soft X-ray region. *SPIE*, Vol. 1140.

Linnenbom, V. J. (1962). *Range-energy plots.* NRL Report 5828. Naval Research Laboratories, Washington, DC.

Mar, B. W. (1966). *Nuclear Science and Engineering*, **24**, 193–9.

Michette, A. G. (1989). *Optical systems for soft X-rays.* Plenum Press, New York.

Price, B. T., Horton, C. C. and Spinney, K. T. (1957). *Radiation shielding.* Pergamon, Oxford.

Profio, A. E. (1979). *Radiation shielding and dosimetry.* Wiley, New York.

Robinson, A. L. (1989). X-ray microimaging for the life sciences. *Synchrotron Radiation News*, **2**, 13–8. (Also Report LBL-27660. Lawrence Berkeley Laboratory, Berkeley, CA.

Samson, J. A. R. (1967). *Techniques of vacuum ultraviolet spectroscopy.* Wiley, New York.

Seltzer, S. M. (1979). Electron, electron-bremsstrahlung and proton depth–dose data for space shielding applications. *IEEE Transactions on Nuclear science*, **NS26**, 4896–4904.

Tada, H. Y., Carter, J. R., Anspaugh, B. E. and Downing, R. G. (1982). *Solar cell radiation handbook*, (3rd edn.), JPL Publication 82-69. Jet Propulsion Laboratory, Pasadena, CA.

Veigele, W. J., Briggs, E., Bates, L., Henry, E. M. and Bracewell, B. (1971). X-ray cross section compilation, from 0.1 keV to 1 MeV, Kaman Sciences Corporation Report to DNA 2433F, Vol. 1, Rev. 1, NTIS No. AD890434. Kaman Sciences Corporation, Colorado Springs, CO.

Williams, C. K., Reisman, A., Bhattcharya, P. and Ng, W. (1988). Defect generation in silicon dioxide from soft X-ray synchrotron radiation. *J. Appl. Phys.*, **64**, 1145–51.

Wilson, M. (1990). Update on the compact synchrotron X-ray source, HELIOS. *Japanese Journal of Applied Physics*, **29**, 2620–4.

Wright, K. A. and Trump, J. G. (1962). *Journal of Applied Physics*, **33**, 687–...

Wyard, S. J. (1952). *Proceedings of the Physical Society*, **A65**, 377.

Ziegler, J. F., Biersack, J. P. and Littmark, U. (1985). *The stopping power and range of ions in solids.* Pergamon Press, Oxford.

Ziegler, J. F. (1991) *The transport of ions in matter*, Software package. IBM Research, Yorktown, NY.

12
Computer methods for particle transport

12.1 Introduction

Earlier discussion indicates the complexity involved in estimating radiation effects within a vehicle, usually a spacecraft. Often, the designer will be provided with environment data in the form of orbit-averaged particle flux spectra for the project under consideration. However, these data can themselves be generated using the environment models described in Chapter 2 and their associated computer programmes. The environment is also often specified in terms of the 'dose–depth' curve giving the dose as a function of depth of shielding in a simple geometry (slab or sphere). The necessary end-point will be the estimation of the radiation dose likely to be received by a particular device at a particular point within the structure. 'Manual' methods were described earlier, but powerful computer codes now serve as tools for the engineer.

An overview of the radiation analysis problem is given in Fig. 12.1. The steps necessary to provide the final dose can be summarized as:

(1) prediction of orbital and mission particle fluxes;
(2) generation of a simple-geometry 'dose–depth' curve from (1);
(3) extrapolation of (2) to complex vehicle geometries, using sector analysis, accounting for shielding afforded by various parts of the real spacecraft.

Simplifying assumptions are made in separating steps (2) and (3). A rigorous analysis would require that the geometry be explicitly considered as the dose is computed; such methods are usually of the Monte Carlo type described below. Manual methods may be useful for a preliminary dose assessment, but the simplifications introduced may introduce undue error at the detailed design stage where highly detailed computer-based sector analysis of the structure of the vehicle can be performed very efficiently.

A brief review is given here of the various computer programs and their use for space radiation. Calculations of single-event upset rates and the problems of reactor and isotope environments are also described.

328 Computer methods for particle transport

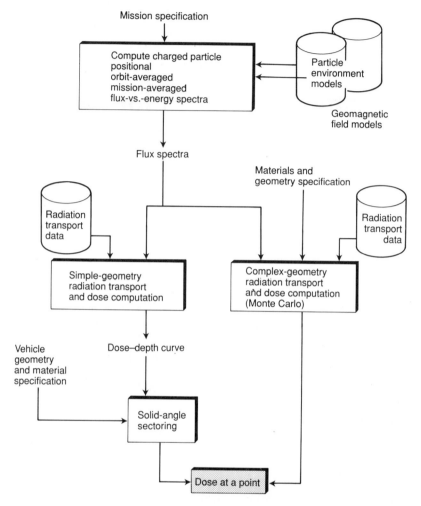

FIG. 12.1 Radiation transport analysis—an overview (Daly 1989). (ESA with permission.)

12.2 Environment calculations

The space environment models described in Chapter 2 are available on computer tape from the WDC-A-R&S (NSSDC) at NASA-GSFC and can be used to produce the required particle flux spectra. To do this, they must be used in conjunction with an orbit generator, giving the spacecraft trajectory and a geomagnetic field program to provide the corresponding B-L coordinates of the trajectory. These steps have been integrated together in the

UNIRAD system (Daly 1986), which also includes programs for dose and equivalent-fluence calculations.

12.3 Dose computation

Over the last two decades, a number of computerized methods have been developed to handle the complete space radiation problem. Some of these are complex and require considerable computer capacity; some employ simplified approaches, making certain reasonable approximations. These methods are in wide use and are recommended especially when considering geometrical and material complexities of spacecraft.

Programs are available in Europe through the program library of the Nuclear Energy Agency (operated by the OECD) . Some programs are also available from ESA and others from NASA via the COSMIC software agency. It should be noted that many of the older programs are not designed for universal application, have poor documentation, and/or are not clearly coded.

12.3.1 Space particle types

Electrons. The major problem to be dealt with in computing the transport of electrons through materials is scattering. Most interactions between energetic electrons and the electrons and atoms of the material through which they are moving involve small energy losses and trajectory deflections. This makes the computational treatment of electron transport difficult.

Protons. Because of their mass, protons do not undergo significant scattering in travelling through a medium. They slow down, losing energy quasi-continuously, in a straight line. Therefore, the well-known range–energy relation and a 'continuous slowing down' approximation can be applied quite simply to their motion. Occasionally, protons produce secondary particles when they collide with atomic nuclei.

Bremsstrahlung (photons). As electrons slow down in a material, they generate bremsstrahlung photons with a distribution of photon energies and directions. Photons subsequently interact with matter through a number of processes (photoelectric, Compton, pair-production), resulting in loss or scattering.

Cosmic rays. Cosmic rays, being highly energetic, are also highly penetrating. They travel in a straight line through a material, producing a cylindrical track of dense ionization. A number of secondary particles are generated by this track, including 'delta rays', which are energetic electrons.

Other particles. The above particle species are the main concern for total-dose and single-event upset problems in space vehicles. A number of secondary particles can be produced in the interactions between protons or heavier ions and the spacecraft material, or in the residual atmosphere in low Earth orbit. These can include secondary protons, neutrons, spallation fragments, and more exotic nuclear particles. Continuous exposure to space radiation can also lead to 'activation' of spacecraft material, which then emits radiation. Finally, on-board nuclear sources, e.g. reactors and thermoelectric generators, can be an important source of radiation. These various 'second-order' radiations must be considered in circumstances where there is a specific sensitivity to them, e.g. in manned flights or when the radiation interferes with instrument detectors.

12.3.2 Monte Carlo techniques

Monte Carlo techniques numerically plot the trajectories of large numbers of particles and predict their interactions in the material through which they are travelling. Interactions usually have a distribution of possible outcomes to which random sampling is applied. For electrons, successive interactions are too numerous to follow individually; instead, attention is given to a small section of the electron's path containing a large number of individual interactions. The net result of all the interactions can be expressed analytically and at the end of each section, the electron energy loss is computed and its direction is altered by random sampling of a scattering distribution. The section length is chosen such that the energy loss in the section is a small fraction of the electron energy. This is the 'condensed history' Monte Carlo technique (Berger 1963). Monte Carlo techniques can also be used to compute the transport and interactions of other particles and their secondaries, including bremsstrahlung and neutrons.

Some Monte Carlo programs include particle transport and interactions in complex geometries, while others consider simple slab geometries. Some commonly used programs and their lead authors include:

ETRAN   : electrons and photons in slab (Berger and Seltzer, NBS)
ITS/TIGER: electrons and protons in various geometries. (Halbleib *et al.*, Sandia)
BETA   : electrons and protons in complex geometries. (Jordan, ART–AFWL)
GEANT-3   : multiparticle complex geometries. (Brun *et al.*, CERN)

The first three are available from the NEA.

Typical outputs of the electron bremsstrahlung codes are:

(1) transmission and back scatter of electrons and photons;
(2) production of bremsstrahlung by electrons and its transport;

(3) spectrum of energy deposition in a thick target by an electron beam;
(4) flux spectra as a function of depth in the material;
(5) deposition of dose as a function of depth of shielding (dose–depth).

TRIM (Ziegler, IBM) is a program for the penetration of ions into solids. Ion energies are from 10 eV to 2 GeV per amu; complex targets of compound materials may be defined with up to three layers. The output of TRIM is a final distribution of the ions and calculation of the phenomena associated with the ion's energy loss. Target atom cascades are followed and redistribution of target atoms is determined. TRIM will run on a PC and has proved very useful in predicting proton and alpha particle damage in silicon devices with complex structures, such as charge-coupled devices.

12.3.3 Methods using a dose 'look-up table': SHIELDOSE

If one is always dealing with a simple (e.g. slab) geometry, then—for an incident particle of a single energy—the generated dose per unit fluence at a given depth of shielding will be the same. Seltzer has created a large data set containing the dose per unit of incident fluence as a function of depth of aluminium shielding and particle energy. The total dose at a given depth of shielding for a given incident particle spectrum is then found by summing the contributions from each particle energy, considering the incident flux at that energy and the time duration. Seltzer's data set (1979) is created with the ETRAN Monte Carlo code, considering all the relevant physics, and his SHIELDOSE program (Seltzer, 1980) can rapidly give the dose for any arbitrary input spectra. Protons are treated according to their range–energy relation, with nuclear reactions neglected. Figure 12.2 shows the three simple geometries considered by SHIELDOSE. If one is not interested in performing a geometrical analysis, the appropriate geometry to choose will depend on the location of the particular dose point. However, the slab case may be too optimistic in practice, since it represents a situation where there is relatively good shielding in most directions. The widespread use of the slab model stems from the fact that it is easier to compute doses with slabs than with spheres. The sphere case is more conservative because, in a slab, there are longer paths through the material in directions away from the normal. For use in subsequent sectoring, the sphere case should be used, since slant paths in a laterally infinite slab are meaningless when considering a small solid angle about a particular direction. This programme is interfaced with the environment spectra and included in ESA's UNIRAD system.

12.3.4 Methods using a straight-ahead approximation

Simpler and shorter alternatives to full Monte Carlo codes exist where analytical methods have been developed for specific application to spacecraft. These include:

332 Computer methods for particle transport

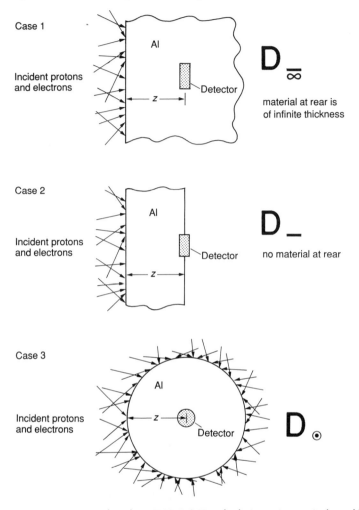

FIG. 12.2 Geometries considered in SHIELDOSE calculations; isotropic broad-beam fluxes of protons and electrons are assumed incident on aluminium targets, and absorbed dose is calculated for small volumes of 'target' or 'detector' materials Al, H_2O, Si, and SiO_2. Case 1, semi-infinite medium; case 2, finite-thickness slab; case 3, solid sphere (after Seltzer 1980). Reprinted by permission of the author.

CHARGE (Yucker and Lilley, McDonnell Douglas, available from NEA).
SHIELD (Davis and Jordan for JPL, available from COSMIC).

These simpler codes assume in the first instance that particles travel in straight lines and, according to the basic range–energy relations, lose energy

12.3 Dose computation

continuously. Corrections are applied to account for the angular effects and the 'degraded' spectrum is computed at a number of depth intervals in the shielding. These methods have the advantage over SHIELDOSE that they can treat the effects due to material differences. They can model laminations of materials in simple (slab, sphere) geometries. Differences between materials are accounted for by normalizing the thickness to density to give g cm^{-2} (or by converting to the equivalent aluminium thickness).

12.3.5 CHARGE program

In the case of electrons, an attempt is made to account for the effect of their angular scattering by applying 'transmission coefficients' derived from Monte Carlo runs. For example, CHARGE uses the empirical relations developed by Mar (see Chapter 11). Bremsstrahlung is treated in CHARGE by using the Koch and Motz (1959) model for generation with transmission based on an exponential attenuation model. 'Build-up' factors are applied to account for bremsstrahlung angular distribution effects. As

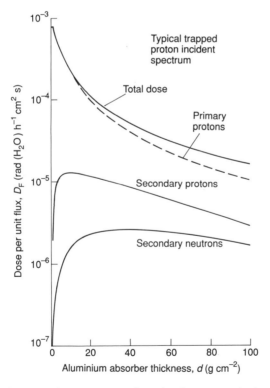

FIG. 12.3 Secondary particle generation: dose-depth curves calculated by CHARGE; showing the effect of secondary particles generated by the passage of typical trapped protons in space through thick aluminium shielding.

indicated above, proton-slowing can be treated well by straight-ahead methods.

The ability of CHARGE to treat laminations of different materials has been used in an ESTEC study (Daly and Adams 1984), where the program was assessed and then applied to the problem of predicting the shielding efficiency of various laminations of high-Z and low-Z materials. CHARGE can also optionally calculate the dose from secondary protons and neutrons. Typical dose–depth curves from CHARGE are shown in Fig. 12.3. This shows the dose in water behind very large thicknesses of aluminium shielding ($100 \, \text{g cm}^{-2}$ or about 37 cm). At 'normal' thicknesses (i.e. a few millimetres), the dose from secondary protons is almost two orders of magnitude lower than that due to primary protons. The incident proton spectrum in this example is 'a typical trapped-proton spectrum'. Other proton-generated particles, e.g. muons, pions, electrons, or positrons, are not treated by CHARGE owing to insufficiency of data. Alsmiller (1964, 1967) has shown that the doses due to muons and photons can be ignored if present understanding is reliable. Given the appropriate fluence-to-response conversion functions, CHARGE can compute doses, dose equivalents for biological effects, and activation response.

12.3.6 Sector analyses

Application of one of the simple-geometry techniques can produce the dose versus depth relation, which can be combined with solid-angle sectoring to produce an approximate dose at a point in a representative model of the spacecraft geometry. Often, the engineer is provided with this dose–depth curve as a specification of the environment. The choice of geometry for the dose–depth calculation sectoring is only an approximation, since the angular scattering of electrons and angular distributions of bremsstrahlung are not explicitly treated.

Manual methods for sectoring calculations were described previously, but computer programmes available for the task include:

DOSRAD, an ESABASE analysis module (Ferrante *et al.* 1984; available from ESA)

SIGMA II, developed for JPL (Davis and Jordan 1976; available from COSMIC)

SIGMA I/B (Jordan and Yucker, available from NEA)

MEVDP, AFWL (Lilley and Hamilton, available from NEA)

RADATA/RADSCAN (Jordan 1990; may be available from BAe)

As an ESABASE module, DOSRAD has the advantage of the use of the ESABASE 'pre- and post-processing' visualization utilities for geometry checking and 3D display of results. The spacecraft geometry is defined in terms of simple shapes and the dose–depth curve taken automatically from

12.3 Dose computation

the SHIELDOSE program in the UNIRAD system. DOSRAD, in common with other sectoring methods, employs ray-tracing, in which a large number of rays are followed out through the geometry; intersections with structure materials are found and the total shielding thickness along each ray is computed. The elemental solid angles around the rays are used to 'weight' the interpolated dose values from the (4π) dose–depth curve at the appropriate depth values and an integration is performed over all rays and solid angles. Thus, a highly detailed sectoring of complex spacecraft geometries is possible.

SIGMA operates on basically the same principle, but has interesting additional capabilities. Multiple attenuation curves can be input, for example individual radiation-species doses and ion fluences. Dependence on material of the attenuation of the various input radiation species can be partially accounted for by specifying different thickness-normalizing factors for each material and radiation type. The usual tracing of the rays, accounting for their slant paths through materials, is augmented by a 'minimum path' estimation to give the minimum shielding a multiply scattered electron would encounter. Parametric shielding calculations are possible in which the sensitivity of the dose to additional 'spot' shielding is computed. The COSMIC package containing SIGMA II also contains the shield optimization program SOCODE and the SHIELD program described above. However, in common with other older programs, SIGMA with its limited documentation is quite difficult to use. The original SIGMA, still available through the NEA, was extremely difficult to use for the definition of complex geometries, since it was necessary to define the shield structure in terms of bounding-surface quadratic equations, and voids also had to be explicitly defined. The SIGMA II program, available through COSMIC, has improved geometry definition facilities and no longer requires void specification.

RADADATA/RADSCAN is a PC-based program for sector analysis, analysing 648 sectors. Up to 20 components on 20 PCBs can be handled per run. The output includes component identity, orientation, and position versus dose limits, and highlights components outside their radiation limits. A sector map of component shielding may also be produced.

12.3.7 Comparisons

While it is not the objective of this brief review of computer methods to make rigorous comparisons between the various approaches, a number of specific instances where comparison has been possible are given.

CHARGE validation examples. The CHARGE documentation contains a number of comparisons made by way of validation of the code. Agreement with other calculations of primary proton dose is good. Secondary nucleon doses are reasonable when compared with Monte Carlo results. Electron doses computed by CHARGE were found to be higher than with other Monte Carlo methods, the disagreement being within a factor of 2 at normal shield thicknesses.

Bremsstrahlung dose: CHARGE and 'manual' methods. The CHARGE documentation includes tabulated data for dose in water behind aluminium shielding in an elliptical polar orbit for which a trapped-electron spectrum is given. The bremsstrahlung dose calculated by the manual approximation is 0.6 rad(H_2O) per month. The dose calculated by CHARGE is between 0.4 and 0.5 rad(H_2O) per month, depending on shield thickness. This correlation is considered to be reasonable.

Electron dose: CHARGE, SHIELD, and BETA. Davis and Jordan (1976) compared the two 'simplified' programs and the BETA Monte Carlo program in the calculation of an electron depth-dose curve for the Jupiter environment. The flux is isotropic upon a spherical aluminium shield with the dose point at the centre. Excellent agreement is obtained for shield thicknesses of up to 4 g cm^{-2} (15 mm of Al). Divergence thereafter is probably due to the weakness in CHARGE with electrons greater than 10 MeV energy.

Electron transmission: BETA, ETRAN, and experiment. The BETA documentation contains a number of comparisons between experimental electron and bremsstrahlung transmission data which show very good agreement. ETRAN results have also been reported (Berger and Seltzer 1968) and these too agree very closely with experimental data.

CHARGE and SHIELDOSE. The results of the Daly and Adams study mentioned in Section 12.3.5 showed that:

1. In spherical geometry, there is good agreement between proton and bremsstrahlung doses.
2. In slab geometry, the CHARGE bremsstrahlung dose is 2 to 3 times higher.
3. SHIELDOSE predicts electron doses 2 to 3 times higher than CHARGE in spherical geometry, but agrees well in slab geometry. This is probably due to the solid-sphere geometry of SHIELDOSE, with account taken of back-scattered electrons.
4. SHIELDOSE should be used as a tool for routine analysis, with CHARGE available for situations where materials are important.

12.4 Single-event upset prediction

The first SEU prediction program was 'CRIER' (cosmic ray induced error rate), by Pickel and Blandford (1980). In 1982, Adams and co-workers at

NRL produced CREME (cosmic ray effects in microelectronics) based on a series of NRL reports; it was updated in 1985. The upset calculation is functionally equivalent to CRIER but, in addition, there are programs for deriving the ion environment, considering geomagnetic shielding, 'weather' conditions, material shielding, etc. SEU prediction programs require device test and technology data as input parameters. At its simplest, the test data include threshold LET and cross section and the technology data require definition of the sensitive volume, or volumes, within a device. The technology data are best obtained from the device manufacturer. To obtain the necessary information by 'reverse engineering' is quite an extensive task.

The CREME programs are a group of FORTRAN routines that calculate differential and integral energy and LET spectra of cosmic rays incident on the electronics inside any spacecraft in any Earth orbit. Input parameters for running these programs describe the interplanetary and magnetospheric weather conditions, the spacecraft's orbit, the shielding surrounding the electronics, and the characteristics of the device under consideration. Input data files contain tabulations of stopping powers and ranges of cosmic ray nuclei in aluminium and silicon and geomagnetic cutoffs. The program's output files contain energy and LET spectra and single-event upset error rates. Modifications have been introduced by Daly to allow the direct introduction of the experimentally measured upset cross section versus LET curve to account for the fact that the upset rate does not saturate rapidly at a unique LET. In other words, the bistables in a device have a spread in sensitivity.

The SPACE RADIATION (Letaw, Severn Communications Corporation 1989) program is based on CREME but includes trapped protons. The user may specify the orbit and the shielding material. The output includes SEU rates and absorbed dose. In the future it will include the electron environment and dose calculation. The program runs on a PC.

The CUPID programme (Clemson University Proton Interaction in Devices, McNulty 1989). This covers proton-induced spallation reactions and determines SEU cross section as a function of the critical charge for a sensitive volume of defined dimensions.

12.5 Neutrons, gamma-rays, and X-rays

The foregoing sections have been concerned mainly with the use of computer programs to determine doses in spacecraft. There are also many computer methods for dealing with terrestrial radiation, such as the shielding of nuclear assemblies, accelerators, and X-ray machines. Extensive computer libraries are maintained for such purposes at nuclear facilities such as the US National Laboratories. A prime example is the Radiation Shielding Information Center at Oak Ridge, Tennessee.

In terrestrial facilities, the calculation of radiation shielding has the major function of reducing the exposure of persons in the neighbourhood of radiation sources, or 'biological shielding'. Within reactors, shielding may also have the function of protecting the vessel from excessive heating or radiation damage. Unlike the space environment, there is little limitation on the mass available to attenuate the radiation. The role of the computer program is to determine the most efficient geometry of shielding and combination of materials to minimize the dose to persons or sensitive parts. In power reactors and fusion machines, the problem is greatly complicated by the presence of large coolant ducts, along which radiation can 'stream' (see Chapter 2). In remotely controlled ground-based vehicles (see Section 2.6 and 15.3) shield weight is at a premium, as it is in space, but the terrestrial environment is less easily attenuated then the space environment. A special feature of nuclear material calculations is the fact that the emission rates and energies of sources such as fuel or reactor cladding will vary strongly with time once a reactor is powered-down.

Among references which deal with neutron, X-ray and gamma shielding comprehensively are Kimel (1966), Chilton *et al.* (1984), Lamarsh (1989), and Paić (1988).

12.6 Conclusions

This chapter has provided a brief review of the many computer methods which are now available for performing analyses of particle fluxes and radiation doses in simple and complex models of spacecraft structures. When using simple geometries for routine dose computation, the sphere case is preferable to the slab case unless the use of the latter can be justified by the location of a specific component in the spacecraft. Shielding in a single-slab model is relatively good in most directions. The use of the programs described in this chapter allows an assessment to be made of the environment and doses in spacecraft, considering the mission and structural materials and geometry. Their use is recommended as an integral part of the detailed spacecraft design phase.

The ESABASE/RADIATION module is designed as an easy-to-use tool for calculating mission radiation environments, using the UNIRAD suite of programs and the resulting dose in a complex model of the spacecraft geometry using the DOSRAD sectoring program; see Fig. 12.4. Alternatively, CHARGE or SHIELD can be combined with SIGMA to compute doses in complex geometries. Single-event upset sensitivity may be assessed with the CREME program together with device data, or other programs now becoming available.

For a review of the wide range of computer methods for dealing with neutrons, gamma-rays, and other radiation not associated with space, the reader is referred to other sources.

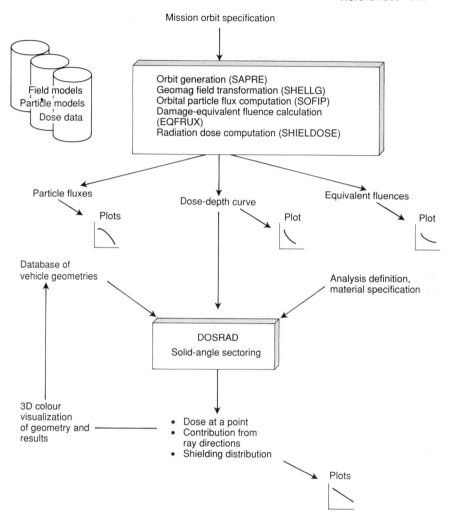

FIG. 12.4 Architecture of the ESABASE/RADIATION application (Daly 1989). (ESA with permission.)

References

Adams, J. H. (1986). Cosmic ray effects in microelectronics, Part IV, NRL Memorandum Report 5901. Naval Research Laboratories, Washington, DC.

Alsmiller, R. G. (1964). Report ORNL-3714. Oak Ridge National Laboratory, Oak Ridge, TN.

Alsmiller, R. G. (1967). *Nuclear Science and Engineering*, **27**, 158.

Berger, R. J. (1963). Monte Carlo calculation of the penetration and diffusion of fast charged particles. In *Methods in Computational Physics*, Vol. 1, *Statistical Physics*. Academic Press, New York.

Berger, R. J. and Seltzer, S. M. (1968). *Penetration of electrons and associated bremsstrahlung through aluminium targets*, NASA Special Publication SP-169. NASA, Washington, DC.

Chilton, A. B., Shultis, J. K. and Faw, R. E. (1984). *Principles of radiation shielding*. Prentice-Hall, Englewood Cliffs, NJ.

Daly, E. J. (1986). The UNIRAD system for radiation environment prediction, ESTEC Technical Note. ESA, Noordwijk, Netherlands.

Daly, E. J. (1989). *The radiation environment; the interaction of radiation with materials; computer methods*. ESA (1989).

Daly, E. J. and Adams, L. (1984). *Radiation doses in spacecraft—assessment of computer programs*, ESTEC EWP 1389. ESA, Noordwijk, Netherlands.

Davis, H. S. and Jordan, T. M. (1976). Improved space radiation shielding methods, JPL Tech Memo. Jet Propulsion Laboratory, Pasadena, CA.

ESA (1989). *Radiation design handbook*. ESA Document PSS-01-609.

Ferrante, J. G., Coffinier, P., Aube, B. and de Kruyf, J. (1984). Spacecraft systems engineering and geometry modelling; the ESABASE–MATVIEW approach. *ESA Journal*, **84(4)**.

Jordan, G. C. (1990). PC based radiation dose calculation for electronic components. In Proceedings of Space Environment Analysis Workshop, ESA WPP-23. ESA, Noordwijk, Netherlands.

Kimel, W. R. (ed.) (1966). *Radiation shielding*. US Govt Printing Office, Washington, DC.

Koch, H. W. and Motz, J. W. (1959). *Reviews of Modern Physics*, **31**, 920.

Lamarsh, J. R. (1983). *Introduction to nuclear engineering*. Addison-Wesley, London, and Reading, MA.

McNulty, P. J., Beauvais, W. J., Adbel-Kader, W. G., El-Teleaty, S. S., Mullen, E. G. and Reay, K. P. (1991). *IEEE Transaction on Nuclear Science*, **NS38**, 1642–6.

Paić, G. (1988). *Ionizing Radiation: Protection and dosimetry*. CRC Press Inc, Boca Raton, FL.

Pickel, J. C. and Blandford, J. T. (1980). Cosmic ray induced errors in MOS devices. *IEEE Transactions on Nuclear Science*, **NS27**, 1006–15.

Seltzer, S. M. (1980). SHIELDOSE—a computer code for space shielding radiation dose calculations, NBS Tech. Note 1116. National Bureau of Standards, Gaithersburg, MD.

Seltzer, S. M. (1979). Electron, electron-bremsstrahlung and proton depth–dose data for space shielding applications. *IEEE Transactions on Nuclear Science*, **NS26**, 4896–904.

13
Radiation testing

13.1 Introduction

The sensitivity of semiconductor devices to radiation is often very variable and it is therefore impossible to use theory alone to predict the effect on a device of a given exposure to radiation. Actual irradiation tests must then be an integral part of the evaluation of a device and, sometimes, tests must be performed on each batch of devices. The simulation of radiation effects in the laboratory is quite difficult to achieve and the results thereof are often not those expected during planning.

This chapter is aimed at providing guidance on the design of adequate test programmes. It reviews:

(1) the radiation sources which may be used together with their advantages and disadvantages;
(2) the main types of measurement which can be made, together with suitable implementation methods.

Reference is also made to the standard test procedures developed in the USA and Europe, and extensive source books such as Attix (1986), Johns and Cunningham (1974) Lamarsh (1983) Profio (1976, 1979) and Kerris (1989).

13.2 Radiation sources

13.2.1 Simulation of radiation environments

Space. The differences between radiation conditions in space and their simulation in the laboratory are frequently quite large. The incident space irradiation consists of a complex, mixed particle environment which, as discussed earlier, is altered and made even more complex by passage through spacecraft enclosures. The dose is delivered steadily over a long period of time — often several years. Most of the time, we can obtain radiation beams from machines at only a number of discrete energies and often it is difficult to modify the rate at which the machine will deliver the radiation. Bearing these facts in mind, we must therefore modify the raw short-term results of 'simulation' tests when converting them into predictions of space radiation conditions. We may also wish to monitor the irradiated device at intervals of minutes, days, and months after exposure and introduce factors to allow for qualitative differences in space and laboratory radiation as well as the 'dose-rate effects' discussed in Chapter 3.

For single-event phenomena the upsets due to high-energy *protons* can be adequately simulated-since a number of accelerators can achieve the appropriate energies up to 300 MeV. Simulation of *galactic cosmic rays*, however, can be achieved only by using ions of similar linear energy transfer (LET) but significantly lower energy (tens or hundreds of MeV). The mechanisms of upset may be influenced by the different ionization track structure at different energies (Stapor *et al.* 1988), and this is currently under investigation by a number of workers.

Military. For the military environment, the problem is the converse of that in space simulation. The need is to simulate a very short pulse of intense radiation. This is usually achieved using pulsed sources of bremsstrahlung and electrons, such as flash X-ray machines and linear accelerators (LINACs). Specialized-purpose pulsed and steady-state reactors are used for neutron testing.

Nuclear industry. For the nuclear industry the problem is often one of acceleration factors. The appropriate test environment is available in a nuclear reactor, but the usual requirement is to demonstrate an extended life (typically 30 years) in a reasonably short test time. Most nuclear industry testing is carried out in high-flux reactors (HFRs), the prime subjects being fuel rod and cladding materials, and pressure vessel steels (see for example Fabry and Lowe 1984). For nuclear fusion reactors, a possible power source of the future, the problems will be broader. A broad range of equipment will be exposed and the confinement vessel will contain exotic metals such as niobium and lithium.

Despite the above-mentioned problems, a surprisingly wide range of devices can be tested suitably by the use of reactors, accelerators and high-activity gamma-ray-emitting isotopes, e.g. using the well-known cobalt-60 hot cell or 'irradiator'. However, we must always classify carefully the physics of the effect which we are aiming to simulate: whether it is classed as a total-dose ionization effect, permanent bulk-damage effect, single-event upset, or transient effect of some other kind.

High-energy physics. Some radiation testing for future high-energy physical research accelerators has been performed at points near the experimental stations of existing accelerators (Tavlet *et al.* 1991). However, to achieve ten-year doses in a shorter time, it may be necessary to test in reactors, gamma-ray sources and other accelerators (see e.g. Schönbacher *et al.* (1990)).

13.2.2 Gamma-rays

The source most commonly used for simulation of ionization effects in silicon components or materials is Co-60, which emits photons of energy 1.173 and 1.332 MeV and has a half-life of 5.27 years. It is utilized for industrial irradiation, sterilization, radiotherapy, and biological research; well-designed irradiators are therefore widely distributed geographically. Several standard commercial irradiation cells are on the market and contract irradiation facilities are often available. Cs-137 (photon energy 0.662 MeV) may gain increased acceptance owing to its 30 year half-life and less severe shield requirements.

It has been amply proven that ionization due to gamma-rays provides a useful simulation of the ionization due to penetrating electrons and protons in the space radiation spectrum. For all practical purposes, one rad(Si) deposited by gamma rays produces the same quantitative response in SiO_2 films with respect to charge-trapping and interface-state creation as does the same amount deposited by high-energy space protons, electrons, and bremsstrahlung. It should be noted that gamma-rays cannot normally be used for bulk-damage simulation (see Chapter 3).

Cobalt-60 is produced from inactive Co-59 by heavy neutron irradiation in a reactor. In a typical irradiator, a cylinder of Co-60, sealed in a steel jacket, is placed in a thick lead shield or concrete cell. A large number of electronic samples, arrayed in sockets in circuit boards, can be placed near the

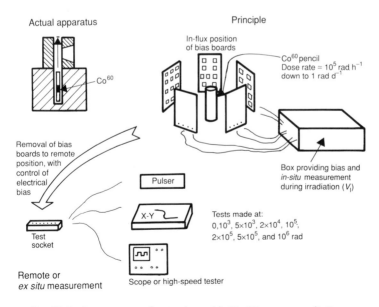

FIG. 13.1 Arrangement for testing with Co-60 gamma-radiation.

source and their response to the radiation can be monitored continuously by means of wires leading out of the cell. In some commercial irradiators, designed for the irradiation of chemicals or animals, the whole exposure takes place in an enclosed structure which can stand freely in the corner of the laboratory. Because gamma-rays are so penetrating, circuit boards can be stacked.

A source of medium strength will have an activity of about 1000 Ci (3.7×10^{13} Bq). Some typical configurations and dose rates are shown in Figs 13.1 and 13.2. Given a room of several metres length, it can be seen that dose rates can be varied from over 10^5 rad h^{-1} 10^3 Gy h^{-1} (allowing the accumulation of mission doses in under an hour) to, say, 100 rad h^{-1} 1 Gy h^{-1}. The latter rate is only about 30 times a typical space dose rate (a high-radiation orbit may average 2×10^5 rad per 10 years, or about 3 rad h^{-1}) and, if real-time conditions are desired, shielding by a few centimetres of lead or steel can produce a further reduction. Alternatively, a source of lower activity can be used at a closer distance (Hardman et al. 1985).

Although the spectrum of a radioactive source is considered to be a line spectrum, this spectrum is modified by scattering within the source itself, the source housing, and the walls and ceiling of the source room. Such scattering can lead to a significant low-energy component (zero to 0.5 MeV) which may lead to severe dose enhancement (Kelly et al., 1983).

13.2.3 X-rays: steady-state and pulsed

Being similar to gamma-rays, X-rays will simulate the space environment by inducing ionization. Even low-energy X-rays, provided they can be introduced into the active region of the device and the doses correctly estimated, can be effective. The construction of X-ray machines is adequately dealt with in Johns and Cunningham (1983). X-rays were first used on oxide-passivated silicon devices by the staff of Bendix and RCA (Larin 1968; Poch and Holmes-Siedle 1969). The tube voltages were 150 and 250 keV respectively. The response of devices was found to be identical to that given by 1 MeV and 125 keV electrons, so long as the base widths of the transistors were low enough that bulk damage was minimal. The use of a 150 kV radiographic set, with and without filtration and the response of a number of device technologies have been studied for the purpose of space simulation testing by Adams and Thomson (1980). Kelliher (1985) performed careful dosimetry of a 320 kV X-ray set, using actual CMOS circuits as a cross-reference. The use of energies in the 10 keV range on unpackaged wafers is described later.

X-rays are generated when a beam of electrons bombards a target, usually of a high-Z metal such as tungsten or copper. For high beam currents, the target must be cooled; the power supply is large if high beam currents are desired. The electron beam, colliding with the target, excites a 'white spectrum' of bremsstrahlung X-rays (actually peaked broadly at about half the beam energy), upon which the K and L peak emissions of the target

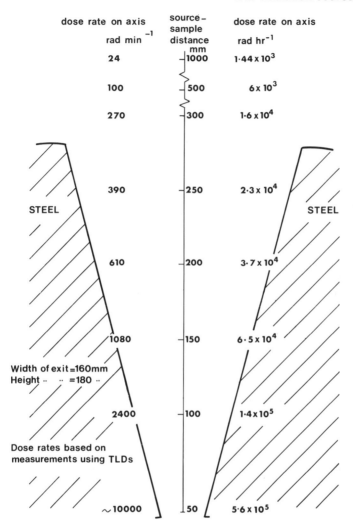

FIG. 13.2 Cross section of steel jaws of Co-60 cave, with dose rates corresponding to position on axis; doses in rad(H_2O), source activity ~1000 Ci (37 T Bq).

metal are superimposed. For tungsten, the main peak is 59.3 keV (K line) and for copper 8.04 keV (K line). The L peak for tungsten is 8.396 keV. It is normally desirable to filter these out and also the lower-energy white radiation, so as to avoid too much influence of the encapsulation on the dose penetrating to the active silicon device. Dose rates at 10 mA in Adams's 150 kV source (tungsten X-rays) were approximately 10^3 rad min^{-1} (10 Gy min^{-1}) at a distance of 330 mm and about 10^2 rad min^{-1} (1 Gy min^{-1}) at

860 mm. The attenuation of this beam by a 0.3 mm Kovar lid is about threefold.

The use of X-rays requires care, but can be recommended for identification of sensitive technologies and the irradiation of limited numbers of devices. The main advantages are the low cost, the wide distribution of X-ray equipment, and the high safety standards available in such equipment. The care which is required revolves around the accurate administration of dose. In order to produce repeatable penetration power in an X-ray beam, the power-supply voltage and tube current must be stable. The degree of filtration must be kept constant, because this controls the X-ray photon energy spectrum. Both the degree of package penetration and the response of dosimetry are very sensitive to changes in energy spectrum.

There has been much interest in recent years in the use of a low-energy (10 keV) X-ray tester for total-dose testing at wafer level (Palkuti and LePage 1982). This uses the tungsten L line; a collimated source is fitted with a shutter and incorporated into a wafer prober. This allows the irradiation of a single selected die on a wafer. A typical dose rate from such a source is in the range 10^5–10^6 rad(Si) min^{-1} (10^3 – 10^4 Gy(Si) min^{-1}). There have been some difficulties and controversy in correlating this source with Co-60 owing to different recombination and dose-enhancement effects. The general consensus is that this is a valid test technique, provided that careful correlation studies are carried out for the semiconductor technology of interest (Dozier *et al.* 1983, 1985; Benedetto *et al.* 1988; Fleetwood *et al.* 1988*a* and *b*). Unlike gamma isotope facilities, X-irradiation equipment has not been fully standardized for component testing. In view of the limited beam angle and relatively high absorption coefficient of kilovolt X-rays, isotope sources are preferred for bulky equipment, a large throughput of parts and high accuracy of result.

The flash X-ray (FXR) generator is a source used for military transient testing. The FXR produces a highly intense pulse of bremsstrahlung with a pulse duration in the range 50–200 ns and a peak pulse dose rate in the range 10^9–10^{12} rad s^{-1} (10^7 – 10^{10} Gy s^{-1}). This is achieved by an energy storage and pulse-forming network discharged by means of a cold cathode tube. An intense pulse of electrons is directed to a high-atomic-number target (typically tungsten or tantalum) to produce bremsstrahlung X-rays. The accelerating voltage is typically 10 MeV with a few mA of current per pulse X-ray pulses may also be generated using a LINAC with a high-atomic-number target powered by klystrons generating pulses of electrons in the range 1 to 5 A, 4 to 40 MeV (Kerris 1989).

13.2.4 Electrons: steady-state and pulsed

All electron beams act as a source of ionization, but only the high-energy machines (particles of energy considerably greater than 0.1 MeV) will produce displacement damage in semiconductors. Consequently, one of the

13.2 Radiation sources

most generally useful sources is the van de Graaff generator, especially because this machine can be designed to operate at any particle energy between 0.1 and 10 MeV. An electron beam is accelerated by the field between earth (the target) and an electrode charged to a very high static potential. The charging is accomplished by means of a moving belt which carries charge from a DC generator to the insulated 'head' electrode. Potentials of 10 million volts can be produced, but with the more common machines, 1 million volts is the limit. Electrons released in the head are accelerated away from it down an evacuated column and emerge through a titanium vacuum window as a beam of about 2 cm diameter. Devices can be placed in this beam. At 1 MeV, the electrons can travel several centimetres in air without great loss in energy, so that irradiation can be performed in air. Typical device encapsulation (e.g. 0.3 mm Kovar IC lids) extracts energy and scatters 1 MeV electrons heavily. Therefore, there is always uncertainty about the dose received at the chip of an encapsulated device. Some engineers object to removing the encapsulation for the purposes of testing. The dose rate can be varied by altering beam current and beam focus or sweeping the beam. Beam current can often be varied from 10 nA to 10 μA, which, in a 20 mm diameter beam, yields particle fluxes from about 2×10^{10} to 2×10^{13} cm^{-2} s^{-1}. These fluxes correspond to dose rates from about 600 to 600 000 rad s^{-1} (6 to 6000 Gy s^{-1}).

TABLE 13.1 Typical factors for converting electron fluence to dose[a]

Cgs units	Beam diameter 1.13 cm Area 1 cm^2	Beam diameter 2cm Area 3.14 cm^2	Beam current
Flux (cm^{-2}s^{-1})	6.2×10^{10}	1.97×10^{10}	10 nA
Dose rate (rad(Si) s^{-1})	2.07×10^3	6.57×10^2	10 nA
Flux (cm^{-2}s^{-1})	6.2×10^{13}	1.97×10^{13}	10 μA
Dose rate (rad(Si) s^{-1})	2.07×10^6	6.57×10^5	10 μA
SI units	Beam diameter 0.013 m Area 10^{-4}m^2	Beam diameter 0.02 m Area 3.14 $\times$ 10^{-4}m^2	Beam current
Flux (m^{-2}s^{-1})	6.2×10^{14}	1.97×10^{14}	10 nA
Dose rate (Gy(Si) s^{-1})	20.7	6.57	10 nA
Flux (m^{-2}s^{-1})	6.2×10^{17}	1.97×10^{17}	10 μA
Dose rate (Gy(Si) s^{-1})	2.07×10^4	6.57×10^3	10 μA

[a] For 1 MeV electrons incident normally, and with no cover on device (3×10^7 cm^{-2} = 1rad(Si))

Table 13.1 gives the fluence-to-dose conversion factors for electrons. Since the dose rates in space are in the range 10^{-5} to 10^{-3} rad s^{-1}, the acceleration of test rate here is over 10^6 times. In many cases, this is acceptable, but it is inconvenient in that, because beam currents cannot be controlled below a few nanoamperes, the dose rate cannot be lowered further.

Resonant transformer accelerators yield electron beam currents up to 1 mA in the 1 to 3 MeV range. Since the peak annual fluences encountered in space with $E > 0.5$ MeV are 3×10^{14} cm^{-2}, it can be seen that test exposures take only a few minutes.

A linear accelerator (LINAC) provides intense pulses of electrons of higher energies, typically 4 to 40 MeV in rapid, square pulses. Electrons fired from an electron gun in a waveguide pick up RF energy and are accelerated from a few keV to several MeV. Average currents are again the microampere range, but instantaneous dose rates can be as high as 10^{10} rad s^{-1} (10^8 Gy s^{-1}).

A typical experimental arrangement such as would be used for any high-energy electron (or proton) beam exposure is shown in Fig. 13.3. It was developed at RCA for solar cells and a 1 MeV van de Graaff generator and gives very high dose uniformity (Wysocki 1963). Scanning of an array of samples can be achieved magnetically, thus avoiding movement of the samples, and it has been shown that doses can be administered in this way with suitable accuracy. However, owing to the pulsed nature of the irradiation, other workers prefer to use a scattering foil to spread the beam. Another source of electrons of interest for special experiments on low dose rates is the isotope Sr-90/Y-90. It emits beta rays over a spectrum not dissimilar to that of space. Rates can be achieved such that a typical 1 year dose can be accumulated in several months, i.e. the acceleration factor is less than

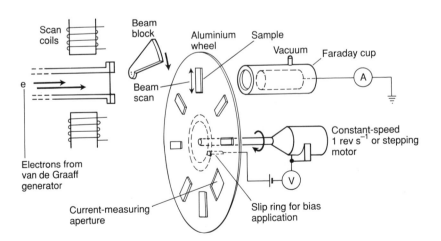

FIG. 13.3 Typical arrangement for 1 MeV electron-beam radiation testing.

10. However, the dosimetry for such a continuous beta spectrum is difficult. The DERTS laboratory in Toulouse has carried out a number of experiments using this source, including low-dose-rate experiments on MOS devices.

Unfocused beams of electrons in the kilovolt range are easy to form with a kilovolt power supply and a simple electron gun (television tube or 'home-made' gun). Such beams can be used to irradiate thin films, and beam currents of milliamperes can be obtained. Another type of kilovolt beam, which is of lower current but of much higher precision, is that used in the scanning electron microscope (SEM) (Galloway 1977). Typical energy is 35 keV and the beam can be precisely aimed and rastered over a selected microscopic area of a semiconductor device (uncapped, of course). Currents are usually in the nanoampere range, but the irradiated area can be as small as a tenth of a micrometre square. Thus, the local dose can be made very high (many megarads per second). If the beam is rastered over the whole of the chip and beam-blanking is employed, dose rates as low as a few kilorads per second may be achieved.

The SEM method is worthy of consideration because the beam is already used in the imaging mode for quality control inspection of semiconductor chips at low dose levels. The opportunity of adding a 'sacrificial, high-dose irradiation on a selected area of the chip is therefore an economical complement to the usual inspection. The problem is dosimetry, because the precise thickness of the 'passivation' on wafers is not always known. A recent Canadian development, the 'SEMFET' (Thomson *et al.* 1990) attempts to address these problems by the use of an MOS dose-measuring device and a computer interface for calculation of the attenuation of the different layers on the chip.

13.2.5 Protons

This section deals mainly with displacement damage induced by protons in encapsulated silicon components (see Chapters 2 and 6). The range in aluminium of a 15 MeV proton is about 1.5 mm (0.060 inch). For surface coatings and windowless optoelectronic components such as X-ray imaging charge-coupled devices (CCDs), energies down to 1 keV must be considered. Special facilities are available at number of laboratories for their exposures at low energies. For the acceleration of protons to energies above 15 MeV, the most common machine is the cyclotron. In this instrument, high-frequency currents applied to two D-shaped electromagnets supply energy to a beam of hydrogen ions injected into a circular 'race track'. The trajectory of the particles is an outward spiral and particles can be picked off at an exit tube. A typical energy for a nuclear research cyclotron would be 20 MeV and fluxes of the order of 10^{13} cm^{-2} h^{-1} are achieved. As the damage efficiency of protons in silicon falls off with increasing energy, 20 MeV is suitable for devices. The spectrum in space penetrating a compartment will be richest in this energy range, lower energies being attenuated greatly by the intervening absorber.

Many devices have been tested using protons at 10 MeV (Boeing Seattle USA and AEA Harwell UK), 100 MeV (McGill University, Canada), 16.8 MeV (Princeton University, NJ, USA), and 22 and 40 MeV (NASA Langley, VA, USA). In Europe, some work has been done in the 7 to 50 and 200 to 3000 MeV ranges. Proton irradiation of solar cells and bulk silicon was used to construct the 'BGR' curves mentioned in Chapter 11.

Since typical space proton fluences ($E > 15$ MeV) for scientific missions are at most 10^{11} cm^{-2}, the fluences required for testing can be built up on a cyclotron in a fraction of an hour. Even at the peak of the proton belt, annual fluences ($E > 10$ MeV) are about 10^{13} cm^{-2}. Some van de Graaff machines can be converted for accelerating protons by reversing polarity and supplying a source of ionized hydrogen at the head. On some modern machines in the 0.5 to 3 MeV range, this conversion is effected with ease.

Other sources of high-energy protons are occasionally available for irradiation. Proton linear accelerators, operating in the 10 to 100 MeV range, are used as injectors for GeV research accelerators, and cyclotron accelerators operate up to 600 MeV. These higher-energy accelerators are used for investigating proton-induced single-event phenomena. A recent development is the installation of the Space Proton Irradiation Facility at the Paul Scherrer Institute in Villigen, Switzerland. Monoenergetic protons of energies up to 590 MeV are available and continuous proton spectra in the range 10 to 300 MeV can be generated to simulate solar flare and proton belt conditions.

13.2.6 Neutrons: steady-state and pulsed

The bulk defect structures resulting from neutron irradiation bear a family resemblance to those produced by higher-energy protons and electrons. Materials test reactors, which possess beam tubes or hydraulic tubes ('rabbits') or have a swimming-pool design, can be used for exposure of samples to fast neutrons. Military programmes rely heavily on neutron testing, generally carried out using special-purpose, pulsed nuclear reactors. The 'TRIGA' reactor is a light-water-cooled, zirconium-hydride-moderated reactor which can be operated in steady-state or pulsed mode. The peak pulse flux is in the range of 6×10^{17} fast neutrons cm^{-2} s^{-1}. The GODIVA is an unmoderated reactor with a uranium sphere in three 'slices'. The outer slices can be driven inwards or outwards with respect to the centre slice. A neutron pulse is generated by driving a fuel rod through a channel in the centre slice, after bringing the assembly together. The peak pulse flux is in the range 10^{17}-10^{20} n cm^{-2}s^{-1}. For a fuller description of military neutron testing the reader is referred to Messenger and Ash (1986). Baur *et al.* (1981) provide a useful tabulation of test reactors and accelerators suitable for testing fusion reactor components.

13.2 Radiation sources   351

Neutrons are generated by the fission of uranium-235 and have energies spread over the range 0.1 to 3 MeV ('fission spectrum'). However, if the flux of neutrons is allowed to collide with the moderator material or coolant, the neutron energies are reduced to thermal energy of the order of 0.025 eV. This is undesirable for device irradiation because neutrons of energy below 10 keV yield few displacements. Moreover, they can be captured by the device materials, particularly gold and silicon, and produce radioactivity (the devices emerge 'hot') and an inappropriate type of damage. Omnidirectional fast-neutron fluxes in reactor cores are usually well above 10^{15} n cm^{-2}h^{-1}. They are accompanied by isotope gamma-ray doses of the order of 10^6 rad hr^{-1}, which complicates the interpretation of responses. Beams of 14 MeV neutrons from the fusion of deuterium and tritium ions, colliding at a few keV, can be produced either in electrostatic machines, which accelerate deuteron beams at 200 keV, or in various experimental plasma generators. Other neutron sources such as accelerator tubes and isotopes are described by Profio (1976, 1979), Liska *et al.* (1981) and Heikkinen *et al.* (1981).

13.2.7 UV photon beams and other advanced oxide-injection methods

Vacuum ultraviolet (VUV) light gives, qualitatively, exactly the same type of charge build-up in SiO$_2$ films on Si as high-energy radiation. However, the method at present is a research technique suitable mainly for characterizing oxide film technology at the wafer stage. The same comments apply to avalanche injection techniques in which a controlled avalanche breakdown in the semiconductor injects hot electrons or holes into the oxide film. The difficulty is in control, dosimetry, and interpretation. A third advanced method, not so well characterized, is the application of a corona discharge to a bare oxide. These techniques are further described in Chapter 4.

13.2.8 Summary of requirements for steady-state radiation sources

In summary, the desirable features of ionizing radiation test sources for semiconductor devices are:

(1) easy access (for 'in situ' access and rapid sample change);
(2) large area of beam (for large sample throughput);
(3) highly penetrating radiation (avoids encapsulation problem);
(4) unambiguous dosimetry;
(5) high stability (reduces the cost of dosimetry);
(6) safety in operation;
(7) low capital cost;
(8) repeatability of doses from one facility to another;
(9) flexibility in dose rate.

352 Radiation testing

Isotope sources rate highly under all of the above criteria and service a very wide range of the space radiation tests required for electronic and optical devices. Their universal acceptance as standard radiation sources for total-dose ionization effects is envisaged. In a number of cases, bulk displacement damage may be important (e.g. solar cells, thyristors) and here, particle irradiation is essential. Devices sensitive to single-event upset constitute another special group which also requires particle irradiation (heavy ions).

13.3 Cosmic-ray upset simulation: heavy ions

Single-event upset (SEU) simulation requires a source of energetic heavy ions with linear energy transfer (LET) values ranging from about 1 to about 45 MeV mg^{-1}cm^2. While the lower LET values are used to investigate the behaviour of a device around threshold LET, the higher values are used to establish the limiting cross section or saturated error rate. Table 13.2 gives a range of commonly used ions and their LET in silicon. Care must be taken in choosing LET and energy to ensure that adequate penetration into the silicon is achieved. Certain single-event mechanisms such as burn-out of power MOSFETs occur at a depth of 20 μm or greater. After traversing this depth of silicon, the ions must still have a high-enough LET to trigger the single-event mechanism.

The machine used most frequently for SEU testing is the cyclotron, at accelerating potentials of up to 300 MeV, and using a range of gaseous ion

TABLE 13.2 Table of ions commonly used for SEU testing in the Tandem accelerator, at IPN Orsay (Chapuis 1990)

Ion	Energy (MeV)	Energy per nucleon (MeV A^{-1})	Linear energy transfer (MeV mg^{-1}cm^2)	Range in Si (μm)
^{7}Li	44	6.3	0.45	253
^{11}B	50	4.5	1.6	91
^{12}C	70	5.8	1.9	105
^{14}N	69	4.9	2.8	73
^{16}O	100	6.25	3.05	95
^{19}F	100	5.3	4.3	73
^{24}Mg	125	5.2	6.8	61
^{28}Si	137	4.9	8.9	53
^{32}S	160	5.0	10.8	53
^{35}Cl	145	4.1	12.8	43
^{40}Ca	160	4.0	16.3	39
^{58}Ni	132	2.3	28.7	24
^{127}I	100	0.8	47.5	15.5
^{197}Au	127.5	0.65	59.3	1

sources such as krypton, argon, oxygen, and neon. The device to be tested is mounted in an evacuated target chamber which contains silicon detectors and a Faraday cup for beam-monitoring and has feed-throughs for electrical connection. Generally provision is made for the device to be tilted with respect to the beam so as to allow the path-length of the ion through the device to be varied.

Although often somewhat limited with respect to the maximum LET and penetration achievable, the tandem electrostatic generator is also a suitable source of heavy ions. This generator is a special form of the aforementioned van de Graaff design. The accelerating electrode within the Tandem is halfway down the column and charged to a positive potential. The ion source produces negative ions which are attracted by and accelerated towards the

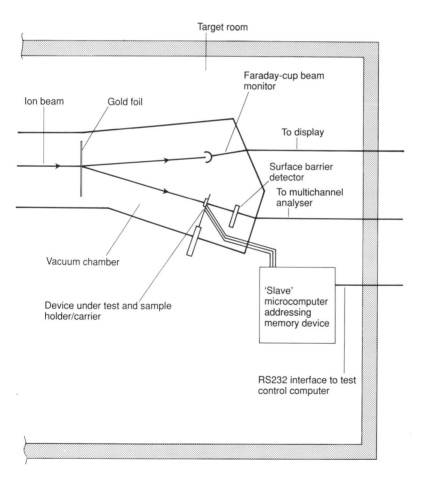

FIG. 13.4 Cyclotron accelerator test configuration (Sanderson *et al.* 1982, with permission of author).

electrode. On nearing the electrode, they are stripped of their charge by foils to a positive charge stage and accelerated away from it towards the mass analyser and the target beam line. The Tandem produces two stages of acceleration, and because the charge state of the ion can be quite high (e.g. oxygen: +5), the overall acceleration energy may be several times the terminal voltage of the machine.

A recent development in SEU testing (Stephen et al. 1983) is the use of fission products from a small radioactive source (1 μCi of californium-252) for the production of upsets. The mean LET of the fission products is 43 MeV mg^{-1} cm^{-2}, which in general is sufficiently high to establish the limiting cross section. The LET can be degraded by the use of foils or an atmosphere of gas, but only reliably to about 15 to 20 MeV mg^{-1} cm^{-2}, which is not low enough to determine the threshold of most of the modern technologies. The main advantages of the CASE system (californium assessment of single-event effects) are its low cost, simplicity, and flexibility. The entire test facility is contained within a simple bell jar and, provided the normal precautions for the handling of radioactive sources are taken, may be used in any laboratory. The system may be interfaced with any test equipment and used for extended periods to accumulate good SEU statistics. A similar bell jar configuration can be used with an americium-241 alpha

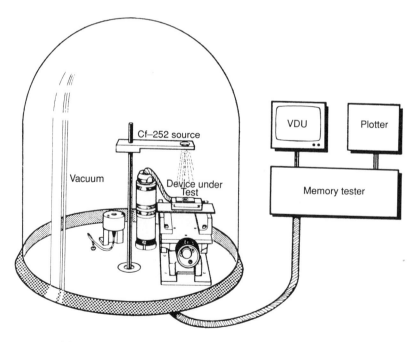

FIG. 13.5 Californium-252 CASE test configuration (Sanderson et al. 1982, with permission of author).

TABLE 13.3 Nuclear characteristics of a Cf-252 source of activity 1μCi (3.7×10^4 Bq) as used in the testing of single-event upsets.

Activity	1 μCi (3.7×10^4 Bq)
Half-life of emission	2.65 years
Omnidirectional neutron emission	4×10^3 n s^{-1} (average energy 2 MeV)
Neutron biological dose rate	0.023 mrem h^{-1} at 1 m
Alpha particle emission	3.1×10^4 particles s^{-1}
Alpha particle energy	5.975–6.119 MeV
Fission fragment emission	10^3 particles s^{-1}
Fission particle energies	80 and 104 MeV
Gamma-ray dose rate	0.002 mR (milliRoentgen) h^{-1} at 1 m

source (LET 0.6 MeV mg^{-1}cm^{-2}), which gives a lower bound to SEU threshold sensitivity. An advanced CASE system has recently been developed (Ackerman *et al.* 1990) using two thin foil detectors and coincidence circuitry to determine the LET of the particle causing upset. Figures 13.4 and 13.5 (Sanderson *et al.* 1982) show schematics of cyclotron and CASE test configurations and Table 13.3 gives the nuclear characteristics of the californium-252 source.

13.4 Dosimetry for testing

Types of dosimeter were surveyed in Chapter 1. In this section we shall consider only the specific application of dosimetry to the testing of silicon devices under radiation and the particular problems involved. The object in this case is to determine the energy deposited in the active region of the silicon chip in rad (Si) or Gy (Si) to within 10 per cent.

Energy-dependence of dosimetric materials. The problem of dosimetry for encapsulated silicon devices is complex, especially when low-energy beams are being used. However, the problems are not insoluble, as the energy absorption physics is well understood and good local dose estimates can be made.

The main problems fall into two fields:

(1) widely differing photon absorption coefficients for Si, Fe, LiF, and H$_2$O in the otherwise useful low-energy region of 30 to 300 keV;
(2) the lack of secondary-electron equilibrium in typical device packages under photon irradiation.

Owing to the strong dependence of the photoelectric absorption effect upon photon energy and atomic weight, small variations in photon beam energy and the composition of a sample can affect both the attenuation of radiation in the device package and the amount absorbed in the active region. These considerations apply particularly to kilovolt X-ray machines. For X-rays in the 30 to 300 keV range, the dosimetry method used must simulate closely the device packaging and structure, while close control must be kept on the penetrating power of the beam.

The energy-dependence problem in LiF dosimetry is adequately explained by the analogy of 'tissue versus bone'; these materials have energy responses not too different from LiF and Si respectively. The energy absorption coefficient changes very rapidly with photon energy for silicon and bone, and much more slowly for LiF, water and tissue. Thus, a shift of a few per cent in photon energy can produce a large disparity in energy absorption between the two groups of materials. This is why control of accelerating voltage is of a high standard in many X-ray sets and must be checked with care in radiation testing. The energy-dependence effect is further complicated by an additional dependence on energy of the thermoluminescent output per unit dose of LiF, which varies by a factor of 1.27 per unit dose between 300 and 50 keV.

It is natural that problems associated with the control and measurement of X-ray dose have received much attention in the field of medical radiology, and methods have been developed to deal with them. The equilibrium question will not be described in detail here. The Bragg–Gray cavity theory concerned is well described in dosimetry manuals (see for example Johns and Cunningham 1974). Briefly, at the discontinuity between two dissimilar materials under irradiation, a secondary-electron spectrum and flux characteristic of the first material persist for some distance until a new equilibrium ('Compton equilibrium' in the megavolt range) is established. Thus, for example, if a small sample such as a silicon chip is irradiated by Co-60 gamma-rays in a steel can (gases being ignored), the silicon receives much of its dose from the Compton electrons generated in the steel. The best dosimeter for measuring rad or Gy (Si) is, not surprisingly, silicon itself (see chapter 1).

13.5 Test procedures for semiconductor devices

13.5.1 Introduction

Having described radiation test facilities, we must now discuss how they shoud be used for irradiating semiconductor devices. The design of a valid space radiation simulation test is not easy. It is important that guidelines for testing are agreed and promulgated, so that these tests are both valid and amenable to comparison. This section does not attempt to present finished guidelines, but discusses the development of procedures and makes comments on them which embody our opinion of acceptable approaches. Winokur, Fleetwood and colleagues at Sandia Laboratories (see e.g. Winokur *et al.*, 1990), Brown and colleagues at US Naval Research Laboratories (see e.g.

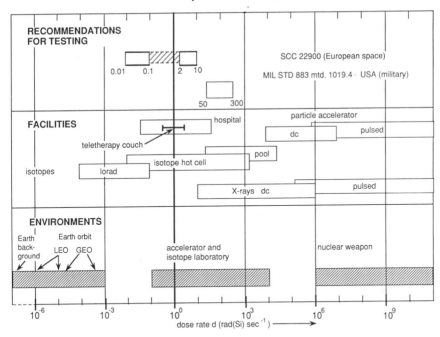

FIG. 13.6 Dose rates and simulation; comparison of the dose rates found in some radiation environments and the rates used in the testing of components for tolerance to cumulative exposure ('total dose').

Brown et al. 1989) and Barnes and colleagues at Jet Propulsion Laboratories (see e.g. Barnes et al. 1991) have been developing, with other groups, a sound physical basis for the radiation testing of integrated circuits. Chapters 3, 4 and 5 have given detailed accounts of the physical issues.

In section 13.2, we noted the differences between the dose rates experienced during use and those used for many 'total-dose' test procedures. Fig. 13.6 demonstrates the scale of these differences, showing that most laboratory facilities and test methods employing them operate in the region above 1 rad s^{-1} (0.01 Gy s^{-1}). The figure shows that specialized facilities giving ultra-low and ultra-high dose rates can be found if necessary. Further comments on time-dependent effects and dose rates in testing are made in Section 13.9.

13.5.2 Objectives

The objectives of a space radiation qualification and test procedure are:

(1) to ensure that the long-lived degradation produced by space radiation lies within an acceptable range;

(2) to produce data which will be of further aid to the electronics designer in estimating the degraded 'end-of-life' characteristics of the piece part;

(3) to produce data showing whether a device is sensitive to single-event phenomena and produce a quantitative estimate for latch-up or soft error rates.

It should be noted that these data enable equipment designers to introduce radiation-tolerance into their designs in two different ways, namely:

(1) by an 'accept/reject' approach, where only the more tolerant devices are accepted for the equipment;

(2) by a 'predict and derate approach', where sensitive devices are accepted and the circuit design allows for quite strongly degraded performance at end-of-life; for single-event phenomena, such an approach may result in the use of latch-up protection and error-correcting codes.

The combination of a standard procedure and data processing method should be such that a test procedure and format are achieved which present the information to advantage and enhance design optimization. An ideal format is the 'growth curve' in which the change of a parameter is plotted versus dose (or time in orbit at a given spacecraft location). On these curves may be noted the 'fixed failure criteria' and the 'stated dose or fluence values' suggested by some authors. The information contained in the full growth curve is more useful than either of the pieces of information mentioned above. For single-event phenomena a similar sensitivity curve is required, showing the variation of error rate as a function of LET and extending from threshold LET for upset to saturated error rate LET. A similar plot is required for proton-induced upsets, where proton energy replaces LET.

13.5.3 Comparison of space with military requirements

It will be noted that not all of the above objectives coincide with those of the test for military environments. In the latter, a single total-dose level is often set (e.g. either the 'tactical environment' with doses in the kilorad range or the 'strategic environment' with doses in the megarad range). In this case, the doses are in reality received in a few short pulses, so that intermediate points on the growth curve are of no interest. As explained elsewhere, the space designer will be considering components which degrade gradually and, for natural space environments, can ignore the transient effects of pulsed doses, while there is no associated neutron damage. On the other hand, the range of semiconductor components used in the two fields is virtually identical and the effects which occur in the two cases are, qualitatively, very similar.

13.6 Radiation response specifications

13.6.1 General

In the many cases where neither time nor funds are available for the radiation-hardening of devices, some hardening of a piece of equipment may be achieved by a rigorous selection of commercially available components. Once a specific device type has been chosen, there is still the serious question of assuring that all units in the batches used perform acceptably under radiation. This field, called 'hardness assurance', has received much attention in military and space work and is a mixture of well-established product assurance techniques and special radiation assessment (Wolicki *et al.* 1985; Messenger and Ash 1986; Adams *et al.* 1989).

13.6.2 Product assurance techniques and special radiation assessment

Institutes in the USA and Europe have developed national and international standards for the assessment of devices under radiation. In Europe, for example, the ESA Space Components Coordination Group (SCCG) has issued a space testing specification. In the UK, a method for qualifying a series of 'radiation-assessed devices' has been circulated by the British Standards Institution (BSI). This was issued by the Ministry of Defence as a draft for BSI's BS 9000 series of electronics assessment methods and is on the table in Europe. In the USA, a similar scheme exists under the ASTM and MIL specification systems. Some references to documents are given in the following sections. As yet, the above schemes have been applied only to the requirements of military and space environments. Although no scheme has been devised for the testing of electronics for use in the nuclear power and commercial industries, the BS 9000 radiation specification is designed for extensions which make it suitable for these industries. An IEEE standard specifies radiation tolerance levels for a class of instruments (1E) used in the nuclear industry see Section 15.3.5.

The body for 'harmonizing' national standards in Europe is CEN (Comité Européen de Normalisation); the section dealing with radiation effects in devices is CECC (CENELEC Electronic Components Committee). Existing standards, after harmonalization, are issued with an added CECC designation (e.g. in Britain, a standard such as BS9000:1992 might be cited as BS CECC 90000:1992). On a higher level still are standards approved by the International Electrotechnical Commission of the International Standards Organization. These may be cited in Britain under titles like BS ISO/IEC 00000;1992. Other national standards such as the German DIN system (Deutsche Institut für Normung) have specifications dealing with radiation safety but not radiation effects on engineering materials. Japanese and Indian space projects generally follow US MIL standards.

13.6.3 ESA–SCC specification (Europe)

The ESA Space Components Coordination Group has developed a specification of radiation test procedures (ESA/SCC Basic Specification No. 22900, proposed Issue 3, dated 1992). The purpose of this specification is to define the testing of semiconductor devices for the effects of total-dose ionization and displacement relevant to the space environment. Cobalt-60 gamma-rays or electron accelerators may be used. This specification also leaves the door open for other sources, provided that correlation with cobalt-60 can be demonstrated. Sample sizes greater than 10 are recommended (see Section 15.6.6).

13.6.4 BS 9000 specification and CECC (Europe)

A draft specification entitled 'Specification of basic requirements for the assessment of semiconductor components for tolerance to high-energy radiation' has been submitted for a place in the BS 9000 series. This is a comprehensive specification of radiation test and device assessment procedures, marking methods for BS 9000 device packages, amendments to BS 9000 data sheets, and manufacturers' quality assessment. The effects considered include those connected with pulsed radiation, total dose, and neutrons. The devices produced to this specification are entitled to the designation 'radiation-assessed devices'. Note that this does not imply 'radiation-hardened'.

13.6.5 MIL specifications (USA)

Radiation standards adopted by the US Department of Defence include the following parts of the MIL-STD system:

Method 1015 (MIL-STD-75): Steady state primary photocurrent irradiation procedure (electron beam)

Method 1017 (MIL-STD-883): Neutron irradiation

Method 1019 (MIL-STD-883): Steady state total dose irradiation procedure (now in fourth revision)

Method 1020 (MIL-STD-883): Radiation-induced latch-up test procedure

Method 1021 (MIL-STD-883): Dose rate threshold for upset of digital microcircuits

Method 1022 (MIL-STD-883): MOSFET threshold voltage

Method 1023 (MIL-STD-883): Dose rate response of linear microcircuits

MIL STD 1547 (USAF): Total dose hardness assurance guideline for CMOS microcircuits in space environments.

13.6.6 ASTM specifications (USA)

Radiation standards adopted by the American Society for Testing and Materials (ASTM), a civil body, include the following:

13.7 Device parameter measurements

E 693 Standard procedure for characterizing neutron exposure in ferritic steels in terms of displacements per atom (DPA).

E 763-85 Method for calculation of absorbed dose from neutron radiation application of threshold-foil measurement data

E668-78: Standard practice for the application of thermoluminescence-dosimetry (TLD) systems for determining absorbed dose in radiation hardness testing of electronic devices

E720-86: Selection of a set of neutron-activation foils for determining neutron spectra used in radiation hardness testing of electronics

E721-85: Standard method for determining neutron energy spectra with neutron-activation foils for radiation-hardness testing of electronics

E722: Characterizing energy fluence spectra in terms of an equivalent monoenergetic fluence for radiation hardness testing of electronics

E820-81: Standard practice for determining absolute absorbed dose rates for electron beams

E1249-88: Practice for minimizing dosimetry errors in radiation hardness testing of silicon electronic devices using Co-60 sources

E1250-88: Methods for application of ionization chambers to assess the low energy gamma component of cobalt-60 irradiators used in radiation-hardness testing of silicon electronic devices

F448-80: Method for measuring steady-state primary photocurrent

F526-81: Method of dose measurement for use in linear accelerator pulsed radiation effects tests

F1192-90: Standard guide for the measurement of single event phenomena (SEP) induced by heavy ion irradiation of semiconductors.

13.6.7 Comparison of standards

Figs 13.7a and b compare flow charts given in two widely-used methods of testing for total-dose effects, namely Method 1019. 4 of MIL-STD-883D and ESA-SCC Method 22900. These standards, developed separately, have been harmonized as far as possible but nevertheless reflect some differences in methodology in the military and space fields.

13.7 Device parameter measurements

This section describes briefly the parameters which are often degraded in widely used silicon devices and also notes some features of their behaviour under irradiation which may require special procedures during testing. Three different desirable sorts of testing are defined: DC and AC parametric testing, and functional testing. The first group discussed here, the DC tests, may form a large proportion of those performed for radiation

(a)

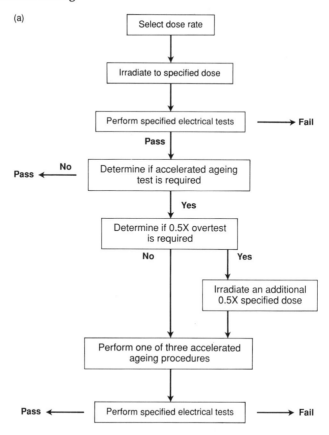

FIG. 13.7a Flow chart(s) for total-dose radiation testing: chart(s) of the steps required in two standards (a) Methods 1019.4 of MIL-STD-883D.

effects Schwank *et al.* (1989) have described criteria for lot acceptance of CMOS devices.

13.7.1 MOS threshold voltage

The MOS threshold voltage (V_T) has been defined in Chapter 4. Briefly, it is that voltage at which a certain, practically measurable, channel current (commonly 10 μA) flows after turn-on by inversion. The inversion point in an n-channel device is about 3 V more positive than the flatband voltage. Some special problems with V_T ('hysteresis', distortion, annealing) are covered by the discussion of C-V plots below. V_T may be measured during irradiation, but it must be noted that:

(1) this disturbs the desired stable oxide field condition;
(2) even at a dose rate of 1 rad s^{-1} (10^{-2} Gy sec^{-1}), photovoltaic effects may interfere.

13.7 Device parameter measurements

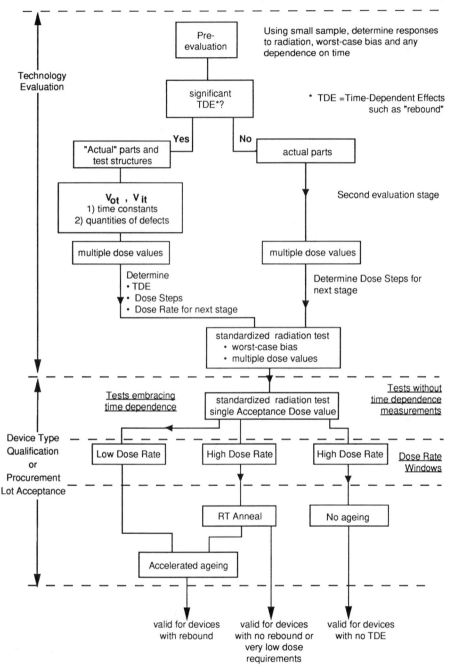

FIG. 13.7b Method 22900 (proposed Issue 3) of the European Space Agency and the Space Components Co-ordination Group. Both include the assessment of time-dependent effects (annealing).

13.7.2 MOS flatband voltage (V_{FB}) and C-V plot

In an MOS capacitor, threshold channel currents do not exist. However, the whole capacitance–voltage curve contains even more information on the state of the semiconductor and interface. The flatband condition (at which surface potential is zero, and hence, no bending of the silicon conduction band) lies on the C-V curve at a point where C/C_0 has a value commonly about 0.8, on the depletion side of the minimum of the plot.

Unfortunately, the capacitance of an LSI transistor gate is usually so small that most C-V sensing circuits cannot measure it. Special fieldplates or specially large transistors have to be fabricated.

Irradiation-induced interface states produce distortions in the C-V curve from which quantitative information can be obtained. One special experimental difficulty caused by irradiation is the production of 'hysteresis' in the C-V plot owing to the generation of 'slow' states.

13.7.3 Quiescent current (I_{ss}) in CMOS logic

The VTNZ effect in CMOS logic (see Chapter 4) leads to a drastic increase in current in the power supply circuit. This is usually measured in the V_{SS} circuit or earth leg of the devices concerned, using a nanoammeter. In LSI circuits, where V_T cannot be measured, the measurement of I_{SS} may be the only method available for detecting the onset of the VTNZ effect.

13.7.4 Leakage currents

The reverse leakage of p–n junctions is usually greatly increased by irradiation, especially when the junction is under bias; the field encourages charge build-up on the surface. Current may rise from picoamperes to milliamperes, thus upsetting the impedance-matching of the test circuit. These currents will be temperature-sensitive, so that standard temperatures are important. Instruments with a wide dynamic range should be used. Dose rates down to 1 rad s^{-1} can give detectable photocurrents; measurements during irradiation therefore require care.

13.7.5 Current gain

The conventional instrument used for measurement of the current gain in bipolar transistors is the oscilloscope curve tracer or parameter analyser. This displays base current and collector current at several different values, and the gain (i.e. the ratio) can be calculated. However, current steps usually cover less than a decade, whereas, as indicated earlier, it is vital to measure radiation-induced degradation of gain over about four decades of collector current.

This illustrates two special features of radiation testing:

1. Measurements of parameters at values far outside those to be used in

application may provide important diagnostic information on which the expert can base a more confident prediction of degradation in use.

2. The routine test instruments may not provide the best form of display or readings for our purpose. For example, the usual set of parameter plots is a wholly inappropriate form; much more suitable would be a computer plot of the change in base current (I_B) versus dose, given for a range of values of I_C or I_E, spread over four orders of magnitude, followed by measurements gained by periodic tracking of the same parameters over several days after irradiation (Brown and Horne 1967). As explained elsewhere, the surface effect on gain is similar to MOS effects, e.g. it can anneal slowly at room temperature and can often be reduced by heating.

13.7.6 Input offset in analog ICs

In integrated analog circuits, only the input and output points are accessible for measurement. The input offset voltage and current are two important parameters in operational amplifiers which degrade very seriously under ionizing radiation.

13.7.7 Noise-immunity and DC switching of logic gates

The switching characteristics of a bipolar or MOS logic gate can be plotted by stepping the input voltage and plotting the output currents or voltages. It is then simple to determine from these curves the loss in noise-immunity.

13.7.8 AC and functional testing

With large-scale integrated circuits, such as memories and microprocessors, it is essential to test the circuits at the required switching rates and loading conditions, and over the full voltage and temperature ranges, because, when a number of integrated circuits operate together, the first functional degradation may be the inability of some section to transmit signals rapidly enough. The same argument, of course, applies to high-frequency communications circuits, where devices may be working near to their frequency cutoff or where the tuning of circuits may drift.

13.7.9 Single-event upset testing

The single-event upset test is essentially a form of functional test which seeks for a change in state of bistable elements. In the case of memories a certain pattern is written into the memory array and the array monitored for any change of content during irradiation. For processors and similarly complex devices, special-purpose software must be written to give access to registers and latches. It is important that different 'patterns' be used, as the bistable elements may have a different sensitivity for changing from one to zero and viceversa. A typical test program would employ 'checkerboard' and 'inverse checkerboard', as well as 'all ones' or 'all zeros'. The techniques used for monitoring of the device are extremely important and should be fully reported, using software flow-charts as appropriate. In the testing of complex devices a number of software and hardware precautions need to be taken, including latch-up detection and protection as well as watchdog timers. (Harboe-Sorensen *et al.* 1988).

13.7.10 Measurement of transient photocurrent

The origin and nature of transient photocurrents has been explained in Sections 3.5, 4.6, and 5.6. Measurement requires fast circuits having controlled RC properties (e.g. coaxial cable). Figure 13.8 shows the simplest possible circuit for the measurement of the transient response of a single diode. A reverse bias, V_R, is applied to the junction via a limiting resistor, R_1 establishing a baseline voltage value on the oscilloscope. The transient photocurrent is conducted by the terminating resistor R_T, and the corresponding IR drop is recorded at node N as a voltage waveform. If the response is off-scale, then a shunt resistor R_2 can be added. The response of a calibrated p-i-n diode, in the same beam, is normally recorded simultaneously on the second channel of the oscilloscope.

13.8 Engineering materials

The procedure for testing and qualifying materials other than semiconductors will, in general, be simpler. The demands made on engineering materials are usually less severe, except in the case of materials for use in a reactor environment. For space use, thermal control and optical materials require careful testing. The general approach for materials testing is to define the critical mechanical, electrical or optical properties to be monitored and to test to a higher radiation level than the specification limit in order to obtain a true 'worst case' result.

In cases where radiation induces chemical reactions, such as in organic materials (Chapter 10) and labile inorganic materials (especially water and salts), then one should expect that dose rate and ambient gases will have an effect on the result and that gases may be evolved. National standards organizations have developed test methods for nuclear engineering materials (see Section 13.6).

13.9 Time-dependent effects and post-irradiation effects

For many modern semiconductor technologies, the problem the problem of time-dependent effects (TDE), sometimes called post-irradiation effects (PIE), is becoming increasingly important. As noted in Chapter 4, TDE/PIE may be beneficial or deleterious (the effects may constitute recovery or 'reverse recovery'). These effects are a particular problem for space environment testing and derive from the highly accelerated dose rate used for testing compared with the actual environment. At the conclusion of a radiation test,

13.9 Time-dependent effects and post-irradiation effects

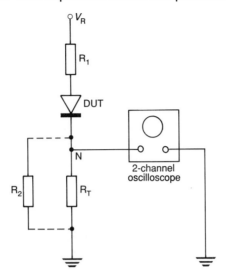

FIG. 13.8 The measurement of transient effects in a diode.

devices may recover or degrade over extremely long periods of time, even under storage conditions. If TDE/PIE are not adequately taken into account, then erroneous conclusions may be drawn regarding the acceptability of devices for a given environment.

As Fig. 13.6 suggests, most high-activity isotope sources have a limited range of dose rates at which they can be used for radiation test. In Europe, these are centred around a rate of 10 rad/sec (0.6 krad/min, 36 krad/hr; 864 000 rad/day). The exposure is about 100 000 times faster than the space dose rate stated above. Serious error could be expected if such exposure rates were not followed by measures to assess time dependence of the radiation effect.

High-activity sources are an essential part of the test facilities needed for space part testing; however, it is best to use them at the lower end of their dose-rate capability. A dose rate of 100 krad/day (1 kGy/day), (i.e. slightly above 1 rad/sec or 10^{-2} Gy/sec) seems a practical and desirable rate to reduce the worst effects of time-dependence of damage states. To deal with the TDE testing problem, an accelerated aging test is coming into use (Fleetwood *et al.* 1989). Two specifications (MIL-883 Method 1019.4 and SCC22900) address the problem by calling for a 168 hour 100°C bake after radiation exposure to 1.5 times the specified dose. In the authors' opinion, this is a good technique for assessing TDE/PIE. However, there should be limitations as to the technologies to which it is applied. Softer MOS technologies may not be so severely affected by rebound effects. In this case, $Q(ot)$ may completely dominate $Q(it)$, so that heat treatment may only relax Q(ot) and not greatly enhance Q(it). In this case, while the aging is indeed accelerated, the result may no longer be a worst case test. The physical factors are explained in Section 4.6; similar considerations appear to apply to bipolar integrated circuit technology, as discussed in Chapter 5.

13.10 Conclusions

Radiation testing is not a simple art and should be approached by the use of exploratory experiments which will be useful in uncovering any pitfalls. Test design is best assigned to an expert, since test facilities are expensive, dosimetry is complex, unusual parasitics are often produced by radiation, and statistics may be difficult. Extracting the correct prediction from a small sample of commercial devices requires considerable skill. This chapter has not attempted to describe a perfect set of facilities or a standard form of test. The problems mentioned prevent rigid standardization at the moment. The descriptions given are intended to record the state of the art, with some recommendations and warnings.

It is unfortunate that details of many of the radiation-effects tests performed never reach publication and are lost to general use. Various authorities including JPL and ESA are attempting to rectify this by placing test reports in a data bank, using a standard format for reporting (Johnson 1990). IEEE is encouraging 'data workshops' at its conferences. A list of radiation-effects data banks in Europe and the USA is given at the end of this chapter.

References

Ackerman, M. R., Browning, J. S., Hughlock, B. W., Lum, G. K., Tsacoyeanes, W. C. and Weeks, M. D. (1990). Advanced development of the Spectrum Sciences model 5005-TF single event test fixture, Sandia report SAND90-2007, UC140. Sandia National Laboratories, Albuquerque, NM.

Adams, L. and Thomson, I. (1980). The use of an industrial X-ray source for electronic components radiation effects work. *IEEE Transactions on Components, Hybrids and Manufacturing Technology*, (3), 144–9.

Adams, L., Harboe-Sorensen, R., Daly, E. J. and Holmes-Siedle, A. G. (1989). Evaluation, simulation and hardening for space radiation effects. In RADECS '89, Université de Montpellier, France.

Attix, F. H. (1986). *Introduction to radiological physics and radiation dosimetry*. Wiley, New York.

Baur, J. M. *et al.* (1981). Radiation hardening of diagnostics for fusion reactors. *General Atomics Report No. GA-A 16614*. GA Technologies, San Diego, CA.

Barnes, C. E., Shaw. D. E., Rax, B. and Agarwal, S. (1991). Post irradiation effects (PIE) in integrated circuits. *Proc. RADECS '91 Conference*, Montpellier, September 9–12 1991, *IEEE Conference Record*. IEEE New York.

Benedetto, J. M., Boesch, H. E. Jr. and McLean, F. B. (1988). Dose and energy

dependence of interface trap formation in cobalt-60 and X-ray environments. *IEEE Transactions on Nuclear Science*, NS-35, 1260-4.
Brown, D. B., Jenkins, W. C. and Johnston, A. H. (1989). Application of a model for treatment of time dependent effects on irradiation of microelectronic devices. *IEEE Transactions on Nuclear Science*, **NS36**, 1954-62.
Brown, R. R. and Horne, W. E. (1967). Space radiation equivalence for effects on transistors, NASA CR-814. US Dept of Commerce, Washington, DC.
Chapuis, T. (1990). Mise en place d'une ligne d'irradiation aux ions lourds a l'Institut de Physique Nucléaire d'Orsay: presentation de l'installation, Report RA/DP/QA/CE/90-002. CNES, Toulouse.
Dozier, C. M. and Brown, D. B. (1983). The use of low energy X-Rays for device testing: a comparison with Co-60 radiation testing. *IEEE Transactions on Nuclear Science*, **NS30**, 4382-7.
Dozier, C. M. *et al.* (1985). Defect production in S_{1O_2} by X-ray and Co-60 radiations. *IEEE Transactions on Nuclear Science*, NS-32, 4363-8.
Fabry, A. and Lowe, A. Jr. (1985). Workshop: LWR-PV physics, dosimetry, damage correlation and materials problems, in Genthon, J. P. and Röttger, H. (eds) Reactor Dosimetry, *Proceedings of 5th ASTM-Euratom Symposium on Reactor Dosimetry*, Geesthacht September 24-18 1984, 999-1000.
Fleetwood, D. M. *et al.* (1988a). Comparison of enhanced device response and predicted X-ray dose enhancement effects on MOS oxide. *IEEE Transactions on Nuclear Science*, **NS-35**, 1265-71.
Fleetwood, D. M., Winokur, P. S. and Schwank, J. R. (1988b). Using laboratory X-ray and cobalt-60 irradiations to predict CMOS device response in strategic and space environments. *IEEE Transactions on Nuclear Science*, NS-35, 1497-505.
Fleetwood, D. M., Winokur, P. S., Riewe, L. C. and Pease, R. L. (1989). An improved standard total dose test for CMOS space electronics. *IEEE Transactions on Nuclear Science*, **NS36**, 1963-70.
Galloway, K. (1977). Important considerations for SEM total dose testing. *IEEE Transactions on Nuclear Science*, **NS24**, 2066-70.
Harboe-Sorensen, R., Adams, L. and Sanderson, T. K. (1988). A summary of SEU test results using californium-252. *IEEE Transactions on Nuclear Science*, **NS35**, 1622-8.
Hardman, M., Farren, J., Mapper, D. and Stephen, J. H. (1985). Low level radiation testing of microelectronic components, ESA Contract Report 5313/83. AERE, Harwell.
Heikkinen, D. W. and Logan, C. M. (1981) RTNS-II [Rotating-target neutron source]: present status. *IEEE Transactions on Nuclear Science*, **NS28**, 1490-3.
Johns, H. E. and Cunningham, J. R. (1974). *The physics of radiology*. Charles C. Thomas, Springfield, IL.
Johnson, S. (1990). The ESA radiation effects database. In Proceedings of ESA Electronic Components Conference. ESA SP-313. ESA, Noordwijk, Netherlands.
Kelliher, K. (1985). Calibration of an on-site X-ray equipment as a total dose simulator, Internal report. British Aerospace, Bristol.
Kelly, J. G. *et al.* (1983). Dose enhancement effects in MOS ICs exposed in typical Co-60 faclities. *IEEE Transactions on Nuclear Science*, **NS30**, 4388-93.

Kerris, K. G. (1989). Source considerations and testing techniques in Ma, T. P. and Dressenderfer, P. V. (eds.) *Ionizing radiation effects in MOS devices and circuits*. Wiley, New York.

Lamarsh, J. R., (1983). *Introduction to nuclear engineering*. Addison Wesley, Reading, MA.

Larin, F. (1968). *Radiation effects in semiconductor devices*. Wiley, New York.

Liska, D. and Machalek, M. D. (1981). FMIT – the fusion materials irradiation facility. *IEEE Transactions on Nuclear Science*, **NS28**, 1304–7.

Messenger, G. C. and Ash, M. S. (1986). *The effects of radiation on electronic systems*. Van Nostrand Reinhold, New York.

Palkuti, L. J. and LePage, J. J. (1982). X-ray wafer probe for total dose testing. *IEEE Transactions on Nuclear Science*, **NS29**, 1832–7.

Poch W. J. and Holmes-Siedle, A. G. (1969). *A prediction and selection system for radiation effects on bipolar transistors. IEEE Transactions on Nuclear Science*, **NS15**, 213–9.

Profio, A. E. (1970). *Experimental reactor physics*. Wiley, New York.

Profio, A. E. (1979). *Radiation shielding and dosimetry*. Wiley, New York.

Schönbacher, H. and Tavlet, M. (1990). Radiation sources for material testing in Europe. *Report No. TIS-CFM/IR/90-02*. CERN, Geneva.

Schwank, J. R., Sexton, F. W., Fleetwood, D. M., Shaneyfelt, M. R., Hughes, K. L. and Rodgers, M. S. (1989). Strategies for lot acceptance testing using CMOS transistors and ICs. *IEEE Transactions on Nuclear Science*, **NS36**, 1971–80.

Sanderson, T. K., Mapper, D. and Stephen, J. H. (1982). Effects of space radiation on advanced semiconductor devices, ESA Contract Report CR(P)1801 (AEA Harwell). ESA, Noordwijk, Netherlands.

Stapor, W. J. *et al.* (1988). Charge collection in silicon for ions of different energy but same linear energy transfer (LET). *IEEE Transactions on Nuclear Science*, **NS35**, 1585–9.

Stephen, J. H., Sanderson, T. K., Mapper, D., Farren, J., Harboe-Sorensen, R. and Adams, L. (1983). Cosmic ray simulation experiments for the study of single event upsets and latch-up in CMOS memories. *IEEE Transactions on Nuclear Science*, **NS30**, 4464–9.

Tavlet, M. and Florian, M. E. L. (1991). PSAIF – Facilité d'irradiation an PS-ACOL du CERN. *RADECS' 91*, Montpellier, September 9–12 1991.

Thomson, I., Mackay, G. and Haythornthwaite, R. (1990). Advances in SEM beam microirradiation. In Proceedings of ESA Electronic Components Conference, ESA SP-313, pp. 447–50. ESA, Noordwijk, Netherlands.

Winokur, P. S. Fleetwood, D. M. *et al.* (1990). Implementing QML for radiation hardness assurance. *IEEE Transactions on Nuclear Science*, **NS37**, 1795–1805.

Wolicki, E. A., Arimura, I., Carlan, A. J., Eisen, H. A. and Halpin, J. J. (1985). Radiation hardness assurance for electronic parts: accomplishments and plans. *IEEE Transactions on Nuclear Science*, **NS32**, 4230–6.

Wysocki, J. (1963). Radiation studies on GaAs and Si devices. *IEEE Transactions on Nuclear Science*, **NS10**, 60–5.

Data compilations

Among the data banks established for the collection and dissemination of radiation test data of electronic components are the following:

Hahn–Meitner Institut, Berlin. *Data compilation of irradiation-tested electronic components*, HMI-B353 (loose-leaf computer-printed data sheets). Total dose.

KAMAN-TEMPO, Santa Barbara (USA). Incorporates the data bank initiated by US Army (HDL), Adelphi, MD.

NASA (USA). Compilation (space application). Available through the Product Assurance Group, Goddard Space Flight Center, Greenbelt, MD.

RADATA (USA). Compilation (space application). Available through Jet Propulsion Laboratory, Pasadena, CA, and accessible via DARPANET or SPAN electronic mail system. Total dose and SEU.

RADFX, ESA-ESTEC, Noordwijk. *Electronic components radiation effects data base*. Compilation (space application). Includes tests performed by a number of European and US authorities. Total dose, SEU, latch-up, and proton upsets. Software developed by Spur Electron Ltd., Havant, England.

SIRE (UK). Compilation on EPIC computer database. Available to military contractors only through UK Ministry of Defence (AWD), London.

14
Radiation-hardening of semiconductor parts

14.1 General

'Radiation-hardening' derives from military terminology. The term has acquired a wide variety of meanings, depending on the user group and radiation environment concerned. That environment may present a variety of levels of total ionizing dose, transient dose rate, neutrons, or single-event upsets. Each combination may require a different approach to hardening. A scientific-satellite designer may consider a level of 100 krad (1 k Gy) tolerance; to be 'hard', a military designer may apply the term 'hard' only to megarad tolerance if it also includes tolerance of transient effects. We must be careful, therefore, to qualify the term 'hardening' or 'hardened' with a definition of the environment involved.

A component common to most radiation environments is ionizing dose. Much of the following discussion will be concerned with hardening against so-called 'total dose', a problem which is important for the complementary metal-oxide-semiconductor technology discussed in Chapter 4 especially in the space environment (Adams *et al.* 1992). As a rule, when referring to a 'process' or 'baseline process', we imply the fabrication processes used on the production line by manufacturers of commercial, large-scale CMOS silicon integrated circuits such as microprocessors, memories, logic, and analogue functions. The unit product, an encapsulated silicon chip, often goes by the prosaic name of a 'part', a term in reliability engineering. The present chapter should be read with reference to some of the sections in of Chapter 4, and the literature reviewed therein. Other aspects of hardening will be briefly discussed.

14.2 Methodology of total-dose hardening

A certain methodology is required for successful radiation-hardening. By successful we mean not only a technology which behaves acceptably under test and operation, but also one for which there is a good understanding of the reasons for the response of the unhardened forms, as well as a good understanding of the various modifications which may lead to the achievement of hardness. There should be confidence that the hardened process

14.2 Methodology of total-dose hardening

is stable and repeatable, and that the hardness level achieved is relatively insensitive to normal processing variations. Radiation-hardening must be fully supported by mathematical simulation and physical analysis and each proposed change needs to be evaluated for its effect on overall performance and impact on other areas of process and design. In semiconductor devices, there are few, if any, changes made for radiation-hardness which do not have an impact in other areas of performance.

Ideally one would start the design of a radiation-hard silicon device with a 'clean sheet of paper', that is a free choice of technologies. More often, we are faced with an existing technology and design which satisfies the application requirements, but falls short of the radiation requirements. The funding for a hardening programme may be limited and the manufacturer, while interested in a hardening programme, generally requires that such a programme should cause minimal disruption to his existing production. The market for 'rad-hard' products is quite limited and there is little financial incentive to set up a 'rad-hard' production line for a given product, in parallel with normal commercial production. There is currently a natural wish to compromise i.e. to develop processes and designs which combine tolerance to radiation *and* suitability as commercial products.

The key to the successful hardening of existing semiconductor processes and products lies in the characterization of the response of the baseline process. This is not easily achieved using functional IC designs (e.g. gates, amplifiers, memories), even using unconventional measurement techniques with which one might be able to extract some parametric information, such as threshold voltage or substrate photocurrents. Quantitative assessment of oxides and interfaces, which is essential for proper characterization, requires the use of special test structures and measurement techniques (Johlander 1990).

Typical test structures include various types of MOS transistor and capacitor using both gate and field oxides, and gated diodes. Test structures may also include functional blocks, such as sense amplifiers and latches, which are essential elements in a number of designs. The process, design rules, and layout of typical designs must be studied for the existence of parasitic elements, particularly involving field oxide and polysilicon or metal conductor patterns. These parasitic elements may require the fabrication of representative field oxide FETS. This is particularly important for modern technologies which use thin, and consequently tolerant, gate oxides, but frequently suffer leakage problems in field oxide isolation areas.

Typical measurement techniques include high-frequency and quasi-static capacitance–voltage (C–V) characteristic plotting, sub-threshold characterization, deep-level transient spectroscopy (DLTS), and $1/f$ noise, as well as standard parametric measurements such as threshold voltage, mutual conductance, and leakage current.

Radiation assessment should be performed using a range of dose rates

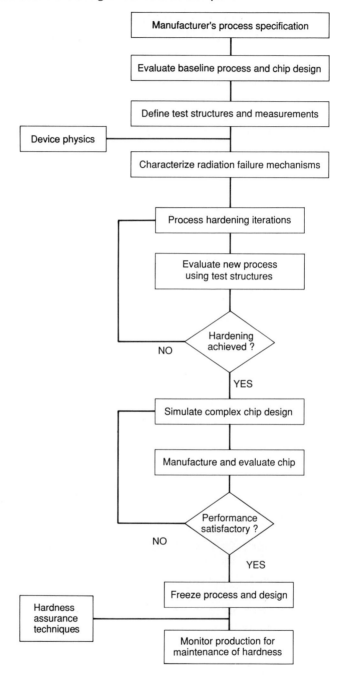

FIG. 14.1 Methodology for the radiation hardening of a semiconductor device manufacturing process.

and a number of different bias conditions (including the passive or unpowered condition). Post-irradiation effects (PIE) should be evaluated using high-temperature (100°C) bakes with interim measurements.

Once a radiation-hard process has been developed, then this needs to be tested for process variability: from lot to lot, from wafer to wafer, and across a wafer. These statistics are important for the setting of lot-acceptance test levels and prediction of yield for a given hardness level. Once in production, a 'rad-hard' process requires continuous monitoring, in order to verify the stability of the process and insensitivity to process variations. Figure 14.1 summarizes a methodology of radiation hardening.

14.3 Hardening of a process

14.3.1 Introduction

The hardening against total ionizing dose from the process point of view is primarily a question of understanding, modifying, and optimizing oxides and interfaces. Chapter 4 emphasized the roles of oxide charge Q_{ot} and interface charge Q_{it} in controlling the electrical operation of a device. These are directly related to the number charge of traps N_{ot} and N_{it}. Device response is governed by the combined effect of Q_{ot} and Q_{it} which are in competition with each other in n-channel devices and support each other in p-channel devices. The objective of radiation hardening is to minimize both Q_{ot} and Q_{it} as well as to control the manner in which each component grows or relaxes during and after radiation exposure.

The threshold voltage shift of an n-channel transistor may be quite small under radiation, but the absolute values of Q_{ot} and Q_{it} may be quite high. This gives apparent hardness which may not be maintained in a different dose rate environment, and may give serious post-irradiation effects (see Sections 4.6 and 13.9).

A large number of processing steps (several hundred) are involved in fabrication a typical MOS integrated circuit and many of these steps can influence the radiation hardness of a device. The most important factors are those which affect the charge trapping characteristics of the oxide or interface. These factors may be summarized as:

(1) material preparation and cleaning
(2) gate oxide growth:
 (a) thickness
 (b) temperature
 (c) ambient
(3) gate oxide anneal:
 (a) temperature
 (b) ambient

(4) gate electrode:
   (a) material
   (b) deposition conditions
(5) subsequent high-temperature processing:
   (a) interlayer dielectrics
   (b) passivation
   (c) packaging.

14.3.2 Material preparation and cleaning

Material preparation includes the perfection and purity of starting material. Silicon defects such as dislocations and stacking faults, as well as contamination and impurities, can all have a deleterious effect on radiation hardness. The wafer surfaces should be oriented in the (100) plane, and surface damage should be minimized by care in polishing. The addition of fluorine prior to gate oxidation by the use of dilute HF in the final cleaning step has been found to improve radiation-hardness.

14.3.3 Oxide growth

As will be appreciated from sections 4.4–4.5, the thickness of the gate oxide has a multiple impact on total-dose radiation hardness. The greater the volume of oxide, the larger the number of holes that can be trapped. Furthermore, the oxide thickness directly controls the shift for a given radiation-induced charge. The amount of trapped charge also depend on the trapping efficiency of the oxide (A-value). The relation between threshold voltage V_t shift and oxide thickness is a power law, t_{ox}^n, where n has been found to lie between 1 and 3. The value of n varies with growth ambient and the electric field applied. 'Wet' oxides grown in pyrogenic steam (made by combustion of hydrogen gas) are found to have a higher value of n than 'dry' oxides grown in dilute dry oxygen.

The use of a thinner gate oxide gives a vital means of improving radiation-hardness. The improvement can be by a factor of between two and ten, and is frequently the first step in a radiation-hardening programme. The reduction of oxide thickness must be accompanied by due consideration of yield and reliability aspects, and the increase of transconductance conferred. 'Ultra-thin' gate oxides (15–30 nm) used in some modern VLSI technologies have been found to be considerably harder than would be predicted by the considerations given earlier. In fact, such oxides have been found to be 'megarad-hard'. It is thought that this extreme hardness is the result of trapped-hole annhilation by electron tunnelling from the oxide–silicon interface and the polysilicon gate. The temperature used for oxide growth has a strong influence on radiation-hardness. The optimum temperature depends on the growth ambient and generally lies in the region of 850°C for wet oxides and 1000°C for dry oxides. Additive and diluting gases (chlorine, nitrogen, and argon), added to the oxygen gas flow, may also affect radiation-hardness. Argon is generally believed to be beneficial, nitrogen to be detrimental (Derbenwick and Gregory 1975).

14.3.4 Oxide anneal

After gate oxide growth, a high temperature anneal in an inert gas or vacuum is performed in order to relieve stress in the oxide/silicon interface and reduce the defect density in the oxide. For commercial processing the anneal would commonly be performed at a temperature around 1100°C in nitrogen. For radiation-hard processing, high-temperature processes subsequent to gate oxidation must be minimized; temperatures and times should be kept as low as practicable. The optimum annealing temperature is in the range 850–950°C for a duration of ten to several tens of minutes. The use of a forming gas (10 per cent hydrogen, 90 per cent nitrogen) as an annealing ambient is found to be beneficial compared with the normal nitrogen alone.

14.3.5 Gate electrode

Common gate electrodes are films of aluminium or polycrystalline silicon (polysilicon); more recently, refractory materials including metals and silicides have been evaluated as alternative gate electrodes. Polysilicon is now the common gate electrode technology, as it allows the implementation of a 'self-aligned' technology. However, this requires that the gate insulator is grown early in the process, which makes it vulnerable to degradation of hardness by subsequent high-temperature processing steps. In fact, it is processing steps rather than material characteristics which are responsible for differences in radiation-hardness, with metal gate being better than polysilicon gate technologies. There are indications that polysilicon gate technologies are beginning to show similar hardness levels to metal gate; this may be due to the advent of lower processing temperatures in general. Refractory metals and metal silicides, such as tungsten and tantalum, offer lower resistivity than doped polysilicon and are of interest for modern VLSI technology (see also chapter 4). These high-atomic-number materials may lead to dose-enhancement effects in certain radiation environments where low-energy components dominate. This is due to the generation of Compton electrons in the high-atomic-number material, which then penetrate into the low-atomic-number gate insulator. In general, refractory materials seem to improve radiation-hardness: interface states are reduced as well as oxide trapped charge.

14.3.6 Modified gate insulators

A number of hardening investigations have been performed using alternative gate insulators which have been modified either by the doping of the oxide or the use of 'sandwich' structures. A wide range of implanted dopants including aluminium, chromium, and nitrogen have been used with varying degrees of success, improved hardness under certain conditions often being accompanied by degraded hardness under other conditions. Dual-layer insulators with phosphosilicate glass or nitride layers on top of the oxide have been evaluated, again with varying degrees of success. Dunn and Wyatt (1989) reported on a megarad-hard gate oxide using reoxidized nitrided oxide. Aluminium oxide was evaluated intensively as a gate insulator and showed some promise (Dennehy et al. 1969), but has not been developed except for special-purpose, mainly low-temperature, applications (see Hughes in Ma 1989). Despite all of these efforts, it would appear that the best gate insulator is a defect-free, clean, low-temperature-grown silicon dioxide (see, for example Borkan's review (1977)).

14.3.7 Field oxide hardening

Field oxides have been mentioned earlier as being an important factor in radiation hardness. In fact, field-oxide-induced leakage tends to be the dominant radiation response in modern MOS technologies (see Chapter 4). This is due to the existence of parasitic field oxide transistors (FOXFETS), which may have a high threshold voltage (about 20 V) but are very sensitive to charge build-up, owing to the thick (several hundred nanometres) unhardened oxide. Similar effects may be found in bipolar devices using recessed field oxide isolation. The number of potentially troublesome parasitic structures can be minimized by careful layout, and the remaining structures can be hardened by doping of the underlying silicon by implantation in order to increase the threshold voltage. A hardened field oxide was reported by Adams et al. (1977). Few details of the processes involved are given, owing to the high level of security classification placed on this topic. Nishioka et al. (1990) hardened a field oxide by means of a fluorine implant, and Watanabe (1985) used a double layer of thick CVD SiO_2 on top of a thin thermal oxide with heavy doping of the silicon to prevent inversion.

14.3.8 Other processing steps

A number of processing steps, particularly those involving high temperatures, may affect radiation-hardness. 'Self-aligned' structures involve ion implantation through an oxide mask. The damage induced in the oxide by this process increases the number of trapping centres. A radiation-hard process requires that damaged oxide be stripped off and replaced by regrowing a pure oxide. Most technologies employ intermediate insulating dielectrics between layers. The temperature and ambient used for reflow and densification of intermediate dielectrics has an effect on radiation-hardness. The use of glasses melting at less than 900°C and a nitrogen or forming-gas ambient is recommended (Winokur et al. 1985). Polyimide is becoming widely used as a planarizing thin-film dielectric; there is currently no evidence that the use of this material affects radiation-hardness in any way (see Chapter 10).

In-process radiation damage may result from plasma etching steps, X-ray lithography, X-rays generated during thermal evaporation of metals, and UV generated during sputtering. Passivation layers, which are the final processing step, can affect radiation-hardness. Plasma-enhanced chemical-vapour-deposited silicon nitride has been found to be worse than the common phosphorus-doped glass (Anderson 1977). The final manufacturing step, packaging, can also affect radiation-hardness. High-temperature die bonding and package sealing may degrade hardness. The ambient gas used during sealing has been shown to affect hardness. In one case, the use of a forming-gas ambient gave a fourfold improvement compared with nitrogen alone.

14.4 Hardening for total dose by 'layout'

The influence of previously discussed parasitic FOXFETs requires careful design and layout in order to minimize the number of FOXFETS. Guard bands may be required in certain areas, and substrate bias may also be employed to bias 'OFF' any parasitic FOXFETs (Rockett 1988). Threshold voltages may be modified in order to accommodate radiation-induced shifts better. n-channel thresholds can be increased from a typical 0.7 V to about 1.0 V and p-channel thresholds can be reduced by about the same amount. Such threshold voltage adjustment can give up to a twofold improvement in radiation-tolerance.

The relative hardness of logic functions requires careful consideration. For example, the CMOS 'NOR' gate uses a series configuration of p-channel devices and parallel n-channels. Since n-channel thresholds decrease (increasing leakage) and p-channel thresholds increase (reducing drive) as a result of radiation, the 'NOR' gate is vulnerable to ionizing radiation. The 'NAND' qate configuration with n-channels in series and p-channels in parallel is likely to be a more tolerant design.

SOS/SOI technologies feature an 'edge' transistor where the gate electrode crosses the edge of an epitaxial silicon island. A parasitic transistor with a thick oxide is formed at this point. This may give leakage problems for the same reasons as a FOXFET does. The employment of closed-geometry devices with annular gates can overcome this problem, but this design is wasteful in terms of surface area required. Also, in SOS/SOI technologies, the body of the device is frequently left 'floating' at an undefined potential. This potential may shift as a function of radiation, and hence hardness can be improved by connecting the body to the source electrode. Some other considerations involving integrated device geometry in relation to radiation response were discussed in sections 4.10, 5.5 and 6.14 with respect to total dose effects in MOS, bipolar, and optical devices respectively.

14.5 Hardening against transient radiation

14.5.1 Pulsed gamma rays

The effects of pulses of ionization in MOS and bipolar ICs was discussed in Chapters 4 and 5. 'Topological adjustment' of the mask layout to reduce photocurrents in the output nodes of MOS inverters and gates has been demonstrated (Dennehy *et al.* 1971). However, dielectric isolation has been more widely used (see Chapter 4). For bipolar integrated circuits, attempts to reduce the effects of the currents induced by gamma-ray pulses, were discussed in Chapter 5; dielectric isolation is an expensive way of hardening bipolar circuits at the chip level, but this improvement by itself is not sufficient to meet the needs of all military equipment. Section 15.3.6 goes further into the considerations guiding hardening decisions for equipment exposed to pulsed gamma rays.

14.5.2 Single-event phenomena

1. *Latchup*

The latch-up mechanism in CMOS devices is due to the activation of a parasitic pnpn or npnp thyristor structure and has been described in Sections 3.6 and 3.7. To harden against latch-up requires elimination of the thyristor structure or assurance that the conditions required for 'turn on' of the thyristor are never satisfied. Elimination of the thyristor structure can be achieved by the use of insulating substrates: SOS or SOI. These technologies are rather specialized and tend to be restricted to the military and space markets.

For the broader range of CMOS technologies, hardening against latch-up may be achieved by the use of a lightly doped epitaxial layer on a heavily doped (low resistivity) substrate. The low-resistivity substrate degrades the gain of the parasitic transistors and limits base–emitter voltages by reducing the distributed resistance across the base–emitter junctions. The substrate also acts as an effective charge collector. Optimization of the resistivity and thickness of the epitaxial layer is important in order to achieve adequate immunity to latch-up. Experiment indicates that a layer thickness of 7 μm or less is desirable. Latch-up has been noted in devices using 12 μm epitaxy, and there was significant temperature-dependence of latch-up sensitivity (Smith *et al.* 1987). Certain aspects of layout have an influence on latch-up sensitivity. Minimum diffusion spacings are important. The use of 'butted contacts' (source–substrate/well), in close proximity, allows frequent strapping of the substrate, limiting the potential difference across the junctions and reducing latch-up sensitivity (Song *et al.* 1987; Chatterjee *et al.* 1987).

2. Upset

As we noted in Chapter 3, the most important parameters in SEU sensitivity are those of critical charge and charge collection. The critical charge is determined primarily by the physical dimensions of the device: the decreasing dimensions of advanced technologies leads inevitably to a reduction in critical charge. The techniques for SEU hardening are therefore concerned with reducing charge-collection volume or increasing the *effective* critical charge by circuit techniques to modify the time constants of response.

The charge-collection volume is minimized by reducing charge-funnelling. This is achieved by the use of insulating substrates (SOS/SOI) or 'tailored' substrates which can terminate an ion track and collect the charge as a 'prompt' photocurrent. 'Tailored' substrates use thin epitaxial layers on top of a highly doped, low-resitivity substrate. The 'shunting' effect of the substrate must be optimized to ensure that the photocurrents do not perturb circuit operation. Floating or high-impedance nodes are susceptible to SEU and should be avoided in hardened circuits; this restricts the use of resistor-load, four-transistor, memory cells. The six-transistor cell has an 'ON' transistor at each data node which couples the node to a supply voltage. This stabilizes the logic state against perturbations (Rockett 1988). The critical charge of sensitive nodes can be increased by increasing capacity. Current sourcing and sinking capability of transistors can be increased to compensate for the effect of increased capacity on circuit speed. This solution is rather unattractive since it uses large areas of silicon. Time constants play an important part in the SEU mechanism. The switching time of bistable elements can be increased so as to prevent a fast SEU pulse from switching the logic state. Resistor decoupling between cells to slow regenerative feedback response of a bistable element allows discrimination between a fast SEU pulse and longer, legitimate, 'write' signals. Polysilicon resistors have a negative temperature coefficient which leads to a slower 'write' response at low temperatures. The 'read' operation is unaffected by intracell resistance because the bit-flip dynamics of the cell are not involved (Rockett 1988).

Resistor hardening involves a number of process problems; the process control of high-value resistors is difficult. Various other types of decoupling are being studied, including diodes, depletion mode transistors, and resistor–capacitor combinations.

Bipolar integrated circuits are comparatively sensitive to SEU. This is due to the intrinsic sensitivity of bipolar transistors and diodes to the charge injection processes of SEU (Zoutendyk *et al.* 1984). The isolation junction to the substrate is a particularly sensitive region, owing to the large active volume of the collector–substrate junctions. There are various design approaches to minimizing the substrate effect, including insulating substrates (bipolar-SOI) and epitaxy on highly doped substrates. A gated feedback element (e.g. a diode 'OR' gate) may also be used to isolate the affected nodes (Berndt 1988).

Error detection and correction is a way of hardening against SEU. This is more likely to be applied at system level although such techniques can be, and are, applied 'on chip' for highly complex VLSI functions. The additional area of silicon required for the additional circuitry makes it unattractive for routine application to standard devices.

14.6 Hardening of parts other than silicon

Most other chapters have provided discussion of radiation-tolerant forms of devices where they are available. The tolerance may be a fortuitous property of the material, as in ultrapure silica optical materials or gallium arsenide, or highly contrived as in silicon–on–sapphire devices. Research on the hardening of parts has only been intensive in the case of active silicon components, many types of solar cell, and some optical components.

The fact that III–V compounds give lower rates of recombination per damaging particle than for silicon has been one of several motivations for the development of solar cells and ICs in these materials. However, the higher carrier mobility and nature of the forbidden energy gap have had equal influence in promoting the development of electronic devices. The same applies to II–VI semiconductor compounds (Davey and Christou 1981). However, ultra-fast GaAs ICs have been developed by defence departments with special urgency because of their general environmental tolerance. Military agencies have engaged in sophisticated research to promote hardening (see e.g. Campbell, *et al.* 1986, and Tabatabaie-Alavi 1986). A special application is in radar signal generation and analysis. Tolerance to single-event upsets is discussed by Bland *et al.* (1988).

14.7 Conclusions

The foregoing is an overview of total-ionizing-dose and single-event hardening techniques and should be used as a guide to a range of research publications. It should be stressed that there is no clear and simple route to a radiation-tolerant silicon integrated circuit. What works for one fabrication process may not work for another, and there are many complex interactions within individual processes and designs. We have attempted to highlight the most important factors and those process changes which should bring improved hardness. The main point is that radiation-hardening as a procedure must be approached in a methodical fashion and with a good understanding of the response mechanisms involved.

References

Adams, J. R., Dawes, W. R. and Sanders, T. J. (1977). A radiation-hardened field oxide. *IEEE Transactions on Nuclear Science*, **NS24**, 2099–101.

Adams, L. and Holmes-Siedle, A. (1992). Radiation hardness assurance of space electronics. *Nuclear Instruments and Methods in Physics Research*, **A314**, 335–44.

Anderson, R. E. (1978). Post gate plasma and sputter process effects on radiation hardness of metal gate CMOS integrated circuits. *IEEE Transactions on Nuclear Science*, **N-25**, 1459–64.

Berndt, D. F. (1988). Designing hardened bipolar logic. *Proceedings of IEEE*, **76(11)**, 1490–6.

Bland, S. W. and Galashan, A. F. R. (1988). *GaAs reliability and radiation hardness*. Final Report of ESTEC Contract 6658/86/NL/MA(SC) Standard Telecommunications Company, Harlow.

Borkan, H. (1977). Radiation hardening of CMOS technologies – an overview. *IEEE Transactions on Nuclear Science*, **NS-24**, 2043–6.

Campbell, A. B. *et al.* (1986). Particle damage effects in GaAs JFET test structures. *IEEE Transactions on Nuclear Science*, **NS-33**, 1435–41.

Davey, J. E. and Christou, A. (1981). Reliability and degradation of active II–V semiconductor devices (in Howes, M. J. and Morgan, D. V. (ed), Reliability and Degradation). pp. 237–300. Wiley and Sons, London.

Dennehy, W. J., Holmes-Siedle, A. G. and Zaininger, K. H. (1969). Digital logic for radiation environments. *RCA Review*, **30**, 668–708.

Dennehy, W. J., Brucker, G. J., Borkan, H. and Holmes-Siedle, A. G. (1971). Topologically adjusted MOS integrated circuits. *IEEE NSREC '71*, Durham N.H., July.

Derbenwick, G. F. and Gregory, B. L. (1975). Process optimization of radiation hardened CMOS circuits. *IEEE Transactions on Nuclear Science*, **NS22(6)**, 2151–8.

Dunn, G. J. and Wyatt, P. W. (1989). Reoxidized nitrided oxide for radiation hardened MOS devices. *IEEE Transactions on Nuclear Science*, **NS36**, 2161–8.

Hughes, H. L. Historical perspective (1989) in T. P. Ma and P. V. Dressendorfer, *Ionizing radiation effects in MOS devices and circuits*. pp. 47–86. Wiley, New York.

Johlander, B. (1990). Test structures and measurement techniques for radiation hardening. In Proceedings of the ESA Electronics Components Conference, ESA SP-313. ESA, Noordwijk, Netherlands.

Nishioka, Y., Itoga, T., Ohyu, K., Kato, M. and Ma, T. P. (1990). Radiation effects on fluorinated field oxides and associated devices. *IEEE Transactions on Nuclear Science*, **NS37**, 2026–32.

Rockett, L. R. (1988). Designing hardened bulk/epi CMOS circuits. *Proceedings of IEEE*, **76(11)**, 1474–83.

Smith, L. S., Nichols, D. K., Coss, J. R. and Price, W. E. (1987). Temperature and epi thickness dependence of the heavy ion induced latch-up threshold for a 16K static RAM. *IEEE Transactions on Nuclear Science*, **NS34**, 1800-2.

Song, Y., Vu, K. N., Coulson, A. R., Lizotte, S. C., Cable, J. S. and Miscione, A. M. (1987). Parametric investigation of latch up sensitivity in !.25 micron CMOS technology. *IEEE Transactions on Nuclear Science*, **NS34**, 1431-7.

Tabatabaie-Alavi, K. *et al*. (1986). Application of GaAs/(Ga, Al) As super-lattices to dose rate hardening of GaAs MESFETS. *IEEE* 1986 *GaAs IC Symposium*, 137-140.

Watanabe, K., Kato, M., Okabe, T. and Nagata, M. (1985). Radiation hardened devices using a novel thick oxide. *IEEE Transactions on Nuclear Science*, **NS32(6)**, 3971-4.

Winokur, P. S., Erret, E. B., Fleetwood, D. M., Dressendorfer, P. V. and Turpin, D. C. (1985). Optimizing and controlling the radiation hardness of a Si-gate CMOS process. *IEEE Transactions on Nuclear Science*, **NS32**, 3954-60.

Zoutendyk, J. A., Malone, C. J. and Smith, L. S. (1984). Experimental determination of single event upset (SEU) as a function of collected charge in bipolar integrated circuits. *IEEE Transactions on Nuclear Science*, **NS31**, 1167-74.

Further reading for Chapter 15

Becquet, M. C. (ed.) (1992). *Teleoperation: numerical simulation and experimental validation*. Kluwer, Dordrecht.

Coenen, S. and Decréton, M. (1993). Feasibility of optical sensing for nuclear robotics in highly radioactive environments. *IEEE Trans. Nucl. Sci.*, in press.

Decréton, M. (1992). Position sensing for advanced teleoperation in nuclear environment. In Becquet (1992).

Horne, R. (1992). Practical experience using teleoperated technology; teleoperated devices used in an accelerator complex. In Becquet (1992).

Raimondi, T. (1992). Transporters for teleoperations in JET. In Becquet (1992).

15
Equipment hardening and hardness assurance

15.1 Introduction

The introduction of tolerance to radiation ('radiation-hardness') into large electronic systems is one of the major tasks to which this Handbook will be put. The essentials of the task are shown in Figure 15.1. This diagram may appear simple but, if the radiation environment is severe, the task can — judging by experience in Europe and the USA — require a large amount of effort. The skills required, mainly subspecialities of physics and engineering, are those outlined in this handbook. In the case of a severe environment, if 'radiation-hardening' is to be successful, it has to be very thorough. In particular it has to permeate all stages of the development of a piece of equipment, including the earliest system engineering and technology development. Clearly then, radiation-effects modelling should start months or years before equipment is built and continue into assembly and deployment. It clearly has to be well managed. The US Department of Defense and Department of Energy have taken this message seriously and invested large amounts in teams of radiation-hardening experts, working directly with major industrial members of contractor teams (see e.g. Holmes-Siedle 1974a, 1974b). Interplanetary space projects, for example to the high-radiation planet Jupiter, demonstrate the same sense of thoroughness (van Lint et al. 1977). The serious effect of a late start is stressed by Messenger and Ash (1986): '... the cost [of radiation hardening] ... is relatively modest when [it] is incorporated at the inception of the development ... [but] ... eleventh-hour add-ons to existing systems ... are expensive, unduly extensive and ... manifest reduced confidence ...'. Nuclear power projects, previously paying most attention to metals and ceramics, are now devoting similar attention to electronics (see e.g. Baur et al. 1981).

In the above fields, a major variable is the degree of penetration of which the radiation is capable. For example, in nuclear fuel handling, the most severe effects come from gamma-rays; given the normal weight constraints, it is difficult to protect mobile electronic gear adequately by shielding, since it requires many centimetres of material to attenuate gamma photon beams significantly (see, for example Chilton et al. 1984). The same applies to mobile military equipment in the nuclear gamma–neutron environment. By

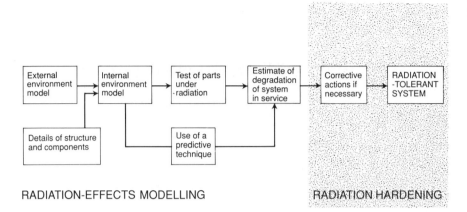

FIG. 15.1 Essentials of improving tolerance of a system to radiation: in modelling, information on the raw external environment is combined with information on structure and components—say, of a vehicle—to give a detailed internal environment model, which can vary with position in the vehicle. Using prediction or tests, the degradation of performance of the system during its mission is estimated. If necessary, corrective actions are taken (see text). The product is a more radiation-tolerant or 'hardened' system.

contrast, much though not all space radiation is heavily attenuated. The electron component is stopped by a few millimetres of material, leaving only a 'tail' of penetrating components (see Appendix E). Accelerator radiation is often in the form of secondary X-rays. Rules for the hardening of equipment usually require a judicious balance of shielding versus circuit alteration; therefore the design approach turns on the distinctive differences in penetration described above. We divide our discussion of equipment hardening into practices suitable for 'penetrating' and 'non-penetrating' radiation. However, we can first mention some elementary principles for inculcating tolerance to radiation effects which are applicable to both types of environment.

15.2 Elementary rules of hardening

15.2.1 General

As was seen in the bar chart of Fig. 3.17, there is a wide spread of radiation-tolerance amongst materials which go into electronic equipment. Materials with high tolerance include all metals, many ceramics, and inert gases. The tolerance to radiation of semiconductors, optics, and polymers is 'fair' to 'very poor', depending on part design and subsystem design. To the 'fair' class belong bipolar ICs, rectifying diodes, and 'hardened' MOS circuits; to the 'very poor' class, devices such as commercial MOS circuits, analogue devices, power transistors, solar cells, and charge-coupled devices.

Device degradation predictions must be prepared early in the design process so as to influence circuit design and other hardening procedures. Given timely warning, circuit designers can accommodate degradation problems. Judicious part selection can often increase the tolerance of equipment to radiation for little increase in weight.

Chapter 14 refers heavily to the radiation-hardening of metal-oxide semiconductor integrated circuits. This technology usually presents the most important problems for aerospace microelectronics in a radiation environment having mission doses above 1000 rad (10 Gy). For other devices, the question of hardening is dealt with in the chapters on device physics.

15.2.2 Measures at systems level

15.2.2.1 *Prediction and statistics*

The degradation of semiconductor devices upon irradiation exhibits an important degree of scatter. Thus, for predicting and providing remedies for the degradation, the accuracy of predictions will have to be dealt with in rigorous fashion, using statistical methods, often covered by the term 'hardness assurance' (see later). The theory used here is similar to that used in other statistical calculations in product assurance for microelectronics.

15.2.2.2 *System theory*

After analysis and prediction of the degradation of specific parts, the next step is to use circuit simulation and control theory to predict the exact effects at subsystem or system level. System theory may also evolve original design solutions, for example in the improvement of fault tolerance despite the multiple failure modes caused by the radiation environment. This may require special sensors to warn of radiation-induced failure.

15.3 Robots, diagnostics and military vehicles in penetrating radiation

15.3.1 Manipulators for nuclear plant

In the nuclear industry, there is a requirement to employ machines which possess a degree of autonomy. The presence of man in certain radioactive locations can be tolerated only for a very short time. The classic examples are:

(1) 'the cavity' in light-water fission reactors in which surveillance and maintenance of the pressure vessel are carried out;
(2) 'hot cells' in spent-fuel processing facilities.

Less structured but more dramatic is the case of the disaster environ-

388 Equipment hardening and hardness assurance

ment; at Chernobyl (see Chapter 2), men literally sprinted to spent fuel fragments, worked until their total dose allowance was exhausted, and then sprinted out again. Clearly, the aim in the future must be to allow machines, especially manipulators, to do more, and to require a minimum in the way of operator commands, that is, machines become 'semi-autonomous tele-operators' (Larcombe *et al.* 1984; Abel *et al.* 1991). The intended result is to make the tasks of man less risky and to increase the profitability of plant

FIG. 15.2 An underwater colour camera system which has to work near intense radiation sources: the camera is used for inspection and handling operations on spent uranium fuel rods of the BR3 reactor at CEN-SCK, a nuclear research institute at Mol, Belgium. The photograph shows the colour camera, Type CE1/VKC/50 accompanied by lamps, a telescopic arm, and pan-tilt support. The camera is positioned above some fuel elements. The image sensing element is a charge-coupled device (CCD). Because of the sensitivity of CCDs to radiation, the camera is barred from going too close to spent fuel elements. European groups are developing CCD-based camera systems which will tolerate higher doses and dose rates. Photograph by courtesy of M. Decreton, CEN/SCK. (Reproduced with permission.)

15.3 Robots, diagnostics and military vehicles

used in mining, accident management, reactor maintenance, decommissioning, and fuel handling and reprocessing. Types of remote teleoperator include long-reach arms and 'snakes' for inspection and welding, manipulators attached to gantries for work on heavy loads and large objects, and mobile robots. One of the most intense source of radiation encountered in this field is spent fuel, emitting gamma-rays; another source is activated reactor parts, also gamma-emitters. Even though other particles are present, the dominant radiation environment is the highly penetrating gamma-rays. Figure 15.2 shows a CCD camera for inspecting fuel elements.

For teleoperators requiring a degree of autonomy, it may be impossible to avoid placing electronic devices in the 'hot' area. For example, sensors for measuring distance, force, and angle are required. The data must be sent to an operator in a 'cold' area. These may be separated by leak-tight concrete and metal walls, which forbids many kinds of communication link, especially multicore cables. An RF link or a single-way cable is typically required. Thus, a multiplexer must be placed in the 'hot' area, along with the sensors (Leszkow and Decreton 1991). A preamplifier may also be required. If RF is used, then communications circuits are required. It may also be an advantage to include automatic control loops in the 'hot' area. Finally, shielding from gamma-rays by even one order of magnitude may impracticable because of weight (about 100 mm of steel would be required). Sensors themselves also may contain radiation-sensitive elements.

In one EEC-sponsored programme, the dose-rate requirement was 10^3 rad h^{-1} (10 Gy h^{-1}) and the lifetime dose specifications ranged from 10^5 to 10^8 rad (10^3 to 10^6 Gy), depending on the application (Lauridsen 1991). From previous chapters, it is clear that, without selection of tolerant devices and special design measures, some subsystems of a teleoperator might fail catastrophically because of the exposure values discussed above. This must be recognized in virtually all projects for hostile nuclear plant environments. In some programmes, radiation-hardening has been introduced early into the design process (see for example Holmes-Siedle, 1985; de Nordwall and Schaller 1984; Leszkow and Decreton, 1991; Lauridsen 1991).

Particles likely to cause bulk damage are not usually present in missions concerned with handling gamma-emitters, although for machines servicing accelerators, spallation neutrons and other particle showers may be present. For the gamma case, surface effects in semiconductors and ionization effects in optics, reducing transmission throughout the whole material, are the most likely radiation effects for this environment. Less frequent are the bulk damage effects of Compton electrons (see e.g. Chapters 5 and 6).

In the case of a very high dose rate, it is also possible that radiation-induced photocurrents or photovoltages will be present. The dose rates will be high enough to cause significant noise in sensors, especially light-operated pickup devices such as solid-state cameras. The rates in the cases described above will not normally be high enough to cause logic upsets.

15.3.2 Hardening of a robotic vehicle

The problem of the robot in radiation is pictured in Fig. 15.3. Most mobile robots contain four major subsystems as shown. Each communicates its state of performance via sensors of position and other parameters. The control computer outside the active area compares these parameters with a 'nominal process model' stored in memory. Faults are displayed as 'residuals' which may cause 'decision logic' to give an alarm. The presence of a radiation environment introduces new elements into the control problem. Radiation-induced degradation contributes new types of fault to the robot and to the process model. In the figure, inputs dealing with radiation effects are marked A and B. If the degradation of specific components has been properly modelled, the measurement of radiation dose allows the computer to calculate the expected state of the circuits and compare it with the actual state. An element of hardness—awareness of incipient failure—has been introduced. Awareness of the precise effects of radiation at system or subsystem level also supplies the engineer with the necessary information to improve radiation-tolerance, not only by substituting parts but also by changing circuit design or system operation to combat the degradation. The use of radiation measurement in a 'health monitor' subsystem is discussed further under 'preventive replacement', below.

Simple circuits which will tolerate doses greater than 10^8 rad (10^6 Gy) and dose rates greater than 100 rad s^{-1} (1 Gy s^{-1}) have been constructed by the use of discrete bipolar transistors (Leszkow and Decreton 1991). The degradation of most integrated circuits at these dose values means that few available ICs can be employed. This is practically true despite major efforts over the last 20 years to achieve tolerance to several tens of megarads (see for example Dennehy *et al.* 1969; Dupont-Nivet *et al.* 1991). The performance of charge-coupled device sensors degrades both by virtue of dose-rate 'fogging' and degradation of device parameters (see Chapter 6).

Radiation-tolerant TV cameras have been designed using bipolar circuits and vidicon sensor elements (see Chapter 6). Sharp and Dumbreck (1991) have developed selection methods for solid-state videocameras. The best selected work until a 20 krad (200 Gy) dose is reached; work is in progress to improve the radiation-tolerance of solid-state imager systems in cameras designed for specific tasks in nuclear plants (Emeriau 1991).

15.3.3 Preventive replacement and fault detection

The consequences of the failure of a machine in nuclear plant are acknowledged to be so serious that strategies for avoiding failures are widely used in equipment practice. In particular, the scheduling of replacements in a timely manner—preventive maintenance—is a good way of avoiding catastrophic events. The case of a mobile robot failing due to the VTNZ

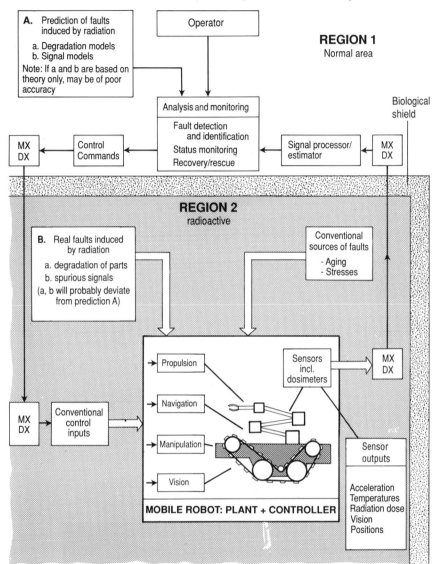

MX − DX = Multiplexer − demultiplexer

FIG. 15.3 Control diagram for a robotic system in a radiation environment. Subsystems communicate their state of performance and positions to a control computer, which compares these parameters with a 'nominal process model'. Faults generate 'residuals' and possibly an alarm. Radiation-induced degradation contributes new types of fault to the model. The 'radiation health monitor' allows the computer to calculate the expected state of the circuits. (Sinha and Holmes-Siedle unpublished work.)

392 Equipment hardening and hardness assurance

effect in a microprocessor (see Chapter 4) in a narrow passageway is a prime example of such an event. Although serious, the event combines the favourable elements of

(1) predictability;
(2) limitation to a small group of key components;
(3) small size of failed circuits.

Therefore, the avoidance of such failures is quite possible and is a typical application of the predictive techniques prescribed in other chapters. However, feature (1) requires accurate knowledge of the dose at the part. It is also, of course, better backed up by decision-making circuitry which recognizes incipient failure (more parts). Garlick (unpublished work) has developed the concept of an 'exo-brain'; that is, a small circuit module containing the part due to fail and which can be replaced using manipulators when the warning is given.

For accurate knowledge of doses, the present authors and co-workers have recommended the use of on-board dosimetry (Holmes-Siedle and Adams 1986; Holmes-Siedle *et al.* 1990). For fault detection ahead of failure, Sinha and others (unpublished reports 1992) note that advanced computer technologies will be able to handle the problem of fault detection and preventive replacement in a better way than has hitherto been possible. For example, in remote control, centralized alarm systems normally detect faults after the event and are designed to hand over control to an operator, who then proceeds to rectify the fault manually. A more appropriate approach is to exploit the use of a computer external to the 'hot cell' to determine the status of the critical subsystems such as the communications, computation and actuator/controller interfaces. By careful design of the control and communications architecture, forward-looking fault detection can be embedded within these systems. It is therefore proper to include a 'radiation health monitor' system as a building block to be incorporated in the automated equipment as necessary. The device can be thought of as the equivalent, for the robot, of a radiation badge for humans.

15.3.4 Remotely controlled maintenance of fusion reactors

In fusion reactors, two classes of equipment will be exposed to radiation: permanently installed equipment (see for example Holmes-Siedle 1974, 1980) and mobile maintenance equipment. The permanent equipment will be present during ignition and hence exposed to both neutrons and gamma rays (see e.g. Baur *et al.* 1981; Holmes-Siedle *et al.* 1984 for the case of diagnostic equipment). The maintenance equipment will handle radioactive parts during power-down and may be exposed only to gamma-rays. The hardening problem will vary greatly with the application, but rules similar to those given above for robotics will often apply.

15.3 Robots, diagnostics and military vehicles

15.3.5 Instruments and detectors in power reactor and accelerator facilities

a. Nuclear

Subsequent to the TMI accident, regulatory authorities have placed a radiation dose specification on instrumentation in reactor facilities (see e.g. Johnson *et al.* 1983). At the instigation of the US Nuclear Regulatory Commission, a high dose specification was set for the design of vibration monitors for pressure-relief valves on the primary coolant system of US reactors. The piezoelectric monitors give a warning when loss of coolant causes vibration in the valve. The dose specification used was 200 Mrad (2 MGy). To meet this hardness specification in the case of a preamplifier, designers have used discrete bipolar transistors in a circuit which uses feedback to compensate for gain degradation and avoids the use of large bypass capacitors which may leak under irradiation (Binkley 1982).

Another specification is for radiation doses received in routine service in a nuclear building (IEEE 1974). Equipment designated 'Class 1E' is replaced when estimated exposure exceeds an established maximum in a range below 1 Mrad (10 kGy). This is the same 'preventive replacement' strategy that was mentioned above for robots. Area radiation monitors are distributed throughout a nuclear facility to estimate the radiation dose and dose rates likely to be suffered by personnel and equipment. However, the uncertainty attached to the total dose at a particular location some distance from an area monitor will be fairly high.

Many advanced instruments contain microprocessors and memories and therefore have fairly low radiation-tolerance unless design for degradation is carried out. Holmes-Siedle *et al.* (1990) have also recommended that equipment in reactor facilities should carry individual RADFET dosimeters as monitors for total dose. This will give assurance that key MOS devices (memories, processors) in critical equipment do not exceed the allowed total dose. The authors suggest that such monitoring will repay its cost, since accurate knowledge may allow an *extension* of the time between replacements of the equipment. For equipment which is mobile in non-uniform radiation fields, individual dosimeters indeed represent the only way to monitor such 'wear-out' with confidence.

b. Accelerator

Designers of high-energy physics facilities (see section 2.7) have predicted serious radiation damage problems in front-end electronics (see section 4.9), detector diodes and CCDs (see sections 6.8 and 6.12), and scintillator materials (see section 8.5). Particles and secondary photons leaving the main beam cause a mixture of ionization and displacement damage which may cause failure many times over within the life of the accelerator.

15.3.6 Military systems

A nuclear explosion emits neutrons, gamma-rays, X-rays, thermal energy, radioactive debris, and radiofrequency energy (electromagnetic pulse, EMP). The highest dose rate is the prompt gamma pulse, emitted in a microsecond or less. The other effects are strongly dependent on whether the explosion is inside or outside the atmosphere. Further details were given in Chapter 2. The effects can be divided into transient and long-lived. Shielding of vehicles against gamma-rays and neutrons is rarely effective, given the weight budgets in question. For example, a thickness of 20 mm of steel is needed to halve the flux of 1 MeV gamma photons (see Chapter 11). By contrast, electromagnetic shielding of equipment is an essential part of the design of most military equipment.

The practices used for managing neutron/gamma effects vary greatly with the equipment. Messenger and Ash (1986) divide equipment into strategic and tactical types. The former is exemplified by long-distance bombers and satellites; and the latter by machines operating on the earth's surface or near to it, such as ships, short-distance aircraft and tanks. A major difference between the two types is the absolute level of the radiation damage likely to be caused; although exact figures are classified information and vary with the mission, it can be said that tactical levels are often based on levels lethal to humans (order of 10^3 rad (10 Gy) and 10^{12} fast neutrons cm^{-2}) and that strategic levels, where unmanned vehicles are involved, may be a thousand times higher than the above.

Transient photocurrents in semiconductor components cause many of the most difficult problems, and some special methods are required to allow a system to pick itself up after the pulse. 'Reset' implies that computers

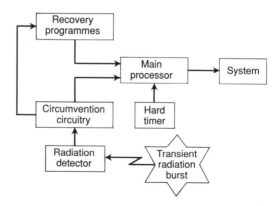

FIG. 15.4 Circumvention of pulsed radiation effects. A detector senses the rising edge of the radiation pulse; decision circuitry maintains the computer in a state of suspension for as short a time as possible before recovery.

15.3 Robots, diagnostics and military vehicles

are restarted after a gamma-induced upset. 'Circumvention' implies that a detector gives warning of the pulse; circuit measures are taken to hold voltages within a desired range during the disturbance and then programmes continue afterwards without restart (Retzler 1973; see also Ricketts 1972; Rudie 1980; Levy and Friebele 1985; Messenger and Ash 1986). This also requires the presence of a timer which is not upset by the radiation, as in Fig. 15.4. Strategic requirements for tolerance to neutrons and total dose may be severe. Device-hardening research funded recently in the USA has addressed these degradation problems intensively, as described in Chapters 4 and 14.

The effects of pulsed neutrons and gamma-rays on a circuit can be predicted from a knowledge of the radiation waveform and radiation fluences. The circuit analysis required is however quite complex, and network analysis programmes have been developed (see for example Malmberg and Cornwell 1964; Bowers and Sedore 1971; Lindholm 1971) to predict the effect of currents and voltages induced by the junctions within a circuit.

Transient circuit effects include:

(1) the 'scrambling' of logic and memory;
(2) spurious or swamped output signals;
(3) loss of power regulation;
(4) four-layer pnpn effects, or 'latch-up' (see Crowley *et al.* 1976);
(5) burnout of junctions and conductors from excessive photocurrent or latch-up currents
(6) transient reduction and subsequent 'rapid annealing' of gain (see below) in transistors and four-layer switches;
(7) transient conductivity in insulators, especially capacitors and substrates (Pigneret 1985);
(8) luminescence in optoelectronic devices;
(9) internally generated electromagnetic pulses from photoemission of electrons from surfaces in vacuum or low-pressure gases;
(10) induced currents from the penetration of electromagnetic radiation into the system.

These short-lived effects leave behind them any of the multitude of long-lived degradation effects described elsewhere. The amounts of silicon bulk damage and oxide charging produced immediately after the radiation pulse are normally much greater than those long-lived values such as are observed after low-rate irradiation. The low-rate levels noted in earlier chapters are, in fact, the long-lived product of an initial high defect or charge concentration which has then undergone a 'rapid anneal' to an equilibrium level (Srour *et al.* 1984). Larin (1964), Ricketts (1972), and Pigneret (1985) give examples of transient output waveforms from solid-state circuits.

15.3.7 General guidelines for hardening against pulsed gamma rays and neutrons

Ricketts (1972) gives a list of prescriptions for circuit design, intended to counter the effects of neutron–gamma pulses listed above to increase tolerance to neutron/gamma pulses, as in Table 15.1. These are circuit measures and do not include the commonsense initial stage of selecting devices with high tolerance and the subsequent measures such as circumvention, EMP suppression, and the reduction of neutron/gamma fluxes by layout (see Thatcher 1965; Retzler 1973, Halpin *et al.* 1976; Levy and Friebele 1985; Friebele 1991; and later in this chapter on spacecraft layout).

Messenger and Ash (1986) give an authoritative discussion of the hardening requirements for an intercontinental missile re-entry vehicle (5×10^{14} n cm^{-2}, 10^6 rad(Si), 10^{12} rad(Si) s^{-1}) and the hardening problems of the guidance computers of the Missile-X and Trident missiles (the current US ICBMs). To harden for these environments is probably the most challenging exercise in the radiation-effects field. Although worse environments may exist one component at a time (see below), the combination of a large system, limited power, serious consequences of failure, high dose rates, and serious gamma and neutron effects probably makes ICBM equipment a uniquely difficult case.

Specifications for the survivability of military system exist in abundance but, except for a few general documents, are of restricted circulation. Such NATO documents bear the numbers STANAG 4145 and 4328. Some documents in the NATO Allied Engineering Publications (AEP) Series, published by the NATO International Staff, Defence Support Division

TABLE 15.1 Guidelines for circuit design: equipment, techniques, and properties to be utilized (after Ricketts 1972)

Passive devices (especially inductors and transformers)	Minimal requirements for ultra-stable voltages, currents, and frequencies
Tunnel diodes	Potting with high-ionization-energy materials
Magnetic memories	
Protective devices to prevent burnout	Minimum impedance levels
High ambient temperature	Negative feedback
High injection levels	Voltage clamping
Maximum base drive for transistors in saturation	Current-limited designs
	Temperature compensation
Minimum-fanout designs	Cancellation of spurious currents
Minimum capacitance values	Time sequencing/delay
Maximum supply voltage to MOS devices	Direct-coupling stages
	Maximum bias stabilization
Maximum gain margin	Saturated logic
Maximum noise margin	

(TSP), present guidelines to survivability under nuclear weapon effects (termed NATO NWE Survivability Documents) but are restricted in circulation. Some US Department of Defense documents, such as MIL Handbooks Nos. 279 and 280 and DNA 5909F present hardness assurance guidelines for semiconductor devices and microcircuits, with respect to total dose and neutron hardness respectively. Related test specifications are MIL-STD-883 methods (see Chapter 13).

15.3.8 When only vacuum electronics will do

When electronics intrudes close to high-level sources, conditions may be such that solid-state devices are not adequate for the mission. Examples include: very high ionizing doses for some of the isotope handling requirements; neutrons and gammas in fusion power reactors, vehicle-borne fission reactors, and underground nuclear weapon tests; and particles in the target chamber of accelerators. On these occasions, it may be necessary to revert to vacuum electronics, using conventional thermionic tubes and photocells of various kinds. Another prescription for this case is the use of 'vacuum microelectronics' such as the 'vacuum FET' and other miniature, micromachined hot or cold electron sources. Some types employ silicon micromachining (Araragi and Nuraki 1992; Gray 1990; Spindt 1976). Others are based on ceramics and metals (Lynn 1985). These devices have been developed for operation at temperatures for which junction devices do not operate well, for example in engine and reactor control; or for use in ultrafast circuits such as microwave electronics for military decoys. Since electronic transport in solids is minimized, degradation under radiation is also minimized. We estimate that, if the parts do not include micromachined silicon, then conditions as high as 10^{13} rad (10^{11} Gy), 10^{22} n cm^{-2} and 10^{13} rad s^{-1} can be withstood. At these values, there will be some swelling of materials and uncontrolled electron emission. The above discussion indicates that the radiation-tolerance outstrips most of the conceivable applications and that the devices will remain in specialized, high-temperature applications, except perhaps, in the future, fusion power reactors (Holmes-Siedle et al. 1984).

15.4. Equipment in non-penetrating radiation: space, X-rays, and beta-rays

15.4.1 Introduction

In previous sections, we have described how to predict the degradation for various types of semiconductor devices in a space environment, given a calculated dose level. In other sections, we have described how to calculate the attenuation of the external space dose by materials given in aluminium equivalent thickness values. For unmanned spacecraft, the launch envelope, weight, and power budget are dominant; given the sensitivity of dose

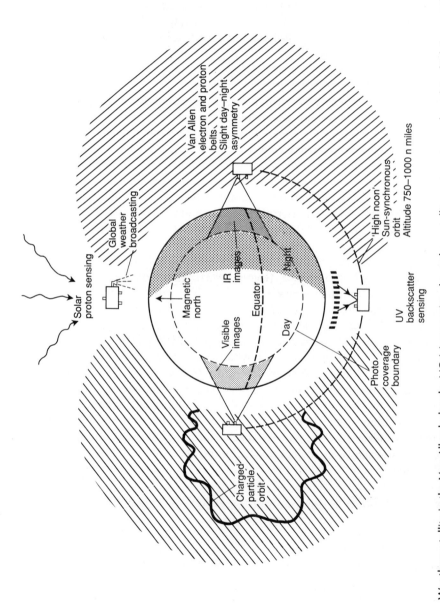

Fig. 15.5 Weather satellite in the Van Allen belts: the NOAA operational weather satellite system, in a polar orbit which 'grazes' the bottom edge of the Van Allen belts; altitude cannot be reduced, or coverage is affected (Holmes-Siedle and Poch 1971).

15.4 Equipment in non-penetrating radiation

value to the amount of absorber around a component, the layout of equipment and tolerance to space radiation are seen intuitively to be critically connected. Thus, for a high-radiation mission, a precise 'radiation-effects engineering' process has to be brought to bear on equipment design practice. The aim of this discipline, sometimes enshrined in procedures (ESA 1989) is to produce a space vehicle of maximum capability (for example with the maximum number of communication channels or scientific experiments) which will survive for a maximum period of time at minimum penalty (attributable to the radiation environment) in cost and launch weight. Holmes-Siedle and Poch (1971) describe the application of predictive methods and shielding analysis to the TIROS weather satellite mission (see Fig. 15.5). This section introduces some design rules which may be applied and also gives some examples of more recent practice in analysing existing spacecraft.

Because of the ease of stopping some part of space radiation, all components of a spacecraft can be thought of as shielding one another. All mass surrounding a spacecraft component can be regarded as 'shielding' or protection, even though that mass serves some other primary, usually structural, purpose. We will describe this type of protection as 'built-in'. Material added primarily for shielding will be called 'add-on' shielding. In this context, a neutral term used here for radiation-stopping material is 'absorber'. Because of its ambiguity, the term 'screening' is not recommended.

The ultimate aim of the design practice described in this section is to use 'built-in' shielding such that the need for 'add-on' shielding is minimized. In other words, layout has a fundamental importance in the design of a radiation-tolerant spacecraft. The concept is emphasized in the diagram shown in Fig. 15.6 (Holmes-Siedle 1965). If one had complete control of spacecraft layout, one would build a sphere with the microelectronics at the centre. This, of course, has not happened since Vanguard, launched in 1957.

Circuit rules which applied to all radiation environments were given earlier. Some schemes which apply specifically to spacecraft include the wide use of redundancy. Redundancy can be 'on-chip' (gate functions reassigned periodically) or 'cold spare redundancy' of subsystems. The object is to allow 'time off', for active recovery of charge build-up (Adams and Holmes-Siedle 1991).

The use of high-density integrated circuits allows larger numbers of functions to be shielded with a small mass. However, this technology is sometimes less tolerant to total-dose effects than the corresponding low-density integrated circuits and — more important — can be very sensitive to single-event upset.

15.4.2 Typical spacecraft configurations and materials

15.4.2.1 *General*

An unmanned spacecraft normally has a very stringent mass budget. The detailed layout of this sparse amount of mass has quite a strong influence

400 Equipment hardening and hardness assurance

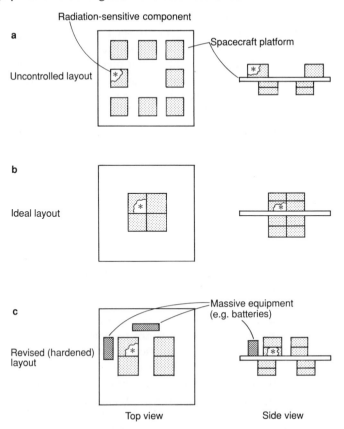

FIG. 15.6 'Hardest' layout for an unmanned spacecraft. Effect of spacecraft geometry on the radiation-tolerance of the system for constant weight: (a) typical one-platform spacecraft, (b) ideal layout for mutual radiation shielding (possibly 10 times longer life than (a)); (c) suggested compromise in which some box covers are thickened and boxes are clustered (Holmes-Siedle 1965).

on the radiation dose reaching the silicon chips which are the focus of our interest. For example, if the system designers opt for a spinning satellite, the solar array will be of the wrap-around type and partially shield all components inside it. The shielding effect of a typical wrap-around solar array will be the equivalent of 3 to 4 mm of aluminium. As we shall see from space dose versus depth curves, this thickness gives a significant attenuation of the space fluxes; if the solar cells were not wrapped around the equipment as in the typical 'paddle' arrangement, then tens of kilograms of add-on shielding would be needed to achieve the same attenuation.

Built-in shielding is also provided by equipment platforms, box covers, and circuit boards. For example, an integrated circuit in the *centre* of a stack

15.4 Equipment in non-penetrating radiation

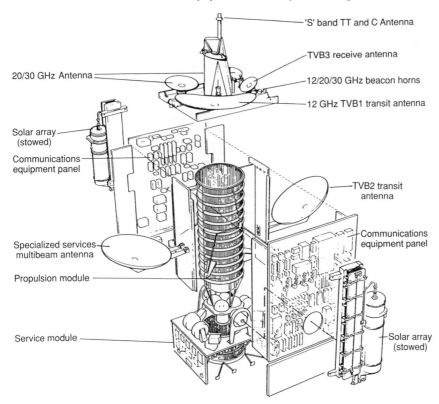

FIG. 15.7 Olympus spacecraft (courtesy European Space Agency): a modern stabilized satellite with solar cell paddles furled; multiple platforms do not supply much mutual protection, and mass is concentrated in the central column.

of printed-circuit boards may be exposed to only one-tenth of the dose received by the same circuits on the *uppermost board* of a stack. Similarly, equipment boxes near the *edge* of a platform receive more dose than those near the centre. The highly significant shielding effect of densely packed circuit boards was confirmed by the distributed RADFET dosimetry on the CRRES and METEOSAT spacecraft (Ray *et al.* 1992; Soli *et al.* 1992; Adams *et al.* 1991*b*). Mullen and Gussenhoven (1991), who designed CRRES experiments, comment that, for deeply buried parts, passive board shielding can be as effective as about an inch (25 mm) of aluminium.

Figure 15.7 shows the layout of the Olympus large communication satellite, which operates in a geostationary orbit. The structure is 'openwork' compared with Fig. 15.6; equipment is secured to four platforms attached to a central cylinder which in turn is joined to the launcher by a conical transition section. The solar array paddles provide no protection.

402 Equipment hardening and hardness assurance

To calculate the dose levels at a given point within a given box, all radiation-absorbing masses present in the satellite have to be taken into account. These, of course, constitute an extremely complex array of masses, but we must calculate as closely as possible how they contribute to radiation-stopping. This is done by 'sector analysis', as described in Chapter 11.

15.4.2.2 *Properties of typical spacecraft materials*

A space vehicle is composed of a large number of small components of widely varying materials. Appendix C includes properties of commercial materials used by one spacecraft contractor in spacecraft subsystems and a table of polymers likely to be used in spacecraft. One of the uses of these tables is that it is often unpractical to assign a solid angle to every component in the spacecraft, and some form of 'homogenization' of the small parts into an equivalent solid slab of representative atomic number will be necessary for a 'manual' calculation of dose. Degradation of these materials is rarely of major importance except for the outer skin of a spacecraft; the degradation of structural materials is unlikely to present problems in the natural space environment. As seen in Appendix E, radiation doses more than 2 mm from the surface are likely to be in the kilorad range — well below the damage threshold for most structural materials.

15.4.2.3 *Spacecraft structure as a radiation stopper*

The analysis of typical designs of geostationary spacecraft shows that the radiation-shielding power of spacecraft often resides mainly in the central structure and not in the equipment platform. The reason why it is difficult to assess the true protection given by the built-in mass is the fact that the mass is divided between a few large structures and a large number of small components. A few important large masses (e.g. batteries) can be used as shadow shields. Ideally, electronic boxes should be 'modular' as in the scheme proposed for the ESA On-Board Data Handling system (Gough 1978) and modular shields for radiation should also be available. Spacecraft electronics box design techniques in the 1990s have developed so that covers are thin foils of aluminium with stopping powers of tenths of millimetres.

15.4.3 Add-on shielding

5.4.3.1 *Introduction*

If built-in mass on the spacecraft cannot be arranged so as to protect all sensitive components, then — as a last resort — some 'add-on' absorber may have to be judiciously added. The use of add-on shielding should be regarded as a last resort after other approaches have been exhausted, essentially because of the high price of payload per kilogram and the high 'revenue' which useful payload can earn.

The first aim of add-on shielding is to interpose a few millimetres of any suitable material between the device of interest and the external environment.

15.4 Equipment in non-penetrating radiation

If the array of devices to be shielded is small, we can save weight by enclosing the array in a compact shield rather than build the same thickness on to the outside of the equipment. This is the idea of 'local' shielding: simply to obtain a given dose reduction in a given volume for the minimum weight penalty. For instance, a single integrated circuit would best be protected by a blob of filled plastic applied directly to the package or by using thicker Kovar for the covers. We will call this type of shield a 'spot shield'.

The particle-scattering property of materials has some dependence on the atomic weight. This is weak in the case of protons and strong in the case of electrons. Thus, the choices of atomic weight for an add-on shield in a proton-dominated orbit such as that of Exosat might differ from that for an electron-dominated orbit such as the geostationary case.

Even small solid angles can admit large amounts of radiation if the absorber in the path of that radiation is thin (say, less than 1 mm). The minimum weight of add-on shielding is obtained by small dense local shields (i.e. large solid angle subtended by small mass). For trade-off purposes, a figure of merit for shielding in units such as 'rads per gram' should be utilized.

15.4.3.2 *An example of shield weight trade-off*
For long-term geostationary missions, as shown in Fig. 15.8, the control of radiation-tolerance of devices is critical. In the example illustrated, a sensitive component is placed inside a 5 litre cubic aluminium box with a 4.5 mm wall thickness. The time required for the maximum acceptable dose, $D_{A(max)}$, to be reached is shown as a function of added shielding. For a device with a $D_{A(max)}$ value of 10^3 rad (10 Gy), it is impossible to add sufficient shielding to make the device survive the desirable 7-year mission. Even with thick shields (3 kg), it will barely survive half the mission. Raising $D_{A(max)}$ to 3×10^3 rad (30 Gy) allows 7-year survival by applying less than 1.5 kg of shielding; when $D_{A(max)}$ is raised to 5×10^3 rad (50 Gy), only about 0.5 kg will be necessary. The 'hardening' has occurred at the best point in the system, namely in the device. The cost penalty of 'hardening' to this rather low level may amount only to that incurred by the radiation testing involved. The cost saving in weight reduction or life enhancement may be many times the cost of a well-designed series of laboratory tests.

Figure 15.9 shows similar curves for an equipment box in a highly elliptical scientific orbit. Here, unlike the geostationary case, the most sensitive device can survive the mission with 2 kg of shielding while, with $D_{A(max)} = 10^4$ rad (100 Gy), scarcely any special absorber is necessary. The curves shown are, of course, only examples and specific to one particular component location on one space vehicle. However, they present the combined information on sector analysis, testing results (including recovery), physics theory, and circuit analysis in a concise form that the system engineer will find useful.

In summary, 'add-on shields' can be classified as 'local', 'whole-board' and

404 Equipment hardening and hardness assurance

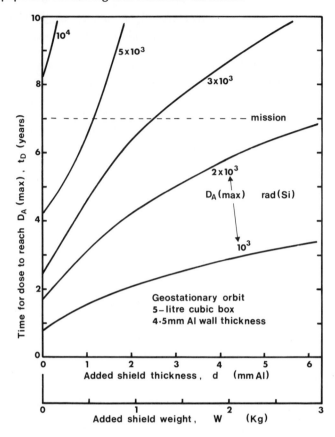

FIG. 15.8 Trade-off of shielding vs. life, geostationary orbit: the time taken to reach the 'maximum acceptable dose', $D_{A(max)}$, for certain MOS devices (see Chapter 4) as a function of thickness or added weight of shielding for values of $D_{A(max)}$ in the low kilorad range. The component is within a 5 litre cube with initial wall thickness of 4.5 mm aluminium; the 'add-on' material is uniform.

'whole-box'; each may be needed at different times. Modularity in these shields is useful, but not essential. The shields used must be simple and easily applied. There is a strong incentive to apply the shielding only to the volume which really requires it (e.g. only one area of a printed-circuit board); this arises from the cubic relation between weight and linear dimension. In different cases, the shield may be poured or moulded polymer, aluminium or a very dense metal. Attachment to the device lid by adhesive has been found acceptable for some spacecraft. Device packages using layered shields have been designed (Schmid et al. 1985)

Shield weight trade-offs can also be made in the case of solar cells, where cover glass thickness is traded against end-of-life power of the array

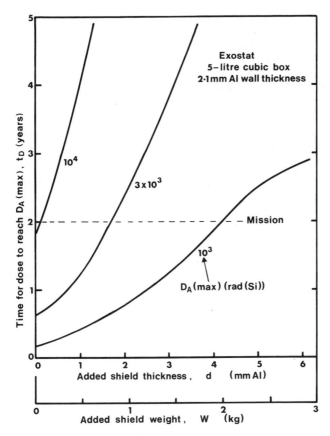

Fig. 15.9 Trade-off of shielding vs. life, elliptical orbit: as for Fig. 15.8.

(Tada *et al*. 1982). The subject of shielding for humans in spacecraft is covered in a separate body of literature (e.g. Barnea *et al*. 1987). The case of the European manned spacecraft, Hermes, is described by Comet (1991).

15.5 On-board radiation monitoring

The predictive methods described in this handbook inevitably involve a large number of approximations and may thus be subject to quite large percentage errors in predicting radiation exposure at a given dose point and the resulting device degradation. In spacecraft, just as thermal control calculations are ultimately verified by the installation of thermocouples, and telemetry of the temperatures, so it is desirable to install small radiation monitors in equipment boxes both to check predictions and to provide operations staff with guidance on the status of equipment during the mission (see Adams *et al*. 1991*a* and *b*; Sections 1.5 and 15.4.2). A similar argument applies in robots military equipment and accelerator experiments, since the radiation environment will be very difficult to predict and only an integrating detector will give a reliable reading of the accumulated dose.

15.6 Hardness assurance

15.6.1 General

This final section covers the science, or perhaps the art, of hardness assurance, which we broadly define as the management of radiation effects. This section is drawn from experience derived mainly from the field of civil/scientific space activities, but the general approach is applicable to many terrestrial problems, such as robotics in the nuclear industry. We do not address the military assurance field specifically. The concept of hardness assurance derives from the military field, where it is a well-established discipline (see for example Messenger and Ash 1986), and many civil standards derive from military ones. Having said this, we point out that military procedures are widely referred to in other chapters. There, the reader will find that the military scenario is unique and requires specialized assurance rules. The 'first line' of military defence is a hardened system built using hardened components; certain elements of the military 'threat' associated with thermonuclear weapons are unique; the costs of system failure are largely unacceptable. We thus acknowledge the efforts of the military community, which laid the foundations of hardness assurance, and led to a good understanding of many of the basic physical mechanisms of radiation damage. However, we cannot directly transfer the management techniques and engineering approaches into the space and terrestrial sectors. In the latter sectors we are primarily concerned with ensuring the survival of what are essentially commercial systems built with commercial components; the cost of failure is more easily quantified and accepted. It follows that, at least for space projects, a somewhat different approach from that of hardened military systems has evolved but absorbed the experience gained in the latter.

15.6.2 Hardness assurance defined

Hardness assurance impinges on all areas of a programme: design, manufacturing, product assurance, and procurement. Hence hardness assurance should be fully integrated into a project life-cycle from the feasibility study phase onwards. The later in a project that hardness assurance is considered, the more costly is the impact if problems are uncovered.

The basic steps in a hardness assurance programme are:

(1) definition and analysis of the 'threat', the 'threat' being the radiation environment of the system, subsystem or component;
(2) definition of the system and/or component response by test and analysis;

(3) definition of 'countermeasures';
(4) implementation and verification of 'countermeasures';
(5) monitoring of the efficiency of 'countermeasures' during the mission;
(6) the transfer of experience to subsequent hardness assurance programmes.

Items (5) and (6), despite being extremely important, are probably the most neglected elements in hardness assurance programmes. The difficulty is that projects tend to operate on internal lines only and under extreme pressure. The 'fixing' of a project problem tends to be a limited exercise. Even if the 'fix' would be of interest and use to other projects, a project organization cannot afford the time to disseminate the information. Frequently the mechanisms for such dissemination are lacking.

15.6.3 Management of hardness assurance

There are many ways of managing the process of hardness assurance described above. The chosen technique is largely a function of the existing management structure within a laboratory or agency and how people communicate. The options are a centralized radiation-effects management or a devolved management, in which subsystem engineers are responsible for radiation effects in the same way as they are responsible for other environmental constraints. In either case, the most effective approach is by means of 'requirements and reviews'. Requirements are reflected in a 'radiation control plan' which defines the external environment, the techniques to be used to convert this into an environment at component location, and the radiation design margin required. Subsystem engineers are then responsible for performing the necessary actions to ensure that requirements are met. For the review process it is recommended that radiation be part of the normal preliminary design review (PDR) and critical design review (CDR) process. Certain intractable radiation problems may require the attention of a 'radiation control committee'. Robinson (1987) recommends that such meetings and bodies should be limited in membership to technical personnel, otherwise they tend to take on a life of their own.

In discussing hardness assurance management we need to be aware how space project hardware comes into being. Robinson (1987) identifies three categories which cover all eventualities:

(1) new design, new hardware;
(2) inherited design;
(3) inherited hardware.

The previously described approach of 'requirements and review' applies to all three categories but, assuming that the hardware has not previously been characterized for radiation, the changes imposed by radiation are progressively greater in scope. In the the first category, *'ab initio'* management is likely to be smooth; in the second category, design changes are likely; in the third category, design changes, component changes and added radiation shielding may all be needed.

15.6.4 Databases

Hardness assurance requires effort dedicated to databases, otherwise vital experience generated during a project is lost to future projects. The importance of an effective database has been recognized by military authorities, NASA, ESA, and terrestrial projects such as the EEC's TELEMAN (robotics for the nuclear environment).

Each project starts by obtaining some benefit from existing data and it is then the responsibility of each project to add to the database for the benefit of its successors. The authors have been, and are, active in attempting to ensure some coordination and commonality between databases (see Chapter 13). With the current status of information technology it is not difficult to transfer data between databases, assuming that agreement on test methods, reporting formats, and data verification is obtainable. A considerable amount of effort is required, but is fully justified on the grounds of economy and efficiency of management.

15.6.5 Parts procurement and radiation design margins

Messenger and Ash (1986) note that most of the difficulties in the radiation-hardening of systems lie with the electronic portions of the system and that the major emphasis of a hardness assurance programme is electronic parts assurance. The hardness of any electronic system will only be as good as the hardness of the individual electronic parts. The 'traditional' approach to parts hardness assurance, as derived from military programmes, requires: process and production control during manufacture, including the monitoring of in-process controls; hardness assurance verification testing by manufacturer and contractor; and a statistical approach to determining parts hardness capability (Messenger and Ash 1986).

Pease *et al.* (1990) note that the 'traditional' approach assumes that a radiation effects database exists, that technologies and parts are mature, and that the mechanisms for radiation-induced failure are well understood. They further comment that the 'traditional' approach is more difficult for space equipment projects than for military because:

(1) the production requirements are small and the production cycle is short;

(2) some aspects of the environment are difficult to simulate;

(3) the required system survivability and confidence levels are high;

(4) the budget for testing and hardening tends to be low.

We note that, in the 1980s, a common approach to parts hardness assurance for space projects emerged from a number of laboratories. This

15.6 Hardness assurance

approach is based on the categorization of parts, control of parts selection by means of a preferred parts list (including radiation criteria) and the application of radiation verification tests performed on samples from delivered batches of parts. Parts are categorized in relation to a project requirement which combines estimates of the internal radiation environment with a rationally chosen value for a radiation design margin (RDM).

Dose values for the internal environment are set using a number of assumptions regarding shielding. Even before final spacecraft layout is decided, it can be agreed that all components can if necessary be protected by material of some sort up to a mass equivalent to say, 15 mm of aluminium. Another project planned for exactly the same external environment may decide that protection can be guaranteed with only an equivalent of 8 mm of aluminium. Similarly, one project may decide on a conservative RDM of 3, while another may decide that an RDM of 1.5 is justified. The differences may be due to perfectly good reasons based on technology, economics or risk.

The radiation tolerance values used for parts are derived from data-bases, manufacturers' data and, occasionally, generic data (i.e. data based on related families of parts). The use of generic data with any degree of confidence requires the use of a conservative RDM, generally between 5 and 10.

Four categories for the acceptability of parts are commonly used:

(1) reject;
(2) accept with caveat that a lot acceptance test is required;
(3) accept without caveat;
(4) no data: evaluation tests required.

It should be possible to transfer category (4) parts to one of the other three categories on completion of evaluation tests. However, because of known batch-to-batch variations, a part placed in category (3) requires a similar RDM to that advised for 'generic' data. This may not be a constraint: an evaluation test which gives a hardness level of 100–150 krad against an internal environment of 20 krad would probably not require lot acceptance tests, but lot testing would be required if the hardness level were 30–40 krad (300–400 Gy).

There is no general agreement as to the degree of circuit complexity which should be subjected to an active radiation test (i.e. what constitutes 'a part'). ESA experiments have found that some hybrid circuits survive to dose values higher than would be expected from the part failure criterion for the CMOS circuits of known sensitivity. Clearly the circuit network has been designed with greater tolerance to, for example, leakage than assumed in the part criterion.

A new approach to parts hardness assurance is based on the 'qualified manufacturers list (QML)' philosophy currently under discussion in the

USA. This is proposed as an alternative to the 'qualified parts list (QPL)' used for the qualification of products and suppliers in some high-reliability military and space application. The QML approach is based on the quality of a manufacturer's capability to produce high-reliability products by following a defined set of process and 'design rules' with consistency. The QML approach seeks to eliminate 'class designations' as used previously in quality assurance and radiation-hardness assurance procedures. The European equivalent to QML is known as 'capability approval'. The concept of QML and the reasons for its introduction have been described by Alexander (1990) and may be summarized as follows:

1. A manufacturer may be prepared to adopt 'total quality management' (TQM).
2. Design authorities are increasingly prepared to use application-specific integrated circuits (ASICs).
3. Deficiencies in 'QPL' have been identified: qualification takes a prohibitively long time and testing and documentation requirements for the QPL approach are prohibitively expensive.

The QML approach would require that tolerance to radiation form part of all design and processing aspects of most high-technology components. Certain measurable parameters must then be identified which have a direct bearing on the ultimate hardness of a product. These parameters may require the use of 'surrogate' or test structures and chips (see Chapter 14). Winokur *et al*. (1990) studied the Sandia CMOS process from the QML point of view and found the QML approach to be technically feasible. A number of changes in the normal 'Quality Control' approach were required and major importance was attached to the identification of simple 'test structures' providing near-perfect models of the changes which jointly led to the radiation failure of a complex design. The concept was proved but a significant amount of research was required. The general impression in the industry appears to be that QML hardness assurance could well be a general solution but that it may impose an unacceptable burden on a manufacturer. It is our opinion that it may be some considerable time before the QML approach to hardness assurance may be shown to be viable, and still further time will be required to see how many manufacturers decide to 'sign up' for the procedure. In the early 1990s, technology evaluation, design verification, and lot acceptance procedures will continue to be used.

15.6.6 Economics of hardness assurance

15.6.6.1 *Programme costs*

We tend to agree with Messenger and Ash (1986) that the costing of a radiation hardness assurance programme can be a 'vexing topic'. It is difficult to find quantitative data on hardness assurance costs. Hardness assurance is a

15.6 Hardness assurance

multidisciplinary activity extending over the areas of design, product assurance, engineering, procurement, and test. Because of this wide spread of activities there are many hidden costs, that is, ones which are difficult to trace.

Robinson (1987) suggests the following 'mission cost ratios' for hardness assurance:

Proof of concept	1
6 months mission	1.1–2
1 to 3 year mission	3–4
Mission >3 years	4–5

For the last two categories these costs are broken down into 'sub-programmes' as a percentage of total programme cost as follows:

Cost of parts purchased	5–10 per cent
Quality assurance	10–15 per cent
Tests	5–10 per cent

For the programmes with which the authors have been associated, the total product assurance costs have been in the range of 12 to 15 per cent of the programme cost and hardness assurance is possibly a third of this. This leads to an estimate for hardness assurance in the region of 5 per cent of programme cost, that is, a cost about one seventh of Robinson's estimate.

The disparity in these estimates comes from dissimilarity in programme types. Robinson (1987) was discussing JPL programmes which are long-duration interplanetary missions with severe radiation requirements. Our experience has been with programmes having radiation requirements in Earth orbit and, possibly, employing less sensitive component technology. The disparity also underlines the problem of defining exactly what activities are properly charged to a hardness assurance budget. In our case much of the background research, development of test methods and, indeed, a significant amount of component testing does not appear as a programme cost but is considered to be 'infrastructure'. We believe that in JPL much of what we know as 'infrastructure' is considered as programme cost.

Using the propositions of Robinson (1987) and our own experience we have constructed Fig. 15.10 as a qualitative view of trends in hardness assurance costs. The curve shows a fairly rapid increase in cost when moving from a 'proof of concept' or flight experiment to a project intended to operate for 3 years reliably. The increase is a combination of analysis, engineering, and test costs. The cost increase in going from a 3 year project to a project of 5–10 years is not large since, in our experience, the major increase is in the parts selection effort, satisfied usually without more testing. Projects requiring more than ten years' operation tend to be in the 'big league' such as interplanetary, multi-user communications or scientific observatory projects. The users often demand very high reliability and hence a higher Radiation

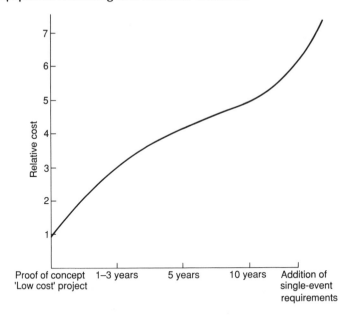

FIG. 15.10 Cost of hardness, short and long missions: a model for the relative costs of hardness assurance applied to an unmanned spacecraft; absolute values of cost vary with project and country. Costs increase with the required mission duration.

Design Margin—a form of overdesign to compensate for uncertainty. Any mission which demands high immunity to single-event upset and latch-up incurs a significant increase in engineering and test costs. In Fig. 15.11 we have included an extension to the cost curve under the name of 'retrofit'. Any late changes in component, sometimes called 'retrofits', can cause a dramatic cost increase occasioned by engineering, testing, and requalification. Occasionally, the design and manufacture of special semiconductor parts may be required.

In the development of equipment, the costs of hardness assurance depend not only on the level of assurance but also on the point in development—the phase—at which hardness assurance is introduced into a programme. This is the message of Fig. 15.11: hardness assurance is the least costly when introduced as part of technological research and feasibility study. A space project may be studying the feasibility of a mission in the radiation belts, or a terrestrial project may be discussing the use of robotics in a very severe environment. Radiation response modelling, done in good time, may divert the project from embarking on an impossibly expensive enterprise to a feasible one.

Up to 'engineering model' (EM) using a normal 'military' quality level for the selection of parts, the costs of introducing hardness in the first

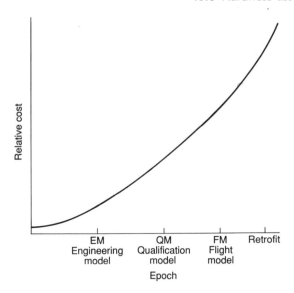

FIG. 15.11 Cost of hardness—expensive if late: a model for the relative costs of hardness assurance applied to unmanned spacecraft. Absolute values of cost vary with project and country. Costs increase if the requirements are introduced at later stages.

paper designs are tolerable and such changes during development are normal. The costs of introducing hardness assurance increase rapidly once the EM design is frozen, if only because justification for, and documentation of, a change in a 'frozen' design requires significant efforts. Costs increase even more steeply after the qualification model (QM) is built, as significant changes require extensive retesting.

15.6.6.2 *Testing costs*

The cost of radiation testing of parts can be costed straightforwardly. A test performed under gamma rays as part of an evaluation and selection of parts for a piece of space equipment costs in the region of a few thousand US dollars, maybe as high as $10 000 if there is significant engineering effort connected with the test. A 'lot acceptance' test under gamma rays would be in the range of $2000 to 5000 per lot. The costs of 'lot acceptance' for standard 'rad-hard' parts are often absorbed in the manufacturer's unit cost. In the case of a specially requested lot acceptance, the cost is strongly influenced by the level specified and the amount of parameter degradation allowed (this determines the manufacturing yield). This point is not always properly appreciated, and it is important that the dose level and parameter degradation are properly determined by reference to design constraints and are pronounced feasible by the parts manufacturer.

414 Equipment hardening and hardness assurance

Too severe a requirement for lot acceptance can lead to unnecessarily high costs.

A single-event test on a moderately complex part, using a particle accelerator, could cost in the region of $50 000 to 150 000. Such a test requires significant engineering effort as well as a test team to work at a high-energy ion accelerator site for one or two weeks. Due to the high cost of an ion beam facility, it makes sound economic sense to collect candidate parts for sone time in order to have a high throughput when the test campaign begins.

15.6.6.3 Sample sizes

In hardness assurance management, one is frequently faced with the problem of defining design margins and selecting sample sizes for test. The larger the sample, the greater the assurance of hardness but also the greater are the costs of acquisition and testing. There exists a statistical relation between the following variables:

(1) the sample size chosen for testing a part;
(2) the design margin and confidence level chosen;
(3) the spread expected in a device parameter when affected by radiation.

A formal parts characterization criterion (PCC) for the likelihood of failure under a given mission irradiation condition has been developed by a US space parts working group (Pease 1992). This links mathematically the three variables listed above. A mean dose for failure, R_F, and its standard deviation are determined from a lot of size n devices, tested to failure. From this, a variable S is calculated. Desired confidence factors are established, which are combined in a variable K. The part criterion is then a number

$$PCC = e^{Ks}$$

The specification dose, R_S, is established. Then, if the expression

$$\frac{R_F}{PCC \times R_S}$$

is greater than one, there is a critical danger of failure in the mission (Pease 1992). Above a PCC value of one, the expected danger decreases. But, as noted, K is sensitive to sample size, n, up to a value between 10 and 20. Reducing n below 10 gives high values of PCC and hence of expected danger. This fact will influence experimenters to test at least 10 samples (see e.g. Section 13.6.3) but will also allow him to calculate confidence levels when sample numbers are for some reason smaller e.g. when very costly devices are to be tested. Robinson (1987) explained the statistical basis of the calculation of the probability of failure of a part under mission radiation, based on a given sample size and design margin, RDM. RDM is the nominal failure dose, R_F, of the part, divided by the mission specified dose, R_S. The ambition of the designer is always to have a good margin i.e. for the value of RDM to be much greater than 1. The recipe for restful nights is an RDM value well above 2.

15.7 Conclusions

The practices recommended here for inculcating radiation-tolerance in equipment require advanced physical modelling techniques, precise engineering procedures, and firm assurance procedures. The degree to which these procedures should be used in an equipment project can be measured by the severity of the raw radiation environment, the desired reliability of the system, and the requirement of that project for radiation-sensitive technologies. The balance of device/circuit design versus shielding will depend on whether the radiation is highly penetrating — as in isotope handling or military environments — or readily attenuated, as in space.

In this chapter we have attempted to summarize the essential features of the radiation hardening of large assemblies of equipment and how effective control of radiation effects, that is hardness assurance, can be implemented. There is no clear path to success. As we have shown throughout this book, radiation effects present extremely complex problems, embracing device physics, design, engineering, product assurance, procurement, and project management. Following the very effective European Space Agency practice, it is often cost-effective for a product assurance authority to issue a set of guidelines for hardness engineering (see ESA 1989). To quote Robinson (1987), following Curran*: *'Hardness Assurance can be achieved, but it is a bit like liberty:- The price of liberty is eternal vigilance'*.

References

Abel, E., Brown, M. H. *et al.* (1991). The NEATER telerobot. In Proceedings of the ANS 4th Topical Meeting on Robotics, .
Adams, L. and Holmes-Siedle, A. G. (1991*a*). La survie de l'electronique dans l'espace. *La Recherche*, **22**, 1182-9.
Adams, L., Daly, E. J., Harboe-Sorensen, R., Holmes-Siedle, A. G., Ward, A. K. and Bull, R. A. (1991*b*). Measurements of SEU and total dose in geostationary orbit under normal and solar flare conditions. *IEEE Transactions on Nuclear Science*, **NS38**, 1686-92.
Alexander, D. R. (1990). Implications of qualified manufacturers list reliability procedures for radiation hardness assurance. In IEEE Nuclear and Space Radiation Effects Conference Short Course, Reno, NV. IEEE, New York.

---

*John Philpot Curran, in his *Speech on the right of election of the Lord Mayor of Dublin*, 10 July 1790: 'The condition upon which God hath given liberty to man is eternal vigilance; which condition if he break, servitude is at once the consequence of his crime, and the punishment of his guilt'.

Araragi, M. and Nuraki, C. (1992). Planar field emission devices with three dimensional gate structures. *Journal of Micromechanics and Microengineering*, **2**, 5-9.

Barnea, G., Berger, M. J. and Seltzer, S. M. (1987). Optimization study of electron-bremsstrahlung shielding for manned spacecraft. *Journal of Spacecraft*, **24**, 158-61.

Baur, J. F. et al. (1981). Radiation hardening of diagnostics for fusion reactors, General Atomics Report GA-A16614. GA Technologies, San Diego, CA.

Binkley, D. M. (1982). A radiation hardened accelerometer preamplifier for 2×10^8 rad total dose. *IEEE Transactions on nuclear science*, NS-29, 1500-7.

Bowers, J. C. and Sedore, S. R. (1971). *SCEPTRE: a computer program for circuit and systems analysis*. Prentice-Hall. Englewood Cliffs, NJ.

Chilton, A. B., Shultis, J. K. and Faw, R. E. (1984). *Principles of radiation shielding*. Prentice-Hall. Englewood Cliffs, NJ.

Comet, B. (1991). Evaluation of the radiation risk of HERMES astronauts. In Proceedings of Radiation Effects in Electronic Components and Systems, Paper IP2. Université de Montpellier, France.

Crowley, J. L., Junga, F. A. and Schulz, T. J. (1976). Technique for the selection of transient radiation hard junction isolated circuits. *IEEE Transactions on Nuclear Science*, **NS23**, 1703-8.

Dennehy, W. J., Holmes-Siedle, A. G. and Zaininger, K. H. (1969). Digital logic for radiation environments. *RCA Review*, **30**, 668-98.

De Nordwall, H. J. and Schaller, K. H. (1984). A Community view of robotics for nuclear engineering. In Proceedings of International Conference on Research for Environmental Extremes, CESTA, Marseilles.

Dupont-Nivet, E. et al. (1991). A hardening technology on SOI for analog devices. In Proceedings of Radiation Effects in Electronic Components and Systems, Université de Montpellier, France.

Emeriau, J. (1991). TELEMAN RATOCA Project No. TM34. Progress Report to EEC. Sofretec, Bezons, France.

ESA (1989). *Radiation design handbook*, ESA Document PSS-01-609 ESTEC, Noordwijk, The Netherlands.

Friebele, J. (1991). Photonics in the space environment. In IEEE Nuclear and Space Radiation Effects Conference (Short Course), San Diego, CA. IEEE, New York.

Gough, B. I. (1978). Modular packaging for on-board units. *Proceedings of the International Conference on Spacecraft On-Board Data Management*, Nice, 24-27 Oct. ESA SP-141, European Space Agency, Noordwijk, The Netherlands.

Gray, H. F. (1990). Vacuum microelectronics – hope for ultra-radiation hardness. *Invited talk, IEEE NSREC '90*, Reno, Nevada.

Halpin, J. J. et al. (1976). Guidelines for nuclear effects survivability for tactical Army systems, Report No. HDL-PRL-76-2. Harry Diamond Laboratory, Adelphi, MD.

Holmes-Siedle, A. G. (1965). Space radiation – its influence on satellite design. *RCA Engineer*, **10**, 100-5.

Holmes-Siedle, A. G. (1974a). Radiation effects in the fusion power programme. *Nature*, **251**, 191-6.

Holmes-Siedle, A. G. (1974b). Radiation-induced ionization effects in solids – a review of research problems. *Proceedings of the IEEE*, **62**, 1196-207.

Holmes-Siedle, A. G. (1980). Radiation effects in the Joint European Torus experi-

ment: guidelines for preliminary design. Report R857/2. Fulmer Research Laboratories Ltd, Stoke Poges, UK.
Holmes-Siedle, A. G. (1985). Radiation effects in remote handling equipment. *Nuclear Industry Requirement for Robots*, (ed. I. D. Salter and D. P. McCabe). Taylor Hitec Ltd, Chorley, UK.
Holmes-Siedle, A. G. and Adams, L. (1986). RADFETS: a review of the use of metal-oxide-silicon devices as integrating dosimeters. *International Journal of Radiation Physics and Chemistry*, **28**, 235-44.
Holmes-Siedle, A. G. and Poch, W. J. (1971). The design of a weather satellite for the radiation environment — a case history. *Journal of the British Interplanetary Society*, **24**, 273-88.
Holmes-Siedle, A. G., Engholm, B. A., Battaglia, J. M. and Baur, J. F. (1984). Damage calculation for devices in the diagnostic penetration of a fusion reactor. *IEEE Transactions on Nuclear Science*, **NS31**, 1106-12.
Holmes-Siedle, A. G., Leffler, J. S., Lindgren, S. R. and Adams, L. (1990). The RADFET System for real-time dosimetry in nuclear facilities. *Proc. 7th Annual ASTM — Euratom Symposium on Reactor Dosimetry, Strasbourg, August 27-31 1990*. (Kluwer, Dordrecht 1992) pp. 851-9.
IEEE (1974). *Qualifying Class 1E Equipment for Nuclear Power Generating Stations*, IEEE Standard 323-1974. Institute of Electrical and Electronics Engineers, New York.
Johnson, R. T., Thome, F. V. and Craft, C. M. (1983). A survey of aging of electronics with applications to nuclear power plant instrumentation. *IEEE Transactions on Nuclear Science*, **NS30**, 4358-62.
Larcombe, M. H. and Halsall, J. R. (1984). *Robotics in nuclear engineering*. Graham and Trotman, London.
Larin, F. (1964). *Radiation effects on semiconductor devices*. Wiley, New York.
Lauridsen, K. (1991). The TELEMAN/ENTOREL project. Proc. RADECS '91, Montpellier, France, Sept 9-12 1991. *IEEE Conf. Record*, Cat. No. 91 THO 400 2, Vol 15, pp. 317-19.
Leszkow, P. and Decréton, M. (1991). Radiation tolerant techniques of multiplexing analogue signals in a nuclear environment. Proc. RADECS '91, Montpellier, France, Sept 9-12 1991. *IEEE Conf. Record*, Cat. No. 91 THO 400 2, Vol 15, pp. 254-8.
Levy, P. and Friebele, E. J. (ed.) (1985). *Critical reviews of technology: optical materials in radiation environments*, SPIE Vol 541. Society of Photooptical and Instrumentation Engineers, Bellingham. WA.
Lindholm, F. (1971). Device characterization for computer analysis of large circuits. *IEEE Transactions on Nuclear Science*, **NS18**, 206-9.
Lynn D. K. *et al.* (1985). Thermionic integrated circuits: electronics for hostile environments. *IEEE Transactions on Nuclear Science*, **NS32**, 3996-4000.
Malmberg, A. F. and Cornwell, F. N. (1964). NET-1 network analysis program, Report LA-3119. Los Alamos Scientific Laboratory, NM.
Messenger, G. C. and Ash, M. S. (1989). *The effects of radiation on electronic systems*. Van Nostrand Reinhold Co., New York.
Mullen, E. G. and Gussenhoven, M. S. (1991). *Results of space experiments: CRRES*. Philips Laboratory Report, PL/PHP. Hanscom AFB, MA, USA.
Pease, R. L., Johnston, A. H. and Azarewicz, J. L. (1990). Radiation testing of semiconductors for space electronics. In IEEE Nuclear and Space Radiation

Effects Conference Short Course, Reno, NV. IEEE, New York.

Pease, R. and Alexander, D. (1992). Hardness assurance for space microelectronics *International Journal of Radiation Physics and Chemistry*, Special issue on space radiation environment and effects, to be published.

Pigneret, J. (1985). The chopped X-ray source concept: an efficient method for TREE analysis and control. *IEEE Transactions on Nuclear Science*, NS32, 4270-5.

Price, B. T., Horton, C. C. and Spinney, K. T. (1957). *Radiation shielding.* Pergamon Press, Oxford.

Ray, K. P., Mullen, E. G., Stapor, W. J., Circle R. R. and McDonald, P. T. (1992). CRRES dosimeter results and comparison using the Spacerad dosimeter and p-channel dosimeters. Paper PD-1, *IEEE NSREC '92*, New Orleans, July 1992.

Retzler, J. P. (1973). *General hardness design guidance for design engineers.* Litton Guidance and Control Division, Woodland Hills, CA.

Ricketts, L. V. (1972). *Fundamentals of nuclear hardening of electronic equipment.* Wiley, New York.

Robinson, P. A. (1987). Packaging testing and hardness assurance. In IEEE Nuclear and Space Radiation Effects Conference Short Course, Snowmass, IEEE, New York.

Rudie, N. J. (1980). *Principles and techniques of radiation hardening.* Western Periodicals Publication Vol. 3, N. Hollywood, CA.

Salter, I. D. and McCabe, D. P. (1985). *Nuclear industry requirement for robots.* Taylor Hitec Ltd, Chorley, UK.

Schmid, J. P., Strobel, D., Mullis, J., Merker, M., Spratt, J. and Featherby, M. (1985). Microcircuit packages incorporating radiation shields for use in space-craft electronics. In Proceedings of International Society for Testing and Failure Analysis (ISTFA), pp. 111-3.

Sharp, J. and Dumbreck, A. (1991). A standard method for the radiation testing of solid-state sensor based, CCTV cameras. In Proceedings of Radiation Effects in Electronic Components and Systems, pp. 9-12. Université de Montpellier, France.

Soli, G. A., Blaes, B. R., Buehler, M. G., Ray, K., and Lin, Y-S. (1992). CRRES microelectronic test chip orbital data. Paper D-6, *IEEE NSREC '92*, New Orleans, July 1992.

Spindt, C., Brodie, I., Humphrey, L., and Westerberg, E. R. (1976). *Journal of Applied Physics*, 47, 5248.

Srour, J. R., Long, D. M., Millward, D. G., Fitzwilson, R. L. and Chadsey, W. L. (1984). *Effects on and dose enhancement of electronic materials.* Noyes Data Corp., Park Ridge, NJ.

Tada, H. Y., Carter, J. R., Anspaugh, B. E. and Downing, R. G. (1982). *Solar cell radiation handbook*, (3rd Edn), NASA-JPL Publication 82-69. TRW Systems Group and Jet Propulsion Laboratory, Pasadena. CA.

Thatcher, R. K. and Kalinowski, J. J. (1969). *TREE handbook.* Battelle Memorial Institute, Columbus, OH.

van Lint, V. A. J., Raymond, J. P., Hart, A. R. and Price, M. L. (1977). *Radiation design handbook for the Jupiter probe*, Report No. MRC/SD-R-14. Mission Research Corp., San Diego. CA.

Winokur, P. S. *et al.* (1990). Implementing QML for radiation hardness assurance. *IEEE Transactions on Nuclear Science*, NS37, 1794-1805.

16
Conclusions

The effects of radiation on electronic devices and materials, especially those occurring in LSI devices and optics, provide the engineer with a wide range of practical and theoretical problems. Our aim has been to point out the problems and lead the engineer towards a balance of solutions, set in a review of the theory of radiation effects in solids. At the same time, we have not hesitated to mention management aspects when appropriate, and extend the review so as to serve as a grounding for students. Summary graphs, tables and calculation methods have often originated in projects sponsored by ESA, the EEC, the US Department of Defense and NASA; these were often developed spontaneously as aids in establishing the significance of a radiation environment problem and communicating such problems to management and colleagues.

Because of the fast-moving nature of technology, all we can provide here is a 'snapshot' of the field. Much of the basic material, however, does not change. This basic material represents some 80 per cent of the book and provides a reference text. We hope to have introduced some new ideas in the field of hardness assurance management, the techniques of which are required in many areas of technology development. In this field the space community have borrowed heavily from their military colleagues. We have appropriately modified and developed their techniques for space and they can now perhaps also be applied in the fields of high-energy physics, fusion, processing, and robotics.

Descriptions of the environment, its problematic aspects, and possible solutions have been couched, as far as possible, in simple language, with copious references to original sources, other textbooks and useful further reading. A large fraction of the graphical material is original and gives comparisons of data from several publications or unpublished data. Wherever possible, improvements in analytical methods, developed by the authors and colleagues (especially co-workers at the European Space Agency), have been presented here, with due acknowledgement. In less pretentious words, we have attempted to bring together all the bits and pieces of useful information that we have gleaned in our lives and present these, with our views, in an accessible form.

Among the bits and pieces are clarification of some difficult questions of transmission of particles into electronic boxes. Approximate methods for calculating dose–depth curves are given, with the caveat that they useful for preliminary investigations, but that computer approaches are recommended for detailed equipment design; the latter is acknowledged in the set of dose–depth curves prepared by Eamonn Daly for orbits of importance ('reference missions') and presented in Appendix E. Overall, the aim is to provide in written form our own personal bridges between the mass of scientific reports and the learner. The learner may be a student, an engineer or a manager.

The problem of predicting the responses of advanced electronic devices to radiation is a challenging one because the device physics involved is complex and the field is in continual development. Improvements are still needed in engineering models for the prediction of degradation or pulsed response. For example, a better version of the the Simple Engineering Model for charge buildup in MOS devices is needed. In future, the designer must have professional design tools which allow him greater accuracy in prediction and a clearer choice between circuit alteration, added shielding, and the cost penalties attendant on hardened devices. By force of circumstance, designers will probably have to use many devices that are available only in an unhardened form. Thus, the sections on radiation testing will be useful to system engineers and management.

Looking forward, an increased formal interaction between radiation-effects experts with equipment project groups is foreseen. New trends in microelectronics include stacked geometries of nanometre dimension in silicon and optoelectronic devices with active polymers contributing to the integration of sensors with microcircuits. New trends in nuclear industrial engineering design include a major increase in the use of remotely controlled or robotic machines to increase smartness in 'hot' locations, including unpredictable situations such as Chernobyl and, further away, in full-power reactors based on nuclear fusion. New trends in space include the use of orbital or lunar stations, increased usage of on-board data processing and sensors, and spaceborne power reactors. The goal of uniform radiation-tolerance in space systems (especially interplanetary probes and operational satellites) is gradually being attained in the space field, after major efforts over the past three decades.

The authors hope to pass on the understanding and use of these achievements to developing fields. It is hoped that radiation workers apply that understanding to ingenious engineering techniques, novel research, and better management policies.

Appendix

Appendices A to F

These appendices are issued as technical information, are not a licence to use information, and not a part of a contract. In particular, while all information furnished is believed to be accurate and reliable, no liability in respect of any use of the material is accepted by the authors or Oxford University Press.

A. Useful general and geophysical data

TABLE A1 Conversion factors, physical properties, and constants

Conversion factors

1 year	$= 5.2596 \times 10^9$ min $= 3.15576 \times 10^7$ s
1 day	$= 1.440 \times 10^3$ min $= 8.640 \times 10^4$ s
1000 Å	$= 0.1\ \mu m = 100$ nm
1 mm	$= 0.03937$ in $\simeq 40$ thou
0.001 in $= 1$ mil or thou $= 25.4\ \mu m$	
1 m^3	$= 10^6$ cm^3
1000 cm^3	$= 10^{-3}$ m^3
1 L	$= 1000$ cm$^3 = 0.220$ gal (UK)
1 g cm^{-3}	$= 1000$ kg m^{-3}
1 ev	$= 1.602 \times 10^{-19}$ J
1 MeV	$= 1.602 \times 10^{-13}$ J
1 J	$= 10^7$ erg
1 cal	$= 4.187$ J
1 eV per molecule	$= 23.1$ kcal mole^{-1}
Electronic charge, q	$= 1.602 \times 10^{-19}$ Coulomb
1 Coulomb	$= 6.241 \times 10^{18}$ electrons
1 μA cm^{-2}	$= 6.241 \times 10^{12}$ electrons cm^{-2} s^{-1}
1 N	$= 10^5$ dynes
1 mm Hg	$= 133.322$ N m^{-2}

Physical properties

Permittivity of free space, ϵ_0	$= 8.86 \times 10^{-14}$ F cm^{-1}
	$= 8.86 \times 10^{-12}$ F m^{-1}
	$= 55.4$ electronic charges V^{-1} um^{-1}
Permeability of free space, μ_0	$= 1.26 \times 10^{-6}$ H m^{-1}
Velocity of light, c	$= 2.997925 \times 10^8$ m s^{-1}
Electron rest mass, m_e	$= 9.11 \times 10^{-31}$ kg
Proton rest mass, m_p	$= 1.67 \times 10^{-27}$ kg

Appendix

TABLE A1 Contd

Constants

Boltzmann constant, k		$= 1.381 \times 10^{33}$ J K^{-1}
		$= 6.62 \times 10^{5}$ eV K^{-1}
kT at room temperature		$= 0.0259$ ev
Planck constant, h		$= 6.63 \times 10^{-34}$ J s^{-1}
Avogadro constant, N_A		$= 6.022 \times 10^{23}$ mol^{-1}

TABLE A2 Frequency, wavelength, and energy

Energy × wavelength, $E\lambda = 1.239\,854 \times 10^{-6}$ eV m
Energy + wavenumber, $E/\nu = 1.986 \times 10^{-25}$ J m
Wavenumber + energy, $\nu/E = 8.065\,46 \times 10^{5}$ eV^{-1} m^{-1}
Frequency + energy, $f/E = 2.417\,966 \times 10^{14}$ Hz eV^{-1}

Wavelength of photon of energy:
  1 eV (infrared) $= 1.239\,854$ μm
  10.2 eV (Lyman α) $= 121.554\,31$ nm

Energy of photon of:
  wavelength 1 μm (infrared) = 1.239 854 eV
  wavenumber 50 000 cm^{-1} $= 6.199\,927$ eV

Wavelength of photon of wavenumber:
  50 000 cm^{-1} = 200 nm
  10 000 cm^{-1} =   1 μm
     50 cm^{-1} = 200 μm

Wavenumber of photon of energy 1 eV = 8065.46 cm^{-1} = 806 546 m^{-1}.

TABLE A3 Geophysical and orbital parameters and conversion factors

Average radius of Earth, R_E	= 3959 statute, miles
	= 3438 nautical miles
	= **6371.315 ± 0.437 km**
Altitude of Geostationary Orbit	= 22 284 statute miles
	= 19 360 nautical miles
	= **35 863 km**
Geocentric Distance of Geostationary Orbit	= 6.629 R_E
	= 42 234 km

Being circular and equatorial, the geostationary orbit is a special case of the geosynchronous orbit (period 24 hours).

1 statute mile	= **1.609344 km** = 0.869 nautical miles
1 nautical mile	= 1.852 km = 1.151 statute miles
1 astronomical unit	= 1.496 × 10^{11} m
Gravitational acceleration	= 0.980665 m s^{-2} = 32.1741 ft s^{-2}

B. Useful radiation data

TABLE B1 Radiation units and data

Absorbed dose
1 rad   = 100 erg g^{-1} = 6.25 × 10^{13} eV g^{-1} = 10^{-2} Gy = 1 cGy
1 Mrad = 10^8 erg g^{-1} = 6.25 × 10^{19} eV g^{-1} = 10 kGy
1 Gy   = 1 J kg^{-1} = 100 rad
1 kGy   = 10^5 rad
1 MGy   = 10^8 rad
10^{20} eV g^{-1} = 1.6 Mrad = 16 kGy

Exposure
1 roentgen (R) = 86.9 erg g^{-1} (air) = 2.58 × 10^{-4} C kg^{-1} (air)
1 roentgen of 1 MeV photons = 1.95 × 10^9 photons cm^{-2}
This fluence of 1 MeV photons deposits:
 0.869 rad in air
 0.965 rad in water
 0.865 rad in silicon
 0.995 rad in polyethylene
 0.804 rad in LiF
 0.862 rad in Pyrex glass (80% SiO$_2$)

Biological dose units (traditional and SI)
 0.75 mrem = 7.5 μSv
 1 mrem   = 10 μSv
 2.5 mrem = 25 μSv
 100 mrem = 1 mSv
 1 rem   = 10 mSv
 5 rem   = 50 mSv
 100 rem   = 1 Sv

Biological Quality Factors
X-, gamma-, beta-rays   1
Slow neutrons   5
Fast neutrons and protons   10
Alpha particles   20
HZE particles   20

Radioactivity
1 becquerel (Bq) = 1 disintegration s^{-1} 1 pCi = 37 Bq
1 curie (Ci) = 3.7 × 10^{10} Bq = 37 GBq   1 Bq = 0.0270 pCi

Radioactivity in food, water, etc.
Quantity per unit volume 1 pCi l^{-1} = 3.7 × 10^4 Bq m^{-3}
Quantity per unit mass   1 pCi g^{-1} = 3.7 × 10^4 Bq kg^{-1}

A 1 Ci point source emitting one 1 MeV photon per disintegration gives an exposure of 0.54 R h^{-1} at 1 m
A 1 Ci Co-60 source gives 1.29 R h^{-1} at 1 m.
Photon fluence at 1 m from a 1 Ci point source = 1.059 × 10^{-9} cm^{-2} (assuming 1 gamma-ray per disintegration)

SI units
Recommended SI units are:
Absorbed dose: the gray (Gy) = 1 J kg^{-1}
Exposure: the coulomb per kilogram (C kg^{-1}, no name given)
Radioactivity: the becquerel (Bq) = 1 disintegration s^{-1}
Biological dose: the sievert (Sv) = absorbed dose (Gy) × Quality Factor

TABLE B2 Energy absorption versus photon energy for air[a]

Energy (MeV)	Mass energy absorption coefficient (μ_{en}/ρ) air ($cm^2 g^{-1}$)	Photon fluence per roentgen ($10^8\ cm^{-2} R^{-1}$)
0.05	0.0406	267
0.10	0.0234	232
0.50	0.0296	36.7
1.0	0.0278	19.5
5.0	0.0174	6.24
8.0	0.0152	4.47
10.0	0.0145	3.75

[a] From H. E. Johns and J. R. Cunningham, *The physics of radiology*, Courtesy of Charles C. Thomas, Publisher, Springfield, IL, 1969, Table A3.

TABLE B3 Typical performance figures for high-energy radiation sources

Source	Beam current	Typical energy deposition
Ultraviolet		
250 W mercury arc	–	$10^7\ erg\ cm^{-2} s^{-1}$ over 2 cm spot ($5.2 > h\nu > 3.4$ eV)
100 W hydrogen arc	–	$10^4\ erg\ cm^{-2} s^{-1}$ over 2 cm spot $12 > h\nu > 9$ eV)
X-Rays		
40 keV	20 mA	10^6 rad h^{-1} over 1 cm spot
Gamma-rays		
Cobalt-60 cells	–	10^5 rad h^{-1} over $>10^3\ cm^3$
Spent fuel rig	–	10^6 rad h^{-1} over $>10^3\ cm^3$
Electrons		
4 MeV LINAC	20 μA	5×10^8 rad h^{-1} in 5 cm spot
0.5 MeV VdG	10 μA	5×10^9 rad h^{-1} in 2.5 cm spot
100 keV TEM	20 μA	10^8 rad s^{-1} over 2 cm spot or 10^{12} rad s^{-1} over 20 μm spot
30 keV SEM	0.1 nA	10^2 rad s^{-1} over $10^{-1}\ cm^2$ or 10^8 rad s^{-1} over $10^{-6}\ cm^2$
3–15 keV Betaprobe	0.5 μA	10^9 rad s^{-1} in 1 mm spot
Fast neutrons		
Test reactor	–	$10^{14}\ cm^{-2} s^{-1}$ over 100 cm^3
Ions		
3 MeV protons	2 μA	$10^{13}\ cm^{-2} s^{-1}$ over 1 cm spot
0.4 MeV protons	0.2 μA	$10^{12}\ cm^{-2} s^{-1}$ over 1 cm spot
Ions 0.005–3 Mev	1 μA	$10^{13}\ cm^{-2} s^{-1}$ over 1 cm spot

TABLE B4 Typical photon energies and wavelengths

	Energy $h\nu$ (eV)	Wavelength, λ (nm)
Gamma rays		
Co-60 gamma	see Table B5	
1 MeV photon	10^6	1.2398541×10^{-3}
X-rays		
100 keV photon	10^5	1.2398541×10^{-2}
Tungsten $K_{\alpha 1}$, 59 keV	5.931824×10^4	2.0901×10^{-2}
Hydrogen lines		
Lyman α line	10.198785	121.56681
H_2 160.8 nm line	7.7105	160.8
Mercury lines		
Hg 253.7 nm	4.8871	253.7
Hg 312 nm	3.9739	312
Hg 365 nm	3.3969	365
Hg 436 nm	2.8437	436
Visible light		
Violet	3.10–2.92	400–424
Blue	2.92–2.52	424–491
Green	2.52–2.16	491–575
Yellow	2.16–2.12	575–585
Orange	2.12–1.92	585–647
Red	1.92–1.77	647–700
Krypton line (primary wavelength standard)	2.046 706	606.78021
Near Infrared		
1 eV photon	1.0	1239.8541 (1.24 μm)
1 μm photon	1.2398541	1000.00 (1 μm)
CO_2 line	0.116967	10 600.00 (10.6 μm)
Thermal energy	0.025	49 594.16 (49.6 μm)
Far infrared		
100 μm photon	0.012398 541	100 000.00 (100 μm)
HCN line (44.859 cm^{-1})	0.0055764	222 292.07 (222.292 μm)

TABLE B5 Radioisotopes useful in radiation experiments: main emission energies[a]

Nuclide	Half-life (years)	Photons		Particles	
		Energy (MeV)	Fraction emitted (%)	Energy (MeV)	Transition probability (%)
Co-60	5.27	1.173 1.333 (av. 1.25)	99.86 99.98	0.318 1.491	99.1 0.1
Ir-192	0.526 (192 d)	0.296 0.308 0.316 0.468 0.604 0.612	29.6 30.7 82.7 47.0 8.2 5.3	0.530 0.670	42.6 47.2
Cs-137	30	0.662 0.032 −0.038 (Ba-137 K X-rays)	85.1 8	0.512 1.174 plus internal conversion electrons 0.65 MeV	94.6 5.4
Sr-90 + daughter	28	0.54	100	–	–
Y-90	0.176	2.27	100	–	–
Kr-85	10.6	0.15 0.67	0.7 99.3	0.51 –	0.7 –
Cf-252	2.65	–	– Neutrons Alphas Fission fragments	– : 2 MeV : 5.9–6.1 Mev : 80 and 104 Mev	–

[a] The Radiochemical Centre, Amersham, UK 1977

TABLE B6 Practical ranges (R_p) of electrons in aluminium[a]

Particle Energy (MeV)	Practical range (g cm^{-2})	(mm)	Particle energy (MeV)	Practical range (g cm^{-2})	(mm)
0.001	0.000012	4.446×10^{-5}	**1.0**	**0.42**	**1.56**
0.003	0.000038	1.408×10^{-4}	1.25	0.52	1.92
0.005	0.000080	2.964×10^{-4}	1.40	0.60	2.22
0.010	0.00017	6.299×10^{-4}	1.50	0.65	2.41
0.030	0.0015	0.0056	1.75	0.80	2.96
0.050	0.0041	0.015	2.0	0.95	3.52
0.100	0.0135	0.050	2.15	1.00	3.705
0.125	0.020	0.074	2.5	1.20	4.45
0.250	0.058	0.214	3.0	1.45	5.37
0.300	0.078	0.289	3.5	1.75	6.48
0.375	0.110	0.408	4.0	1.9	7.04
0.500	0.165	0.611	4.5	2.3	8.52
0.540	0.20	0.741	5.0	2.6	9.63
0.625	0.22	0.815	5.5	2.8	10.37
0.750	0.25	0.926	6.0	3.2	11.86
0.78	0.2969	1.000	7.0	3.7	13.70
0.97	0.40	1.48	8.0	4.2	15.56
1.0	**0.42**	**1.56**	9.0	4.7	17.41
			10.0	5.2	19.27

[a] Collected from curves in V. Linnenbom, NRL Report 588, 1962; C. J. Calbick, *Physics of thin films*, **2**, 63–145, 1964

428 Appendix

TABLE B7 Selected values of range of protons in aluminium[a]

Energy (MeV)	Range (g cm^{-2})	Range (mm)	Energy (MeV)	Range (g cm^{-2})	Range (mm)
0.1	0.00019	0.0007	4	0.034	0.126
0.3	0.00073	0.0027	10	0.163	0.604
0.5	0.00143	0.0053	13.1	0.270	1.00
			15	0.340	1.26
1	0.0039	0.0145	19.3	0.539	2.00
3	0.021	0.078	25	0.809	3.00
5	0.049	0.182	29	1.080	4.00
10	0.163	0.604	30	1.130	4.2
30	1.13	4.187	37	1.619	6
50	2.80	10.37	40	1.888	7
100	9.20	34.09	43	2.159	8
			46	2.429	9
			48	2.699	10
			65	4.048	15
			74	5.397	20
			100	9.200	34.1

[a] Source as for Table B6

TABLE B8 Range of alpha particles in Silicon[a]

Energy (MeV)	Range (10^{-3} gm cm^{-2})	Range (μm)
0.5.	0.445	1.92
1	0.784	3.38
2	1.63	7.02
3	2.70	11.65
5	5.50	23.7
10	15.9	68.5
20	49.7	214.0
30	98.9	426.2
100	805.4	3470

[a] Collected from curve in Ziegler, J. F. (ed.) 'The stopping and ranges of ions in matter' Vol 6 (1980) and TRIM programme.
The density of silicon assumed was 2.321 gm cm^{-3}.

B. Useful radiation data

TABLE B9 Total mass attenuation coefficients (μ/ρ, cm^{2}g^{-1}) (coherent scattering included) of selected materials for photons of energies 0.01 to 100 MeV suitable for calculations according to Lambert's law for narrow geometry[a]

Photon energy (MeV)	H$_2$ (8.988 × 10^{-5})[b]	Be (1.85)	C (2.25)	N$_2$ (1.250 × 10^{-3})	O$_2$ (1.429 × 10^{-3})	Mg (1.74)	Al (2.6989)
0.01	0.385	0.593	2.28	3.57	5.78	20.8	26.3
0.05	0.335	0.156	0.187	0.187	0.213	0.329	0.369
0.10	0.294	0.133	0.152	0.150	0.156	0.169	0.171
0.50	0.173	0.0773	0.0872	0.0871	0.0873	0.0864	0.0844
1.0	0.126	0.0565	0.0637	0.0636	0.0637	0.0628	**0.0613**[c]
5.0	0.0505	0.0235	0.0271	0.0274	0.0278	0.0287	0.0284
10.0	0.0325	0.0163	0.0196	0.0202	0.0209	0.0231	0.0231
50.0	0.0141	0.0102	0.0142	0.0156	0.0169	0.0222	0.0230
100.0	0.0116	0.00992	0.0145	0.0163	0.0179	0.0241	0.0251

Photon energy (MeV)	Si (2.42)	Fe (7.86)	Cu (8.93)	Sn (7.29)	W (17.1)	Pb (11.34)	U (18.7)
0.01	34.2	173.0	224.2	141.6	95.5	133.0	178.0
0.05	0.437	1.94	2.62	10.70	5.91	7.81	1.11
0.10	0.184	0.37	0.461	1.68	4.43	5.40	1.91
0.50	0.0875	0.084	0.0836	0.0946	0.136	0.161	0.193
1.0	**0.0635**	**0.0599**	0.0589	0.0578	0.0654	**0.0708**	0.0776
5.0	0.0297	0.0314	0.0318	0.0354	0.0407	0.0424	0.0445
10.0	0.0246	0.0298	0.0308	0.0385	0.0464	0.0484	0.0506
50.0	0.0252	0.0382	0.0410	0.0588	0.0760	0.0804	0.0850
100.0	0.0275	0.0432	0.0465	0.0677	0.0881	0.0934	0.0984

Photon energy (MeV)	Air (1.205 ×10^{-3})	H$_2$O (1.00)	SiO$_2$ (2.20–2.32)	Perspex (PMMA) (1.19)	Polyethylene (0.91–0.97)	Bakelite (2.23)	Pyrex glass (2.23)
0.01	4.99	5.18	1.90	3.25	2.01	2.76	1.71
0.05	0.208	0.227	0.318	0.208	0.209	0.200	0.302
0.10	0.154	0.171	0.169	0.164	0.172	0.161	0.166
0.50	0.0870	0.0968	0.0874	0.0941	0.0995	0.0921	0.0870
1.0	0.0636	**0.0707**	0.0636	0.0687	0.0727	0.0673	0.0633
5.0	0.0275	0.0303	0.0287	0.0292	0.0305	0.0286	0.0284
10.0	0.0204	0.0222	0.0226	0.0211	0.0215	0.0206	0.0222
50.0	0.0161	0.0167	0.02028	0.0151	0.0142	0.0147	0.0201
100.0	0.0168	0.0172	0.0224	0.0154	0.0142	0.0150	0.0215

[a] Reprinted courtesy of the National Institute of Standards and Technology, from NSRDS-NBS 29, *Photon cross sections, attenuation coefficients, and energy absorption coefficients from 10 keV to 100 GeV*, by J. H. Hubbell, 1969. Not copyrightable in the United States.
[b] Density (ρ, g cm^{-3}) at 20°C
[c] Values for commonly used materials are highlighted at an energy of 1 MeV–close to that of Co-60 gamma photons

C. Useful data on materials used in electronic equipment

TABLE C1 Densities and chemical formulae of commercial plastics (in order of increasing density)

Name	Density[a] (g cm^{-3})	Typical formula	Example of trade name and properties
Polypropylene	0.90	$(C_3H_5)_n$	Grace
Polyethylene		$(CH_2)_n$	
High-density	0.941–0.965		
Medium-density	0.926–0.941		Xylonite H.F.D. 4201
Low-density	0.910–0.925		
Polystyrene	1.04–1.08	$(C_6H_5CHCH_2)_n$ $(C_7H_8)_n$	
Polyurethane	1.088	$(C_{10}H_{10}N_2O_4)_n$	Thiokol Solithane 113 (d = 1.073)
Acrylonitrile–butadiene-styrene (ABS)	1.06–1.08	$(C_{15}H_{19}N)_n$	
Epoxy	1.12	$(C_{18}H_{24}O_3N_2)_n$	Shell Epikote 828
Nylon	1.13–1.15	$(C_{12}H_{24}O_3N_2)_n$	Nylon 6.6 (d = 1.13)
Poly (vinylidene chloride)	1.15–1.80	$(C_2H_3C_l)_n$	
Poly (methyl methacrylate) (PMMA)	1.18–1.20	$(C_5H_8O_2)_n$	Perspex, Lucite
Phenol–formaldehyde	1.3–1.6	$(C_7H_7O)_n$	Bakelite
Polysulphone	1.41	$(C_{27}H_{22}SO_4)_n$	Union Carbide P1700
Polyester	1.5–2.1	$(C_{24}H_{12}O_{12})_n$	Mylar
Silicone	1.08–2.8	$(C_2H_6SiO)_n$	Dow Corning 93–500 (d = 1.08)
Polytetrafluoroethylene (PTFE)	2.1–2.3	$(CF_2)_n$	Teflon
Epoxy–Glass-Fibre Composite	2.2	–	Fortin FR4

[a] R. C. Weast, *CRC handbook of chemistry and physics*, 1972–3, pp. C764–74
J. Brandrup and E. H. Immergut (ed.), *Polymer handbook*, Interscience, 1966, Table IX-5
N. A. Waterman (ed.), *Fulmer materials optimiser*, Fulmer Research Institute, 1977

C. Useful data on materials used in electronic equipment

TABLE C2 Radiation absorption effectiveness of various materials used in electronic equipment boxes

Material	Comercial name or type number	Function	Typical thickness (mm)	Typical density (g cm^{-3})
(a) Organic				
Polyurethane	Thiokol Solithane 113	Coating	0.1–1.0	1.073
Silicone	Dow Corning 93–500	Encapsulant gasketing	0.1–5	1.08
Epoxy	Shell Epon 828	Adhesive, coating or PCB	0.1–5	1.1
PTFE	BS 2848	Cable sleeving	0.1–5	2.1
Nylon	Nylon 6, 6	Bushings, spacers	0.5–5	1.13
Polysulphone	Union Carbide P1700	Connector mouldings	0.5–5	1.24
PVC	3M	Adhesive tape, cable sleeving	0.1–5	1.4
PVF	Kevlar	Cable sleeving	0.1–5	–
Polyester	Du Pont Mylar	Dielectric	0.025–0.1	–
Phenolic	Bakelite	Connector	0.5–5	1.6
DAP	Type 10256	Transistor pads	1–2	–
Neoprene	Neoprene	Gasketing	0.1–5	1.25
(b) Inorganic (non-metal)				
Beryllia	Nat. Beryllia Corp.	Heat-sink	0.1–1	3
Fibreglass	Fortin RF 4 epoxy–glass laminate	PC board		2.1–2.5
Borosilicate glass	Corning 7740	Encapsulant	0.1–5	2.23
Mica	EAC 699	Isolation shim	0.1–1	2.6–3.2
Alumina	Various ceramics, Linde sapphire	IC header and lid, epitaxial substrate	0.1–1	3.9–3.97
Silicon	Monsanto	Electronic chip	0.2–2.0	2.33
Silica	GEC Crystal or Spectrosil B	Electronic crystal substrate, optical window	0.1–5	2.65
Gallium arsenide	Monsanto or metals Research	Electronic chip	0.1–1	5.32
Molybdenum disulphide	Moly-ITC-Bond	Lubricant	0.01–0.1	4.80
Zinc oxide	–	Pigment	0.01–0.1	5.6

TABLE C2 Contd

Material	Comercial name or type number	Function	Typical thickness (mm)	Typical density (g cm^{-3})
(c) Alloys				
Brass	Cu70Zn30	Structures	1–10	8.5
Kovar	Fe54Ni29Co17	IC lids, connectors, wires	0.25–1.0	8.2
Solder	Pb64Sn36	Solder joints	0.1–2.0	9.43
Stainless steel	Fe63Cr25Ni12	Bolts, vessels	1–10	7.5
Monel	Ni67Cu28Fe (etc.)	RF gaskets	0.1–1	8.8
Mumetal	Fe18Ni75Cu5Cr2	Magnetic shields, coil cores	0.1–5	8.58
(d) Composites				
Fibreboard	Fortin FR 4 Micaply 3M Cu-clad	PCB, box wall	0.1–5	2.2 (not incl. Cu)

(e) Metals	Atomic number	Atomic weight	Function	Typical thickness (mm)	Typical density (g cm^{-3})
Beryllium	4	9.01	Structure	1–10	1.848
Magnesium	12	24.32	Structure	1–10	1.738
Aluminium	13	26.98	Structure	1–10	2.699
Titanium	22	49.7	Bolts, sheets	0.1–5	4.507
Chromium	24	52.01	Plating	0.01–0.1	7.20
Iron	26	55.85	Bolts, sheets	0.1–5	7.87
Cobalt	27	58.94	Structure	1–10	8.85
Nickel	28	58.69	Structure	1–10	8.90
Copper	29	63.54	Cladding, cable, structure, heat sinks	0.01–10	8.93
Silver	47	107.88	Plating	0.01–0.1	10.50
Tin	50	118.7	Plating	0.01–0.1	7.31
Tantalum	73	180.88	Shielding	0.1–5	16.65
Platinum	78	195.23	Plating, wire	0.001–0.01	21.45
Gold	79	197.2	Plating, wire	0.001–0.05	19.32
Lead	82	207.2	Solder	0.1–1	11.34

D. Test data: radiation response of electronic components

TABLE D1 Degradation under total ionizing dose exposure: **transistors**

Type	Technology	Manufacturer	Dose for 25% degradation[a] (krads)	I_c or V_{gs}
2N2222	Low-power npn	SGS(F)	40	1 mA
2N2222		MOT(F)	12	1 mA
2N2219		SGS(F)	65	1 mA
2N2222A		AEG(D)	25	0.25 mA
2N2369		SGS(F)	180	0.2 mA
2N2484		SGS(F)	20	1 mA
2N2880	High-power npn	Solitron	20–75	20 mA
2N2880		Unitrode	40	20 mA
2N2905	Low-power pnp	SGS(F)	35	1 mA
2N2905A		SGS(F)	30	1 mA
2N2907		SGS(F)	60	1 mA
2N2907A		SGS(F)	35	1 mA
2N2907A		MOT(F)	100	1 mA
2N2920	Dual npn	SGS(F)	21	1 mA
2N3439	High-power npn	MOT(F)	170	10 mA
2N3501		MOT(F)	12–20	10 mA
2N3637	Low-power pnp	MOT(F)	41	10 mA
2N3700	Low-power npn	SGS(F)	35	1 mA
2N3749	High-power npn	Solitron	70	50 mA
2N3810	Dual pnp	SGS(F)	35	2 mA
2N3811		MOT(F)	120–150	1 mA
2N5005	High-power pnp	SGS(F)	>400	1 A
2N5004		SGS(F)	380	0.5 A
2N5154	High-power npn	SGS(F)	80	2.5 A
2N5664		Unitrode	>500	1 A
2N5664B		Unitrode	200	1 A
2N5666	Low-power npn	Unitrode	180	4 mA
2N5672	High-power npn	SGS(F)	180	1.5A
2N6764	n-channel FET	IR(GB)	30–60	1 V
2N6766			50–70	1 V
2N6782			60–65	1 V
2N6784			60	1 V
2N6786			80	1 V
2N6796			60	1 V
2N6798			60–90	1 V
2N6800			80	1 V
2N6802			85	1 V
2N6804	p-channel FET	IR(GB)	50–60	4 V
2N6806			20	4 V
2N6849			30–40	4 V
2N6851			20	4 V

[a] Dose in krad at which 25 per cent degradation in gain (h_{FE}) occurs.
Manufacturers' nationality: D = Germany F = France GB = Great Britain

434 Appendix

TABLE D2 Dose for onset of malfunction under total ionizing dose: **memories**[a]

Type	Manufacturer	Quiescent leakage (krad)	Functional failure (krad)
CMOS dynamic random-access memories (DRAM)			
MB814100-10PSZ 4M × 1	Fujitsu	8	8
HM514100ZP8 4M × 1	Hitachi	–	10
MT4C1004C 4M × 1	Micron	20	13–20
MCM514100Z80 4M × 1	Motorola	13	11–15
D424100V-80 4M × 1	NEC	15	18–22
M514100-80J 4M × 1	OKI	–	18–22
KM41C4000Z-8 4M × 1	Samsung	12	8
HYB514100J-10 4M × 1	Siemens	8	8
TMS44100DM-80 4M × 1	Texas Inst	15	18–25
TC51400Z-10 4M × 1	Toshiba	13	11–14
TC514100J-10 4M × 1	Toshiba	–	12–15
EDI441024C100ZC 1M × 4	EDI	52	35–75
MCM4C1000P 1M × 1	Mitsubishi	–	59–71
D421000C-10 1M × 1	NEC	14	16–30
HYB511000A-70 1M × 1	Siemens	–	11–12
TMS4C1024-12N 1M × 1	Texas Inst	4	5–6
SMJ4C1024-12J 1M × 1	Texas Inst	4	6
TC511000AP-10 1M × 1	Toshiba	–	15
TC514100J-10 256K × 4	Toshiba	–	12–13
D424256C-80 256K × 4	NEC	–	16–22
M5M44C256P 256K × 4	Mitsubishi	–	40–100

D. Test data: radiation response of electronic components 435

TABLE D2 Contd

Type	Manufacturer	Quiescent leakage (krad)	Functional failure (krad)
CMOS static random-access memories (SRAM)			
MT5C1008C 128K × 8	Micron	50	30–50
88128C100CM 128K × 8	EDI (Mitsubishi)	–	10–13
EDI88130H5CM 128K × 8	EDI (Philips)	7	45
88128CS35CC 128K × 8	EDI (Sharp)	10	7–10
HM628128L–10 128K × 8	Hitachi	–	15–18
HMS1832SMLP–15 32K × 8	Hybrid Inc	–	8–10
71256S55DB 32K × 8	IDT	–	6–8
71256RE 32K × 8	IDT	35	32–38
71256–L55P 32K × 8	IDT	–	10
7164RE 8K × 8	IDT	30	80
MB81C81A–45 256K × 1	Fujitsu	15	38–71
EDH8832C–15 32K × 8	EDI (Mitsubishi)	–	7–8
HM62256LP–10 32K × 8	Hitachi	–	9–10
D43256AC–10LL 32K × 8	NEC	–	16–20
TC55257P–10 32K × 8	Toshiba	–	12–15
CY7C198–35PC 32K × 8	Cypress	13	>35
CY7C187–35PC 64K × 1	Cypress	16	35–35
IMS1601 64K × 1	INMOS	–	8–10
MCM6287P35 64K × 1	Motorola	4	9–11
CY7C185–35PC 8K × 8	Cypress	12	30
P4C164–35CC 8K × 8	Performance	12	32–70

Table D2 Contd

Type	Manufacturer	Quiescent leakage (krad)	Functional failure (krad)
CMOS static random-access memories (SRAM)			
CY71C128–35PC 2K × 8	Cypress	–	10
HM1 65162–2 2K × 8	HMS	–	10
P4C116–35DC 2K × 8	Performance	10	50
CMOS programmable read-only memories (PROM)			
HN27C256EG 32K × 8 UV–EPROM	Hitachi	Not measured	13–15
M5M27CC256K12 32K × 8 UV–EPROM	Mitsubishi	–	12.7
D27C256AD–15 32K × 8 UV–EPROM	NEC	–	20–23
28C256 32K × 8 EEPROM	SEEQ	–	10 writing 37 reading

[a] Collected by European Space Agency

D. Test data: radiation response of electronic components 437

TABLE D3 Observations of single-event upset and latch-up under beams of **heavy ions**.[a]

Device type	Organization	Manufacturer	SEU Threshold LET (MeV mg^{-1} cm^2)	Cross Section per bit (cm^2)	Comments
CMOS static random-access memories (SRAM)					
HM6504	4K × 1	Harris	5	2E-6	Latch-up at LET 13
HM6516	2K × 8	Harris	4	2E-6	Latch-up at LET 15
HM65262	16K × 1	Harris	3	3E-6	
HMS65641	8K × 8	MHS	2.5	3E-6	Latch-up at LET 50
MT5C2568	32K × 8	Micron	<3	3E-6	Latch-up at LET 15
IMS1600SL	64K × 1	INMOS	<3	3E-6	Latch-up at LET 15
D43256	32K × 8	NEC	3	1.5E-6	
MB84256	32K × 8	Fujitsu	<3	4E-6	Latch-up at LET 25
HM6116	2K × 8	Hitachi	5	5E-6	Latch-up at LET 13
HM6264	8K × 8	Hitachi	5	1.5E-6	Latch-up at LET 13
71256	32K × 8	Hitachi	4	4E-7	
HM628128	128K × 8	Hitachi	4	6E-7	Latch-up at LET 80
CXK58225	32K × 8	Sony	9	4E-7	Latch-up at LET 25
CXK581000P	128K × 8	Sony	3	8E-8	Latch-up at LET 55
MSM8128SLMB	128K × 8	NEC	4	6E-8	Latch-up at LET 85
51C98	16K × 4	INTEL	3	1E-6	Latch-up at LET 40
TC5516	2K × 8	Toshiba	6	3E-6	Latch-up at LET 27
HC6116KSH-T	2K × 8	Honeywell	25	1E-6	
HC6364	8K × 8	Honeywell	>90	—	
MT5C1008C-25	128K × 8	Micron	2	1E-6	
88128C100CM	128K × 8	EDI (Mitsubishi)	2	2E-7	
CXK58100P-10L	128K × 8	Sony	2	1E-7	
88128CS35CC	128K × 8	EDI (Sharp)	—	2E-7	Cf252 test (LET 43)
88130H-45CM	128K × 8	EDI (Philips)	—	2E-7	Cf252 test (LET 43)

Table D3 Contd

Device type	Organization	Manufacturer	SEU Threshold LET (MeV mg^{-1} cm^2)	Cross Section per bit (cm^2)	Comments
CMOS static random-access memories (SRAM)					
HM628128L-10	128K × 8	Hitachi	2	3E-7	
71256-S55DB	128K × 8	IDT	2	8E-7	
71256 RE	32K × 8	IDT	2	8E-7	
7164 RE	8K × 8	IDT	2	5E-7	
CYC189-15DC	16K × 4	Cypress	1	3E-9	Latch-up at LET 8
CMM6167	16K × 4	RCA	15	6E-9	SOS
MAS9167	16K × 1	MEDL	>72	–	SOS
7187	64K × 1	IDT	–	–	Latch-up at LET 6
71256	32K × 8	IDT	2.5	1E-6	Latch-up at LET 15
S32KX8	32K × 8	Seiko	3	5E-7	
OW5962	32K × 8	Omni Wave	8	2E-7	
P4C188L	16K × 4	Performance	5	1E-7	
MB814100-10PSZ	4M × 1	Fujitsu	1	7E-7	
HM5141002PB	4M × 1	Hitachi	2	1E-7	
MT4C1004C	4M × 1	Micron	2	3E-7	
D424100V-80	4M × 1	NEC	1	3E-7	Latch-up at LET 40
KM41C400Z-8	4M × 1	Samsung	2	4E-7	Latch-up at LET 22
HYB514100J-10	4M × 1	Siemens	1	5E-7	Latch-up at LET 11
TMS44100DM-80	4M × 1	Texas Inst	1	3E-7	
TC514100Z-10	4M × 1	Toshiba	1	5E-7	

D. Test data: radiation response of electronic components   439

Other integrated circuits

Part	Size	Manufacturer			Notes
28C256L EEPROM	32K × 8	SEEQ	10	1E-8	
28C256-50 EEPROM	32K × 8	SEEQ	3–10	1E-9	
28C256 EEPROM	32K × 8	SEEQ	>120	–	
27HC642 UV EPROM	8K × 8	ATMEL	12	1E-9	
CY7C263 UV EPROM	8K × 8	Cypress	10	–	Latch-up at LET 15
HM6617 PROM	2K × 8	Harris	–	2E-10	Cf252 (LET 43)
HS-1-6617RH PROM	2K × 8	Harris	19	1E-9	
LL7320Q	Gate array	LSI Logic	–	–	Latch-up at LET 25
LRH9320Q	Gate array	LSI Logic	30	5E-6	
LRH91000	Gate array	LSI Logic	50	3E-8	Latch-up at LET >100
XC3042	PLA	Xilinx	6	1E-6	Latch-up at LET 13
XC2064	PLA	Xilinx	6	2E-6	Latch-up at LET 12
ACT1010	FPGA	Actel	25	5E-6	
ACT1020	FPGA	Actel	25	5E-6	
ACT1020	FPGA	Actel	22	2.3E-6	
ACT1020B	FPGA	Actel	25	2.3E-6	
EP910JC-40	PLA	Altera	4	–	Latch-up at LET 15

[a] Collected by European Space Agency

TABLE D4 Observations of single-event upset and latch-up under beams of **protons**.[a]

Device type	Organization	Manufacturer	SEU Threshold energy (MeV)	Cross section per bit (cm²)	Comments
CMOS static random-access memories (SRAM)					
HM1 6504-2R	4K × 1	Harris	<40	5E-15	
MM1 6504 SH11	4K × 1	MHS	–	9E-14	
HM1 6516-9	2K × 8	Harris	<40	1.5E-13	
HM 6116L1-2	2K × 8	Hitachi	<40	2E-12	
TC 5516 AP-2	2K × 8	Toshiba	<40	1E-14	
MAS 6116F	2K × 8	MEDL	–	<2E-15	SOS
HM1 65162-2	2K × 8	MHS	–	5E-13	
CY7C167-35DC	16K × 1	Cypress	Latch-up at 200 MeV		
HM1 65262-2	16K × 1	MHS	–	6E-13	
SMJ61CD16LA	16K × 1	Texas Inst	Latch-up at 200 MeV		
MB8464-15	8K × 8	Fujitsu	–	3E-13	
HM 6264LP-15	8K × 8	Hitachi	<40	7E-14	
D4363C-20L	8K × 8	NEC	–	9E-15	

D. Test data: radiation response of electronic components 441

Part	Size	Manufacturer	Latch-up at 60 MeV	Latch-up at 60 MeV
D4464-15	8K × 8	NEC		
TC5564PL-15	8K × 8	Toshiba		5E-13
IMS1600S55	64K × 1	INMOS	–	
MT5C1008C-25	128K × 8	Micron	30	2E-13
88128C100CM	128K × 8	EDI (Mitsubishi)	30	1E-13
88130H45CM	128K × 8	EDI (Philips)		
HM628128L-10	128K × 8	Hitachi	30	1E-13
CXK581000P-10L	128K × 8	Sony	30	3E-15
71256 S55DB	32K × 8	IDT	<50	2E-13
71256 RE	32K × 8	IDT	<50	3E-13
7164 RE	8K × 8	IDT	<40	1E-13
MB814100-10PSZ	4M × 1	Fujitsu	30	1.5E-13
MT4C1004C	4M × 1	Micron	30	8E-14
MCM514100Z80	4M × 1	Motorola	30	1.5E-13
D424100V-80	4M × 1	NEC	30	4E-13
M514100-80J	4M × 1	OKI	30	7E-14
KM41C4000Z-8	4M × 1	Samsung	30	7E-14
HYB514100J-10	4M × 1	Siemens	30	3E-13
TMS44100DM-80	4M × 1	Texas Inst	30	2E-13
441024C100ZC	1M × 4	EDI	<50	3.5E-14

[a] Collected by European Space Agency

TABLE D5 Total ionizing dose and single event upset data on some **microprocessors**. (LET in MeV per mg per cm^2. Cross section in cm^2 per bit).

Device	Threshold LET for SEU	Limiting cross-section (×10^{-6})	Total dose (krad)	Latch-up LET	Proton cross-section
Intel 8086	2	5	–	–	–
Harris 80C86	2	4	5–7	–	–
SGS Z80A	3	10	–	–	–
Zilog Z80	4^1	15	–	–	–
Harris 80C85 (Rad-hard)	35	100 per device	–	–	–
Harris HS 82C54RH	22	–	–	–	–
Harris MD 82C54	10	2	–	–	–
MHS 80C31L	3	1	–	–	–
Intel 80386	6	0.1–100 (register dependent)	(CHMOS IV)15; 24 cycled	28	–
Mot 68020^2	12	2000 per device	7	12	10^{-8} per device

D. Test data: radiation response of electronic components 443

Mot 68882[3]	–	–	–	27	–
Texas DSP TMS 320–C25	12	1 (RAM)	5	–	10^{-14} (at 800 MeV)
Analog Dev.[4] ADSP 2100A	8	2	>25	12–19	5×10^{-14} (at 200 MeV)
Transputer family					
T414	2–3	2 (RAM)	15–50[5]	–	–
T800	–	–	3[6];40[7]	–	2×10^{-12} (at 100 MeV)
T425	–	–	14–44 (batch dependent)	–	–
T222	–	–	3–4[6]; 40[7]	–	–
'RISC' Processors					
Siemens R3000A-33-AE	5	2	–	–	–
IDT R3000AE-25G	7	1	–	26.9	–
LSI LR3000AHC-25	5	1	–	59.7	1[6]
Performance PR3000A-25	5	1	–	–	–

[1] For some internal elements, the threshold is 0.7; [2] proton threshold 20 MeV; [3] similar to 68020; [4] latch-up cross-section 10^{-5} per device, proton latch-up 200–800 MeV; [5] dose rate dependent. From 1988 onwards, 5 krad (internal RAM failure); [6] internal RAM failure; [7] internal RAM disabled.

444 Appendix

E. Dose–depth curves for representative satellite orbits

Compiled by Daly (1989). Reproduced by permission of the European Space agency from reference ESA (1989).

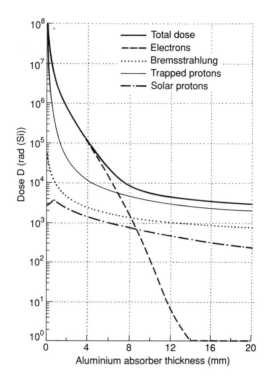

FIG. E1 Geostationary transfer orbit, 200 × 35 876 km (used to transfer geostationary and interplanetary satellites to geosynchronous altitude): 4π dose at centre of aluminium spheres; 1 year, one large flare (Daly 1989).

Daly, E. J. (1989) *The radiation environment: Dose-depth curves for various reference missions.* In ESA (1989).
ESA (1989). *Radiation design handbook.* ESA Document PSS-01-609.

E. Dose–depth curves for representative satellite orbits 445

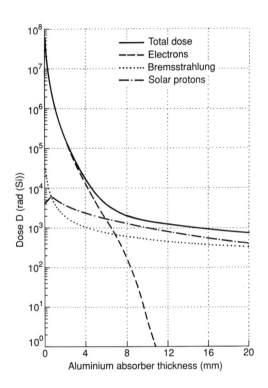

FIG. E2 Geostationary orbit, 35 786 × 35 786 km (used for communications satellites): 4π dose at centre of aluminium spheres; 1 year, one large flare (Daly 1989).

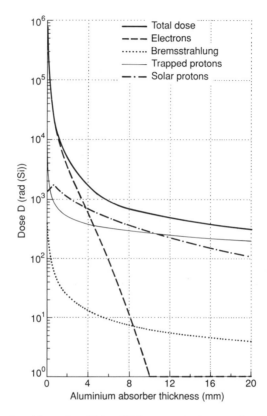

FIG. E3 Low Earth orbit, polar, 784 × 784 km, 98° inclination (used for Earth observation satellites and some types of communications and meteorological satellites): 4π dose at centre of aluminium spheres; 1 year, one large flare (Daly 1989).

E. Dose–depth curves for representative satellite orbits

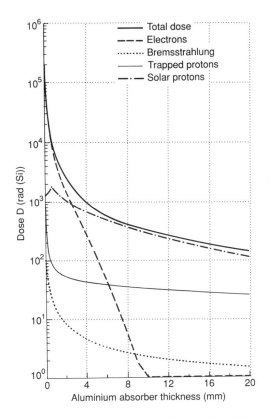

FIG. E4 Low Earth orbit, 500 × 500 km, 28° inclination (used for manned missions, e.g. space station): 4π dose at centre of aluminium spheres; 1 year, one large flare (Daly 1989).

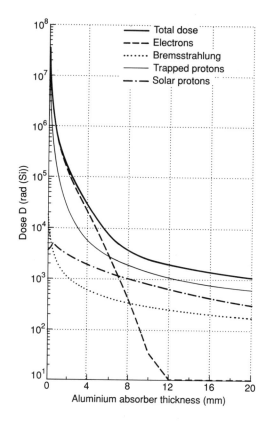

FIG. E5 'Molniya' orbit, 1250 × 39 105 km, 63° inclination (used for communications satellites for coverage of high latitudes): 4π dose at centre of aluminium spheres; 1 year, one large flare (Daly 1989).

E. Dose–depth curves for representative satellite orbits   449

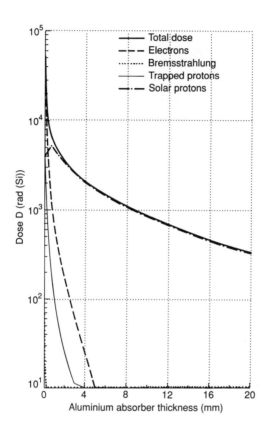

FIG. E6 Typical interplanetary mission (e.g. Mars; cometary rendezvous) with one transfer orbit before injection: 4π dose at centre of aluminium spheres; 2 years, one large flare (Daly 1989).

F. Degradation in polymers

Several authors have listed degradation in polymers in the form of bar charts giving stages of degradation versus dose. One set* is given here for reference (Figs. F1–F3) and an analysis is given in Table F1. Section 10.3 contains further references, discussion of polymer degradation and comments on the use of these charts. More advanced formats for comparison of polymers, for example that of the International Electrotechnical Commission, are discussed in Section 10.3.

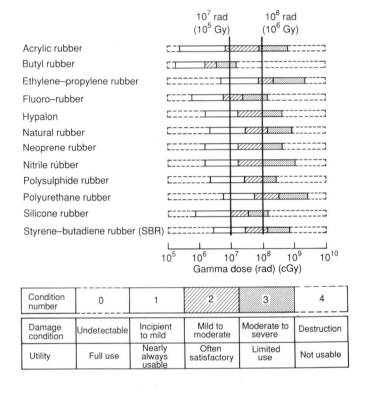

FIG. F1 Radiation-tolerance of elastomers.

---

*Van de Voorde, M. H. and Restat, C. (1972). Selection guide to organic materials for nuclear engineering, Report No. CERN 72-7. CERN, Geneva.

F. Degradation in polymers 451

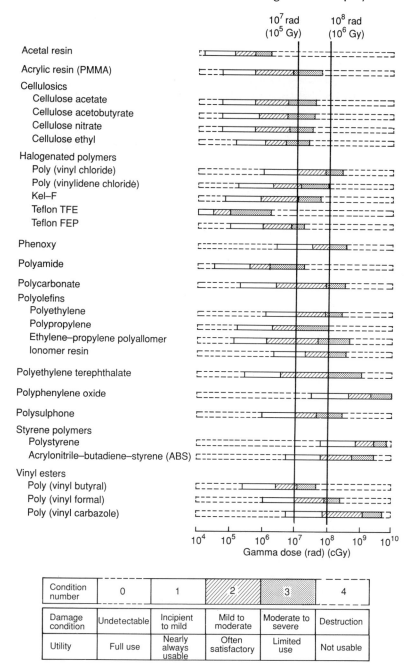

FIG. F2 Radiation-tolerance of thermoplastic resins.

Appendix

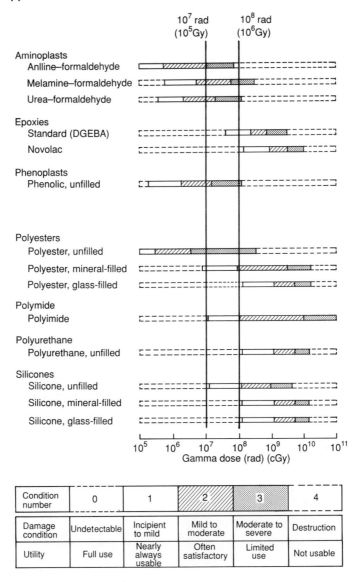

FIG. F3 Radiation-tolerance of thermosetting resins.

F. Degradation in polymers 453

The implications of the charts are brought out by listing the damage caused and remaining utility for various types of plastics at two chosen dose values, 10^7 rad (10^5 Gy) and 10^8 rad (10^6 Gy). This is done in Table F1 below.

TABLE F1 Analysis of data on polymers in Figs F1–F3, showing numbers of types of polymer in conditions 0–4 after irradiation to the doses indicated

COND-ITION	DAMAGE LEVEL	UTILITY	TYPE OF POLYMER (unfilled)						Dose unit
			Elastomer		Thermo plastic		Thermo setting		
			10^7 (10^5)	10^8 (10^6)	10^7 (10^5)	10^8 (10^6)	10^7 (10^5)	10^8 (10^6)	rad Gy
0	undetectable	full use	0	0	2	0	8	5	
1	incipient to mild	nearly always usable	8	0	8	2	1	3	
2	mild to moderate	often satisfactory	3	5	7	3	3	1	
3	moderate to severe	limited use	1	6	7	10	2	4	
4	destruction	not usable	0	1	2	11	0	1	

Author index

Numbers in brackets refer to references at the end of each chapter. Page references are also given for co-authors where the main reference may only be the main author *et al.*

Abbey, A. 218, 222 [230]
Abdel-Kader, W. G. 337 [340]
Abel, E. 389 [415]
Ackerman, M. R. 355 [368]
Acton, L. 308 [324]
Adams, J. H. 24 [44, 45], 336 [339]
Adams, J. R. 378 [383]
Adams, L. 3 14, [14], 24, 27 [45, 46], 57, 59, 63, 68 [92], 99, 107, 111, 119, 121, 131, 143, 144, 145, 147 [154, 157, 158, 160, 162], 215, 218, 222 [230], 334 [340], 344, 354, 359, 366 [368, 369, 370], 372 [383], 392, 393, 400 [415, 417]
Adloff, J-P. 37 [45]
Agarwal, S. 119, 122 [161], 357 [368]
Agullo-Lopez, F. 49 [90], 240, 241 [257]
Alberts, W. G. 50 [90]
Albus, H. P. 167 [198]
Alexander, D. R. 408, 416 [415, 418]
Ali, S. M. 277 [288]
Allen, D. J. 261 [272]
Allinson, N. M. 217 [277]
Allman, M. 190 [197]
Allsop, W. E. 217 [227]
Alsmiller, R. G. 334 [339]
Andersen, J. N. 215 [227]
Anderson, R. E. 379 [383]
Angell, C. A. 247 [259]
Anghinolfi, F. 37 [45], 145 [155]
Anspaugh, B. E. 203, 206 [232], 265 [274] 315 [326], 405 [418]
Apodaca, L. 269 [274]
Araragi, M. 397 [416]
Arimura, I. 359 [370]
Armbruster, P. 145 [158]
Arnold, C. 287 [289]
Arnold, G. 247 [257]
Ash, M. S. 31 [46], 50, 52, 54 [92], 131 [160], 164, 168, 185, 196 [198], 201, 208 [231], 259, 264 [274], 348, 356 [364], 385, 393, 395, 396, 406, 408, 410 [417]
Aspell, P. 37 [45]

Attix, F. H. 3, 8, 9, 12 [14, 104, 45], 302, 304 [325], 341 [368]
Aube, B. 334 [340]
Aubuchon, K. G. 100 [155]
Augier, P. 216, 218 [228]
August, L. S. 120, 122 [156]
Azarewicz, J. L. 408 [418]

Bailey, W. E. 117, 149 [155]
Baker, D. N. 26 [45]
Baker, W. D. 218 [230]
Balk, P. [50]
Barbe, D. F. 218 [230]
Barbier, M. 71 [90]
Barengoltz, J. 54, 76 [94], 174 [199]
Barkas, W. H. 292, 297 [325]
Barnea, G. 405 [416]
Barnes, C. E. 110, 119, 122 [155, 161], 201, 209, 212 [228, 232], 357 [368]
Barry, A. L. 213 [228]
Barsis, E. H. 251, 254 [258]
Bartko, J. 238 [239], 261 [273]
Bartolotta, A. 11, 12 [14]
Bates, L. 306 [326]
Battaglia, J. F. 28, 32 [45], 392, 397 [417]
Baur, J. M. 28, 32 [45], 353 [368], 385, 392, 397 [416, 417]
Baxendale, J. H. 276 [288]
Beauvais, W. J. 337 [340]
Becquet, M. C. [384]
Belchek, E. H. 279 [288]
Bender, G. E. 60, 62 [90, 92], 221 [229]
Benedetto, J. M. 153 [155], 237 [239], 346 [368]
Benjamin, J. D. 149 [155]
Berg, N. 272 [272]
Berger, M. J. 292, 295, 296, 297, 311 [325], 330, 334 [339, 340], 405 [416]
Berndt, D. F. 187 [197], 382 [383]
Beynel, P. 279, 283 [288]
Bezuglii, V. D. 256 [257]
Bhattacharya, P. 308 [326]
Biersack, J. P. 292 [326]

Bingefors, N. 149 [155]
Binkley, D. M. 393 [416]
Bion, T. 148 [155]
Blackburn, D. L. 237 [239]
Blaes, B. R. 400 [418]
Blake, J. B. 26 [45]
Blamires, N. G. 121 [155]
Bland, S. W. 262 [273], 382 [383]
Blandford, J. T. 336 [340]
Blice, R. 187, 191 [198]
Blood, P. 58 [92]
Bloss, W. L. 262 [273]
Blower, N. 24 [46]
Boesch, H. E. 105, 117, 122, 127, 149 [155, 160], 346 [368]
Bolt, R. O. 277 [288]
Borel, G. 146 [157]
Borkan, H. 378, 380 [383]
Borrego, J. M. 272 [273]
Bosnell, J. R. 100 [157], 236
Boudenot, J-C. 216, 218 [228]
Bouquet, F. L. 283 [288]
Bourgoin, J. L. 50, 54, 59 [90, 93]
Bowers, J. C. 395 [416]
Boyd, A. W. 9 [14]
Braasch, R. 209 [230]
Bracewell, B. 306 [326]
Bradley, P. 234 [239]
Brady, F. T. 149 [155]
Brashears, S. S. 120, 122 [156]
Braünig, D. 98 [155], 183 [197], 207, 208, 212 [228, 232]
Brewin, T. B. 43 [45]
Brews, J. R. 102, 125, 127 [160]
Briere, M. A. 212 [228]
Briggs, E. 306 [326]
Brodie, I. 397 [418]
Brotherton, S. D. 235 [239]
Brown, D. B. 68, 69 [91], 108, 110, 117, 119, 125 [155, 161], 308 [325], 344, 355 [369]
Brown, M. H. 389 [415]
Brown, R. R. 174 [197], 365 [369]
Brown, S. F. 207 [231]
Brown, W. L. 59 [90], 316 [325]
Browning, J. S. 355 [368]
Bruce, A. J. 201 [228]
Brucker, G. J. 59 [90], 99, 120, 122 [155, 156, 157, 162], 174, 178, 184 [197], 205, 208, 210, 213 [228], 263 [273]
Buchanan, B. 208 [232], 260 [274]
Bücker, H. 311 [325]
Buehler, M. G. 54 [90], 262 [272], 400 [418]
Bull, R. A. 24 [46], 145 [154], 405 [415]
Buller, J. F. 149 [159]
Burke, E. A. 59, 60, 62, 63, 77 [90, 94], 152 [157], 215 [229], 267 [273]
Burt, D. 214, 215 [228]
Busi, F. 276 [288]

Buskirk, F. R. 267 [273]

Cable, J. S. 380 [383]
Campbell, A. B. 260, 267 [273], 382 [383]
Campbell, F. J. 283 [288]
Campbell, M. 37 [45]
Canfield, L. R. 210 [229], 209, 308 [326]
Carlan, A. J. 359 [370]
Carrol, J. G. 277 [288]
Carson, R. F. 213 [228]
Carter, J. R. 203, 206 [232], 265 [274], 317 [326], 405 [418]
Carter, P. M. 76 [90]
Catlow, C. R. A. 49 [89], 240, 241 [257]
Cellier, F. E. 153 [159]
Chadsey, W. L. 55, 59 [94], 153 [161], 395 [418]
Chadwick, K. H. 9 [14]
Chaffin, R. J. 272 [273]
Chang, J. M. 261 [273]
Chapiro, A. 277 [288]
Chapuis, T. 144 [156], 352 [369]
Charlesby, A. 277 [288]
Chatterjee, A. 277 [288]
Cheng, L. J. 58 [92]
Cherne, R. D. 149 [159]
Chew, N. G. 149 [155]
Chilton, A. B. 291, 293 [325], 338 [340], 385 [416]
Chin, L. M. 14 [14]
Chiu, K. Y. 55 [93]
Chlebek, C. 215, 216, 218 [230]
Chow, W. W. 213 [228]
Chrisey, D. B. 267 [273]
Christian, S. M. 99 [156]
Christou, A. 382 [383]
Chu, D. D. 148 [156]
Chudzicki, M. J. 272 [273]
Chung, J. 150 [160]
Circle, R. R. 401 [418]
Clarke 217 [228]
Clay, P. G. 275 [288]
Cloeren, J. M. 271 [274]
Clough, R. L. 277, 283 [289]
Coakley, P. G. 145 [159]
Coche, M. 11, 12 [14]
Coenen, S. [384]
Coffinier, P. 334 [340]
Cohen, B. L. 44 [45]
Colella, N. J. 145 [159]
Combs, W. 189 [197]
Comet, B. 405 [416]
Compton, W. D. 50 [93], 240 [259]
Coninckx, F. 11, 12 [14], 37 [46]
Conrad, E. E. 60 [90]
Cooley, W. C. 203 [228], 246 [257], 266 [273], 292 [325]
Cooper, M. S. 178 [197]

Cope, A. D. 210 [228]
Coppage, F. N. 261 [272]
Corbett, J. W. 50, 54, 58 [90, 92]
Corbett, W. T. 148 [159]
Cornwell, F. N. 395 [417]
Coss, J. R. 212 [231], 261 [273], 380 [383]
Coulson, A. R. 380 [383]
Crabb, R. L. 203, 206 [228]
Craft, C. M. 393 [417]
Crawford, J. H. 50, 57 [90, 91, 93], 240, 243 [257, 258]
Crouzet, S. A. 210 [228]
Crowley, J. L. 395 [416]
Cucinotta, F. A. 42 [45]
Cuevas, A. 203 [228]
Cullis, A. G. 149 [155]
Cunningham, J. R. 3, 9 [14], 64 [92], 303, 306 [326], 341, 344, 356 [369]
Curtis, O. 54, 55 [90], 203 [229]
Cusick, J. 143 [156]

Dale, C. J. 58, 59, 60, 61, 62, 63, [90, 91, 93, 94], 215, 221 [229, 231]
Daly, E. J. 17, 19, 24 [46], 142, 143 [154, 158], 295, 297, 300 [325], 329, 334 [340], 361 [368], 405 [415]
Danchenko, V. 120, 122 [156]
Dantas, A. R. V. 190 [197], 212, 231, 261 [274]
Darwish, M. N. 153 [156], 238 [239]
Dauphin, J. 281 [289]
Davey, J. E. 382 [383]
Davies, W. R. 124 [161]
Davis, H. S. 334 [340]
Dawes, W. L. 153 [156]
Dawes, W. R. 77 [90], 378 [383]
Dawson, L. R. 211 [233]
Deal, B. E. 67 [91], 103 [156], 217, 218 [229]
DeBruyn, J. D. 205 [229]
Debusschere, I. 205, 216, 217 [229]
Decréton, M. [384] 389, 390 [417]
De Keersmaecker, R. 128, 137 [162]
De Kruyf, J. 334 [340]
DeLancey, W. M. 122 [159]
Delaney, C. F. G. 9 [14]
Dellin, T. A. 152 [161]
DeMartino, V. R. 269 [274]
Dennehey, W. 45 [90], 99, 100, 129, 131, 149 [156, 157], 174 [197], 378, 380 [383], 390 [416]
De Nordwall, H. J. 389 [416]
Derbenwick, G. F. 115, 117, 127 [157, 161], 378 [383]
Desko, J. C. 153 [156]
Dickerson, M. H. 39 [46]
Dienes, G. J. 50 [91], 270 [273]
DiMaria, D. J. 70 [91]

Dolan, R. 208 [232], 260 [274]
Dole, M. 277 [289]
Dolly, M. C. 153 [156, 157], 238 [239]
Donovan, R. F. 166 [197]
Douglas, S. 207 [232]
Downing, R. G. 203, 206 [232], 317 [326], 405 [418]
Dozier, C. M. 308 [325], 344 [369]
Dragnić, I. 37 [45]
Dragnić, Z. 37 [45]
Dressendorfer, P. V. 50 [93], 95, 100, 102, 103, 107, 108, 111, 113, 117, 124, 125, 135, 140 [157, 159, 160, 161, 162], 379 [384]
Drexhage, M. 253 [258]
Dries, L. J. 267 [273]
Dropkin, H. 272 [272]
Dumbreck, A. 390 [418]
Dunn, G. J. 378 [383]
Dupont-Nivet, E. 146 [157], 390 [416]
Dyer, C. S. 71 [91]

Egusa, S. 276, 278, 279, 283 [289]
Eisen, H. 201, 208 [299], 359 [370]
Ekelof, T. 149 [155]
Elliott, T. 214, 215, 218, 220, 221, 222, 223, 224 [230], 308 [324]
El-Teleaty, S. S. 337 [340]
Emeriau, J. 390 [416]
Emily, D. 187, 190 [198]
Emms, C. G. 100 [157]
Engholm, G. A. 28, 32 [45], 392, 397 [417]
English, T. C. 209 [230]
Enlow, E. W. 189, 195 [198]
Ensell, G. 111, 119, 121 [158]
Eriksson, C. 149 [155]
Erret, E. B. 379 [384]
ESA 99, 397, 415 [416]
Evans, B. D. 243, 246, 254 [257]
Evans, G. A. 201 [229]
Evans, L. G. 71 [91]
Evans, R. D. 291 [325]
Ewing, R. L. 287 [290]

Fabry, A. 342 [369]
Faccio, F. 145 [157]
Facius, R. 311 [325]
Fahrner, W. R. 98 [155], 208 [228]
Faith, T. J. 205 [228]
Fan, J. C. C. 149, 152 [157]
Fang, P. H. 120, 122 [156]
Farren, J. 347, 354 [369] [370]
Faw, R. E. 291, 293 [325], 338 [340], 385 [416]
Featherby, M. 405 [418]
Feigl, F. J. 68, 70 [91]
Ferrante, J. G. 334 [340]
Finch, E. C. 9 [14]

Fitzwilson, R. L. 55, 59 [94], 153 [161], 395 [418]
Flanagan, T. M. 50, 54 [94], 218, 221 [232], 286 [290]
Fleetwood, D. M. 68 [91], 108, 113, 128 [155, 157, 161], 308 [325], 346, 356, 361, 367 [369, 370], 379 [384]
Flood, D. J. 205 [229]
Florian, M. E. L. 37 [46], 345 [370]
Fonderie, V. 127, 134 [162]
Fossum, J. G. 115, 117 [157]
Fowler, W. B. 68 [91]
Freeman, R. F. A. 19 [45], 107, 108, 110, 125 [157, 158], 201, 217, 218, 220 [229]
Friebele, E. J. 68 [91], 216 [229, 231], 242, 247, 250, 252, 253 [258], 395, 396 [416, 417]
Friedman, A. L. 130 [157]
Fritsch, D. 217 [228]

Gabbe, J. D. 59 [90], 316 [325]
Gaebler, W. 98 [156], 208 [228]
Galashan, A. F. 262 [273], 382 [383]
Galloway, K. F. 153 [159], 197 [198], 237 [239], 349 [369]
Galushka, A. P. 211 [230]
Garraway, A. 148 [159]
Garth, J. 152 [157]
Gauthier, M. K. 190 [198], 207, 208, 212 [231], 261 [274]
Gehlhausen, M. 59 [94]
Geilinger, J. E. 279 [288]
George, K. P. 40 [45]
Gereth, R. 54 [94]
Germano, C. A. 203 [229]
Gerson, J. D. 58 [92]
Gibson, W. 58 [92]
Giddings, J. 148 [159]
Gigas, G. 54, 76 [94], 174 [199]
Gillespie, G. H. 201 [229]
Gingerich, M. E. 242, 250, 253 [258]
Giroux, R. A. 98 [159]
Gladstone, D. J. 14 [14]
Glaser, M. 145 [157], 211 [230]
Glasstone, S. 71 [91]
Goesele, U. 149 [160]
Goldhammer, L. J. 24 [45]
Golob, J. E. 255 [258]
Goltz, P. 310 [325]
Goodwin, C. A. 153 [156], 238 [239]
Gopal-Ayengar, A. R. 40 [45]
Gordon, B. M. 308 [326]
Goswami, J. N. 24 [45]
Gough, B. I. 402 [416]
Gover, J. E. 196 [198]
Grabmayr, G. 225 [229]
Gray, H. F. 397 [416]

Greening, J. R. 9 [14]
Greenwell, R. H. 201 [299]
Greenwood, L. R. 28 [46]
Gregor, R. B. 190 [198]
Gregory, B. L. 81 [91], 115, 117 [157], 185 [198], 377 [383]
Griffin, P. J. 59, 60, 61, [91, 92], 212 [229]
Griscom, D. L. 68, 69 [91], 248 [258]
Grobman, W. D. 307, 308 [325]
Groom, D. E. 37 [45], 211, 248 [229]
Groombridge, I. 100, 120, 127 [157, 158]
Grove, A. S. 95, 102 [157], 200 [229]
Grunthaner, F. J. 70 [91], 108 [157]
Gudiksen, P. H. 39 [46]
Gummel, H. K. 205 [231]
Gussenhoven, M. S. 401 [417]
Gutmann, R. J. 272 [273]

Haelbich, R. P. 307, 308 [325]
Hall, G. 211 [229]
Halpin, J. J. 359 [370], 396 [415]
Halsall, J. R. 388 [417]
Hamman, D. J. 277 [289]
Hammond, N. D. A. 71 [91]
Hammoud, A. N. 284, 285 [289]
Hanes, M. H. 238 [239], 261 [273]
Hanks, C. L. 277 [289]
Harari, E. 100 [158]
Harboe-Sorensen, R. S. 24 [46], 141, 143, 144, 145 [155, 157, 158, 162], 353, 359, 366 [368, 369, 370, 405, 415]
Hardman, M. 342 [369]
Harper, C. A. 89 [91]
Harper, L. 103, 121, 124 [162]
Hart, A. R. 385 [418]
Hartman, J. A. 63 [94]
Hartmann, R. A. 215 [232]
Hash, G. L. 80 [93], 261 [272]
Hatano, H. 134, 149 [158]
Hauser, J. R. 166 [197]
Hawthorne, R. A. 211 [231]
Haythornthwaite, R. 347 [370]
Heijne, E. 36 [46], 133, 145 [157, 158], 211 [230]
Heikkinen, D. W. 351 [369]
Heimbach, C. 60 [92]
Helms, C. R. 67, 69 [92]
Henderson, B. 68 [92], 108 [162], 240 [258]
Henry, E. M. 306 [326]
Hensler, J. R. 243 [258]
Hevey, R. 127
Heyns, M. 127, 134 [162]
Higbie, P. R. 26 [45]
Hiraoka, H. 281 [289]
Hirata, M. 57 [92]
Hite, L. R. 149 [155]

Ho, P. T. 260 [274]
Holck, D. K. 261 [272]
Holland, A. 57, 59, 63 [92], 215, 217, 218, 221, 222, 224 [230]
Holmes-Siedle, A. G. 3, 14 [14], 19, 24, 27, 28, 32 [45, 46], 48, 57, 59, 63, 68, 81 [92], 99, 100, 107, 108, 110, 111, 119, 120, 121, 122, 125, 126, 127, 129, 131, 133, 134, 145, 149, 154 [155, 156, 157, 158, 161, 162], 174, 182, 184 [197, 198], 205, 215, 217, 218, 220, 222 [228, 229, 230, 231], 247, 248, 251, 252 [258], 278 [289], 344, 359 [368, 369], 372, 378, 380 [383], 385, 389, 390, 392, 393, 397, 400, 405 [415, 416, 417]
Hopkinson, G. R. 63 [92], 215, 216, 217, 218 [230]
Horne, R. [384]
Horne, W. E. 174 [197], 365 [369]
Horton, C. C. 291, 304 [326], 386 [418]
Hrabal, C. A. 49, 50 [92]
Hu, C. 150 [160]
Huang, A. 225, 226 [230]
Huang, C. Y. 267 [273]
Hubbell, J. H. 291 [325]
Hughes, G. W. 115 [158]
Hughes, H. L. [94], 98, 99, 127 [158, 159], 378 [383]
Hughes, K. L. 108, 125 [161], 362 [370]
Hughes, R. C. 111, 117 [159], 287 [289]
Hughlock, B. W. 355 [363]
Humphrey, L. 397 [418]
Hurlson, R. E. 167 [198]
Hwang, J-M. 238 [239]

ICRP 7 [14], 39 [46]
IEC 282, 283 [289]
Ilie 256 [258]
Innes, B. 149 [155]
Itoga, T. 378 [383]

Jahns, J. 225, 226 [230]
Janda, R. J. 203 [228], 246 [257], 266 [273], 292 [325]
Jander, D. R. 225 [231]
Janesick, J. R. 214, 215, 216, 218, 220, 221, 222, 223, 224 [230], 308 [325]
Janousek, B. K. 262 [273]
Janssens, G. 212 [232]
Jarron, P. 37 [45], 145 [157, 159], 211 [230]
Jenkins, T. M. 13 [14]
Jenkins, W. C. 110, 117, 119 [156], 355 [369]
Jensen, L. H. 205 [229]
Jernberger, A. 148 [156]
Jha, R. 37 [45]
Johlander, B. 57, 59, 63 [92], 146 [159], 215, 218, 222 [230], 373 [383]

Johns, H. E. 3, 9 [14], 64 [92], 303, 306 [325], 341, 344, 356 [369]
Johnson, H. [162]
Johnson, G. H. 197 [198]
Johnson, R. T. 27 [46], 393 [417]
Johnson, S. 368 [369]
Johnston, A. H. 110, 117, 119, 124 [155, 159], 357 [369], 408 [417]
Jones, K. W. 307 [325]
Jordan, G. C. 334 [340]
Jordan, T. M. 120, 122 [156, 162], 334 [340]
Jorgesen, L. 148 [159]
Junga, F. A. 267 [273], 395 [416]
Jupina, M. A. 108 [159]

Kakuta, T. 26 [46]
Kalinowski, J. J. 73 [94], 286 [290], 396 [418]
Kamth, G. S. 206 [231]
Kashmitter, J. L. 145 [159]
Kato, M. 113 [160], 378 [383, 384]
Katsueda, M. 113 [160]
Kawanishi, S. 276 [290]
Keen, J. M. 149 [155]
Kelleher, A. 137
Kelliher, K. 344 [369]
Kelly, J. G. 59, 60, 61 [91, 92], 212 [229], 342 [369]
Kerns, S. E. 55, 76 [92, 159]
Kerr, K. A. 148 [159]
Kerris, K. C. 103, 121, 124 [162], 341, 346 [370]
Khivrich, V. I. 211 [230]
Kidd, J. 262 [273]
Kilian, W. T. 211, 231]
Killiany, J. M. 217 [230]
Kim, W. S. 148 [159]
Kimel, W. R. 338 [340]
Kimerling, L. C. 58 [93]
King, C. 143 [156]
King, J. C. 271 [273]
Kinnison, J. D. 145 [159], 271 [274]
Kitazaki, K. S. 63 [94], 215 [232]
Klebesadel, R. W. 26 [45]
Klein, G. W. 277 [289]
Knoll, G. F. 3, 13 [14], 27 [47], 210 [230], 255 [258, 273], 291, 295 [325]
Knudson, A. R. 122 [162], 261, 267 [273]
Kobayashi, M. 256 [258]
Koch, E. E. 307 [326]
Koch, H. W. 333 [340]
Koehl, P. 277 [288]
Koga, R. 142, 143, 146 [156, 159]
Kohno, I. 276 [290]
Kolasinski, A. 143 [156]
Konozenko, I. D. 211 [230]
Korde, R. 209, 210 [230], 308 [326]

Kosier, S. L. 153 [159]
Koyanagi, M. 152 [159]
Kraner, H. W. 211 [230, 231]
Krantz, R. J. 262 [273]
Kraus, D. 211 [230]
Kreidl, N. J. 243 [258]
Krishnakumar, B. 284, 285 [289]
Krull, W. A. 149 [155, 159]
Kruyf, J. 334 [340]
Kurtz, S. 287 [289]

Laghari, J. A. 284, 285 [289]
Lai, D. 24 [45]
Lamarsh, J. R. 37 [46], 338, 340, 341 [370]
Lancaster, G. D. 225 [231]
Lanford, W. A. 76 [94]
Lang, D. V. 58 [93]
Lange, R. 39 [46]
Larcombe, M. H. 388 [417]
Larin, F. 50, 54 [92], 164, 166, 168, 185, 186 [198], 208 [230], 344 [370], 395 [417]
Lauridsen, K. 389 [417]
Lawrence, R. 127
Lazo, M. S. 59 [91], 212 [229]
Leadon, R. E. 50, 54 [94], 220, 221 [232], 286 [290]
Lefevre, H. 67 [92]
Leffler, S. 14 [14], 27 [46], 392, 393 [417]
Lehmann, B. 212 [228]
Lehovec, K. 262 [274]
Lelis, A. J. 117, 122 [159]
Lell, E. 240, 243, 248 [258]
Lemeilleur, F. 211 [230]
Lenahan, P. M. 70 [94], 108 [159, 160]
Leopold, W. 129, 149 [156]
LePage, J. J. 349 [370]
Leray, J-L. 146, 150 [160]
Leslie, S. G. 238 [239], 261 [273]
Leszkow, P. 389, 390 [417]
Levchuk, I. V. 211 [230]
Levy, P. W. 201 [231], 243, 247 [258, 273], 395, 396 [417]
Li, S. S. 149 [156]
Li, Z. 210 [229, 231]
Lin, Y. S. 400 [418]
Lindgren, S. R. 14 [14], 27 [46], 392, 393 [417]
Lindholm, F. 395 [417]
Linnenbom, V. J. 292, 296 [326]
Liska, D. 351 [370]
Littmark, U. 291 [326]
Lizotte, S. C. 380 [383]
Loferski, J. J. 205 [231]
Logan, C. M. 354 [369]
Long, D. M. 55, 59 [94], 153 [161], 395 [418]
Loo, R. Y. 206 [231]
Looney, L. D. 255 [258]

Lopez-Cotarelo, M. 147 [160]
Lowe, A. L. 59, 60 [92], 217 [228], 342 [369]
Lowis, R. 236 [239]
Lowry, L. 119, 122 [161]
Lowry, M. E. 225 [231]
Ludlam, T. 211 [230]
Luera, T. L. 59, 60 [91, 92], 212 [229]
Lum, G. K. 355 [368]
Lynn, D. K. 397 [417]
Lyons, P. B. 248, 255 [258]

Ma, T. P. 50, [93], 95, 100, 102, 140 [160], 378 [383]
McCabe, D. P. [418]
McCall, R. 13 [14]
McCarthy, K. 217, 221, 222 [230]
Machalek, M. D. 353 [370]
McDonald, J. C. 9 [14]
McDonald, P. T. 401 [418]
McGarrity, J. M. 105, 117, 122, 127 [155, 160], 269, 271 [274]
McGuire, R. E. 24 [45]
Mackay, G. 347 [370]
McLaughlin, W. L. 9 [14]
McLean, F. B. 105, 117, 122, 140 [155, 156, 159, 160], 346 [269]
McNett, J. 145 [159]
McNulty, P. J. 337 [340]
McNutt, M. J. 221 [231]
McWhorter, P. J. 103, 120, 124, 152 [160, 161, 162]
McWright, G. M. 225 [231]
Maerker, R. E. 27 [46]
Magee, J. L. 277 [288]
Magno, R. 266 [273]
Magorrian, B. G. 217 [227]
Maier, P. 279, 283 [289]
Maisch, W. G. 267 [273]
Maldonado, J. R. 307, 308 [325]
Mallen, W. 187 [198]
Malmberg, A. F. 395 [417]
Malone, C. J. 79 [94], 381 [384]
Mapper, D. 143 [158], 353, 354, 355 [370]
Mar, B. W. 318, 322 [326]
Marini, G. 256 [258]
Marshall, P. W. 59, 60, 61, 62, 63 [90, 91, 93, 94], 215, 221 [229, 231]
Martin, K. E. 178, 182 [198], 207, 212 [231], 261 [274]
Maruyama, X. K. 267 [273]
Maserjian, J. 70 [91], 108 [157]
Mattern, P. L. 251, 254, 257 [258]
Mattsson, S. 144 [160]
Matzen, W. T. 211 [231]
Maurel, J-M. 147 [160]
Maurer, R. H. 271 [274]
Mayer, B. 277, 280 [289]

Maxseiner, R. 212 [228]
Meddeler, G. 37 [45]
Medvedev, G. 39 [46]
Meisenheimer, T. L. 128 [160]
Menick, R. L. 260 [274]
Merker, M. 405 [418]
Messenger, G. C. 31 [46], 50, 52, 54, 55, 76 [93], 131 [160, 167], 168, 170, 178, 185, 196, 197 [198], 201, 208 [231], 266 [274], 350, 358 [370], 385, 393, 395, 396, 406, 408, 410 [417]
Meyer, W. E. 221 [231]
Michette, A. G. 306, 307 [326]
Milinchuk, V. K. 276, 277, 284 [289]
Miller, A. 9 [14]
Miller, D. L. 58 [93]
Miller, S. L. 120 [160]
Miller, W. M. 120 [160]
Millward, D. G. 55, 59 [94], 153 [161], 395 [418]
Miscione, A. M. 380 [383]
Mistry, K. B. 40 [45]
Mitani, K. 149 [160]
Mitchell, E. W. 242, 248 [259]
Mnich, T. M. 148 [159]
Moghe, S. B. 272 [273]
Mooney, P. M. 58, 59 [93]
Mork, C. 149 [155]
Morrison, M. 308 [325]
Motz, J. W. 333 [340]
Mueller, G. P. 261 [273]
Mullen, E. G. 337 [340], 401 [417, 418]
Muller, R. 141 [158]
Mullis, J. 405 [418]
Murray, J. R. 131 [160]

Naber, J. A. 50, 54 [94], 220, 221 [232], 286 [290]
Nagata, M. 378 [384]
Nargornaya, L. L. 256 [257]
Naruke, K. 134, 149 [158]
Neighbours, J. A. 267 [273]
Nelson, C. M. 248 [258]
Newall, D. M. 260 [274]
Ng, W. 308 [326]
Nichols, D. K. 207 [231], 380 [383]
Nicollian, E. H. 102, 125, 127 [160]
Niishi, M. 276 [290]
Ning, T. H. 70 [93]
Nisenoff, M. 267 [273]
Nishioka, Y. 378 [383]
Nolthenius, H. J. 28 [47]
Norgett, M. J. 59 [93]
Norton, J. R. 271 [274]
Nothoff, J. K. 261, 262 [274]
Nowlis, R. N. 189 [198]
Nuraki, C. 397 [416]

Oberg, D. L. 237 [239]
O'Brien, M. C. 242 [259]
Occelli, E. 211 [230]
Ochoa, A. 80 [93]
Ohyu, K. 378 [383]
Ojha, A. 209 [230]
Okabe, T. 113 [160], 378 [384]
Oldham, T. R. 117, 122, 133 [159, 160]
Oliver, G. D. 13 [14]
O'Loughlin, M. J. 262 [273]
Onori, S. 11, 12 [14]
Ormond, R. 150 [160]
Ornelas, J. R. 203 [232]
Osbourn, G. C. 210 [232]
Othmer, S. 55 [93]

Paić, G. 9 [14], 338 [340]
Paige, E. G. 242, 248 [259]
Palkuti, L. 150 [160], 190 [198], 344 [370]
Paulsson, M. 149 [155]
Pease, R. L. 128 [157], 187, 189, 191 [197, 198], 367 [369], 408, 416 [417, 418]
Pells, G. P. 243 [259]
Pelose, J. R. 260 [274]
Pelous, G. 207 [232]
Pepper, M. 108 [162]
Peterson, E. L. 61, 71 [93]
Peterson, R. I. 225 [231]
Phillips, D. C. 279, 283 [289]
Pickel, J. C. 77 [93], 336 [340]
Pigneret, J. 395 [418]
Pinner, S. H. 277, 278, 283 [289]
Platteter, D. G. 187, 191 [198]
Poch, W. J. 128, 133 [161], 184 [198], 216 [231], 344 [370], 397 [417]
Pons, D. 59 [93]
Pool, F. 215, 218, 220, 221 [230]
Posnecker, K. U. 211 [230]
Powell, R. J. 115, 127 [158, 161]
Price, B. T. 291, 304 [326, 418]
Price, M. L. 385 [418]
Price, W. E. 190 [198], 207, 212 [231], 261 [274], 380 [383]
Primak, W. 247 [258]
Profio, A. E. 3, 5 [14], 291, 304 [326], 341, 351 [370]

Quayle, A. 217 [227]

Radlein, E. 246 [259]
Rai-Choudhury, P. 238 [239], 261 [273]
Raimondi, T. [384]
Rappaport, P. 205 [231]
Ravel, M. K. 217 [231]
Rax, B. 119, 122 [161], 357 [368]
Ray, K. P. 401 [418]
Raymond, J. P. 59, 61, 71 [93], 187 [198], 385 [418]

Razouk, R. 127 [159]
Reay, K. P. 337 [340]
Reddy, R. E. 24 [45]
Reichmanis, E. 281 [289]
Reinheimen, A. L. 217 [231]
Reisman, A. 122, 128 [162], 308 [326]
Remmerie, J. 128, 137 [162]
Renardy, J. 211 [230]
Rendell, R. W. 108, 125 [161]
Restat, C. 278 [290, 450]
Retzler, J. D. 178 [197], 395, 396 [418]
Ricketts, L. V. 265, 269, 270 [274], 395, 396 [418]
Riewe, L. C. 113, 128 [157], 367 [369]
Robinson, A. L. 308, 310 [326]
Robinson, M. T. 59 [92]
Robinson, P. A. 78 [93], 407, 411, 414, 415 [418]
Rockett, L. R. 379, 381 [383]
Rodgers, M. S. 125 [161], 362 [370]
Roesch, W. C. 3 [14]
Roeske, F. 225 [231]
Rogers, S. C. 166 [198]
Rogers, V. C. 50, 54 [94], 220, 221 [232], 286 [290]
Rombeck, F. J. 142 [158]
Roncin, J. C. 207 [232]
Roper, G. B. 236 [239]
Rosati, A. 11, 12 [14]
Rose, M. A. 196 [198]
Rosen, C. A. 208 [228]
Rosenzweig, W. 59 [90], 205 [231], 316 [325]
Rossi, G. 145 [157]
Rouse, G. V. 149 [159]
Roy, T. 216, 218, 219 [231]
Rudie, N. J. 76 [93], 260, 268, 270 [274], 395, 415 [418]

Sah, C. T. 58 [94]
Saito, H. 57 [91]
Sakaue, K. 134, 149 [158]
Saks, N. S. 68, 69 [91], 108, 125 [161], 216, 217, 218 [230, 232]
Salter, I. D. [418]
Samson, J. A. R. 307 [326]
Sander, H. H. 185 [198]
Sanders, T. J. 378 [383]
Sanderson, T. K. 143 [158], 353, 354, 355, 366 [370]
Sangster, D. F. 276 [290]
Sansoe, C. 143 [158]
Santiard, J. C. 37 [45]
Sartori, L. 71 [93]
Sarvandam, K. V. 40 [45]
Sasuga, T. 276 [290]
Sathy, N. 40 [45]
Schaeffer, D. L. 145 [159]

Schaller, K. H. 389 [416]
Schlesier, K. 131, 149 [161]
Schmid, J. P. 405 [418]
Schmitt, E. J. 142 [158]
Schneider, W. 50 [89]
Schönbacher, H. 11, 12 [14], 37 [46], 256 [259], 279, 283, 284 [288, 290], 340 [370]
Schreurs, J. W. H. 243 [259]
Schrimpf, R. D. 153 [159], 189, 197 [198]
Schuler, R. H. 277 [289]
Schulman, J. H. 50 [93], 240 [259]
Schulz, M. 67 [92]
Schulz, T. J. 395 [416]
Schwank, J. R. 103, 108, 124, 125 [161, 162], 346, 362 [369, 370]
Schwarzmann, A. 208 [228]
Scott, T. M. 144 [161]
Sedore, S. R. 396 [416]
Seehra, S. 236 [239]
Seguchi, T. 276 [290]
Seltzer, S. M. 71 [91], 291, 292, 295, 296, 297 [325, 326], 330, 331, 334 [340], 405 [416]
Seran, H. 145 [158]
Sexton, F. W. 80 [93], 108, 124, 125, 149 [161], 362 [370]
Shaneyfelt, M. N. 108, 125 [161], 362 [370]
Shapiro, P. 120, 122 [156], 261 [273]
Sharp, J. 390 [418]
Shaw, C. M. 148 [159]
Shaw, D. C. 119, 122, 155 [161]
Shaw, D. E. 357 [368]
Shedd, W. 208 [232], 260 [274]
Shelby, R. 268 [273]
Sheng, S. L. 206 [231]
Shuegraf, K. K. 187 [198]
Shultis, J. K. 291, 293 [325], 338 [340], 385 [416]
Shutte, N. M. 26 [47]
Sigel, G. H. 243, 246, 254 [258]
Silverman, J. P. 277, 281 [290], 307, 308 [325]
Simons, M. 166 [197]
Sinclair, W. K. 42 [47]
Singmin, A. 41 [46]
Sinha, P. 391
Sinton, R. A. 201 [228]
Sirois, Y. 256 [259]
Sivo, L. L. 190 [198]
Sjolund, A. 149 [155]
Skoog, C. D. 251, 254 [258]
Slifkin, L. M. 50 [90], 240, 243 [257]
Slusark, W. J. 236 [239]
Smith, L. J. 269 [274]
Smith, L. S. 79 [94], 380, 381 [383, 384]
Smits, F. M. 205 [231]
Snidow, N. L. 59, 60 [92]

Snow, E. H. 167 [198]
Snyder, E. S. 152 [161]
Sokel, R. J. 80 [93]
Soli, G. A. 214, 221, 222, 223, 224 [230], 400 [418]
Song, Y. 380 [383]
Spencker, A. 205 [232]
Spindt, C. 397 [418]
Spinney, K. T. 291, 304 [326, 418]
Spratt, J. 170 [198], 405 [418]
Springer, P. 187 [198]
Srour, J. R. 55, 59, 63, 79 [93, 94], 153 [161], 215 [232], 269, 271 [274], 395 [418]
Stanley, A. G. 182, 184, 187, 190 [198], 207, 211, 212 [232]
Stanley, T. D. 113, 149 [161]
Stapor, W. J. 261 [273], 345 [370], 401 [418]
Starchik, M. I. 211 [230]
Stasiak, J. W. 70 [90]
Stassinopoulos, E. G. 19 [47], 120, 122 [156, 161, 162]
Stearns, D. G. 217 [232]
Stecklin, R. A. 187 [198]
Stein, H. J. 54, 56 [94]
Stephen, J. H. 353, 354, 355 [370]
Stevenson, G. R. 37 [47], 133 [162]
Stewart, T. B. 203 [232]
Stofel, E. J. 203 [232]
Stolarz-Izycka, A. 279, 283, 284 [290]
Strahan, V. 130 [157]
Strobel, D. 405 [418]
Stroud, J. S. 243 [259]
Sukuragi, S. 256 [258]
Suli, M. 58 [92]
Sullivan, W. H. 287 [290]
Summers, G. P. 59, 60, 62 [90, 92, 94], 221 [229], 267 [273]
Sundaram, K. 40 [47]
Sung, J. J. 150 [162]
Suter, J. J. 271 [274]
Svensson, C. 67 [94]
Svensson, G. K. 14 [14]
Swallow, A. J. 277 [290]
Swanson, R. M. 201 [228]
Sweetman, J. D. 152 [161]
Sweigard, E. L. 28 [49]
Sze, S. M. 95, 97, 102 [162], 262 [274]
Szondi, E. J. 28 [47]

Tabatabaie-Alavi, K. 382 [384]
Tada, H. Y. 203, 206 [232], 265 [274], 317 [326], 405 [418]
Tavlet, M. 37 [47], 279 [290], 340 [370]
Taylor, T. L. 149 [155]
Thatcher, R. K. 73 [94], 286 [290], 396 [418]
Thome, F. V. 393 [417]
Thomlinson, J. 144 [162]

Thompson, M. W. 49 [94]
Thomson, I. 41 [47], 212 [232], 344, 349 [368, 370]
Tindall, W. E. 225 [231]
Titus, J. L. 153 [156, 157], 197 [198], 238 [239]
Torrens, I. M. 59 [92]
Totterdell, D. H. J. 121 [155]
Townsend, P. D. 47 [89], 240, 241 [257]
Treece, R. K. 148 [159]
Tremere, D. A. 167 [198]
Trew, J. W. 269 [274]
Troeger, C. L. 261, 262 [274]
Trombka, J. I. 71 [91]
Trump, J. G. 318 [326]
Tsacoyeanes, W. C. 355 [368]
Tsao, S. S. 149 [162]
Tucker, R. F. 243 [259]
Tundall, W. E. 225 [231]
Tupikov, V. I. 276, 277, 287 [289]
Turfler, R. M. 187, 191 [198]
Turner, J. E. 37 [47], 276, 277 [290]
Turpin, D. C. 103, 124 [161, 162], 379 [384]

Underwood, C. I. 24 [46]
UNEP 37, 40 [47]

Vandenbroeck, J. 127, 134 [162]
Van de Voorde, M. H. 276, 278, 280, 283, 285 [290, 450]
Van Gunten, O. 120, 122 [156, 162]
Van Lint, V. A. J. 50, 54, 76 [94], 174 [199], 207, 220, 221 [232], 286 [290], 385 [418]
Van Vonno, N. 148 [162], 189, 190 [199]
Vavilov, V. S. 205 [232]
Veigele, W. 306 [326]
Villard, L. 147 [160]
Vineyard, G. H. 50 [90], 270 [273]
Vook, F. L. 56 [94]
Vranch, R. L. 108 [162]
Vu, K. N. 380 [383]

Wagemann, H. G. 98 [155], 207, 208 [228, 232]
Walker, J. W. 58 [94]
Wallmark, J. T. [162]
Walters, M. 122, 128 [162]
Ward, A. K. 24 [46], 145 [154], 405 [415]
Ward, J. 24 [46]
Warlaumont, J. M. 307, 308 [325]
Waskiewicz, A. E. 130 [157]
Watanabe, K. 378 [384]
Watkins, G. D. 58 [93, 94]
Watkins, L. M. 251, 254 [258]
Watts, S. J. 215, 216, 217, 219 [231, 232]
Waxman, A. S. 99 [163]
Webster, W. M. 169 [199]

Weeks, M. D. 358 [368]
Weeks, R. A. 248 [258]
Weidwald, J. D. 217 [232]
Weinberg, I. 206 [232]
Wenger, C. 201, 208 [229]
Wert, L. J. 237 [239]
Wertheim, G. K. 53, 57 [94]
Wertz, J. E. 69 [91], 240 [258]
West, R. H. 255 [259]
Westerberg, E. R. 397 [418]
Wiczer, J. J. 210 [232]
Widdows, S. 236
Wigmans, R. 257 [259]
Wilkins, B. R. 76 [90]
Williams, C. K. 308 [326]
Williams, R. 125 [162]
Wilski, H. 284, 285 [290]
Wilson, M. 307 [326]
Winokur, P. S. 102, 107, 108, 113, 117, 121, 124, 125, 128, 140 [155, 157, 161, 162], 308 [325], 346, 356, 367 [369, 370], 379 [384], 410 [418]
Wintle, H. J. 286 [290]
Wirth, J. L. 166 [199]
Witham, H. S. 70 [94]
Wojick, R. 212 [228]
Wolfson, R. 39, 40 [47]

Wolicki, E. A. 59 [91, 94], 356 [370]
Wong, J. 247 [259]
Wong, T. Y. 187 [198]
Woolf, S. 152 [157]
Wooten, D. 148 [159]
Wright, D. 217, 219 [231]
Wright, K. A. 318 [326]
Wrobel, T. F. 80 [93], 261 [272]
Wyard, S. J. 302 [326]
Wyatt, P. W. 378 [383]
Wysocki, J. T. 205 [232], 348 [370]

Yamada, W. E. 262 [273]
Yaqi, H. 26 [46]
Yip, K. L. 68 [91]
Yu, A. Y. L. 167 [198]
Yu, C. H. 150 [162]
Yu, C-Y. 150 [162]

Zaininger, K. H. 99, 100, 110, 125, 131, 149 [157, 158, 163], 378 [380], 390 [416]
Zakharia, M. 119, 122 [161]
Ziegler, J. F. 76 [94], 292, 297, 326, 331 [340]
Zoutendyk, J. A. 79 [94], 381 [384]
Zsolnay, E. M. 28 [47]
Zuleeg, R. 261, 262 [274]

Subject index

A-value (trapping probability) 108, 115, 217, 376
absorbed dose *see* dose
absorber, shielding 291, 394, 399
absorption, optical 240
absorption coefficient *see* attenuation
absorption constant, for X-rays 307–9
absorption effectiveness of various materials 431, 432
AC testing 365
accelerator environment 36, 279
acceptors 7
accident dosimetry 11, 13
activation 330
activity, specific 4
acute exposure 43
additives, in polymers 282
add-on shielding 291, 399–405
advanced MOS 149
Ag *see* silver
air
  attenuation of photons by 429
  ionization chambers 9
  roentgen unit 6
Al *see* aluminium
alkali chloride 240
alkali doping, in glasses 245
alkali halide 50, 242, 248, 256–7
alpha particle 5, 428
aluminium
  attenuation of photons 429
  doping, in glasses 244
  properties 432
  range of particles in 292, 427–8
aluminium oxide *see also* sapphire, ceramic
  as MOS insulator 378
  as TLD 11
americium-241 354
'Amersil' *see* silica
amorphous semiconductor devices 82, 84
amorphous silica 66, 67
analogue circuits 145, 189–90
annealing
  of bulk damage 56, 184
  of charge in MOS devices 118–27
  of colour centres 257
  at cryogenic temperatures 122
  isochronal 56, 184
  isothermal 57
  model 124
  radiation-induced 125
  room-temperature 120
  slope 124
  of surface effects in bipolar technology 184
  thermal 119, 184
  UV 125
annular MOSFET gate for hardness 379
application-specific integrated circuit (ASIC) 146
assurance, hardness 406–418
ASTM specifications *see* standard
astronaut dose limits 42
atomic displacement 50, 83, 99, 165–8, 171, 225
atomic displacement versus shield depth 316–18
atomic number (Z)
  of electrode material 377
  in generation of photons 303
  in stopping of radiation 296, 321–4
attenuation
  coefficient for photons 6, 247, 302–6, 309, 429
  narrow geometry 304
  of particle fluxes 310–24
Au *see* gold
Auger mechanism 55
autonomous system 389
avalanche diodes 207
avalanche injection 127, 351

B *see* boron
back-channel leakage in SOS 149
back scattering 293, 321
bakelite, attenuation by photons 429
barium titanate 264
base region
  diode
    currents 202
    photodiode 209
    solar cell 205
  transistor
    base-to-emitter junction 168

466 Subject index

structure in integrated circuit 189, 191, 195
structure in power device 234–6, 238
structure in small-signal device 164–5, 168
width 169–71, 238
Be *see* beryllium
becquerel 4, 423
beryllia *see* beryllium oxide
beryllium
  attenuation of photons 429
  properties 432
beryllium oxide 11, 431
'BETA' code 320, 334
beta (gain), change in reciprocal *see* gain damage factor
beta particle 5
'BGR' curves 59, 316, 317, 350,
bias dependence
  of CCDs 216
  of MOS 100–3, 237
biological effectiveness 7
biological materials 308–10
biological shielding 338
biological units 7
bipolar
  integrated circuits 187–96, 381
  power transistors 233–5
  transistors 164–98
'Bit-flip' 76
BL coordinates 326
blocking diodes 207
bond scission, organic *see* scission
bond strain 68
border traps *see* interface
boron
  B-V centre 56
  diffusion 74
  dopant 72
borosilicate glass (pyrex), attenuation 429
box cover as shielding 399–405
Bragg–Gray cavity theory 9, 354
brass, properties 432
breakdown, avalanche 127–351
breakdown voltage 234–6
  BV_{CBO} 234
  BV_{DS} 236, 237
bremsstrahlung 293, 299–300, 329, 336, 342
BS 9000 *see* standard
build-up 8, 306, 333
built-in field, oxide 118
built-in shielding, spacecraft 291, 399
bulk charge in oxide films *see* Q_{ot}
bulk damage 50–4, 83, 219, 247 *see also* displacement, vacancy
  in bipolar transistors 165, 168, 172–8, 184, 233
  in CCDs 220–1

constant (K) 59, 172–8
dependence on energy and particle type 58–60, 224
in diodes 201
energy dependence 58–62
equivalent particle fluence 53, 59, 63–4, 223
in MOS transistors 128
in solar cells 205–6
thermal annealing of 184
bulk defect structure 350
bulk semiconductor, transient effects in 73
bulk states in CCDs 221
buried channel, in CCDs 222, 223
buried gate FET 238
burn-out, in power MOS 237, 238
burn-out, transient effect 395
butane 276
butted contacts 380
butyl rubber 285
BVA resonator 271

Ca *see* calcium
caesium iodide scintillator 256
caesium-137 39, 341–3
calcium fluoride and sulphate dosimeter 11
californium-252 142, 143, 144, 354, 355
calorimeter 9
capacitance–voltage plot 102–3, 364, 373
capacitor 269, 270
carbon, attenuation of photons 309, 429
carrier removal 56, 172, 220, 233
'CASE' testing 354, 355
categorization of parts 409
CCDs *see* charge-coupled devices
CECC/CENELEC *see* standards
centigray *see* rad
ceramic capacitor 269
ceramic materials, radiation tolerance 82
  see also oxides, nitrides
Cerenkov radiation 255, 300
cerium oxide 243
CERN 36–7
Cf *see* californium
channel *see specific types*
characteristic emission line, of X-rays 302, 303, 344, 345
'CHARGE' code 34, 332–6, 338
charge collection 381
charge confinement, in CCDs 222
charge-coupled device (CCD)
  annealing of charge transfer inefficiency, model 222–4
  charge confinement in 222
  charge transfer channel in 219
  charge transfer efficiency (CTE) 215, 220, 221, 223, 224

charge transfer inefficiency (CTI) 64, 213, 215, 221, 223, 224
  difference from MOS switch 215
  displacement damage 220-4
  electron bombardment 217
  interface trapped charge 218
  oxide trapped charge 217
  use in astronomy 215
  use in high-energy physics 215-16, 393
  use as video sensor 215, 388
  use in X-ray detection 215, 221
charge deposition 78
charge funneling 381
charge injection into oxide 127, 237, 381
charge injection device 215
charge, interface and oxide see Q_{it} and Q_{ot} respectively
charge state, of ion 35
charge trapping 65-70, 88, 99-108, 223, 236, 343
chemical formulae of materials 430
chemical vapour deposition 251, 378
Chernobyl disaster 39, 44, 278, 388, 420
chlorine
  in colour centres 242-3
  in oxide growth 377
chromium, properties 432
chronic dose 43
circuit board, shielding by 400, 401
circumvention 130, 395
Cl see chlorine
classification of MOS 138
clocking rate, in CCDs 223
cloth 269, 272
cluster defect 220
CMOS see complementary-symmetry logic
Co see cobalt
coating, optical 247
coating, plastic 282
coating, spacecraft external 280
cobalt-60 280, 343
  emissions and use 280, 426
  metal, properties 432
code, computer see individual names of codes
cold-cathode devices 85, 271, 397
collector-base leakage current 182
collisions, atomic 61
colour centres 50, 88, 99, 226, 240-3, 248, 257
combined effects of stress on plastics 283
commercial MOS technology 140-7
complementary-symmetry logic 95-163
composite absorber 322
Compton effect 8, 59, 300, 305, 377, 389
Compton electron 8, 32, 356
computer methods for particle transport 327-40

conductors 89, 270
configuration and materials of spacecraft 399
consequences of degradation 81
controlled fusion reactor 32, 392
conversion factor
  electron fluence to dose 347, 348
  optical loss 250
  physical quantities 421-2
  proton non-ionizing energy loss 62
  radiation units 422-4
copper
  attenuation of photons 429
  properties 432
'Corning 7490' silica 253
Corning type B fibre 253
cosmic rays 16, 19, 329
  anomalous component 24
  co-rotating events 24
  galactic 19
  other sources 24
  simulation 352-5
  solar 19
  terrestrial 22
cost
  of hardness assurance 410
  of testing 413
coulombic elastic scattering 61, 300
countermeasure 406
'CREME' code 337, 338
'CRIER' code 336-7
critical charge 76, 381
critical temperature (T_c) 267
cross linking 282, 285
cross-section 70, 71, 79, 305
CRRES satellite 401
cryogenic MOS 122
crystals
  change of diffraction power in quartz 264
  electro-optic 225
  optical response 241-2
  quartz resonator 270
  radiation effects in silicon 50-64
Cs see caesium
Cu see copper
'CUPID' code 337
curie 4
curing of plastics by radiation 281
current gain 364
current transfer ratio 213
current value
  bipolar transistor collector-base leakage current I_{CBO} 182
  CMOS quiescent current I_{SS} 134
  diode diffusion current I_{diff} 202
  diode forward current I_F 201-2
  diode generation current I_{gen} 202
  diode primary photocurrent i_{pp} 74-5, 128-30

468 Subject index

diode reverse current I_R 201, 207, 208
solar cell short-circuit current I_{SC} 203
cutoff frequency, f_α 170
C–V plot *see* capacitance–voltage
CVD *see* chemical vapour deposition
CVD silica 251
cyclotron 349, 352

damage energy E_{DAM} 61, 62
dark current in CCDs 64, 215, 218, 225
    spikes 221
databases 368, 371, 408
data, general and geophysical 421, 422
data, radiation 423–9
decay chain 4
decay rate 4
decay series 4
deep level in silcon *see also* bulk damage,
    transient spectroscopy (DLTS) 58, 373
defect characterization 58
degradation
    of bipolar transistor gain 164–6, 168–78
    of JFETs 260–2
    of logic performance
    of organics 274
    of polymers 274–83, 450–3
    predictions for devices 108, 178, 221–3, 387
    processes 48
    of transport power 50
degraded spectrum 333
delayed photocurrent 75
delayed radiation 31
delrin 284
delta rays 329
densities of materials 430
depletion 102
depletion region
    CCD 214
    junction 72, 75
    MOS 102
deposition of dose 310, 318
depth–dose curves, representative orbits 444–9
derating 396
design margin, radiation 408, 412
device parameter measurements 361–6
diagnostics for fusion reactors 33, 263–5
dielectric isolation 130–1, 149, 238, 380
dielectric, plastic 284
diffusion length 52, 75, 202, 220
diffusion length damage constant 172
diffusion length degradation 203, 204
digital IC, bipolar 186–7
digital signal processors 145
DIN *see* standard
diode 201–13
    avalanche 207

direct bandgap 210
dosimeter and dose-rate detector 12–13
    Gunn 271
    high-power rectifier 207
    light emitting 212, 213
    low power rectifier 206–7
    microwave 207
    neutron dosimeter 13
    photodiode 209–11
    p-i-n 13, 270
    radiation detector, high-energy 210–11
    reverse leakage 206, 207
    Schottky barrier 211
    Zener 207
direct-coupled FET logic 262
displacement, atomic *see* bulk damage
displacement energy E_d 50
divacancy 51
'donut' structure 195
dopant 56, 72, 378
dose
    computation of 329
    conversion to fluence *see* conversion
    vs. depth plots 310–15
    enhancement of 152, 344, 346, 377
    limits
        for astronauts 42
        for public and workers 39
    at a point, 3D computation 327–8, 339
    total 372–5, 433–4, 442–3
    uniformity of 348
    units of 6–8, 423
    values
        in test and real environments 357
        vs. risks associated 43
dose-rate effects 128–31, 195–7, 342
    in capacitors 270
    in insulators 286
    in optical fibres 253, 255, 256
dosimeter
    calorimeter 8
    dE/dx detector 13
    diode 12
    dye 42
    Faraday cup 10
    Fricke 11
    gammachrome 12
    ionization chamber 8
    MOS (RADFET) 14, 401
    on-board spacecraft monitoring 14, 392, 405–6
    perspex 12
    pigment or paint 12
    polyethylene/hydrogen pressure 12
    proportional counter 9
    RADFET 14, 401
    radiochromic 12, 308
    radiophotoluminescent (RPL) 11

silicon 12
thermoluminescent (TLD) 10
dosimetry
  general 8
  for testing 355–6
DOSRAD code 334, 335, 338
drain-source breakdown 236, 237
DRAM *see* dynamic random-access memory
drawn optical fibre 253
D–T reaction 29, 32, 351
dye, radiochromic 12
dynamic random-access memory 142

E centre in Si 51, 57, 184, 221
E' centre in silicon dioxide 68
ECL *see* emitter–collector logic
economics, of hardness assurance 410
edge transistor, in SOI 379
elastomers 285
electret 283
electrolytic capacitors 270
electromagnetic pulse (EMP) 32
electromagnetic radiation 299–310
electron 5
  belt 17, 329, 444–9
  dose calculation code 336
  fluence to dose conversion 347
  injection 70
  range 293, 427
  transmission coefficient 295, 336
  scattering 295
  sources for testing 346–9
  spectrum 297–8
  trajectory 294
electron spin resonance (ESR) 58
electron volt 5
electro-optic crystals 225
elongation 277
emission line, characteristic 302, 303
emitter (bipolar transistor) 164–6
  nested 192–3
  walled 191
emitter–base surface effects 189
emitter–collector logic 187
energy absorption coefficient 306, 356
energy absorption versus photon energy 424
energy dependence, dosimetric materials 355–6
energy level
  bandgap E_G 51, 69, 72
  conduction band edge E_C 51, 69, 72
  Fermi level E_F 168, 204
  midgap E_{MG} 51
  trap E_T 51
  valence band edge E_V 51, 69, 72
energy loss 296, 299
energy units 5, 421–5
engineering materials testing 366

engineering model, spacecraft phase (EM) 412
engineering polymers 283, 450–3
environment 16
  processing 28
  reactor 26–8
    containment 27
    controlled fusion 32
    mobile 28
    vessel and cavity 27
  robot 33
  space 16
    cosmic ray 19
    electron 19, 444–9
    interplanetary 449
    other planets 25
    proton 19, 444–9
  terrestrial and man made 37
    background 37
    external 37
    extra-terrestrial 37
    fallout 39
    ingested 38
    internal to the body 40
    monazite survey, India 40
    radon 41
  weapon 29
environment calculations 328
epitaxial layer 130, 380, 381
epoxy polymer 276, 278, 283
equilibrium 8, 356
equipment hardening 385–406
equipment, in non-penetrating radiation 397–406
equivalent fluence 83, 203, 206–7
error detection and correction (EDAC) 382
error rate 78
ESABASE code 338, 339
ESA-SCC specification 359, 360
ETRAN code 312, 330, 331
exposure, unit of 6, 430

F *see* fluorine
f_α *see* cutoff frequency
fabrication processes, semiconductor 372
fading 119
F-centres 243
fallout 39
Faraday cup 10
fast interface states 68
fault detection 390–2
Fe *see* iron
Fermi level 168, 204 *see also* energy, Fermi
ferrite 152, 266, 267
FET *see* field-effect transistor
fibre, optical 248–57
  luminescence 253, 255
  radiation-induced loss 248–53

470 Subject index

field-effect transistor
  GaAs 262
  general 84
  Junction 260–2
  logic 262
  MES 262
  MOD 262
  MOS 95–163
  vacuum 397
field emission cathode 271, 397
field oxide
  charge trapping effects 134, 236
  leakage 134–7
  recessed 378
field oxide hardening 378
field-programmable gate array 146–7
fillers, in polymers 278, 282
fission neutrons 351
flare, solar 16, 24, 444–9
flatband condition 102
flatband voltage 104, 217, 364
flash X-ray 196, 342, 346
fluence 6
fluence rate 6
fluence to dose conversion 348
fluences, equivalent 83, 202
fluorine
  in oxide growth 376, 378
  in optical media 242–3, 251–3
flux 6
flux density 6
food irradiation 29, 280
forming gas 377, 378, 379
FOXFET 135, 368, 373–4
free radical 276
frequency to wavelength conversion factor 422
Fricke dosimeter 11
f_T see gain–bandwidth product
functional testing 365
funnelling see charge
fusion neutrons 351
fusion reactor
  environment 32–4, 392
  maintenance 392

gain–bandwidth product, f_T 170–1
gain, bipolar transistor
  damage factor 171, 172
    bulk 165–80
    surface 180–4
gallium arsenide
  field-effect transistor 262, 271
  integrated circuit 382
  laser see laser diode
  light-emitting diodes see light
  on silicon 152
  solar cell 206

gamma dot see dose rate
gamma ray 5, 300–8
  attenuation of 295, 303, 429
  pulse 196, 375, 390, 391
  sources 343
gas multiplication 10
gaseous release 285
gate electrode material 377
gate insulators, modified 377
gate oxide thickness dependence 105, 113
gate voltage 99
Ge see germanium
GEANT code 295, 330
Geiger–Müller counter 10
general and geophysical data 421, 422
generation lifetime 211
geomagnetic cutoff 24
geomagnetic shielding 22, 25, 337
germania-doped quartz 251
germania-doped silica 226
germanium 50, 262, 264, 265, 292
glass
  capacitor 269
  composition, influence of 243
  fibre 248–56
  radiation sensitivity 248
  window 242
glow curve (TLD) 11
gold
  attenuation of photons 309, 429
  properties 432
  scattering of photons 150
gray unit 6, 423 see also rad unit
grippers on teleoperator 278
growth curve 111, 113, 115, 117, 358
growth temperature, gate oxide 376
guard band 135
Gunn diode 272

H see hydrogen
H centre 241
half life 4
Hall effect sensor 266, 267
halogen
  in optical media 242–3
  in polymers 284
hard, definition of 97–8
hardened ('rad-hard') product 147, 373
hardening
  against transient radiation 380–2
  of device by layout 379
  of equipment 385–406
  of semiconductor parts 372–83
  of a semiconductor process 375–9
hardness assurance 406–18
hardness categorization 409
heavy ion
  burn-out, induced by 236

cosmic ray 22
  device, response to 76, 437–43
  linear energy transfer 78
  for testing 352–5
  track in Si 76–7
  transport through matter 299
heavy-metal electrode 150–2, 377
Hg *see* mercury
high-energy physics
  detector diode 210–11
  environment 36
  environment simulation 342–84
high-energy radiation source 424
high-flux reactor (HFR) 342
high-power rectifier diode 208
high-voltage CMOS 153, 238
hole injection 127
hole trapping 99
hot electrons and holes 127
hydrogen
  attenuation of photons 429
  effects in MOS 107, 108

I see current
I *see* iodine
image charge 104
impurity–vacancy interaction 51, 56
incipient failure 390, 392
incipient latchup 129
indium antimonide detector 264
indium phosphide solar cell 206
induced radioactivity 70
infrared
  absorption 242
  detection 262–3
  emission 212
ingested radiation 38
injection at interfaces *see* avalanche
injection level 54, 178
input offset current and voltage in amplifier 365
input voltage, effect on MOS logic 96, 110–12
insulator
  in accelerator 279
  in CCDs 216–20
  in devices, function 82, 86–7
  in MOS device 65–7, 376–8
  polymer as 278–9, 286
  sandwich 150, 377
  in space 280
integrated injection logic 186–7
integrated optical device (IOD) 200
interface (unless otherwise stated, refers to the interface between silicon and grown silicon dioxide; *see also* Q_{it}, oxide)
  border traps 68

  in CCDs 218
  charge traps 106 *see also* Q_{it}
  fast states 68, 99, 108, 124, 127, 137, 138, 184, 215, 217, 221, 225, 236, 377
  location 67
  slow state drift 99, 107
  trapped charge *see* Q_{it}
intermediate dielectric 378
intermittent bias response 111, 113
internal radiation 40
internal spectrum 297
interstitial atom 50
iodine-131 39
ion *see* heavy ion
ionization chamber (gaseous) 9
  thimble 9
  electrometer 9
ionization, consequences of 88
ionization effects in solids 64
ionization in CCDs 216–17
inversion 102
IRAN (Irradiate–Anneal) 120, 184
iridium-192 426
iron
  attenuation of photons 429
  properties 432
iron-55 X-ray 55, 222, 308
irradiation bias 99, 105, 110, 217
isochronal anneal 56
isoflux plot 17
isolation technology 190, 191
isothermal anneal 57
isotope
  cell 343
  properties 426
  in terrestrial environment 37–40
ITS/TIGER code 297, 330

J centre 51, 184
Joint European Torus fusion experiment 32
Josephson junction 267
joule 5
junction field effect transistor (JFET) *see* field effect
junction, p–n
  in bipolar transistor 164
  in diode 201
  general 74–5
  light-doped drain 150
  moderately-doped drain 150
  in MOS device 95–7
  nested emitter 192–3
  transient effect 74–5

K centre 184
kapton 282, 284
kerma 7
knock-on atom 50, 60–4

472 Subject index

kovar 432
krypton-85 426

L centre 241
Lambert's law 304
laser diode, III–V, 212–13
latch-up
  hardening against 380
  response 79, 196, 366, 395
late radiation effects *see* postirradiation effects, time-dependent effects
lateral transistor 79, 187, 190
lattice defect 50
layout for radiation tolerance
  of circuits 379
  of robots 390
  of spacecraft 402
lead
  attenuation of photons 429
  flint glass 243
  properties 432
  zirconate titanate 152, 262
leakage currents *see* current
lifetime damage 202
lifetime, generation 211
lifetime killing 130
lifetime, minority carrier 165–9, 172, 200–2, 211
light-emitting diode 212–13
light guide, optical 248
lightly-doped drain (LDD) 150
linear accelerator (LINAC) 73, 342, 346, 348
linear energy transfer 78, 237, 337, 342, 352–5, 358
linear integrated circuit 189
lithium 342
lithium borate 11
lithium dopant in silicon 205
lithium-drifted detector 13
lithium fluoride 11, 355, 356
lithium niobate 225
local shielding against space radiation 403
logic
  bipolar digital 186
  gallium arsenide FET 262
  MOS families 95–6, 131, 145–7
  programmable 147
long-lived radiation effects 98, 186
  in polymers 281
long wavelength X-ray 306
loss
  energy 295–9, 329–35
  optical 242, 249
lot acceptance 413
low dose rate effects in MOS 128
low power rectifier diode 207, 208
lubricants 280

luminescence, in fibres 253–5
luminescent efficiency 255

magnesium
  oxide 240, 242
  properties 432
magnetic components 266
magnetic rigidity 22
magnetosphere 17
maintenance, remotely controlled 392
majority carriers 54
management of hardness assurance 407, 408
manipulators for nuclear plants 387–9
Mar approximation 324
mass attenuation coefficient 306, 429
mass thickness 292, 318
material
  biological 37, 275
  insulating 82, 86–7
  interaction with radiation 291–325
  magnetic 266
  optical-fibre 251–6
  organic 276
  polymer technology 283–5, 450–3
  semiconducting 84–5
  as shielding 310, 318
  structural 399–402, 430–2
  window 242–7
material preparation and cleaning (for hardening) 376
material type, influence on radiation shielding 318–24
materials data, tabulated 430–2
materials testing 366
maverick device 183, 207
maximum acceptable dose 101, 113, 403
maximum range 292, 295
mechanical sensor 268
memory
  dynamic MOS 141
  ferrite 153, 266
  non-volatile 152
  static 141
  test data 433–40
  technology 141–2
mercury cadmium telluride 264–5
mercury UV line 425
MESFET 262, 271
Messenger–Spratt equation 169
metal halide 242
metal-oxide-semiconductor (MOS) device
  capacitor 102–3, 364, 373
  CCD 214–18
  CMOS 95
  degradation mechanism 102–5
  dosimeter 14, 405
  logic families 145

Subject index   473

metal-gate 377
  power transistor 236–7
  structure 64–7, 100–2
metal silicides as contacts 377
metals, radiation damage 27–8, 49–50, 266
METEOSAT satellite 401
methodology for hardening of semiconductor parts 372–5
MEVDP code 334
Mg *see* magnesium
mica 83
micromachined silicon 397
microprocessor
  response to radiation 442–3
  technology and radiation 142–5
microwave devices 271
microwave diodes 208
military environment, simulation 342
military specifications *see* standard
military systems, hardening 394–7
minimum ionizing radiation 297
minority carrier 52
minority-carrier concentration 73
minority-carrier diffusion length 202, 203, 221
minority-carrier lifetime 52, 74, 116, 165–9, 172, 202–4, 208, 211, 212, 221, 235
  see also lifetime
minority-carrier lifetime damage constant 168, 172
miscellaneous components and hardware 268–72
mission cost ratios 411
mission dose 322–3, 338–9, 450–3
mobility, degradation of carrier 54, 128
modified gate insulators 377
modular packaging and shielding for spacecraft 402
molybdenum disulphide, properties 432
monazite survey in India 40
monel metal, properties 432
monitoring of radiation dose
  in accelerators 393
  in nuclear reactors 393
  on-board satellites 405
Monte-Carlo method 295, 322, 330, 331
MOS *see* metal-oxide-semiconductor
multicomponent optical glass 244
multilayer shielding 318, 322
multiple errors 78
mumetal, properties 432
mylar 83, 278, 284

N *see* nitrogen
Na *see* sodium
NAND gate 379
Nb *see* niobium
n-channel transistor 103

neoprene 285
network formers in glass 244
neutron 5
  activation 70–1
  capture 5, 70, 71
  controlled fusion 33
  dosimetry 361
  fission reactor 27
  secondary 299
  shield 293
  sources for testing 350–1
  weapons 31
neutron damage 54, 391
  in bipolar technology 170, 178
  in capacitors 269
  equivalent fluence 361
  in ferrites 266, 267
  in JFETs 261, 262
  in lasers and light emitters 212
  in microwave devices 272
  in optocouplers 212
  in phototransistors 211
  in power transistors 234, 235, 237
  in quartz crystals 270
  in resistors 268
  in sapphire 243
  in silica 247
  in silicon 57
  in sensors 261–4
  in solar cells 203
  in thyristors 236
neutron-induced defect 50
neutron sources, for testing 350, 351
new optics 225
nickel, properties 432
niobium in fusion reactors 342
nitride *see* silicon
nitrile rubber 285
nitrogen
  attenuation of photons 429
  nitrided oxide growth 377
noise immunity of logic 133, 365
noise in MOS $(1/f)$ 128
non-binding orbital 68
non-ionizing energy loss (NIEL) 60, 63, 223, 224, 265, 267, 317
non-volatile RAM (NVRAM) 150, 266
NOR gate 379
'notch', in CCDs 222
nuclear industry, environment simulation 342
nuclear reactor monitors 393
nuclear scattering 62–3
number transmission coefficient (NTC) 295, 324
nylon 282, 430, 451

O *see* oxygen
Olympus satellite 401

474 Subject index

on-board data handling (OBDH) in spacecraft 147, 402
on-board radiation monitoring 14, 392, 405, 406
operational amplifiers 189–90
optical coatings 247
optical computing 226
optical density *see also* optical loss, fibre, glass, light etc.
   absorption 243
   conversion factors 250
   radiation-induced change 240, 256
optical diode 201
optical fibre, polymeric 254, 277
optical glasses 243–6
optical light guide 248
optical media 239–56
optically triggered thyristor 236
optocoupler 213, 214
optoisolator 212
organic scintillator 256, 279
organics 275–90
oxidation of silicon 376–8
oxide (here taken to mean oxide film grown on silicon unless, otherwise stated)
   annealing, thermal 377
   in CCDs 217
   charge injection into 351
   growth 376
   interface with silicon *see* interface
   isolation by mean of 191–5
   passivation by means of 165, 202, 261
   recessed 191–4
   sandwich structure 377
   thickness dependence of V_T shift 105, 113, 176
   trapped charge *see* Q_{ot}
oxides other than silicon
   of aluminium 243, 378
   of germanium 251
   magnesium 242
   network disrupting 247
oxygen
   attenuation coefficient 429
   in silicon defects 57, 204
oxygen-vacancy centre *see* A centre

paper
   capacitor dielectric 269
   degradation in radiation 272
parameter measurement 361–6
parasitic elements 373, 378, 379
parasitic thyristor 79, 380
parasitic transistor 79, 237, 379
particle damage statistics 63
particle induced defects in optical media 247
particle transport and range 292
parts categorization 409

parts procurement 408
P_b centre 68, 107, 108
p-channel MOS transistors 105
perspex dosimeter 12
phosphorus
   diffusion 74
   in silicon defects *see* E centre
phosphorus-doped glass 379
phosphorus-vacancy centre *see* E centre
photoconduction 209
photocurrent 73–5, 98, 196
   in bipolar technology 74, 166
   in diode 74, 209
   in insulator 286
   in MOSFET 129, 380, 381
   prompt 75
   radiation pulse length 74–5
   in solar cell 74, 203
   transient 361, 394
photodiode 209, 213
photoelectric absorption process, photons 304
photoemission 86, 127
photogeneration 202
photologic 85, 213
photon, high-energy 329
   energy vs. wavelength 425
photoresists 281
phototransistor 211, 213
photovoltaic effect 209
piezoelectric polymer film 268
p-i-n diode 13, 380
pinchoff voltage 261
planar shield 310
planar transistor 164
plasma, solar 16
plastic-coated silica fibre 251–2
plastic dielectric capacitor 267
plastic films 282
plastic, structural 270, 399–402, 430–2
platinum, properties 432
platinum resistance thermometers 264
plutonium 29
p-n junction *see* junction
polar horn 19
polaron 68
polycrystalline silicon
   emitter 189
   gate MOS structure 96–7, 372
   resistor 376
polyethylene
   attenuation 429
   dosimeter, hydrogen pressure 12
   properties 430
polyimide 374
polymer 64, 273–88, 450–3
   degradation under radiation 65, 273–88, 446–9

Subject index 475

grafting 28
properties 430, 431
shielding with 318
polymeric optical fibre 254
polymers 64, 273–88
poly(methyl methacrylate)(PMMA) 12, 254, 281
polyolefin 276
polypropylene 275
polysilicon see polycrystalline silicon
polyurethane 278
poly(vinyl chloride) (PVC) 278, 284
positron 5
postirradiation effect (PIE) 119, 361–2, 370
power CMOS 153
power devices 186, 232–8
    MOSFET 235–6
    'smart' form 237
    transient burnout of 236
practical range 290, 294
preconditioning of quartz crystal 291
prediction model see also degradation
    for bipolar transistor degradation 178, 182
    of CTI in CCDs 222
    of damage in bipolar technology 178
    for equipment hardening 387
    for logic performance 100, 186
    for MOS device degradation 108–10
    for optical fibre loss 248
    for single event upset 336, 337
    for solar cell degradation 205
pressurised water reactor (PWR) 27
preventive replacement 390–2
primary knock on atom (PKAS) 60–4
primary photocurrent see photocurrent
printed circuit board (PCB) 285, 400, 401
process hardening 375–9
processing, industrial 29, 280, 281
processing variables, semiconductor 138
product assurance techniques 359
production of electromagnetic radiation 302
programmable logic device (PLD) 146
programmable memory 436, 439
prompt radiation 31
propane 292
proportional counter 10
protection 42
proton 5
    damage in CCDs 221
    interaction code 'CUPID' 337
    range 293, 428
    solar flare 22
    sources, for testing 349, 350
    space 329
    transfer curve 224
    transport 299
PT see platinum

PTFE 278, 284
Pu see plutonium
pulse radiolysis 276
pulsed neutron effects 187, 188, 395–7
pulsed neutron source, for testing 350
pulsed photon effects 184, 188, 380, 395–97
punch-through 234

Q_{it} 67–70, 103–5, 110, 125, 367–8, 375 see also interface, surface, V_{it}
Q_{ot} 67–70, 103–5, 108–9, 125, 367–8, 375 see also oxide, charge, V_{ot}
qualification model (QM) 413
qualified manufacturers list (QML) 409
quality factor (Q) 7
quartz 226, 242, 248, 251, 269
quartz crystal 270, 271
quiescent current, I_{SS} 364

Ra see radium
rad unit dose, dosimeter
    1 MeV photons in various materials 423
    per gram shielding figure of merit 403
    per roentgen 6, 423
    tissue 13
    water 12
RADATA/RADSCAN code 334, 335
RADFET dosimetry 14, 143, 393, 401
radiation belt see trapped radiation
radiation chemical yield 276
radiation control plan 407
radiation data, tabulated 423–9
radiation design margin (RDM) 408, 412
radiation hardening of metals 69 see also hardening
radiation index (RI), for polymers 277, 278
radiation-induced conductivity (RIC) 270, 286
radiation processing/sterilisation 28, 280, 281
    sources and doses 29
radiation qualification and test procedures 357, 358
radiation testing of devices and materials
    methods 341–71
    qualification and procedures 357–8
    sources 341–55
    tabulated data 433–40
radiation tolerance assessment 359, 373
radiation tolerance of plastics 283–5
radioactivity 4, 70, 423, 426
radiochromic dye dosimeter 12
radioisotope data, tabulated 426
radiolysis 275
radiolytic reactions 276, 277
radiophotoluminescent dosimeter (RPL) 11
radium 3, 41

476 Subject index

radon 41
random walk 66, 122, 124
range of particle 292, 427, 428
ray tracing 335
reaction node in oxide film 70
reactor (fission)
   terrestrial 26
   spaceborne 28
reactor (fusion) 32
'rebound' in MOS devices 107, 120, 124, 137
recessed oxide *see* oxide
recombination, carrier 54, 55, 73, 99, 212
recovery, of MOS 118–24
redundancy 111, 113, 147, 399
refractory metal 377
regulations 42
relative biological effectiveness (RBE) 7
relaxation 118, 119
rem unit *see* sievert, rad
remote handling 278
remotely controlled maintenance 392
requirements and review 407
resistance of insulator during irradiation (R_R) 286
resistivity change, in MOS 128
resistor decoupling, for SEU 381
resistor, response to radiation 270
response to radiation, specification 359
retrofit, impact of 412
reverse annealing 57, 120-1, 124
reverse current *see* current
'RISC' processor 143, 145, 443
risk of radiation to humans 43
robots 33, 227, 387–93
roentgen unit 6, 423, 424
room temperature recovery of MOS device 120
rubber, response to radiation 285
rules, radiation hardening of equipment 386–7
ruptured bond at an interface 276
Rutherford scattering of particle 62

sample sizes for testing 414
sapphire
   optical window 243
   substrate for MOSFET *see* silicon on sapphire
saturated cross section 79
saturation voltage 185
scanning electron microscope 349
scattering 293, 295, 306, 318, 344
SCC specification 359, 360
Schottky barrier 210
Schottky logic 195
scintillator 256–7, 279
scission of chemical bond 282–3

secondary neutrons 71, 299
secondary particle 330
secondary process in displacement damage 61
secondary radiation 297
sector analysis 327, 334, 402
selection principles for bipolar transistors 178
self-aligned structure 377, 378
'SEMFET' 349
semiconductor
   device classification 84-7
   parts, radiation hardening 372-84
sensor *see* transducer
sewage processing 28, 29
shadow RAM 130
shadow shield 402
SHIELD code 332, 338
shield weight tradeoff 403–5
shielding
   by circuit boards 400, 401
   geomagnetic 22
   materials 318
   satellite 399
   of space radiation 311–15
SHIELDOSE code 312, 331, 336
SHIELDOSE/CHARGE comparison 336
Shockley-Read-Hall theory 52, 178, 203
Si *see* silicon
sievert unit 7
SIGMA code 334, 335, 338
silica, fused
   fibres 250–3
silicate glasses 242–53
silicon
   attenuation of photons 429
   bond length 68
   collision model 60–4
   dose rate response equation 75
   dose unit 7, 13, 355, 423
   dosimeters 12
   junction *see* p-n junction
   micromechanisms 268
   properties 431
silicon controlled rectifier 235
silicon dioxide
   fibres 253
   films *see* oxide, metal-oxide-silicon
silicon-gate MOS device *see* polycrystalline silicon
silicon nitride
   film 379
   nitride-oxide silicon (SNOS) memory 152
silicon on insulator 131, 149, 191, 379, 381
silicon on sapphire 131, 149, 276, 379, 381
silver chloride 3
silver, properties 432
SIMOX 149

simulation
  circuit network analysis 395
  of cosmic rays 352-5
  of environmental damage to device
    by charge injection 351-2
    by electron beam 346-9
    by gamma rays 343-4
    by neutron beam 350-1
    by proton beam 349-50
    by UV photons 344-6
    by X-rays 344-6
  of radiation environments 341-6, 352-5, 327-38
single-event phenomena (SEP) 76, 342
single-event phenomena, hardening against 380, 381
single-event upset, prediction of 336, 337
single-event upset testing 352-5, 365, 366
single-event upset data, tabulated 437-43
'slab geometry' 331, 338
slow states see interface
smart power device 238
Sn see tin
snapback 80
SOCODE code 335
sodium acetate 276
'soft latch' 80
soft X-ray 308-10
solar cell 202, 317
solder, properties 432
space environment model 19, 327, 337, 444-9
space particle types 16-24, 329
space proton irradiation facility 350
SPACE RADIATION code 337
spacecraft configurations and materials 399-401
spacecraft structure as radiation stopper 402
spallation neutron 389
spallation reaction 337
specific activity 4
specification see standard
specific radiation induced loss in fibres 249
spectrosil see also silica 253
spherical shield model for spacecraft 312, 331, 338
spot shield for space radiation 403
solar flare 16, 24
South Atlantic anomaly 17
Sr see strontium
stainless steel, formula and properties 432
standard specification for radiation testing
  ASTM 360-1
  BS 9000 359, 360
  CECC/CENELEC 359, 360
  DIN 359
  IEEE class 1E equipment 27, 359, 393
  ESA/SCC 360

MIL STD 883 359, 360
MIL STD 1553 148
MIL STD 1750A 142
static MOS memory 4, 142, 435-6
static induction transistor (SIT) 238, 261
statistics of particle damage 63
steady state radiation sources, for testing 351, 352
sterilization of products 280
stopping number 296
stopping power 296, 299
straight-ahead approximation 331
strain gauge 268
strategic equipment 394
streaming of radiation 338
string, degradation under radiation 269, 272
strontium-90 38-9, 348, 426
structural plastic 272, 284
superconductors 267
super recovery 124
suprasil 251-3
surface damage 165, 180-2, 202
surface effects annealing 184
surface generation 218
surface linked gain degradation 180-2, 187
surface potential 218
swept quartz 270
system level hardening 387

tactical equipment 394
tandem accelerator 352-4
tantalum, density 432
tantalum capacitor 269
teleoperator 389
temperature sensor 266
terrestrial environment 37
test procedures 356-8
test structure for semiconductor hardening 373
testing 340-71
testing costs 413
thermal annealing 184
thermal generation 73
thermionic cathode 271
thermistor 266
thermocouple 266
thermoluminescent dosimetry (TLD) 10, 11
thermonuclear weapon 29
thermoplastics 284
thermosetting plastics 282
'threat', military from radiation 406
Three-Mile Island (TMI) accident 393
threshold voltage 102, 236, 362, 375, 379
thyristor
  device 235, 236
  parasitic structure in Si device 79, 395
time-dependent effect (TDE) see also post-irradiation effect 118-19, 366-8

tin *see also* solder
  attenuation of photons 429
  properties 432
TIROS spacecraft 399
tissue-equivalent material 13
tolerance of plastics 283–5
topological adjustment of ICs 130, 380
total dose *see dose*
total dose hardening methodology 372–5
track structure 342
trade-off of shield weight 403–5
transconductance 107, 236, 261
transducer 33, 262–6
transient
  effects 54, 71, 395
    in bipolar ICs 195
    in bipolar transistors 166
    in capacitors 270, 390
    in circuits 73, 395
    in diodes 201
    in JFETs 261
    in MOS devices 128–31, 237
    in optical fibres 252–4
    in polymers 284, 390
    in resistors 270
    in sensors 265
  hardening of semiconductors 380–2
  photocurrent 74, 366, 394
  sources for testing 346
transistor *see* bipolar, junction, metal-oxide-semiconductor
transistor–transistor logic (TTL) 186–8
transmission coefficient of electrons 295
transport of electrons 293–8
transport of protons and heavy particles 299
transputer 144
trapped charge 66, 220, 367, 368
trapped radiation belt 16–49, 444–9
trapping centre 66, 375, 378
trench isolation 86, 191, 195
TRIGA reactor 261, 350
TRIM code 224, 292, 331
  attenuation of photons 429
  tungsten plug 150–2
tunnelling of electron in oxide 66–9, 118–27
'turnaround' in nMOS transistor 125, 128
TV camera 215, 226, 390

U *see* uranium
ultimate elongation, in polymers 277
ultra-hard technology 117, 187, 190, 376
ultraviolet (UV)
  absorption 242
  to anneal MOS device 125
  vacuum UV beam for testing 306, 351
  wavelengths 425
UNIRAD code 331, 335, 338
  units, definition 5–7, 421–4

UOSAT spacecraft 143
upset *see* single event
uranium
  attenuation of photons
  fission 26, 29
useful general and geophysical data (tabulated) 421, 422
useful radiation data (tabulated) 423–9

V centres 243
vacancy 50, 184
vacuum devices 201, 226, 397
vacuum microelectronics 397
vacuum tubes 271
valence band *see* energy
van de Graaff accelerator 346–8, 350, 353
vapour-deposited fibre technology 251
voltage value, *see also* breakdown
  bipolar transistor collector–emitter saturation voltage $V_{CE(sat)}$ 166, 185–6, 235
  diode forward voltage drop V_F 12, 209, 213
  irradiation bias for MOS device V_I 110–3
  irradiation bias value negative V_I- 110
  irradiation bias value positive V_I+ 110
  irradiation bias value zero $V_I 0$ 111–13
  MOS capacitor flatband voltage V_{FB} 103
  MOSFET capacitor inversion voltage V_i 103
  shift of V_T induced by Q_{it}: V_{it} 113
  shift of V_T induced by Q_{ot}: V_{ot} 113
  threshold voltage of MOSFET V_T
    definition 101–5
    shift, ΔV_T 102
    ΔV_T prediction model 108–10
voltage regulator 238
voltage shift in CCD 220
'V_T of n-channel device at zero' (VTNZ), condition 102, 113, 117, 125, 133, 218, 364, 390

W *see* tungsten
walled emitter 191
water
  attenuation of photons 429
  attenuation of soft X-rays 309
  dose *see* rad
  as shield 293
weapon environment 29
Webster equation 169, 182
weight of shielding, trade-off 403–5
white radiation, X-rays 302
wind, solar *see* plasma, solar
window materials 242
wood, degradation under radiation 269, 272

X-rays 302–10
  generation 299, 307, 321
  sources for testing 344–6

yttrium-90, emission 426

Z *see* atomic number
Zener diode 208

zero bias response 111
zerodur glass-ceramic 247
zinc crown glass 243
zinc oxide, properties 432
zinc sulphide scintillator 264
zone melt recrystallization 149
zones, inner and outer space-radiation *see* trapped radiation

6 12/99
9 5/03